Auditory and Vestibular Research

METHODS IN MOLECULAR BIOLOGY™

***John M. Walker*, SERIES EDITOR**

493. **Auditory and Vestibular Research:** Methods and Protocols, edited by Bernd Sokolowski, 2009
489. **Dynamic Brain Imaging:** Methods and Protocols, edited by Fahmeed Hyder, 2009
485. **HIV Protocols:** Methods and Protocols, edited by Vinayaka R. Prasad and Ganjam V. Kalpana, 2009
484. **Functional Proteomics:** Methods and Protocols, edited by Julie D. Thompson, Christine Schaeffer-Reiss, and Marius Ueffing, 2008
483. **Recombinant Proteins From Plants:** Methods and Protocols, edited by Loïc Faye and Veronique Gomord, 2008
482. **Stem Cells in Regenerative Medicine:** Methods and Protocols, edited by Julie Audet and William L. Stanford, 2008
481. **Hepatocyte Transplantation:** Methods and Protocols, edited by Anil Dhawan and Robin D. Hughes, 2008
480. **Macromolecular Drug Delivery:** Methods and Protocols, edited by Mattias Belting, 2008
479. **Plant Signal Transduction:** Methods and Protocols, edited by Thomas Pfannschmidt, 2008
478. **Transgenic Wheat, Barley and Oats: Production and Characterization Protocols**, edited by Huw D. Jones and Peter R. Shewry, 2008
477. **Advanced Protocols in Oxidative Stress I**, edited by Donald Armstrong, 2008
476. **Redox-Mediated Signal Transduction:** Methods and Protocols, edited by John T. Hancock, 2008
475. **Cell Fusion:** Overviews and Methods, edited by Elizabeth H. Chen, 2008
474. **Nanostructure Design:** Methods and Protocols, edited by Ehud Gazit and Ruth Nussinov, 2008
473. **Clinical Epidemiology:** Practice and Methods, edited by Patrick Parfrey and Brendon Barrett, 2008
472. **Cancer Epidemiology, Volume 2:** Modifiable Factors, edited by Mukesh Verma, 2008
471. **Cancer Epidemiology, Volume 1:** Host Susceptibility Factors, edited by Mukesh Verma, 2008
470. **Host-Pathogen Interactions:** Methods and Protocols, edited by Steffen Rupp and Kai Sohn, 2008
469. **Wnt Signaling, Volume 2:** Pathway Models, edited by Elizabeth Vincan, 2008
468. **Wnt Signaling, Volume 1:** Pathway Methods and Mammalian Models, edited by Elizabeth Vincan, 2008
467. **Angiogenesis Protocols:** Second Edition, edited by Stewart Martin and Cliff Murray, 2008
466. **Kidney Research:** Experimental Protocols, edited by Tim D. Hewitson and Gavin J. Becker, 2008.
465. **Mycobacteria**, *Second Edition*, edited by Tanya Parish and Amanda Claire Brown, 2008
464. **The Nucleus, Volume 2:** Physical Properties and Imaging Methods, edited by Ronald Hancock, 2008
463. **The Nucleus, Volume 1:** Nuclei and Subnuclear Components, edited by Ronald Hancock, 2008
462. **Lipid Signaling Protocols**, edited by Banafshe Larijani, Rudiger Woscholski, and Colin A. Rosser, 2008
461. **Molecular Embryology:** Methods and Protocols, Second Edition, edited by Paul Sharpe and Ivor Mason, 2008
460. **Essential Concepts in Toxicogenomics**, edited by Donna L. Mendrick and William B. Mattes, 2008
459. **Prion Protein Protocols**, edited by Andrew F. Hill, 2008
458. **Artificial Neural Networks:** Methods and Applications, edited by David S. Livingstone, 2008
457. **Membrane Trafficking**, edited by Ales Vancura, 2008
456. **Adipose Tissue Protocols**, Second Edition, edited by Kaiping Yang, 2008
455. **Osteoporosis**, edited by Jennifer J. Westendorf, 2008
454. **SARS- and Other Coronaviruses:** Laboratory Protocols, edited by Dave Cavanagh, 2008
453. **Bioinformatics, Volume 2:** Structure, Function, and Applications, edited by Jonathan M. Keith, 2008
452. **Bioinformatics, Volume 1:** Data, Sequence Analysis, and Evolution, edited by Jonathan M. Keith, 2008
451. **Plant Virology Protocols:** From Viral Sequence to Protein Function, edited by Gary Foster, Elisabeth Johansen, Yiguo Hong, and Peter Nagy, 2008
450. **Germline Stem Cells**, edited by Steven X. Hou and Shree Ram Singh, 2008
449. **Mesenchymal Stem Cells:** Methods and Protocols, edited by Darwin J. Prockop, Douglas G. Phinney, and Bruce A. Brunnell, 2008
448. **Pharmacogenomics in Drug Discovery and Development**, edited by Qing Yan, 2008.
447. **Alcohol:** Methods and Protocols, edited by Laura E. Nagy, 2008
446. **Post-translational Modifications of Proteins:** Tools for Functional Proteomics, *Second Edition*, edited by Christoph Kannicht, 2008.
445. **Autophagosome and Phagosome**, edited by Vojo Deretic, 2008
444. **Prenatal Diagnosis**, edited by Sinhue Hahn and Laird G. Jackson, 2008.
443. **Molecular Modeling of Proteins**, edited by Andreas Kukol, 2008.
442. **RNAi:** Design and Application, edited by Sailen Barik, 2008.
441. **Tissue Proteomics:** Pathways, Biomarkers, and Drug Discovery, edited by Brian Liu, 2008
440. **Exocytosis and Endocytosis**, edited by Andrei I. Ivanov, 2008
439. **Genomics Protocols**, Second Edition, edited by Mike Starkey and Ramnanth Elaswarapu, 2008
438. **Neural Stem Cells:** Methods and Protocols, Second Edition, edited by Leslie P. Weiner, 2008
437. **Drug Delivery Systems**, edited by Kewal K. Jain, 2008
436. **Avian Influenza Virus**, edited by Erica Spackman, 2008
435. **Chromosomal Mutagenesis**, edited by Greg Davis and Kevin J. Kayser, 2008
434. **Gene Therapy Protocols: Volume 2:** Design and Characterization of Gene Transfer Vectors, edited by Joseph M. Le Doux, 2008
433. **Gene Therapy Protocols: Volume 1:** Production and In Vivo Applications of Gene Transfer Vectors, edited by Joseph M. Le Doux, 2008
432. **Organelle Proteomics**, edited by Delphine Pflieger and Jean Rossier, 2008
431. **Bacterial Pathogenesis:** Methods and Protocols, edited by Frank DeLeo and Michael Otto, 2008
430. **Hematopoietic Stem Cell Protocols**, edited by Kevin D. Bunting, 2008
429. **Molecular Beacons:** Signalling Nucleic Acid Probes, Methods and Protocols, edited by Andreas Marx and Oliver Seitz, 2008
428. **Clinical Proteomics:** Methods and Protocols, edited by Antonia Vlahou, 2008
427. **Plant Embryogenesis**, edited by Maria Fernanda Suarez and Peter Bozhkov, 2008
426. **Structural Proteomics:** High-Throughput Methods, edited by Bostjan Kobe, Mitchell Guss, and Thomas Huber, 2008

METHODS IN MOLECULAR BIOLOGY™

Auditory and Vestibular Research

Methods and Protocols

Edited by

Bernd Sokolowski, PhD

University of South Florida, College of Medicine, Tampa, FL, USA

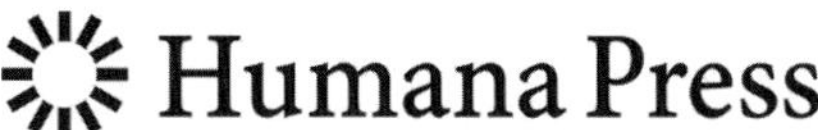

Editor
Bernd Sokolowski
College of Medicine
University of South Florida
Tampa, FL
USA
bsokolow@health.usf.edu

Series Editor
John M. Walker
School of Life Sciences
University of Hertfordshire
Hatfield, Hertfordshire, AL10 9AB, UK

ISBN: 978-1-934115-62-6 e-ISBN: 978-1-59745-523-7
ISSN: 1064-3745 e-ISSN: 1940-6029
DOI 10.1007/978-1-59745-523-7

Library of Congress Control Number: 2008934544

Printed on acid-free paper

9 8 7 6 5 4 3 2 1

springer.com

Preface

Hearing is a sensory modality critical to both language and cognitive development. In its absence, and without sensory input through another modality, such as the manual/visual modality of sign language, cognitive and language development can be severely impaired in the earliest formative years of a child. In its endeavor to discover the mechanisms underlying audition, the field of auditory science has provided rich comparative physiological studies, allowing insights into both the micromechanical and electrochemical world of this system. For many years, the auditory/vestibular sciences have been influenced by the discoveries of electrical engineers and sensory physiologists, who have provided insights into the functions of this dynamic system. The early discoveries in these fields, as well as advancements in microprocessing and materials technologies, provided a means whereby hearing could be regained partly through the use of a bionic device, known as a cochlear implant. Presently, this device and the auditory brainstem implant are the only ones to prosthetically replace brain function.

With the advent of molecular biology tools, such as RT-PCR, the auditory and vestibular fields have made great strides in understanding the genetic basis for various hearing and balance disorders over the past fifteen to twenty years. These technologies permitted the discovery of genes that control inner ear structure and function by overcoming the hurdle of working with small amounts of tissue, as found in the inner ear. The amplification of genes with RT-PCR provided a means to discover gene expression in the small, inner ear endorgans during development, as well as in damaged and normal sensory epithelia in the adult. The use of gene knockout animal models provided the means to verify the effects of genes critical to the development of this system, whereas *in situ* hybridization localized newly discovered gene transcripts. As these technologies continue to broaden the discovery of genes and their regulatory behavior, auditory and vestibular studies have begun to focus on proteins in terms of their interactions, structure, and how these factors relate to function.

In light of the dramatic changes in the auditory and vestibular sciences over these past fifteen plus years, this book describes RNA, protein, and imaging protocols that currently are in use and that have provided insights into genetic regulation, as well as insights into genes and pathogens involved in diseases of the ear. This overview provides a perspective of basic research with both mammalian and non-mammalian animal models, as well as protocols applicable to clinical studies. The chapters in Part 1 include basic protocols of RNA isolation and expression, followed by methods to study cell lineage, gene delivery, and the identification and use of stem cells. This section ends with techniques that are applicable to clinical studies of genes, pathogens, and cancers that lead to hearing loss in humans. Part 2 focuses on the study of inner ear proteins and more specifically on their interactions, including techniques such as the yeast-two hybrid assay, coimmunoprecipitation, plasmon resonance, and protein tagging for mass spectrometry. The final section, Part 3, describes imaging techniques useful for the study of ions, protein-protein interactions, and imaging of proteins at the atomic level.

While the chapters are written by specialists in the auditory and vestibular fields, the techniques described herein will be useful to those exploring genes and proteins in other systems as well, especially where tissues are scarce and where a comparative approach lends itself to discovering the underlying causes of human disorders.

Contents

Contributors

JOHNVESLY BASAPPA • *Department of Neuroscience, Tufts University School of Medicine, Boston, MA, USA; Jean Mayer USDA Human Nutrition Research Center on Aging, Tufts University, Boston, MA, USA*
INNA A. BELYANTSEVA • *Laboratory of Molecular Genetics, Section on Human Genetics, NIDCD, National Institutes of Health, Rockville, MD, USA*
CATHERINE S. BRESEE • *Oregon Hearing Research Center, Oregon Health & Science University, Portland, OR, USA*
JOHN V. BRIGANDE • *Oregon Hearing Research Center, Oregon Health & Science University, Portland, OR, USA*
WILLIAM E. BROWNELL • *Bobby R. Alford Department of Otolaryngology – HNS, Baylor College of Medicine, Houston, TX, USA*
LONG-SHENG CHANG • *Departments of Pediatrics, Otolaryngology, and Pathology, The Ohio State University College of Medicine, Center for Childhood Cancer Research Institute at National Children's Hospital, Columbus, OH, USA*
FRANS P. M. CREMERS • *Department of Human Genetics, Nijmegen Centre for Molecular Life Sciences, Radboud University Nijmegen Medical Centre, Nijmegen, The Netherlands*
BENJAMIN CURRALL • *Department of Biomedical Sciences, Creighton University School of Medicine, Omaha, NE, USA*
DENNIS G. DRESCHER • *Departments of Otolaryngology and Biochemistry, Wayne State University School of Medicine, Detroit, MI, USA*
MARIAN J. DRESCHER • *Department of Otolaryngology, Wayne State University School of Medicine, Detroit, MI, USA*
SUSAN M. DYMECKI • *Department of Genetics, Harvard Medical School, Boston, MA, USA*
GARTH D. EHRLICH • *Center for Genomic Sciences, Allegheny Singer Research Institute, Allegheny General Hospital, Pittsburgh, PA, USA; Department of Microbiology and Immunology, Drexel University College of Medicine, Allegheny Campus, Pittsburgh, PA, USA*
DONNA M. FEKETE • *Department of Biological Sciences, Purdue University, West Lafayette, IN, USA*
GREGORY I. FROLENKOV • *Department of Physiology, University of Kentucky, Chandler Medical Center, Lexington, KY, USA*
ARMIN GIESEKE • *Max-Planck Institute for Marine Microbiology, Bremen, Germany*
PETER G. GILLESPIE • *Oregon Hearing Research Center and Vollum Institute, Oregon Health & Science University, Portland, OR, USA*
SAMUEL P. GUBBELS • *University of Iowa Hospitals, Iowa City, IA, USA*
LUANNE HALL-STOODLEY • *Center for Genomic Sciences, Allegheny Singer Research Institute, Allegheny General Hospital, Pittsburgh, PA, USA; Department of*

Microbiology and Immunology, Drexel University College of Medicine, Allegheny Campus, Pittsburgh, PA, USA

RICHARD HALLWORTH • *Department of Biomedical Sciences, Creighton University School of Medicine, Omaha, NE, USA*

MARGARET C. HARVEY • *Department of Otolaryngology – HNS, University of South Florida, College of Medicine, Tampa, FL, USA*

STEFAN HELLER • *Departments of Otolaryngology – HNS and Molecular & Cellular Physiology, Stanford University School of Medicine, Stanford, CA, USA*

HAKIM HIEL • *Department of Otolaryngology – HNS, Johns Hopkins University, School of Medicine, Baltimore, MD, USA*

HEATHER JENSEN-SMITH • *Department of Biomedical Sciences, Creighton University School of Medicine, Omaha, NE, USA*

JONATHAN J. JUNGWIRTH • *Oregon Hearing Research Center, Oregon Health & Science University, Portland, OR, USA*

THANDAVARAYAN KATHIRESAN • *Department of Otolaryngology – HNS, University of South Florida, College of Medicine, Tampa, FL, USA*

JOSEPH E. KERSCHNER • *Department of Otolaryngology, Medical College of Wisconsin, Milwaukee, WI, USA*

JUN CHUL KIM • *Department of Genetics, Harvard Medical School, Boston, MA, USA*

HANNIE KREMER • *Department of Otorhinolaryngology, Nijmegen Centre for Molecular Life Sciences, Radboud University Nijmegen Medical Centre, Nijmegen, The Netherlands*

OLIVIER LICHTARGE • *Department of Molecular and Human Genetics, Baylor College of Medicine, Houston, TX, USA*

MICHIO MURAKOSHI • *Department of Bioengineering and Robotics, Graduate School of Engineering, Tohoku University, Sendai, Japan*

DHASAKUMAR S. NAVARATNAM • *Department of Neurology, Yale School of Medicine, New Haven, CT, USA*

MICHAEL NICHOLS • *Department of Physics, Creighton University, Omaha, NE, USA*

LIPING NIE • *Department of Otolaryngology, Center for Neuroscience, University of California, Davis, CA, USA*

LAURA NISTICO • *Center for Genomic Sciences, Allegheny-Singer Research Institute, Allegheny General Hospital, Pittsburgh, PA, USA*

KAZUO OSHIMA • *Departments of Otolaryngology – HNS and Molecular & Cellular Physiology, Stanford University School of Medicine, Stanford, CA, USA*

JAMES PAGANA • *Oregon Hearing Research Center and Vollum Institute, Oregon Health & Science University, Portland, OR, USA*

FRED A. PEREIRA • *Bobby R. Alford Department of Otolaryngology – HNS, Huffington Center on Aging and Department of Molecular and Cellular Biology, Baylor College of Medicine, Houston, TX, USA*

MELINDA M. PETTIGREW • *Division of Epidemiology of Microbial Diseases, Yale University School of Public Health, New Haven, CT, USA*

TuShun R. Powers • *Biology Department, New Mexico State University, Las Cruces, NM, USA*
Lavanya Rajagopalan • *Bobby R. Alford Department of Otolaryngology – HNS, Baylor College of Medicine, Houston, TX, USA*
Neeliyath A. Ramakrishnan • *Department of Otolaryngology, Wayne State University School of Medicine, Detroit, MI, USA*
Danielle Rossino • *Department of Biomedical Sciences, Creighton University School of Medicine, Omaha, NE, USA*
Takunori Satoh • *Department of Biological Sciences, Purdue University, West Lafayette, IN, USA*
Pascal Senn • *Departments of Otolaryngology – HNS and Molecular & Cellular Physiology, Stanford University School of Medicine, Stanford, CA, USA; Department of Otolaryngology – HNS, Inselspital, University of Berne, Berne, Switzerland*
Elba E. Serrano • *Biology Department, New Mexico State University, Las Cruces, NM, USA*
Jung-Bum Shin • *Oregon Hearing Research Center and Vollum Institute, Oregon Health & Science University, Portland, OR, USA*
Bernd H. A. Sokolowski • *Department of Otolaryngology – HNS, University of South Florida, College of Medicine, Tampa, FL, USA*
Paul Stoodley • *Center for Genomic Sciences, Allegheny Singer Research Institute, Allegheny General Hospital, Pittsburgh, PA, USA; Department of Microbiology and Immunology, Drexel University College of Medicine, Allegheny Campus, Pittsburgh, PA, USA*
David R. Sultemeier • *Biology Department, New Mexico State University, Las Cruces, NM, USA*
Jonathan C. Thomas • *Yale University School of Public Health, New Haven, CT, USA*
LeAnn Tiede • *Department of Biomedical Sciences, Creighton University School of Medicine, Omaha, NE, USA*
Casilda Trujillo-Provencio • *Biology Department, New Mexico State University, Las Cruces, NM, USA*
Sevin Turcan • *Department of Neuroscience and Department of Biomedical Engineering, Tufts University, Medford, MA, USA*
Ana E. Vázquez • *Department of Otolaryngology, Center for Neuroscience, University of California, Davis, CA, USA*
Douglas E. Vetter • *Department of Neuroscience, Tufts University School of Medicine, Boston, MA, USA*
Hiroshi Wada • *Department of Bioengineering and Robotics, Graduate School of Engineering, Tohoku University, Sendai, Japan*
D. Bradley Welling • *Department of Otolaryngology, The Ohio State University College of Medicine, Columbus, OH, USA*
David W. Woessner • *Oregon Hearing Research Center, Oregon Health & Science University, Portland, OR, USA*
Ebenezer N. Yamoah • *Department of Otolaryngology, Center for Neuroscience, University of California, Davis, CA, USA*

YILING YU • *Department of Developmental Neurobiology, St. Jude Children's Research Hospital, Memphis, TN, USA; Deptartment of Anatomy, Histology and Embryology, Shanghai Medical School, Fudan University, Shanghai, P. R. China*
JIAN ZUO • *Department of Developmental Neurobiology, St. Jude Children's Research Hospital, Memphis, TN, USA*

Part I

Nucleic Acid Protocols

Chapter 1

RNA Isolation from *Xenopus* Inner Ear Sensory Endorgans for Transcriptional Profiling and Molecular Cloning

Casilda Trujillo-Provencio, TuShun R. Powers, David R. Sultemeier, and Elba E. Serrano

Abstract

The amphibian *Xenopus* offers a unique model system for uncovering the genetic basis of auditory and vestibular function in an organism that is well-suited for experimental manipulation during animal development. However, many procedures for analyzing gene expression in the peripheral auditory and vestibular systems mandate the ability to isolate intact RNA from inner ear tissue. Methods presented here facilitate preparation of high quality inner ear RNA from larval and post-metamorphic *Xenopus* specimens that can be used for a variety of purposes. We demonstrate that RNA isolated with these protocols is suitable for microarray analysis of inner ear organs, and for cloning of large transcripts, such as those for ion channels. Genetic sequences cloned with these procedures can be used for transient transfection of *Xenopus* kidney cell lines with GFP fusion constructs.

Key words: RNA, auditory, vestibular, *Xenopus laevis, Xenopus tropicalis*, microarray, cloning, transcriptional profiling, heterologous gene expression.

1. Introduction

Xenopus laevis (*X. laevis*) is a widely used and well-established organism that has contributed to our understanding of embryogenesis and cellular development for well over 50 years *(1)*. In large part, *X. laevis* is a popular experimental species because the stages of *X. laevis* development have been richly detailed from the fertilized egg to the mature adult, and because the organism is extremely easy to breed and maintain in the laboratory *(2)*. However, the relatively long generation time (~18 months) and allotetraploid genome have been an impediment to genetic

Bernd Sokolowski (ed.), *Auditory and Vestibular Research: Methods and Protocols, vol. 493*

DOI 10.1007/978-1-59745-523-7_1 Springerprotocols.com

studies with this species. In recent years, *Xenopus tropicalis* (*X. tropicalis*) has emerged as another member of the genus *Xenopus* that is a superior alternative for molecular genetic analysis and large scale sequencing efforts due to its diploid genome and shorter (~5 months) generation time *(3, 4)*.

Using *X. laevis* and *X. tropicalis* as our experimental systems, we aim to understand the developmental mechanisms that give rise to the uniquely-shaped auditory and vestibular endorgans of the inner ear with their characteristically patterned epithelia comprised of specialized mechanosensory hair cells *(5–7)*. We are especially interested in the specification of the electrical phenotype of hair cells and in determining the cadre of ion channels that typify endorgans of the inner ear *(8–10)*. Our target for molecular investigations is the inner ear of both *Xenopus* species. However, the diminutive inner ears that reside in the *Xenopus* otic capsule provide limited amounts of tissue for RNA isolation. This restriction has posed an additional challenge for the identification of genes that are expressed in the inner ear *(6, 9)*.

The protocols that we have developed and that are presented here permit isolation of high quality RNA from larval and post-metamorphic *Xenopus* inner ears of both species. Furthermore, methods that are applicable for RNA isolation from the *Xenopus* inner ear have been validated as suitable for RNA isolation from other *Xenopus* organs such as brain and kidney, as well as *Xenopus* cell lines. Since the inner ear originates from a neurogenic placode *(11)*, many genes expressed in the inner ear are also expressed in nervous tissue. Inner ears also share common sensitivities with kidneys to antibiotics and other drugs *(12)*. Therefore, these RNA isolation protocols can be used to design experiments that identify genes expressed during development of various *Xenopus* organs, and in the response of organs and cell lines to chemical challenges. The total RNA isolated with our procedures was used in two downstream applications, transcriptional profiling *(13)* and molecular cloning *(9)*. Ion channel genes cloned with these methods can be fused to sequences for a fluorescent reporter molecule such as green fluorescent protein (GFP) and expressed in *Xenopus* kidney cell lines using lipid-mediated transient transfection protocols *(14)*.

2. Materials

2.1. Tissue Preparation

1. *Xenopus laevis* and *Xenopus tropicalis* are obtained from Nasco (Fort Atkinson, WI). *X. laevis* larvae are purchased as a unit of Stage 48–55 tadpoles (cat. no. LM00450MX) and maintained in aquaria until they reach developmental Stage 55–56. Juvenile animals (*X. laevis*, cat. no. LM00453MX and *X. tropicalis*, cat. no. LM00821MX) are used within a week

of arrival. All procedures involving animals are approved by the New Mexico State University Institutional Animal Care and Use Committee.

2. Ethyl 3-aminobenzoate methanesulfonate salt: Prepare a 2% (w/v) stock solution in water (*see* **Notes 1** and **2**). Store the stock solution at 4 °C and discard after one month. Working solutions of 0.2% (larvae) and 0.5% (juvenile animals) are prepared the day of use by diluting the stock solution with 0.01 *M* phosphate buffered saline: 1.5 m*M* KH_2PO_4, 8.1 m*M* Na_2HPO_4, 138 m*M* NaCl, 2.7 m*M* KCl, pH 7.4 (PBS, cat. no. P3813, Sigma-Aldrich, St. Louis, MO).
3. Water treated with 0.1% (v/v) diethyl pyrocarbonate (DEPC, cat. no. D5758, Sigma-Aldrich) (*see* **Notes 1** and **3**).
4. RNase*Zap*® and RNA*later*® (Ambion, Austin, TX).
5. Dissecting tools: Dumont #5 forceps, standard straight fine Iris scissors, scalpel handle #3 (all from Fine Sciences Tools, Foster City, CA); BD Bard-Parker Sterile Scalpel Blades #11 (VWR, West Chester, PA); 60 × 15 mm glass Petri dish (VWR) filled with black wax (Nasco, Fort Atkinson, WI).
6. Olympus SZ61 dissecting stereomicroscope (Leeds Precision Instruments, Minneapolis, MN).
7. Microcentrifuge tubes sterilized by autoclaving, 1.7 mL and 2.0 mL; and 22 gauge needles (VWR).

2.2. Total RNA Isolation

1. RNeasy® Mini Kit (Qiagen, Valencia, CA).
2. 14. 3 *M* β-mercaptoethanol.
3. Kimwipes® (VWR).
4. Brinkmann Polytron PT1200 handheld homogenizer with sawtooth 0.5 mm generator (VWR).

2.3. Total RNA Clean-up

1. The DNA-*free*™ kit, 5 *M* ammonium acetate, 5 mg/mL linear acrylamide, and THE RNA Storage Solution (all from Ambion, Austin, TX).
2. 100% ethanol (Aaper Alcohol, Shelbyville, KY).
3. 70% ethanol is prepared by diluting 100% ethanol with DEPC-treated water.
4. Refrigerated centrifuge (*e.g.*, Beckman Coulter Allegra™ 21R Refrigerated Centrifuge; VWR).

2.4. Determination of Total RNA Quality and Quantity

1. Agilent 2100 Bioanalyzer and Agilent RNA 6000 Nano Kit (Agilent Technologies, Palo Alto, CA).

2.5. Microarray Analysis

1. GeneChip® *Xenopus laevis* Genome Array and GeneChip® One-Cycle Target Labeling Kit (Affymetrix, Santa Clara, CA).

2. gcRMA (Robust multichip averaging) summarization method downloaded from bioconducter.org and executed in the R package (http://www.r-project.org/).
3. Spotfire® DecisionSite® 9.0 for Functional Genomics (Spotfire, Inc., Somerville, MA).

2.6. Molecular Cloning of Genes Identified with Transcriptional Profiling

1. SMART™ RACE cDNA Amplification Kit (Clontech Laboratories, Inc., Mountain View, CA).
2. PicoMaxx® High Fidelity PCR System (Stratagene, La Jolla, CA).
3. Thermal cycler (e.g., TC-512, Techne Inc., Burlington, NJ).
4. Lyophilized gene specific primers, 10 nmole (Operon, Huntsville, AL).
5. SeaKem® LE Agarose (Cambrex Bio Science Rockland, Inc., Rockland, ME).
6. Tris-acetate/EDTA electrophoresis buffer (TAE 1X): 0.04 *M* Tris base, 0.001 *M* EDTA, 1.0 *M* glacial acetic acid. A 50X stock solution is prepared in a glass bottle and stored at room temperature (22–26 °C).
7. Agarose gel with ethidium bromide: Prepare by diluting 10 mg/mL ethidium bromide stock solution (cat. no. E1510, Sigma-Aldrich) to a final concentration of 0. 25 μg/mL in melted agarose gel. Do not include ethidium bromide in the running buffer.
8. 1 kb DNA ladder (New England Biolabs, Inc., Ipswich, MA).
9. S.N.A.P.™ UV-Free Gel Purification Kit and TOPO® XL PCR Cloning Kit with One Shot® TOP10 Electrocomp™ *E. coli* (Invitrogen, Carlsbad, CA).
10. Lambda Bio UV/Vis Spectrometer (Perkin Elmer Corp., Norwalk, CT).
11. Electroporator 2510 and electroporation cuvettes (Eppendorf, Westbury, NY).
12. BigDye® Terminator v3.1 Cycle Sequencing Kit (Applied Biosystems, Foster City, CA).
13. ABI PRISM® 3100 Genetic Analyzer (Applied Biosystems).

2.7. Heterologous Inner Ear Gene Expression

1. pAcGFP1-C1 expression vector (Clontech Laboratories, Inc.).
2. *Xba*I and *Spe*I restriction enzymes; Antarctic phosphatase; and Quick Ligation™ Kit (all from New England Biolabs).
3. *X. laevis* A6 kidney cell line (American Type Culture Collection, Manassas, VA).
4. Cell culture medium per 100 mL: 75 mL of NCTC-109 (cat. no. N1140, Sigma-Aldrich), 10 mL Newborn Calf Serum (cat. no. N4637, Sigma-Aldrich), 2 mL of 200 m*M* L-glutamine (cat. no. G7513, Sigma-Aldrich), and 13 mL of Milli-Q water. Combine medium components, then filter sterilize with a 250 mL Stericup™-GP Filter Unit (Millipore Cor-

poration, Billerica, MA). Passage reagent: Filter sterilized 0.25% Trypsin-EDTA (cat. no. T4049, Sigma-Aldrich). Store aliquots consisting of 10 mL of Newborn Calf Serum, 2 mL of L-glutamine, and 10 mL of 0.25% Trypsin-EDTA at −20 °C.

5. Labconco® Purifier™ Class II Safety Cabinet and VWR Water Jacketed CO_2 Incubator Model 2310 (VWR).
6. BD Falcon culture slides, two- or four-chamber (cat. no. 53106-302 or 53106-304, VWR).
7. Lipofectamine™ 2000 Transfection Reagent (cat. no. 11668-027, Invitrogen).

2.8. Epifluorescence Imaging

1. 3.7% paraformaldehyde is prepared by diluting 10% paraformaldehyde in PBS. The paraformaldehyde should be prepared fresh for each experiment.
2. Hoechst 33342, 10 mg/mL.
3. *SlowFade*® Antifade Kit (Invitrogen).
4. VWR® micro cover glasses No.1, 24 × 50 mm.
5. Clear fingernail polish.
6. CoolSNAP™ HQ CCD camera (Photometrics, Tucson, AZ).
7. Nikon TE2000 epifluorescence microscope equipped with a UV-2A filter cube (UV excitation) and a B-2E/C filter cube (Blue excitation) (A.G. Heinze Inc., Lake Forest, CA).
8. MetaVue™ Imaging System Version 6.0r5 and MetaMorph® Offline Imaging System Version 6.0r5 (Molecular Devices Corporation, Sunnyvale, CA).
9. Adobe® Photoshop® Version 7.0.1 (Adobe Systems Incorporated, San Jose, CA).

3. Methods

Many methods requiring total RNA as a starting material are significantly affected by the quality of the RNA isolated from the tissue of interest. Furthermore, when tissue is limited, the success of methods that dictate a specific amount of starting RNA is hindered because the amount of total RNA that can be retrieved is diminished. Isolation of total RNA from the minuscule inner ear of *Xenopus (5, 7),* for molecular pursuits such as transcriptional profiling and molecular cloning, presents such a challenge. In addition, the total RNA must be of very high quality, with a demonstrable lack of degradation and the presence of transcripts greater than 3 kb.

After numerous attempts to isolate a sufficient amount of high quality total RNA from the inner ear of *Xenopus*, we were able to optimize the method by working quickly through tissue dissections, meticulously cleaning the tools used in the procedure with RNase*Zap*®, bathing the exposed tissue with RNA*later*®, and incorporating the use of the Qiagen RNeasy® Mini Kit.

Additionally, the use of the Agilent 2100 Bioanalyzer for RNA characterization greatly enhanced our capacity to detect and analyze isolated total RNA and to standardize protocols (*see* **Fig. 1.2**). Previously, the use of denaturing gel electrophoresis and a spectrophotometer was standard for assessing the quality and quantity of RNA. This procedure resulted in the loss of μg amounts of total RNA. In contrast, the Agilent 2100 Bioanalyzer can detect and assess the *quantity* of as little as 25 ng/μL total RNA and the *quality* of as little as 5 ng/μL total RNA *(15)*. The Bioanalyzer characterizes the sample with an RNA integrity number (RIN) that can be used to establish a quantitative standard for RNA quality for each application. We consider this type of equipment and quantitative analysis a major contributor to the success and replicability of our methods, especially for transcriptional profiling.

High quality RNA enhances the success of RT-PCR Rapid Amplification of cDNA Ends (RACE) reactions, especially when long transcripts are sought. With our protocols, we typically visualize one sharp band of a PCR product when gene specific primers and the PicoMaxx® High Fidelity PCR system are used to amplify a gene of interest (*see* **Fig. 1.3C**). The PicoMaxx® High Fidelity PCR system uses a blend of *Taq* and *Pfu* DNA polymerases and an exclusive ArchaeMaxx® polymerase-enhancing factor. The system provides high PCR sensitivity and fidelity which facilitates the detection and amplification of low copy number targets up to 10 kb in size. For example, we routinely obtain target PCR products in the 3–4 kb range *(9)*. Using the TOPO® XL PCR Cloning Kit in conjunction with the PicoMaxx® High Fidelity PCR system enabled the cloning of transcripts greater than 3 kb. It is crucial to confirm that clones are full length and of the expected identity by sequencing prior to other downstream applications such as heterologous expression in a cell culture system (*see* **Fig. 1.3D**).

We enlisted the transcriptional profiling approach, using microarray analysis, in order to better understand global gene expression patterns that underlie *X. laevis* inner ear function (*see* **Fig. 1.3A**). Using microarray methods, we identified several genes on the GeneChip® *X. laevis* genome array that are differentially expressed in *X. laevis* inner ear as compared with brain *(13)*. For example, we detected a gene which has been implicated in hereditary deafness, the gap junction protein β2 (GJB2) *(16)*, during analysis of our microarray data. The difference in GJB2 expression levels in *X. laevis* inner ear was nearly 4X that of the brain (*see* **Fig. 1.3B**). The relative GJB2 abundances detected through microarray analysis of inner ear and brain RNA are also apparent when inner ear and brain RNA are used as template in

RT-PCR RACE reactions with GJB2 gene specific primers (*see* **Fig. 1.3C**). Thus, RT-PCR RACE reactions replicate the gene's transcriptional profile between organs as detected with microarrays. We interpret these findings as an indication of the indispensable contribution of high quality RNA to the consistency between replicates and the identification of inner ear specific genes.

In summary, we recommend:

- Work quickly and follow established best practices for handling RNA *(17)*
- Clean tools with RNase*Zap*®
- Bathe tissue with RNA*later*® during surgery
- Use Qiagen RNeasy® Mini Kit
- High quality RNA (RIN>8) is essential; use the Agilent Bioanalyzer or similar equipment to assess RNA quality
- At least 1 μg of high quality RNA (RIN>8) is optimal
- Use total RNA within 72 h for microarray experiments
- 18–20 μg of biotin-labeled cRNA are required before proceeding to the fragmentation and hybridization steps
- Review electropherograms for size distributions of cRNA/mRNA samples; transcripts greater than 3 kb should be present in the sample, especially if ion channels and other membrane transporters and receptors are target genes
- Use the PicoMaxx® High Fidelity PCR System or equivalent
- Use the TOPO® XL PCR Cloning Kit or equivalent

As per standard laboratory procedures, investigators are reminded that they should review chemical MSDS sheets prior to use and ensure that reagents are handled and discarded in compliance with all federal and state regulations. Protocols that use animals in research must be approved by the Institutional Animal Care and Use Committee prior to initiating experiments. Recommended guidelines for the use of *Xenopus* in research are available online *(18–20)*.

3.1. Tissue Preparation

1. Take precautions to eliminate sources of RNase contamination by following rigorous laboratory procedures for working with RNA. Minimal practices include a clean bench, strict adherence to the use of gloves throughout the procedure, a dedicated set of pipettors, and disposable sterile plasticware. Prior to dissections, clean all dissection tools and the Polytron generator with RNase*Zap*® and rinse thoroughly with DEPC-treated water (*see* **Note 4**).
2. Partially immerse stage 55–56 larvae and juvenile (2.5–3.5 cm, 1.0–3.0 g) *X. laevis* or juvenile *X. tropicalis* (1.5–2.5 cm, 0.8–2.0 g) in a solution of ethyl 3-aminobenzoate methanesulfonate salt (0.2% for Stage 55–56 larvae and 0.5% for juveniles) for 10–30 min at room temperature (22–26 °C). The solution should cover the torso of the animal but leave the nostrils of the animal exposed to air.

3. Work quickly during the dissections to minimize tissue exposure to RNases. Pierce larval hearts and decapitate juvenile animals prior to aseptic removal of inner ear tissue. Also, extract brain and kidney tissue from juvenile animals.
4. Use a dissecting stereomicroscope (e.g., Olympus SZ61) for the dissection of tissues.
5. Remove larval inner ears by approaching the tissue from the dorsal side of the animal (*see* **Fig. 1.1A**). Peel the skin away from the area above the brain and otic capsules and remove the soft bone of the skull and otic capsules, using a surgical scalpel and fine forceps. Once the tissue is exposed (*see* **Fig. 1.1B**), bathe it with RNA*later*® to prevent degradation of RNA. Extract the inner ear by grasping the eighth cranial nerve and withdrawing the nerve and its target sensory organs from the otic capsule.
6. Remove juvenile inner ears and brains by approaching the tissue ventrally from the roof of the mouth after removing the lower jaw. To facilitate the dissection, pin the top of the head and the exposed spinal cord to the dissecting dish with 22 gauge needles. Despite the animal size difference, *X. tropicalis* inner ears are only slightly smaller than those of *X. laevis*, and overall the inner ears of the two species are of comparable dimensions (*see* **Fig. 1.1C,D**). Remove the skull and otic capsule bone with a surgical scalpel and fine forceps. Bathe the exposed tissue (*see* **Fig. 1.1D**) copiously with RNA*later*®. Then, extract the inner ear tissue as described in the larval

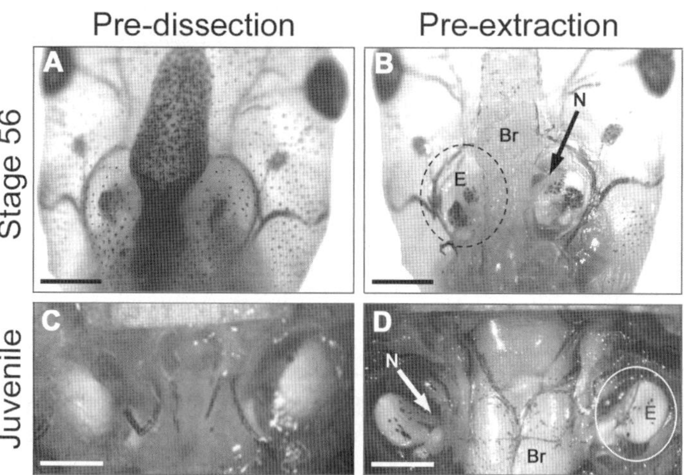

Fig. 1.1. *Xenopus* inner ear and brain dissections. **(A)** Dorsal view of *X. laevis* Stage 56 larva prior to dissection. **(B)** Dorsal view of exposed *X. laevis* Stage 56 inner ear (E) and brain (Br) after the tissue and bone are removed. **(C)** Ventral view of *X. laevis* juvenile upper jaw prior to dissection. **(D)** Ventral view of exposed *X. tropicalis* juvenile inner ear (E) and brain (Br) after tissue and bone are removed. Arrows point to the 8th cranial nerve (N). Anterior (*top*), Posterior (*bottom*). Scale bar = 1 mm.

preparation. After removal of inner ear tissue, sever the brain from all nerve branches and carefully remove with fine forceps.

7. The dissected tissue is placed in a 2.0 mL microcentrifuge tube, containing 200–400 μL of RNA*later*®, immediately after removal from the animal.
8. Use tissue immediately for RNA isolation or store at −20 °C for several months or at −80 °C indefinitely prior to RNA isolation.

3.2. Total RNA Isolation

1. Use the RNeasy® Mini Kit from Qiagen for total RNA isolation.
2. Remove the dissected tissue from the RNA*later*®, blot quickly on a Kimwipe® (from a clean box reserved for RNA work), and immediately place in a pre-weighed 2.0 mL microcentrifuge tube containing 600 μL of Buffer RLT and 6 μL of 14. 3 *M* β-mercaptoethanol.
3. Determine the tissue weight by subtracting the weight of the tube plus Buffer RLT and β-mercaptoethanol from the weight of tube containing the tissue plus Buffer RLT and β-mercaptoethanol.
4. Extract no more than 30 mg of tissue, as recommended by the vendor, for each total RNA isolation (*see* **Note 5**).
5. Disrupt the tissue using a homogenizer such as a Brinkman Polytron with a 0.5 mm generator set at the highest setting. Pulse samples 3–5 times for ~5–10 s until the tissue is completely homogenized and the solution is uniform in texture and color. Cool the tube on ice between pulses.
6. After the tissue is uniformly homogenized, follow the RNeasy® Mini Kit protocol as described in the manual.
7. Elute total RNA in two steps, first with 50 μL and then with 40 μL of RNase-free water provided with the kit.

3.3. Total RNA Clean-up

1. Use the DNA-*free*™ kit from Ambion to purify total RNA sample.
2. Treat the total RNA obtained from **step 7, Section 3.2** with DNaseI to degrade any DNA in the sample. Follow the protocol manual with a DNaseI incubation time of 30 min.
3. Following the DNase inactivation step, transfer the total RNA solution to a clean 1.7 mL microcentrifuge tube and precipitate by adding 0.1X volume of 5 *M* ammonium acetate, 0.015X volume of 5 mg/mL linear acrylamide, and 2X volume of 100% ethanol.
4. Incubate the total RNA precipitation reaction overnight at −20 °C.
5. Recover the total RNA pellet by centrifugation at 10,000 g for 20 min at 4 °C. Discard the supernatant, and wash the pellet with 300 μL of 70% ethanol. Repeat centrifugation for 10 min.

6. Discard the wash supernatant, and air dry the pellet at room temperature for 15 min.
7. Resuspend total RNA in 8–15 μL of THE RNA Storage Solution and incubate in a 37 °C water bath for 5 min.
8. Store the sample at −80 °C or place on ice and analyze for quality and recovery.

3.4. Determination of Total RNA Quality and Quantity

1. Dilute the total RNA sample 1:3 in THE RNA Storage Solution and use the Agilent RNA 6000 Nano Kit to prepare the RNA for the determination of quality and quantity.
2. Follow the vendor's instructions of the Agilent 2100 Bioanalyzer to generate an electropherogram and an RNA Integrity Number (RIN) for each RNA sample. An electropherogram displaying two well-defined peaks of 1.8 kb (18S rRNA) and 4.0 kb (28S rRNA) and a RIN above 8 are indicative of high quality RNA (*see* **Fig. 1.2B1**). For typical yields of total RNA isolated from *Xenopus* tissues *see* **Note 6**. **Figure 1.2** shows representative electropherograms generated by the 2100 expert software.

3.5. Microarray Analysis

1. Use total RNA, obtained in **Section 3.3, step 8,** from juvenile and larval *X. laevis* in microarray experiments within 72 h of tissue dissection. At least 1 μg (5–10 μL) of high quality RNA with a RIN above 8 is required for cRNA synthesis (*see* **Note 7**).
2. Prepare biotin-labeled cRNA for hybridization to the GeneChip® *Xenopus laevis* Genome Array by using the GeneChip® One-Cycle Target Labeling kit, following the vendor's instructions.
3. Determine the amount of biotin-labeled cRNA by analyzing diluted cRNA (1:10 or 1:20) on the Agilent 2100 Bioanalyzer in order to ensure that 18–20 μg of biotin-labeled cRNA are available for the fragmentation and hybridization steps (*see* **Note 8**). Review electropherograms, generated by using the 2100 expert software from Agilent, to evaluate the size distributions of the cRNA samples to be used for transcriptional profiling. For inner ear and brain, expect the sample to contain cRNA that is greater than 3 kb. (*see* **Fig. 1.2C** for a cRNA sample prepared by the BioMicro Center at the Massachusetts Institute of Technology, Cambridge, MA, using the aforementioned technology).
4. After image acquisition, normalize the raw data using the gcRMA (Robust multichip averaging) summarization method. After normalization, analyze all microarray data using the Spotfire® DecisionSite® 9.0 for Functional Genomics software. A heat map that graphically represents the levels of gene expression from one array of *X. laevis* inner ear tissue is shown in **Fig. 1.3A**.

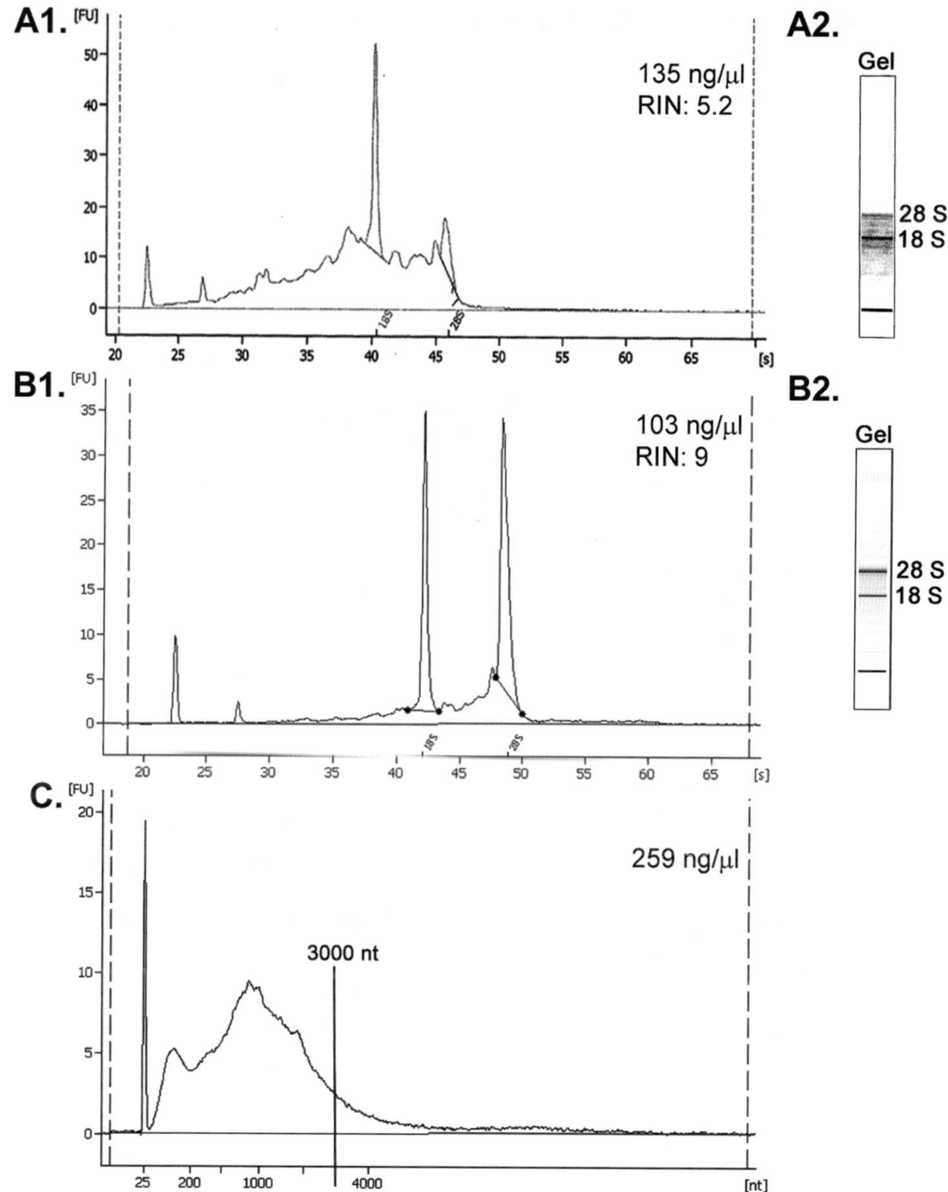

Fig. 1.2. *X. laevis* inner ear total RNA and cRNA evaluated with an Aligent 2100 Bioanalyzer. **(A1)** Electropherogram of degraded total RNA. This electropherogram is typical of RNA isolated from *X. laevis* juvenile inner ear tissue prior to implementation of the Qiagen RNeasy® Mini Kit. The RNA sample had a low (5.2) RNA Integrity Number (RIN). **(A2)** Gel representation of total RNA shown in **A1**. **(B1)** Electropherogram of high quality (non-degraded) total RNA. This electropherogram is typical of RNA extracted from *X. laevis* Stage 56 larvae inner ear tissue with the Qiagen RNeasy® Mini Kit using the procedures described in **Sections 3.2–3.4**. The RNA sample had a high (9.0) RNA Integrity Number (RIN). **(B2)** Gel representation of total RNA shown in **B1**. **(C)** Electropherogram of biotin-labeled cRNA synthesized from the total RNA shown in **B1**. The cRNA of inner ear samples typically includes sequences greater than 3000 nt. Axis units: Abscissa **A1** and **B1**, migration time (seconds, s); Abscissa **C**, nucleotides (nt); Ordinate, fluorescence units (FU).

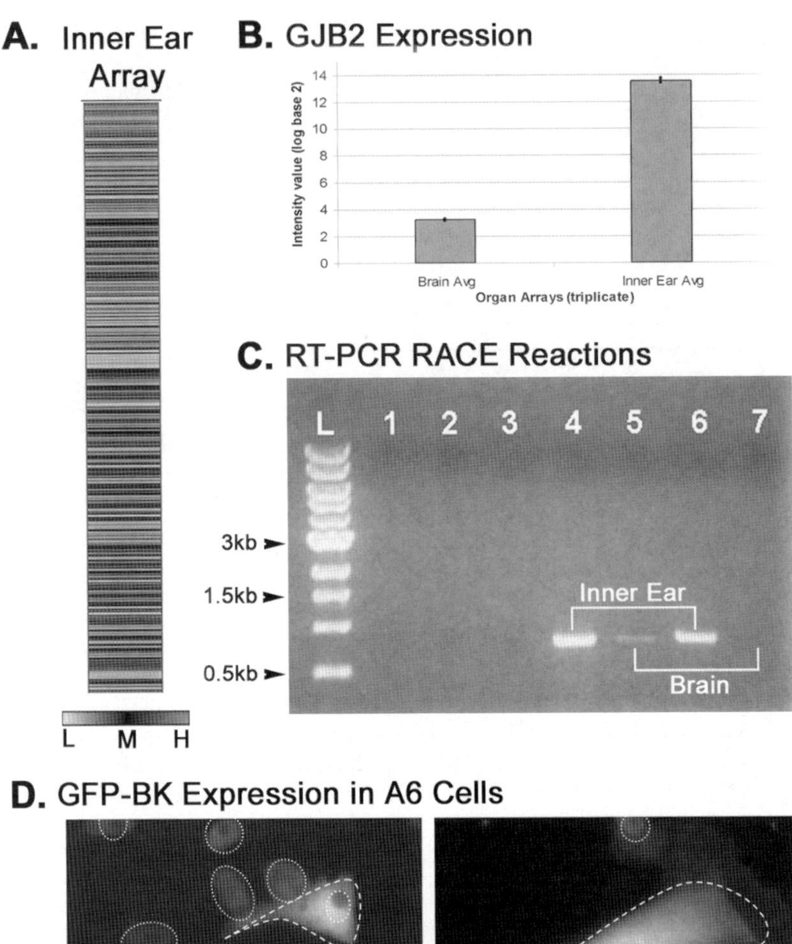

Fig. 1.3. Applications of RNA for transcriptional profiling (**A**, **B**), molecular cloning (**C**), and heterologous gene expression in cell culture (**D**). **(A)** Heat map of inner ear microarray data. *X. laevis* juvenile inner ear total RNA ($\sim 1\,\mu g$) was labeled and hybridized to the GeneChip® *Xenopus laevis* Genome Array. The heat map was constructed using Spotfire® DecisionSite® 9.0 software. The data are organized numerically by probe set ID with the lowest ID number at the top. In this sample, the normalized intensity values (log base 2) for the 15,491 probe sets ranged from 0.8 (low) to 15.22 (high). The intensity color code for expression: L – low, M –intermediate, H – high. **(B)** Microarray detection of the expression of the gap junction protein beta 2 (GJB2) in inner ear and brain. The average and standard deviation values in the histograms were calculated from three different microarray experiments. Low standard deviations characterized the sample replication. **(C)** RT-PCR RACE detection of GJB2 transcript in *X. laevis* juvenile inner ear and brain. Lane 1 – inner ear, minus RT control, GJB2 primer pair 1; Lane 2 – brain, minus RT control, GJB2 primer pair 1; Lane 3 – PCR reaction control, minus cDNA template, GJB2 primer pair 1; Lane 4 – inner ear, GJB2 primer pair 1; Lane 5 – brain, GJB2 primer pair 1; Lane 6 – inner ear, GJB2 primer pair 2; Lane 7 – brain, GJB2 primer pair 2. Lane $\mathrm{L} - 5\,\mu\mathrm{L}$ of NEB 1 kb DNA ladder. cDNA is detected in lanes 4–7. Expression levels correlate with the microarray data shown in **B**. **(D)** GFP-positive *Xenopus* A6 kidney cells transfected with the GFP-BK clone. *Dashed lines* encircle transfected A6 cells (GFP) and *dotted lines* encircle stained nuclei (Hoechst). Scale $\mathrm{bar} = 50\,\mu\mathrm{m}$.

3.6. Molecular Cloning of Genes Identified with Transcriptional Profiling

1. The expression of genes detected with microarray analysis can be confirmed with molecular cloning using RT-PCR RACE protocols.
2. Use *X. laevis* and *X. tropicalis* inner ear or brain total RNA, acquired in **Section 3.3, step 8**, as a template for RT synthesis of first strand cDNA. Prepare 5′-RACE-Ready first strand cDNA in separate reactions, according to the instruction manual of the SMART™ RACE cDNA Amplification Kit, using 1 μg of control RNA from the kit and 1 μg of the following experimental samples: *X. laevis* inner ear or brain total RNA, *X. tropicalis* inner ear or brain total RNA.
3. Run control PCR reactions that amplify the first strand cDNA, prepared from the kit RNA according to the vendor's protocols, to evaluate the success of the 5′ RACE reactions. Analyze control PCR products by agarose gel electrophoresis, as described in the kit manual, before proceeding with the experimental PCR reactions with *Xenopus* samples. If the expected products are observed, use the *Xenopus* first strand cDNA samples for PCR reactions. If products are not detected, the first strand synthesis reactions must be repeated following recommendations outlined in the troubleshooting section of the user manual.
4. Use the 5′-RACE-Ready inner ear or brain first strand cDNA samples (5–10 μL in a 50 μL reaction volume) to amplify second strand cDNA according to the PicoMaxx® High Fidelity PCR System user manual. The thermal cycler parameters are as follows: initial denaturation, 95 °C for 2 min; 35 cycles of 95 °C for 40 s, Tm minus 3 °C for 30 s, 72 °C for 3 min; final extension 72 °C for 10 min. Design gene specific primers to the gene of interest and obtain from Operon (Huntsville, AL). The primer Tm's are specified in the Oligonucleotide Data Sheet provided by Operon and vary by primer pair.
5. Analyze 7 μL of the PCR reaction on a 1.2% agarose gel (0. 25 μg/mL ethidium bromide) in 1X TAE buffer.
6. Isolate the resultant PCR fragments using the S.N.A.P.™ UV-Free Gel Purification Kit, according to the vendor's instructions, until the quantification step. For quantification, analyze 10 μL of the isolated fragment on a Lambda Bio UV/Vis Spectrometer.
7. Clone the purified PCR product using the TOPO® XL PCR Cloning Kit with One Shot® TOP10 Electrocomp™ *E. coli*, according to the kit manual with a 3:1 molar ratio of insert to vector. After the ligation step, use the Electroporator 2510 to transform the electrocompetent cells with a setting of 1400 V.
8. Analyze between 5–10 positive clones by restriction enzyme digestion to confirm the presence of an insert. Sequence

clones with inserts following the BigDye® Terminator v3.1 Cycle Sequencing Kit protocol and analyze sequencing reactions using the ABI PRISM® 3100 Genetic Analyzer.

9. Molecular cloning, using the RT-PCR RACE procedures described above, can be used to confirm the differential expression of genes such as the *X. laevis* EST, GenBank Accession no. BJ076720 (National Institute for Basic Biology Mochii normalized *Xenopus* tailbud library, Clone ID# XL058il6 (3′)). This gene is detected with Affymetrix (Santa Clara, CA) probe set Xl.8924.1.A1_at using the GeneChip® *Xenopus laevis* Genome Array. The Affymetrix annotation states that this probe set targets *X. laevis* connexin 29 (gap junction protein β2, GJB2). Microarray analysis suggests that this gene is expressed at higher levels in the inner ear than in the brain (*see* **Fig. 1.3B**).
10. Design primers to amplify two regions within the 3,292 bp Affymetrix consensus sequence for the Xl.8924.1.A1_at probe set. Use the HMMgene v. 1.1 program *(21)* to determine the putative coding sequence within the consensus sequence for the design of primer pair 1. Use the sequence for the *X. laevis* EST (GenBank Accession #BJ076720) that is arrayed on the chip for the design of primer pair 2.
11. Primer pair 1 amplifies an 804 bp product that includes the entire GJB2 coding sequence. The forward primer (5′-AGTCAGCGCACAGAGACCAA-3′) aligns with the 5′ UTR sequence (31 bp upstream of the translational start site) and the reverse primer (5′-AGCTGACC TGCCACAGTAAC-3′) aligns 7 bp downstream of the stop codon.
12. Primer pair 2 amplifies an 859 bp product within the EST sequence. The forward primer (5′-CGGTCATCATT CAGAGTT-3′) aligns 71 bp upstream of the start of the EST sequence and the reverse primer (5′-ACACTC CAGGAAAACAC-3′) is 24 bp downstream from the end of the EST sequence.
13. Prepare the first strand cDNA from *X. laevis* inner ear and brain total RNA as described previously. The second strand PCR reactions follow the steps outlined in **Section 3.6, step 4** with the following annealing temperatures (Tm): 60 °C for primer pair 1 and 53 °C for primer pair 2.
14. Clone and sequence the resultant PCR products (*see* **Fig. 1.3C**) following the methods described in **Section 3.6, steps 6–7**. Confirm that cloned inserts correspond to the expected GJB2 and EST targets by sequencing. The relative product intensities on the gel parallel the relative transcript levels detected by microarray analysis.

3.7. Heterologous Inner Ear Gene Expression

1. Subclone a TOPO® XL cloned calcium-activated potassium channel (BK) isoform from *X. laevis* inner ear tissue *(9, 14)* that shares 99% nucleotide identity with a posted *Xenopus* spinal cord GenBank sequence (Accession no. AF274053) into the pAcGFP1-C1 expression vector.
2. Use *Xba*I and *Spe*I restriction enzymes to digest 5 μg of the BK TOPO® XL clone, following the vendor's recommended reaction conditions. Isolate the insert using the S.N.A.P.™ UV-Free Gel Purification Kit as described in **Section 3.6, step 6**.
3. Prepare the vector, pAcGFP1-C1, by digesting 1 μg of plasmid DNA with *Xba*I restriction enzyme and then dephosphorylating with Antarctic phosphatase, following the vendor's protocol.
4. Use the Quick Ligation™ Kit to clone the BK isoform fragment into the GFP expression vector, following the vendor's protocol. Confirm positive clones by restriction enzyme analysis and subsequent sequencing reactions using the BigDye® Terminator v3.1 Cycle Sequencing Kit.
5. Culture the *X. laevis* A6 kidney cell line according to the vendor's recommendations and maintain in antibiotic-free media at 26 °C in 5% CO_2.
6. Passage A6 cells at a plating density of 4.5×10^4 cells/cm^2 on two- or four-chamber culture slides 24 h prior to transient transfections.
7. Transfect cells with a final concentration of 2 μg/mL of Lipofectamine 2000 Transfection Reagent and 1 μg/mL of plasmid diluted in serum free NCTC-109 medium, following the vendor's protocol. Transfect cultures with Lipofectamine without plasmid in control experiments.

3.8. Epifluorescence Imaging

1. After 24–48 h, wash cells gently in serum free NCTC-109 medium.
2. Fix cells in 3.7% paraformaldehyde for 10 min at room temperature and wash with PBS (pH 7.4).
3. Expose cells to 1 μg/mL Hoechst 33342 diluted in PBS (pH 7.4) for 10 min at room temperature followed by a PBS (pH 7.4) wash.
4. Prepare slides for imaging by adding AntiFade Solution C from the *SlowFade*® Antifade Kit to the chambers for 15 min at room temperature. Remove the AntiFade Solution C and detach the side walls of the culture chambers.
5. Add a few drops of AntiFade Solution A from the *SlowFade*® Antifade Kit to the slide prior to applying a cover-slip. Seal the cover-slip with clear fingernail polish.
6. Capture images using a CCD camera connected to an epifluorescent microscope (e.g., CoolSNAP™ HQ CCD camera

connected to a Nikon TE2000 epifluorescence microscope with UV-2A, Hoechst and B-2E/C, GFP filter cubes) and the MetaVue™ Imaging System software (*see* **Fig. 1.3D**).

7. Process images offline using appropriate software (e.g., MetaMorph® Offline Imaging System software and Adobe® Photoshop® Version 7.0.1 for figure preparation).

4. Notes

1. Prepare all solutions in Milli-Q water (18 MΩ-cm) with a total organic content of less than five parts per billion.
2. Solutions of ethyl 3-aminobenzoate methanesulfonate salt are buffered to pH 7.0 with sodium bicarbonate (cat. no. S5761, Sigma-Aldrich). Use gloves when preparing and handling solutions of ethyl 3-aminobenzoate methanesulfonate salt. Prepare the stock solution in a chemical fume hood to avoid inhalation of dust.
3. WARNING: DEPC is a combustible explosive and a toxic chemical. It is recommended that users purchase small amounts of DEPC and use all the DEPC as soon as possible. Take appropriate precautions when handling: e.g., always wear personal protective equipment and work in an approved chemical fume hood. Dilute DEPC in water to a final concentration of 0.1%. After adding the DEPC, close the container tightly and vigorously shake. Allow the solution to sit overnight at room temperature and autoclave for 20 min before use. The DEPC-treated water can be stored at room temperature after autoclaving. Solutions requiring DEPC-treated water are so indicated in the text.
4. The Polytron blade can be removed carefully from the outer shaft of the generator, and both the blade and outer shaft must be thoroughly cleaned. RNase*Zap*® is copiously applied and the surfaces are lightly scrubbed with Kimwipes. Rinse the blade and outer shaft several times with DEPC-treated water, rinse once with 70% ethanol, and allow to air dry. (Reminder: 70% ethanol should be made with DEPC-treated water). This cleaning procedure should be repeated between tissue samples.
5. Typically, the amount of tissue recovered from *Xenopus* juvenile animals is 1 mg per inner ear and 10 mg per brain. The amount of tissue recovered from Stage 56 *X. laevis* is 0.7 mg per inner ear. Total RNA is isolated by combining tissue from at least 3 animals.
6. The average yields of total RNA using the Qiagen RNeasy® kit are as follows: *X. laevis* juvenile −1.8 μg per brain and 0.3 μg per inner ear; *X. laevis* Stage 56 – 0.3 μg per inner

ear; *X. tropicalis* juvenile $-1.9\,\mu g$ per brain and $0.2\,\mu g$ per inner ear. Downstream protocols (i.e. Microarray analysis and RT-PCR) typically require at least $1\,\mu g$ of total RNA.

7. The Agilent 2100 Bioanalyzer allows precise optimization of total RNA isolation procedures (*see* **Fig. 1.2**). The Bioanalyzer can detect ng quantities of nucleic acids in $1\,\mu L$ and also quantifies RNA integrity with a number (RIN). Therefore, this instrument is superior to spectrophotometer quantification (260/280 ratios) for evaluating RNA quality.
8. The GeneChip® One-Cycle Target Labeling Kit procedure can successfully label synthesized cRNA from 1–8 μg of total RNA. However, in one reaction of 12 completed with these protocols, a smaller amount ($0.8\,\mu g$) of *X. laevis* inner ear total RNA produced 22 μg of biotin-labeled cRNA.

Acknowledgments

The authors thank Joanna Beeson for technical support with cell culture experiments, and Alicia Arguelles and Erica Koval for their assistance with manuscript preparation. We are grateful to Dr. Charlie Whittaker, of the MIT Center for Cancer Research, and to Manlin Luo and Dr. Rebecca Fry, of the MIT BioMicro Center, for their technical assistance and generous advice. We also acknowledge Dr. Peter Sorger and members of the MIT Cell Decision Processes Center for stimulating intellectual interactions and for facilitating access to essential equipment and computing resources. Funding for this research was provided by awards from the National Institutes of Health (NIGMS-MORE S06GM008136, NIDCD R01DC003292, and NIGMS P50GM068762).

References

1. Tinsley, R. C. and Kobel, H. R. (ed.) (1996) *Biology of Xenopus*. Oxford University Press, Oxford.
2. Nieuwkoop, P. D. and Faber, J. (ed.) (1967) *Normal Table of Xenopus laevis (Daudin). A Systematical and Chronological Survey of the Development from the Fertilized egg Till the End of Metamorphosis*. 2nd Edition. North-Holland Publishing, Co., Amsterdam.
3. Amaya, E. (2005) Xenomics. *Genome Res.* **15**, 1683–1691.
4. Showell, C. and Conlon, F. L. (2007) Decoding development in *Xenopus tropicalis*. *Genesis* **45**, 418–426.
5. Diaz, M. E., Varela, A., and Serrano, E. E. (1995) Quantity, bundle types, and distribution of hair cells in the sacculus of *Xenopus laevis* during development. *Hear. Res.* **9**, 33–42.
6. Serrano, E. E., Trujillo-Provencio, C., Sultemeier, D., Bullock, W. M., and Quick, Q. A. (2001) Identification of genes expressed in the *Xenopus* inner ear. *Cell. Mol. Biol.* **47**, 1229–1239.
7. Quick, Q. A. and Serrano, E. E. (2005) Inner ear formation during the early larval development of *Xenopus laevis*. *Dev. Dyn.* **234**, 791–801.

8. Varela-Ramirez, A., Trujillo-Provencio, C., and Serrano, E. E. (1998) Detection of transcripts for delayed rectifier potassium channels in the *Xenopus laevis* inner ear. *Hear. Res.* **119**, 125–134.
9. Sultemeier, D. R., Provencio-Trujillo, C., and Serrano, E. E. (2007) Cloning and characterization of *Xenopus* inner ear calcium-activated potassium channel α- and β-subunits. *ARO. Abstr.* **30**:740.
10. Gabashvili, I. S., Sokolowski, B. H., Morton, C. C., and Giersch, A. B. (2007) Ion channel gene expression in the inner ear. *J. Assoc. Res. Otolaryngol.* **8**, 305–328.
11. Schlosser, G. and Northcutt, R. G. (2000) Development of neurogenic placodes in *Xenopus laevis*. *J. Comp. Neurol.* **418**, 121–146.
12. Humes, H. D. (1999) Insights into ototoxicity: Analogies to nephrotoxicity. *Ann. N. Y. Acad. Sci.* **884**, 15–18.
13. Powers, T., Trujillo-Provencio, C., Whittaker, C., and Serrano, E. E. (2007) Gene expression profiling of *Xenopus* organs yields insight into the *Xenopus* inner ear transcriptome. *ARO. Abstr.* **30**:741.
14. Sultemeier, D. R., Knight, V. B., Manuelito, S. J., Hopkins, M., and Serrano, E. E. (2005) Heterologous and homologous expression systems for functional analysis of *Xenopus* inner ear genes. *ARO. Abstr.* **28**:28.
15. Agilent Technologies, Inc. (2003) Reagent Kit Guide: RNA 6000 Nano Assay. November 2003 Edition, Germany.
16. Petersen, M. B. and Willems, P. J. (2006) Non-syndromic, autosomal-recessive deafness. *Clin. Genet.* **69**, 371–392.
17. Ambion Inc. (2008) The basics: RNase control. http://www.ambion.com/techlib/basics/rnasecontrol/index.html
18. Nasco. (2008). Animal Protocol for *Xenopus laevis*, African Clawed Frog Colony. http://www.enasco.com/Static.do?page=xen_rcare
19. Grainger Lab. (2001) Grainger Lab *X. tropicalis* Adult Husbandry Protocol. http://faculty.virginia.edu/xtropicalis/husbandry/TropadultcareNew.htm
20. University of Arizona: IACUC Learning Module – *Xenopus laevis*. (2008) Care and Handling of *Xenopus laevis*. http://www.iacuc.arizona.edu/training/xenopus/.
21. Center for Biological Sequence Analysis. (2007) HMMgene v. 1.1 Program. http://www.cbs.dtu.dk/services/HMMgene/

Chapter 2

Synthesis of Biotin-Labeled RNA for Gene Expression Measurements Using Oligonucleotide Arrays

Ana E. Vázquez, Liping Nie, and Ebenezer N. Yamoah

Abstract

Using gene arrays, it is currently possible to simultaneously measure mRNA levels of many genes in any tissue of interest. Undoubtedly, comprehensive measurements of gene expression as part of carefully designed experiments will continue to further our understanding of audition and have the potential to open up new avenues of research. This chapter describes a reliable protocol to prepare high-quality biotin-labeled RNA target, specifically for oligonucleotide array experiments. The procedure includes isolation of high-quality total RNA, synthesis of double-stranded cDNA engineered for in vitro transcription with T7 RNA polymerase, subsequent in vitro transcription in the presence of biotin-labeled ribonucleotides, and fractionation of the RNA to ~ 500 bp fragments, suitable for oligonucleotide array experiments. Because the membranous labyrinth is composed of functionally interdependent cellular structures, which themselves contain numerous, highly differentiated cell types, comprehensive analysis of gene expression in the cochlea is best complemented by immunohistotochemical studies or, if no suitable antibodies are available, by in situ hybridization studies. Either one of these techniques will identify the specific cell types that express the genes of interests.

Key words: Gene expression, oligonucleotide microarrays, gene chips, membranous labyrinth, cochlea.

1. Introduction

Significant improvements in molecular biology methods currently allow precise, simultaneous measurements of a large percentage of transcript levels in a tissue of choice. Optimization of RNA labeling methods, better quality arrays, as well as the development of high-density scanners have made microarray technology an efficient and reliable tool for quantifying levels of gene expression. As a result of these improvements, measurements of gene expression

Bernd Sokolowski (ed.), *Auditory and Vestibular Research: Methods and Protocols, vol. 493*

DOI 10.1007/978-1-59745-523-7_2 Springerprotocols.com

levels are now feasible at the level of the transcriptome, even if only very small amounts of tissue or a few cells are available *(1–3)*. Consequently, array experiments, monitoring changes of gene expression relevant to specific biological processes, are contributing to the progress of numerous research fields. Other molecular methods for measuring gene expression, such as Northern blot analysis or quantitative polymerase chain reaction, preclude the analysis of large numbers of genes. Additionally, Northern blot analysis requires the use of large amounts of RNA and, thus, is not a practical method for small sample sizes.

Presently, a number of platforms are used for global transcript analyses, including cDNA and, short (~ 25 nucleotides) and long (50–70 nucleotides) oligonucleotide arrays. Microarrays of spotted cDNAs are available and can be custom made or spotted on demand. These arrays have an advantage in that they can be designed and spotted, to ensure that complex transcriptomes of specific tissues are well represented or that the array is tailored to the specific scientific questions of interest. Because these arrays are usually prepared on slides by investigators, there is a lower cost associated with them. Collections of cDNAs for auditory research have been compiled into cDNA microarrays *(4, 5)*. For example, spotted cDNA arrays were utilized to study the effects of noise in the cochlea *(6, 7)*.

Oligonucleotide arrays are more costly than cDNA-spotted slides. The commercial oligonucleotide arrays, however, have the advantage that multiple independent measurements are made for each transcript, providing high measurement reliability. It is our experience that the technical variability in experiments, using oligonucleotide arrays, can be almost negligible. For example, high reproducibility was reported using an Affymetrix (Santa Clara, CA) zebrafish genome oligonucleotide array *(3)*. The correlation coefficient for comparisons between replicate samples of zebrafish hair cells and liver gene expression measurements were 0.95 and 0.99, respectively *(3)*, lending high confidence to this methodology. We have focused on the Affymetrix system, but various other biotechnology companies provide high-quality oligonucleotide arrays also. Commercially available oligonucleotide arrays now represent whole genomes of various organisms. For example, the GeneChip Mouse Genome 430A 2.0 Array (Affymetrix) contains most well-characterized genes and expressed sequence tags of GenBank®, dbEST, and RefSeq databases. Oligonucleotide microarrays have been used in studies of hair cell differentiation *(8)* and in studies of the effects of noise in the cochlea *(9)*.

Since changes in gene expression are at the core of many biological processes, such as development, differentiation, response to stress, and apoptosis, the ability to monitor gene expression changes is highly valuable to the understanding of these processes.

This chapter describes the preparation of target biotin-labeled RNA for short (~ 25 nucleotides) oligonucleotide arrays, including the following steps: RNA purification, synthesis of double-stranded cDNA containing a T7 promoter sequence, cDNA purification, in vitro transcription in the presence of biotin-labeled ribonucleotides, and fractionation of RNA. The RNA labeling protocol is based on the priming of cDNA synthesis with an oligo (dT) primer, containing the T7 promoter sequence. Thus, the cDNA obtained can be used for subsequent in vitro transcription with T7 RNA polymerase *(10)*. The T7 polymerase performs a linear amplification, and the in vitro RNA obtained reflects the abundance of each transcript in the initial RNA *(11–13)*.

Although outside of the scope of this chapter, our recommendation is to devote serious effort to the planning of the array experiments prior to tissue collection. The Gene Expression Omnibus (GEO) website, http://www.ncbi.nlm.nih.gov/geo/, is an invaluable resource for gene expression studies. Inner ear and other hearing-related gene expression datasets are currently easily accessed on GEO. Investigators are encouraged to familiarize themselves with GEO and to evaluate the datasets available prior to designing array experiments. Upon completion of the experiments, investigators are urged to submit their results so that they will be available for the benefit of the scientific community.

2. Materials

2.1. Membranous Labyrinth Tissue Collection

1. RNase-free dissection tools. Clean all tools with RNase *AWAY*.
2. RNA *later* RNA stabilizing reagent (QIAGEN Inc., Valencia, CA).

2.2. RNA Isolation

For RNA isolation, we routinely use solutions and columns from the RNeasy Protect Mini Kit (cat. no. 74124, QIAGEN Inc, Valencia, CA).

1. β-mercaptoethanol.
2. RLT buffer (RNeasy Protect Mini Kit).
3. DNase I stock solution: Prepare using RNase-free DNase Set (QIAGEN Inc, Valencia, CA). Before using, dissolve the solid DNaseI (1500 Kunitz units) in 550 μL of RNase-free water and mix gently by inverting the tube only; do not vortex. Divide into 10 μL aliquots and store at −20 °C for up to 9 months.
4. Buffer RDD (RNeasy Protect Mini Kit).
5. Ethanol.
6. Buffer RPE with ethanol added (RNeasy Protect Mini Kit).
7. Tris buffer: 10 m*M* Tris-HCl, pH 8.0.

2.3. Synthesis of Biotin-Labeled RNA Target for Array Analysis

1. MessageAmpII aRNA Kit (cat. no. 1751, Ambion, Inc., Austin, TX) (*see* **Note 1**).
2. 10 m*M* Biotin-11-CTP (Perkin Elmer Life Sciences, Waltham, MA).
3. 10 m*M* Biotin-16-UTP (Roche Molecular Biochemicals Indianapolis, IN).
4. 100% Ethanol.
5. RNase-free water: Prepare by adding 0.1% volume of diethyl pyrocarbonate to nanopure water, shake vigorously to mix well, incubate overnight at 37 °C, and autoclave the next day.
6. 70% ethanol prepared with RNase-free water.
7. 3 *M* NaOAc, pH 5.2, RNase free.
8. 1X Tris-Borate-EDTA (TBE) electrophoresis buffer.

2.4. Fragmentation of In Vitro Transcribed Biotin-Labeled RNA

1. 5X RNA fragmentation buffer; use RNase-free reagents: 200 m*M* Tris-acetate, pH 8.1, 500 m*M* KOAc, 150 m*M* MgOAc. Combine 4 mL of 1 *M* Tris-acetate, pH 8.1, with 0.64 g of MgOAc and 0.98 g of KOAc. Mix thoroughly and filter through a 0.2 μm filter. Aliquot and store at room temperature.

2.5. Gel Electrophoresis to Evaluate the Progression of the Fragmentation Reaction

1. 1X TBE electrophoresis buffer.
2. DNA electrophoresis loading dye.

2.6. Equipment

1. Homogenizer (e.g., PowerGen125, Thermo Fisher Scientific, Pittsburgh, PA).
2. Eppendorf 5804 R centrifuge (Eppendorf of North America, Westbury, NY).
3. Water bath or hot plate.
4. Spectrophotometer.
5. Thermalcycler.
6. Gel electrophoresis apparatus.

3. Methods

3.1. Membranous Labyrinth Tissue Collection

1. Dissect the membranous labyrinth from each cochlea quickly under RNase-free conditions, and place the tissues into 100 μL of RNA*later* (*see* **Notes 2** and **3**).

3.2. RNA Isolation

For RNA isolation, we routinely use the QIAGEN RNeasy Protect Mini Kit, including the in-column DNase digestion specifically as follows:

1. Calculate and prepare the total amount of β-mercaptoethanol/RLT buffer solution needed for all extractions (600 μL of β-mercaptoethanol/RLT buffer solution will be needed for each extraction). Prepare sufficient amount of this solution by adding 10 μL of β-mercaptoethanol per 1 mL of RLT buffer and mix thoroughly by vortexing.
2. Remove the RNA*later* buffer from the tissues.
3. Add 600 μL of β-mercaptoethanol/RLT buffer solution to the tissue (*see* **Note 4**).
4. Homogenize immediately for 40 s using three-fourth of the maximum speed of the PowerGen 125 homogenizer.
5. Centrifuge for 3 min at maximum speed in a microcentrifuge.
6. Remove the supernatant to a new tube.
7. Add 1 volume (600 μL) of 70% ethanol to the lysate and mix immediately.
8. Place the RNease column in a 2 mL collection tube, pipette up to 700 μL of the sample into the column, and centrifuge for 15 s at ≥ 8000 g. Discard the flow-through.
9. Apply the rest of the sample volume by loading aliquots of up to 700 μL, successively, onto the column. Centrifuge after each loading, as in **step 8, Section 3.2**, and repeat until the entire sample is loaded onto the column.
10. Pipet 350 μL of buffer RW1 into the RNeasy column and centrifuge for 15 s at ≥ 8000 g.
11. Add 70 μL of buffer RDD to 10 μL of DNase stock solution and mix by gently inverting the tube. Then, pipette this mix directly onto the RNeasy silica-gel membrane and incubate at room temperature for 15 min.
12. Pipet 350 μL of buffer RW1 into the RNeasy mini column and centrifuge for 15 s at ≥ 8000 g.
13. Place the RNeasy column into a new 2 mL collection tube and add 500 μL of buffer RPE (with ethanol added), and centrifuge for 15 s at ≥ 8000 g to wash the column. Discard the flow-through.
14. Wash with RPE buffer (with ethanol added) again. Repeat **step 13, Section 3.2**, discard the flow-through, and centrifuge again for 1 min at ≥ 8000 g to thoroughly dry the column.
15. Elute the RNA: Transfer the column to a new tube and add 30 μL of warm (50–55 °C) RNase-free water, wait for 1 min, and centrifuge for 1 min as in **step 14, Section 3.2**.
16. Prepare a 1:100 dilution of the obtained RNA in Tris buffer (*see* **Note 5**). Determine the amount of RNA, using a spectrophotometer.

3.3. Synthesis of Biotin-Labeled RNA Target for Array Analysis

We routinely use the MessageAmp™ II aRNA Amplification System (cat. no. 1751 Ambion, Inc., Austin, TX) to synthesize biotin-labeled RNA from the purified RNA obtained above. We recommend following Ambion's method exactly for:

1. Reverse transcription for first-strand cDNA synthesis (*see* **Note 6**).
2. Second-strand cDNA synthesis.
3. Purification of double-stranded cDNA.
4. In vitro transcription to synthesize biotin-labeled aRNA (*see* **Note 7**).

3.4. Quantitation of aRNA Yield

Prepare a 1:100 dilution of the RNA in Tris-HCl buffer and measure the RNA concentration (*see* **Note 8**).

3.5. Fragmentation of In Vitro Transcribed Biotin-Labeled RNA

1. Precipitate the labeled RNA and solubilize the RNA pellet to obtain a 0.64 μg/μL concentration with RNase-free water.
2. Set up fragmentation reactions: 32 μL of RNA (~ 21 μg) and 8 μL of fragmentation buffer (*see* **Note 9**). Set a thermocycler in incubation mode to hold at 94 °C, use a timer, allow fragmentation at 94 °C for 35 min exactly, and place the tube on ice following the incubation.
3. Load 1 μg of fragmented RNA on a 1% agarose gel to evaluate the progress of the fragmentation reaction (*see* **Note 10**).

3.6. Perform Hybridization Experiments to Test Array for Evaluation of Target Quality

It is recommended to further evaluate the quality of the in vitro–transcribed RNA for gene-expression profiling by performing a hybridization experiment to Test3 Arrays. The Test3 Array is an economically priced array. The efficiency of the labeling reaction should be evaluated using these arrays, by inspecting the signal intensity of the housekeeping genes for the organism of interest (*see* **Note 9**). The 3′–5′ signal ratios for genes of the organism of interest should be between 1 and 3.

3.7. Perform Hybridization Experiments to Oligonucleotide Microarray of Choice

The hybridization, washing, staining, and scanning of oligonucleotide arrays are usually performed at a microarray core facility. These protocols are outside of the scope of this chapter. Information about these procedures can be obtain at http://www.affymetrix.com/support/downloads/manuals/expression_analysis_technical_manual.pdf

3.8. Data Analysis, Validation, Data Sharing, and Annotation

Data analysis, validation, data annotation, and sharing are all important components of all array experiments. However, these are outside the scope of this chapter. Briefly, array data analysis may be performed using one of the various software packages available without cost such as DNA-Chip Analyzer software (dChip v.4/14/06, http://biosun1.harvard.edu/complab/dchip/) and BRB array tools from the National Cancer

Institute (NIH) http://linus.nci.nih.gov/BRB-ArrayTools.html. Alternatively, data analysis software packages may be obtainable from the microarray core facilities (*see* **Notes 2, 10, 11**). After completion of array experiments, validation of the data, for a subset of genes of interest to the investigator, should be performed using quantitative PCR, immunohistochemistry, or in situ hybridization. After validation, the data should be submitted to GEO (http://www.ncbi.nlm.nih.gov/geo/) (see **Note 11**).

4. Notes

1. Make sure to store each of the components of the MessageAmpII aRNA at the temperatures indicated. Do not store enzymes in a frost-free freezer.
2. Gene expression experiments must be planned carefully. We recommend becoming familiar and evaluating the array data that is already available at the GEO website, http://www.ncbi.nlm.nih.gov/geo/, which may be relevant to the scientific questions being evaluated. We also recommend that for each experimental condition, tissues be collected and pooled into at least three independent but representative groups (a larger number of replicates should not be required). Isolate total RNA from each of these groups of tissues independently; the RNA samples and thus the resulting arrays within the same group will then be biological replicates.
3. For gene expression studies, the membranous labyrinth of each mouse should be dissected quickly under RNase-free conditions. We routinely place the tissues into 100 μL of RNA*later* (QIAGEN Inc.). Tissues should be stored in RNA*later* at 4 °C overnight and then placed in −20 °C the next morning until further use.
4. Perform **steps** 4–8 continuously and quickly, and always at room temperature. Make sure all centrifugation steps are carried out between 22 °C and 25 °C, not cold. This method yields approximately 2. 5 μg of total RNA from 16 pooled mice membranous labyrinths. The RNA can be stored at −20 °C. However, it is recommended to continue the procedure as soon as possible.
5. For quantitation purposes, dilute the RNA in buffer and not in water, because in water the absorption will not be the same and the measurement will be inaccurate. The OD_{260}/OD_{280} ratio should be between 1.7 and 2. There are more sophisticated methods of obtaining RNA concentration measurements, but some of them require expensive instrumentation. However, it is our experience that spectrophotometer measurements of RNA in buffer are adequate for the purpose of this protocol.

6. The manufacturer (Ambion, Inc., Austin, TX) recommends using 1 μg of RNA. We find that this is not required. Occasionally, we have used 400 ng to obtain high-quality and sufficient target-labeled RNA for more than one array (amounts as low as 250 ng are likely to yield labeled RNA for more than one array). Each array requires 5 μg for the test array and 15 μg of biotin-labeled RNA for the experimental array.
7. Perform in vitro transcription for 14 h.
8. The RNA concentration in the fragmentation reaction should be between 0.5 and 2 μg/μL.
9. Agarose gel electrophoresis and ethidium bromide staining for visualization is sufficient to evaluate the progress of the fragmentation reaction. After fragmentation, a change in the size of the labeled RNA should be apparent. The size of the fragmented RNA should be ~ 200 bp or no larger than the 500 bp DNA marker. We recommend performing a hybridization to a "Test3 Array" and evaluating the detection signal as well as the 3′–5′ signal ratios for the genes of the organism of interest.
10. With any software of choice, it is our recommendation to compute the gene expression levels, the group means and standard errors, by pooling arrays (three arrays of each experimental group are adequate for many experiments). This approach helps to account for measurement inaccuracies and allows for the elimination of outliers (http://biosun1.harvard.edu/complab/dchip). The arrays within the same group are biological replicates.
11. Array data relevant to hearing research is available at GEO (http://www.ncbi.nlm.nih.gov/geo/). Investigators are urged to familiarize themselves with the data already available and relevant to their field of interest, and to submit their results so that they will be accessible to the scientific community.

Acknowledgments

We would like to thank Dr. Kirk Beisel for critical reading of this chapter.

References

1. Eberwine, J., Kacharmina, J. E., Andrews, C., Miyashiro, K., McIntosh, T., Becker, K., Barrett, T., Hinkle, D., Dent, G., and Marciano, P. (2001) mRNA expression analysis of tissue sections and single cells. *J. Neurosci.* **21**, 8310–8314.
2. Kacharmina, J. E., Crino, P. B., and Eberwine, J. (1999) Preparation of cDNA from single cells and subcellular regions. *Meth. Enzymol.* **303**, 3–18.
3. McDermott, B. M., Jr., Baucom, J. M., and Hudspeth, A. J. (2007) Analysis and

functional evaluation of the hair-cell transcriptome. *Proc. Nat. Acad. Sci. USA* **104**, 11820–11825.

4. Pompeia, C., Hurle, B., Belyantseva, I. A., Noben-Trauth, K., Beisel, K., Gao, J., Buchoff, P., Wistow, G., and Kachar, B. (2004) Gene expression profile of the mouse organ of Corti at the onset of hearing. *Genomics* **83**, 1000–1011.
5. Morris, K. A., Snir, E., Pompeia, C., Koroleva, I. V., Kachar, B., Hayashizaki, Y., Carninci, P., Soares, M. B., and Beisel, K. W. (2005) Differential expression of genes within the cochlea as defined by a custom mouse inner ear microarray. *J. Assoc. Res. Otolaryngol.* **6**, 75–89.
6. Lomax, M. I., Gong, T. W., Cho, Y., Huang, L., Oh, S. H., Adler, H. J., Raphael, Y., and Altschuler, R. A. (2001) Differential gene expression following noise trauma in birds and mammals. *Noise Health* **3**, 19–35.
7. Taggart, R. T., McFadden, S. L., Ding, D. L., Henderson, D., Jin, X., Sun, W., and Salvi, R. (2001) Gene expression changes in chinchilla cochlea from noise-induced temporary threshold shift. *Noise Health* **3**, 1–18.
8. Rivolta, M. N., Halsall, A., Johnson, C. M., Tones, M. A., and Holley, M. C. (2002) Transcript profiling of functionally related groups of genes during conditional differentiation of a mammalian cochlear hair cell line. *Genome Res.* **12**, 1091–1099.
9. Kirkegaard, M., Murai, N., Risling, M., Suneson, A., Jarlebark, L., and Ulfendahl, M. (2006) Differential gene expression in the rat cochlea after exposure to impulse noise. *Neuroscience* **142**, 425–435.
10. Van Gelder, R. N., von Zastrow, M. E., Yool, A., Dement, W. C., Barchas, J. D., and Eberwine, J. H. (1990) Amplified RNA synthesized from limited quantities of heterogeneous cDNA. *Proc. Nat. Acad. Sci. USA* **87**, 1663–1667.
11. Li, Y., Li, T., Liu, S., Qiu, M., Han, Z., Jiang, Z., Li, R., Ying, K., Xie, Y., and Mao, Y. (2004) Systematic comparison of the fidelity of aRNA, mRNA and T-RNA on gene expression profiling using cDNA microarray. *J. Biotech.* **107**, 19–28.
12. Feldman, A. L., Costouros, N. G., Wang, E., Qian, M., Marincola, F. M., Alexander, H. R., and Libutti, S. K. (2002) Advantages of mRNA amplification for microarray analysis. *Biotechniques* **33**, 906–912, 914.
13. Polacek, D. C., Passerini, A. G., Shi, C., Francesco, N. M., Manduchi, E., Grant, G. R., Powell, S., Bischof, H., Winkler, H., Stoeckert, C. J., Jr., and Davies, P. F. (2003) Fidelity and enhanced sensitivity of differential transcription profiles following linear amplification of nanogram amounts of endothelial mRNA. *Physiol. Genomics* **13**, 147–156.

Chapter 3

In situ Hybridization Approach to Study mRNA Expression and Distribution in Cochlear Frozen Sections

Hakim Hiel

Abstract

In situ hybridization is well suited to obtaining specific topological information on gene transcripts and thereby to relating such observations to a particular function. In spite of the technical and practical difficulties, the application of molecular biological techniques such as in situ hybridization to the cochlea can provide important insights. However, the rarity of gene products (mRNA and proteins) in the cochlea and its fragile structure require the refinement and adaptation of in situ hybridization methods. The present chapter provides a detailed in situ hybridization protocol adapted to frozen tissue sections collected from adult and neonatal stages of the vertebrate cochlea.

Key words: Cochlea, in situ hybridization, ear, gene expression, autoradiography, digoxigenin-UTP, ^{35}S-UTP, vestibule, riboprobes.

1. Introduction

In situ hybridization involves the use of tagged nucleic acid segments to detect specific mRNAs or DNAs within their preserved cytological or histological surroundings. The technique was first reported by two concurrent studies, John et al. *(1)* and Gall and Pardue *(2, 3)*. Gall and Pardue *(2, 3)* performed in situ hybridization by interacting radioactive ribosomal RNA with DNA from *Xenopus* oocyte nuclei immobilized on a solid or semi-solid matrix. The bound label was then detected by autoradiography. Later, with the immense advancements of nucleic acid biochemistry and molecular cloning techniques, in situ hybridization procedures were refined and applied to a variety of tissues

Bernd Sokolowski (ed.), *Auditory and Vestibular Research: Methods and Protocols, vol. 493*

DOI 10.1007/978-1-59745-523-7_3 Springerprotocols.com

from brain sections to whole-mount organs and even to whole *Drosophila* larvae *(4–10)*.

While this novel technique was applied to the study of gene expression in many tissue types, its application to the inner ear lagged behind. The scarcity of gene products and the delicate and fine structure of the cochlear and vestibular sensory epithelia were an impediment to the quick application of this novel technique to the inner ear. Only in the early 1990s was in situ hybridization used successfully to study gene expression in inner ear organs. Ryan et al. *(5, 6)*, in two successive articles, described the application of this approach to the cochlear end organ. In the first report the authors provided a comprehensive description of in situ hybridization and the effect of decalcification on the preservation of the cochlear mRNA pool; the second report described the expression of non-NMDA glutamate receptor mRNA throughout the cochlear duct.

In situ hybridization provides critical information not only about the expression pattern of mRNA within the whole organ but also about its cellular compartmentalization. Hybridization signals can be visualized by different approaches: radioactive (autoradiography) and non-radioactive (chromogenic reaction or fluorescence). The former approach, while more sensitive and informative about the pattern of distribution of mRNA within an entire organ, has limited spatial resolution due to scatter of the autoradiography signal. The non-radioactive method on the other hand (chromogenic reaction or fluorescence) provides better spatial resolution for cellular mRNA distribution while its lower sensitivity requires a substantial level of expression of the gene product in question. Of course, either approach can only report transcription; immunohistochemistry is required to demonstrate translation into the protein product. The present chapter provides a detailed protocol for in situ hybridization as applied to inner ear tissues, ranging from tissue and probe preparation to hybridization signal detection.

2. Materials

2.1. Tissue Preparation

1. 0.12 *M* phosphate buffer, pH 7.4: Dissolve 16.6 g of sodium phosphate monobasic (NaH_2PO_4) in 800 mL of DEPC water. After adjusting the pH to 7.4, bring the final volume to 1 L with DEPC-treated water.
2. Fixative solution: 4% paraformaldehyde (PFA) in 0.12 *M* phosphate buffer, pH 7.4.
3. 0.5 *M* ethylene di-amine tetra-acetic disodium dihydrate (EDTA), which corresponds to an 18.61% solution, pH 8.0.

4. Decalcification solution: 5–8% EDTA with 2–3% PFA in 0.09 *M* phosphate buffer, pH 7.4–8.0. For a 100 mL working solution, mix 26.9 mL of 0.5 *M* EDTA, pH 8.0, with 73.1 mL of fixative solution.
5. 1 *M* NaOH.
6. 1 *M* HCl.
7. Cryo-protection solution: 30% sucrose in 4% PFA in 0.12 *M* phosphate buffer, pH 7.4.
8. Cryo-sectioning: SuperFrost Plus Slides (Thermo Fisher, Milford, MA), optimal cutting temperature (OCT) medium (Electron Microscopy Sciences, Hatfield, PA), silica gel, and slide boxes that hold 25 slides for tissue section storage at −80 °C or −20 °C.
9. Slide warmer.
10. Iso-pentane.
11. Liquid nitrogen.
12. AbSolve (Perkin Elmer, Shelton, CT): Detergent used to decontaminate glassware, plasticware, and metalware from any RNase or DNase presence. Dissolve 20 mL in 1 L of double-distilled water (ddH_2O) and soak overnight.

2.2. Probe Preparation: In Vitro Transcription

1. 1 *M* Tris-HCl, pH 8.0.
2. 0.5 *M* EDTA, pH 8.0.
3. *D*i-*e*thyl *p*yro*c*arbonate (DEPC)–treated water: Add 0.1% (v/v) DEPC to ddH_2O in autoclavable bottles, shake well, and incubate overnight under a fumehood. The next day, loosen the lids on the bottles and autoclave for 40 min.
4. Gel extraction kit from Qiagen (Valencia, CA).
5. Agarose, molecular biology grade (Invitrogen, Carlsbad, CA).
6. Selected restriction enzymes with their respective buffers (New England Biolabs; Ipswich, MA).
7. 10X digoxigenin labeling mix, which contains untagged UTP, GTP, CTP, and ATP (Roche Applied Sciences, Indianapolis, IN) in addition to digoxigenin-UTP.
8. Control digoxigenin-labeled RNA (Roche Applied Sciences).
9. 3 *M* sodium acetate.
10. 100% ethanol.
11. 4 *M* LiCl.
12. RNase inhibitor or RNA guard.
13. Tris-HCl buffered saline (TBS), pH 7.5: Mix 100 m*M* Tris-HCl with 150 m*M* NaCl.
14. Nylon membrane for blotting (Invitrogen).
15. Tween 20 (Sigma-Aldrich, St. Louis, MO).
16. TBS-$MgCl_2$: 100 m*M* Tris-HCl and 100 m*M* NaCl supplemented with 50 m*M* $MgCl_2$, pH 9.5.
17. Blocking buffer for dot-blot: Mix TBS, pH 7.5, with 0.25% Tween 20 and 5–10% non-fat dry milk.

18. Blocking buffer for tissue sections: TBS pH 7.5 supplemented with 0.25% Triton X-100 and 5–10% sheep serum.
19. NBT (nitro-blue tetrazolium chloride)/BCIP (5-bromo-4-chloro-3-indolyl phosphate, toluidine salt) solution: Dissolve 1 tablet (Roche Applied Sciences) *or* 3.5 mg of NBT and 1.8 mg of BCIP in 10 mL in TBS-$MgCl_2$, pH 9.5.
20. ^{35}S-UTP, 250 μCi with a specific activity of 1,000 mCi/m*M* (20mCi/mL) (General Electric, Piscataway, NJ). ^{35}S-UTP can be substituted with ^{35}S-CTP.
21. Quick Spin Columns for radiolabeled RNA purification (Roche Applied Sciences).
22. RNA polymerases (T7, SP6, or T3) with their respective 10X transcription buffer (New England Biolabs or Roche Applied Sciences).
23. Ribonucleotides: UTP, CTP, GTP, and ATP.
24. UV –cross-linker.
25. Non-fat dry milk (BioRad, Hercules, CA).

2.3. Hybridization

1. 1 *M* Tris-HCl, pH 8.0.
2. 0.5 *M* EDTA, pH 8.0.
3. 20X SSC: 3 *M* NaCl and 0.3 *M* sodium citrate, pH 7.0.
4. Denhardt's solution: Dissolve 1% ficoll, 1% bovine serum albumin (BSA) and 1% polyvinyl pyrrolidone (PVP) in DEPC-water. Split into 500 μl aliquots that can be kept for a year at −20 °C.
5. Salmon or fish sperm DNA (Roche Applied Sciences). Store at −20 °C in 1 mL aliquots. Very stable at −20 °C for 1–2 years.
6. Baker's yeast tRNA: Store at −20 °C in 550 μL aliquots. Use it for up to 6 months.
7. Formamide (Roche Applied Sciences).
8. 1 *M* triethanolamine (TEA), pH 8.0.
9. 5 *M* NaCl.
10. 1 *M* dithiotreitol (DTT).
11. Proteinase K in solution (Roche Applied Sciences).
12. DPX mounting medium (Sigma-Aldrich).
13. RNase A, store at −20 °C in 500 μl aliquots.
14. NBT/BCIP solution (*See* **Section 2.2, step 19**)
15. TBS-$MgCl_2$, pH 9.5 (*See* **Section 2.2, step 14**)
16. TBS, pH 7.5 (*See* **Section 2.2, step 11**)
17. Triton X-100.
18. 100% ethanol.
19. 50% dextran sulfate (Sigma-Aldrich) in DEPC-water: heat at 68 °C for 3–4 h or until dissolved. Store at 4 °C.
20. Hybridization buffer: Combine 10 mL of formamide, 4 mL of 20X SSC, 402 μL of Denhardt's solution, 1 mL of fish or salmon sperm DNA, 504 μL of baker's yeast tRNA, 4 mL

of 50% dextran sulfate, and 100 μL of 0.5 *M* EDTA, pH 8.0. Mix well and let stand on ice for at least 30 min to clear the air bubbles created during mixing. Store at −20 °C, and when in use always keep it on ice.

21. RNase A treatment solution: Mix 25 mL of 5 *M* NaCl; 2.5 mL of 1 *M* Tris-HCl, pH 8.0; 500 μL of 0.5 *M* EDTA, pH 8.0; 500 μl of RNase A (10 mg/mL); and enough ddH_2O to bring the final volume to 250 mL.
22. Hybridization oven or a temperature-controlled slide warmer.
23. Photographic emulsion (Thermo Fisher).
24. Kodak D19 developer: Prepare a 2X stock solution by dissolving 39.15 g in 250 mL ddH_2O. Store in a cool, dark area.
25. 1X Kodak photographic fix solution: Dissolve 44.5 g in 250 mL ddH_2O.
26. Histological mounting medium (e.g., Permount™, Thermo Fisher Scientific, Fair Lawn, NJ).
27. Five-slide plastic slide holders (Electron Microscopy Sciences or Thermo Fisher).
28. Aluminum foil.
29. Colored Tape.

3. Methods

All glassware, plastic, and metalware used for preparing solutions and processing tissue sections need to be RNase and DNase free. All utensils must be washed in detergent-containing bath and autoclaved (*See* **Note 1**). All chemicals used should preferably be of molecular grade.

3.1. Tissue Preparation

3.1.1. Fixation

1. Under the dissecting microscope, remove the stapes from the oval window. With fine forceps, remove the membrane obliterating the round window.
2. Use a 1 mL syringe to perfuse the cochlea gently through the oval or round window with ice-cold fixative solution. Repeat three times (total of 3 mL of fixative solution) for each cochlea.
3. Place the cochleae in the same fixative for post-fixation overnight at 4 °C.
4. Under the dissecting microscope and in ice-cold fixative solution, trim excess connective tissues and bone surrounding the cochlea.

3.1.2. Decalcification

1. Place the tissue samples in the decalcification solution and return to 4 °C under gentle agitation.
2. Monitor the decalcification process every few hours. The bony shell around the membranous tissues should be transparent and rubbery to the touch. The presence of opaque spots on the bony casing indicates that the decalcification process is not complete. If not fully decalcified, put the cochleae back in a fresh decalcification solution and return them to 4 °C under gentle agitation (*see* **Table 3.1** for decalcification times according to the animal age).
3. Once fully decalcified, wash specimens in ice-cold fixative solution twice for 10 min on the rocker. Afterwards, place the specimens in fresh ice-cold fixative and store at 4 °C until further processing.

3.1.3. Cryo-Protection and Cryo-Sectioning

1. Prepare 100 mL of cryo-protection solution and place it on ice.
2. Transfer the specimens to 10 mL of this solution and place on a rocker (or a gentle shaker) at 4 °C until they sink to the bottom of the tube. It usually takes 2–5 days for the specimens to saturate with sucrose (*see* **Note 2**).
3. Proceed with freezing: Use an isopentane bath cooled to −100 to −120 °C with liquid nitrogen. Place and orient the specimen in a plastic or aluminum mold containing OCT medium. Immerse the mold containing the specimen in an isopentane/liquid nitrogen bath for 30–45 sec. Once the specimen is frozen, place it in the cryostat chamber for 1 h to equilibrate before cryo-sectioning. Do not over-freeze the specimen as this may crack the OCT block containing the tissue sample.
4. Cryo-sectioning of the cochleae: Collect onto the histological slides 10–15 μm radial tissue sections, and use only the sections closest to the modiolus. Also, distribute the sections serially over the slides. This is practical when using lower vertebrates such as chick or frog *(11)* (*see* **Note 3**).
5. Place the slides on a slide warmer set at 45–55 °C for 1–2 h to dry the tissue sections and improve their adherence to the slide.

Table 3.1
Describes the decalcification process time course depending on the age of the animals used in the experiments

Post-natal age	P1–P7	P8–P16	P17–P21	> P21
Time in 5–8% EDTA	1–2 h	2.5–4 h	5–8 h	overnight (~ 12–16 h)

6. Place the slides in a slide box that holds 25 slides with desiccant, and tightly seal before storing at −80 °C until use.

3.2. Probe Preparation

The following instructions assume that the user possesses the cDNA of interest and that it is already cloned into a multipurpose vector containing a multiple cloning site (MCS) flanked with phage RNA polymerase T7 and SP6 or T3 promotors such as pGem-T-easy (Promega, Madison, WI) or pBluescript SK (+ / −) (Stratagene, La Jolla, CA). Several considerations require attention when designing and cloning a cDNA for riboprobe synthesis and its subsequent applications. Among these considerations are the size and sequence of the cDNA of interest. It is generally thought that smaller size riboprobes (200–300 bp) have better access to their target transcripts; however, these small size riboprobes provide fewer signaling moieties and so are less sensitive. Longer probes (700–1,200 bp) provide better sensitivity and specificity, therefore achieving higher hybridization rates. Also, the probe's length directly influences the melting temperature (Tm) and, therefore, leads to more stable cRNA–mRNA hybrids that can withstand high stringency washes, resulting in reduced background. Another important consideration is that the selected target sequence of the transcript needs to have the lowest possible homology with other genes.

3.2.1. Linearization

1. The RNA polymerases (T7 and SP6, or T3) will transcribe either cDNA strand independently, leading to the synthesis of anti-sense (complementary) RNA (cRNA) or sense RNA (mRNA copy). The anti-sense RNA will serve as the experimental probe and the sense RNA as the control probe in subsequent experiments such as northern and southern blots, nuclease protection assays, and notably in situ hybridization. Therefore, it is critical to determine, prior to probe synthesis, the orientation of the cloned cDNA in the vector, whether it is inserted $5' - 3'$ or vice versa.
2. Restriction enzyme choice: Choose enzymes that are present in the MCS and that do not cut within the cloned cDNA.
3. Prepare the reaction mixture: 10 μL of plasmid DNA (200–250 ng/μl), 2 μL of 10X reaction buffer, 7.5 μL of DEPC water, and 0.5 μL of restriction enzyme (∼ 10 units). Incubate at 37 °C for 2.5 h. (*see* **Note 4**).
4. Run a 1% agarose gel to check whether the linearization is complete: Load 1–5 μL of the reaction mixture to check the quality of the linearization. A complete linearization should reveal a single clean band of the expected size. If multiple bands per lane are visible, then reincubate the specimen for an additional 1 or 2 h. Add more enzyme if needed. Certain enzymes have a shorter activity life (*see* **Note 5**).

3.2.2. DNA Cleaning from Proteins and Salts (see Note 5)

1. Proteinase K treatment: Add to the linearization reaction 3 μL of 1 *M* Tris-HCl, pH 8.0; 3 μL of 0.5 *M* of EDTA, pH 8.0; 3 μL of proteinase K (0. 5 μg/μl); and enough DEPC water to bring the final reaction volume to 30 μL. Mix well and incubate for 2.5 h at 37 °C.
2. Linearized DNA cleaning: There are numerous ways for cleaning linearized plasmid DNA notably phenol-chloroform extraction, 3 *M* sodium acetate precipitation, or gel extraction. The last method is recommended because it results in a higher yield and a cleaner plasmid DNA, which is critical for the subsequent steps. Several companies offer gel extraction kits (e.g., Qiagen).
3. After running all linearized plasmid DNA on a 1% agarose gel, follow the manufacturer's protocol to extract DNA from agarose until the elution step. To elute the DNA from the column, add 20 μl DEPC water to the column and let stand at room temperature for 2–3 min.
4. Spin at 10,000 g for 1 min at room temperature in a benchtop centrifuge to collect the linearized plasmid DNA in a clean 1.5 mL tube.

3.2.3. Riboprobe Synthesis

Ribonucleotides, RNA polymerases, 10X digoxigenin labeling mix, and ^{35}S-UTP should be kept on ice while preparing the reaction mixture. However, 10X transcription buffers (RNA polymerases reaction buffer) *must be* kept at room temperature. If kept on ice, the spermidine contained in the 10X transcription buffer will precipitate the DNA. The linearized DNA can also be kept at room temperature for the time-span of the reaction mixing.

3.2.3.1. Digoxigenin RNA Labeling

1. Combine 2 μL of 10X digoxigenin labeling mix, 2 μL of 10X transcription buffer, 13 μL of linearized plasmid DNA (0. 5–1 μg), 1 μL of RNase inhibitor, and 2 μL of RNA polymerase (T7, T3, or SP6). Mix well and incubate for 2.0 h at 37 °C. Note 10X digoxigenin labeling mix consists of all four ribonucleotides in addition to the digoxigenin-conjugated UTP.
2. Stop the reaction by adding 2 μL of RNase-free 0.2 *M* EDTA, pH 8.0, to the sample; mix briefly; and place immediately on ice.
3. Precipitation: Add to the reaction mix 2. 5 μL of RNase-free 4 *M* LiCl and 75 μL of −20 °C-chilled 100% ethanol. Mix gently and place the samples overnight at −20 °C or at −80 °C (or at least for 2 h at −80 °C). This step cleans up the riboprobes from proteins, unincorporated nucleotides, and salts present during the synthesis reaction.
4. Centrifuge the samples at 10,000 g at 4 °C for 30 min.

5. Carefully decant the supernatant while keeping an eye on the pellet.
6. Wash the excess salts from the pellet with 100 μL of −20 °C prechilled 70% ethanol.
7. Centrifuge the samples at 3,500 g at 4 °C for 5 min.
8. Carefully decant the supernatant while keeping an eye on the pellet.
9. Air dry the pellet at room temperature for 2–3 min. Do not over-dry the pellet because it will be hard to redissolve. Also, an over-dried pellet becomes translucent and, therefore, hard to see.
10. Resuspend the RNA pellet in 50 μL of DEPC water and 3 μL of RNase inhibitor. Once the RNA pellet is resuspended, keep it on ice. If not in use, store at −20 °C for short-term storage. For longer term storage, reprecipitate in 5 μL of RNase-free 4 *M* LiCl and 125 μL of −20 °C-chilled 100% ethanol and place at −80 °C.

3.2.3.2. Dot-Blot: To Estimate cRNA Concentration and Labeling Efficiency

1. Prepare serial dilutions of the digoxigenin-control RNA and the synthesized digoxigenin-cRNA:
   a. For the control digoxigenin-labeled RNA, prepare four dilutions that contain 10, 1, 0.1, and 0.01 ng/μL.
   b. For digoxigenin-cRNA, prepare four dilutions that are undiluted, 1:10, 1:100, and 1:1,000.
2. Prepare a strip of nylon membrane and blot 1 μL from each dilution of digoxigenin-control RNA and synthesized digoxigenin-RNA in separate rows.
3. Air dry the membrane and cross link it in a UV cross-linker.
4. Wet the membrane briefly with TBS, pH 7.5.
5. Transfer the membrane to 5–10 mL of blocking buffer (TBS, pH 7.5 with 0.25% Tween 20 and 5–10% non-fat dry milk) and incubate at room temperature under gentle agitation for 30 min.
6. Place the membrane in alkaline phosphatase (AP) anti-digoxigenin antibody diluted 1:5,000 in the blocking buffer, and incubate at room temperature under gentle agitation for 45–60 min.
7. Rinse the membrane at room temperature twice for 5 min in TBS, pH 7.5, containing 0.25% Tween 20 under gentle agitation.
8. Rinse the membrane at room temperature under gentle agitation twice for 10 min with TBS-Mg_2Cl, pH 9.5.
9. Visualize the blots by incubating the membrane with AP substrate NBT/BCIP solution for 10–20 min in the dark at room temperature.
10. Stop the reaction by briefly rinsing the membrane with ddH_2O.

11. Estimate the probe concentration by direct comparison of control RNA and synthesized RNA respective dot-blots.

3.2.3.3. Radiolabeled Riboprobe Synthesis

1. Combine 2 μL of 10X transcription buffer, 1 μL of 10 m*M* ATP, 1 μL of 10 m*M* CTP, 1 μL of 10 m*M* GTP, 0–2 μL of 10 m*M* UTP, 1 μL of RNase inhibitor, 3 μL of ^{35}S-UTP (250 μCi, 20 mCi/mL, 20 m*M*), linearized plasmid DNA (~ 1 μg), 2 μL of RNA polymerase (~ 40 units of T7, T3, or SP6), and 0–2 μL of DEPC water. Mix all the components in the order listed above and incubate for 2.0 h at 37 °C. (*see* **Note 6**).
2. Remove unincorporated nucleotides: There are several protocols to purify synthesized radiolabeled RNA from free nucleotides. Presently we suggest two ways:
   a. Lithium chloride precipitation as described above for digoxigenin-labeled probes (*see* **Section 3.2.3.1, step 3**).
   b. Use Quick Spin Columns for radiolabeled RNA purification (Roche Applied Sciences): Follow the manufacturer's step-by-step protocol.
3. If not in use, store reprecipitated, synthesized RNA in RNase-free 4 *M* LiCl and −20 °C-chilled 100% ethanol and place at −80 °C. Radiolabeled probes are not stable because of the slow radiolysis process that occurs during storage. This phenomenon will lead to the breakdown of the radiolabeled probe's chemical bonds, which will translate into high background hybridization signal on the tissue sections. So it is strongly advised to use freshly synthesized radiolabeled RNA within 2–4 days.

3.3. Hybridization

3.3.1. Day 1: Pre-hybridization

1. Post fixation: Transfer the slides from −80 °C to ice-cold 4% PFA in 120 m*M* phosphate buffer, pH 7.4, for at least 15 min.
2. Rinse twice for 10 min at room temperature in pre-hybridization buffer: 100 m*M* Tris-HCl and 50 m*M* EDTA, pH 8.0.
3. Proteinase K treatment: Transfer the slides to a pre-hybridization buffer containing 1–10 μg/mL of proteinase K, depending on the processed tissue.
   a. If processing embryonic tissues, skip **step 3**.
   b. If processing neonatal tissue, use proteinase K at 1–5 μg/mL concentration and incubate at 37 °C for 10–15 min.
   c. If processing adult tissue, use 10 μg/mL proteinase K and incubate at 37 °C for 20–30 min.
4. Rinse briefly in DEPC water at room temperature for 5–10 s to stop proteinase K action.

5. Place the slides into 100 mM TEA solution, pH 8.0, and incubate for 3 min at room temperature.
6. Acetylation: This step reduces the background due to the binding of the probes to molecules containing amine and carboxy groups. (Prepare this solution just before transferring the slides to it.) Transfer the slides to 100 mM TEA containing 0.25% acetic anhydride, pH 8.0, and incubate for 10 min at room temperature.
7. Rinse twice for 2 min in 2X SSC, pH 7.0, at room temperature.
8. Dehydration: Rinse in graded ethanol concentrations at room temperature: 3 min in 70% ethanol, 3 min in 95% ethanol, and twice for 5 min in 100% ethanol.
9. Place the slides in a container with silica gel to air dry at room temperature for 1–1.5 h.

3.3.2. Day 1: Hybridization

1. Prepare the probe solution (*see* **Note 7**).
2. To denature, place the probes (cRNA and the sense RNA) at 83–86 °C for 3–5 min, and then transfer the samples immediately to an iced water bath and incubate for at least 5 min.
3. *For digoxigenin-labeled probes*: Dilute the probe in hybridization buffer to a final concentration of 0. 2–1 μg/mL.
4. *For* ^{35}*S-labeled cRNA*: Dilute the probe in hybridization buffer to a cpm concentration between 8 and 10×10^6 cpm/mL. Supplement the radioactive probe solution with DTT to a final concentration of 10 mM. The addition of DTT prevents the formation of the disulfide bonds between radiolabeled molecules.
5. Place the slides on clean bench-top paper (or a clean paper towel), identify the slides corresponding to each probe (if using more than one), and open and place a cover-slip box nearby.
6. Add 120 μL of probe solution onto each slide. Avoid air bubbles, particularly on or around the tissue sections.
7. Use new gloves (out of the box), and make sure that the gloves' fingertips do not come in contact with anything but the cover slips. Avoid making air bubbles over the tissue sections by carefully and gently placing the cover slips onto the slides.
8. Seal with DPX mounting medium around the edges between the cover slip and the slide. Use disposable plastic or Pasteur pipettes to dispense DPX medium.
9. Place the sealed slides in a hybridization oven and incubate at 56 °C for 14–16 h (*see* **Note 8**).

3.3.3. Day 2: Post-hybridization

1. Prepare ahead of time both RNase A treatment solution and 0.1X SSC bath (supplemented with 10 mM DTT, if using ^{35}S-labeled riboprobes) and place them at 37 °C and 50 °C, respectively.

2. Take the slides from the hybridization oven in batches (usually six at a time). Wait 10–15 s before peeling off the DPX seal (peel it from its proximal corners, being very careful not to tear off the cover slip with it).
3. Place slides, cover slipped, in 4X SSC at room temperature for 10 min or until the cover slips start to come off.
4. Transfer the slides (without cover slips) to a fresh 4X SSC bath at room temperature for 10 min.
5. RNase A treatment: Transfer the slides to RNase A solution and place at 37 °C for 20 min.
6. Rinse twice for 3 min in 2X SSC (supplement with 10 m*M* DTT when using ^{35}S-labeled riboprobe), pH 7.0, at room temperature.
7. Rinse twice for 3 min in 1X SSC (supplement with 10 m*M* DTT when using ^{35}S-labeled riboprobe), pH 7.0, at room temperature.
8. Rinse twice for 3 min in 0.5X SSC (with 10 m*M* DTT when using ^{35}S-labeled riboprobe), pH 7.0, at room temperature.
9. Rinse for 3 min in 0.1X SSC (supplement with 10 m*M* DTT when using ^{35}S-labeled riboprobe), pH 7.0, at room temperature.
10. Rinse one time for 15 min in 0.1X SSC (supplement with 10 m*M* DTT when using ^{35}S-labeled riboprobe), pH 7.0, at 50 °C (*see* **Note 9**).
11. Rinse for 3 min in 0.1X SSC (supplement with 10 m*M* DTT when using ^{35}S-labeled riboprobe), pH 7.0, at room temperature.

3.3.3.1. For Digoxigenin-Labeled RNA Probes

1. Proceed by placing the slides in TBS (100 m*M* Tris-HCl, 150 m*M* NaCl, pH 7.5) containing 0.25% Triton X-100 for 10 min at room temperature.
2. Transfer the slides to the blocking buffer and incubate for 1 h at room temperature.
3. Add to each slide 300 μL of AP-anti-digoxigenin antibody at a dilution of 1:1,000 in the blocking buffer. Place the slides in a humidified chamber for 5 hr at room temperature. A chamber can be made from an empty pipette tip box that is divided into two compartments, separated by the meshed platform used to hold the tips. Fill the bottom compartment with ddH_2O and place the slides on the meshed platform.
4. Rinse the slides twice for 5 min each in TBS, pH 7.5, at room temperature.
5. Equilibrate the slides twice for 10 min at room temperature in $TBS/MgCl_2$, pH 9.5.
6. Prepare NBT/BCIP solution just before applying it.
7. Add 300 μL of NBT/BCIP solution to each slide and incubate in the dark in a humid chamber and at room temperature for 3 h to overnight (*see* **Note 10**).

3.3.3.2. **For ^{35}S-labeled Probes**

Steps 3 to 7 must be accomplished in complete darkness or under limited lighting from the red safelight.

1. Proceed with serial dehydration by transferring the slides through 70% ethanol for 3 min at room temperature, 95% ethanol for 3 min at room temperature, and 100% ethanol twice for 5 min at room temperature.
2. Air dry the slides for at least 30 min.
3. Prepare the photographic emulsion: *This step should be performed under limited lighting from the red safelight.* To 40 mL of ddH_2O warmed to 40 °C, add 40 mL of photographic emulsion (*see* **Notes 11, 12**).
4. Stir gently until the photographic emulsion is completely dissolved (10–20 min).
5. Keep the emulsion under continuous gentle stirring and warm temperatures (35–40 °C).
6. With a steady vertical movement, coat each slide by slowly dipping it in the photographic emulsion.
7. Place the slides vertically in a safe, dark box to dry overnight at room temperature.

3.3.4. Day 3: Visualization and Incubation

3.3.4.1. *For Digoxigenin-Labeled cRNA*

1. Stop the reaction by rinsing the slides in TBS, pH 7.5, twice for 5 min at room temperature (*see* **Note 10**).
2. Mount the slides and observe under the microscope.

3.3.4.2. *For ^{35}S-labeled cRNA*

Both steps must be accomplished in complete darkness or under limited lighting from the red safelight.

1. Place the slides in a box with desiccant silica gel, close the box, seal it with tape, and wrap it in two layers of aluminum foil. Label the box properly.
2. Place the slide box at 4 °C, leaning vertically at approximately a 60° angle with the section of the box containing the desiccant at the bottom, and incubate for 10, 15, 20, or 30 days.

3.3.5. Day $\geq$ 10: Developing ^{35}S-RNA Hybridized Tissue Sections Slides

These steps should be performed in complete darkness or under limited lighting from the red safelight (*see* **Notes 11, 12**). Also it is recommended to initially develop one slide to assess the quality of the auto-radiographic signal.

1. Open the box and select the slide(s) to develop and place it (them) in the five-slide holder. Close the box, seal it, and wrap it back in aluminum foil.
2. Prepare the developer bath by mixing 50 mL of Kodak D19 developer stock solution with 50 mL of ddH_2O. Place the slides in Kodak D19 developer ($\sim$ 18 °C) for 4 min. Agitate gently every 2 min.

3. Rinse briefly in ddH_2O for 10–20 s.
4. Transfer the slides into the Kodak Fixer solution for 5 min.
5. Place the slides for 10 min under running tap water. At this stage turn the regular light back on.
6. While the slides are wet, scrape the photographic emulsion from the back of the slides.
7. Rinse again under running tap water for 5 more min.
8. Air dry and mount in Permount or other histological mounting media.

4. Notes

1. RNase contamination of solutions and material is a major problem in RNA detection experiments. To prevent this, it is strongly recommended to work in an RNase- and DNase-free environment. To fulfill this condition, all glassware, plasticware, and metalware must be cleaned prior and after each experiment by observing the following steps:
   (a) Wear gloves at all times.
   (b) Soak all glassware, plasticware, and metalware overnight in a detergent-containing bath (Absolve, *see* **Section 2.1, step 12**). On the next day, rinse thoroughly with ddH_2O and proceed with autoclaving.
   (c) Glassware and metalware (such as dissection tools) is either autoclaved for 30–40 min or placed overnight in a high-temperature oven (100–120 °C).
   (d) Plasticware is autoclaved for a 20-min cycle.
   (e) All solutions are prepared in RNase-free water (DEPC water), except the solutions destined for use after the hybridization step.
2. Tissue cryo-protection can be achieved by direct infiltration with a 30% sucrose solution or with graded infiltration of the specimens using serially increasing sucrose concentrations (10, 15, 20, 25%) to a maximum of 30%.
3. Depending on the age of the animals and the thickness of the cryo-sections, mammalian cochlea yield 72–120 radial sections spread equally over 10–15 histological slides. The cochlear duct of lower vertebrates (chick or frog), since they are straight cylinders, will generate 200–250 cross-sections distributed serially and uniformly over 20–24 slides *(11)*.
4. Plasmid-cloned cDNA linearization and cleaning, in preparation for in vitro transcription, results in 20–60% loss of DNA quantity. Therefore, it is recommended to start with 4–5 times the required amount for in vitro transcription, which is 1 μg per reaction.

5. The linearization of the plasmid-cloned cDNA should be complete, so that only a single clean band of the expected size should be visible on the gel. Also, the linearized DNA needs to be thoroughly cleaned. These two quality control steps are essential because they will directly reduce the efficiency of the in vitro transcription in terms of quality and quantity of the probe synthesized.
6. During probe synthesis, UTP concentration is the limiting factor. The specific activity of the radioactive probe depends fundamentally on the ratio between ^{35}S-UTP/unlabeled UTP. In the absence of unlabeled UTP, the synthesis produces truncated-size probes with high specific activity. However, when unlabeled UTP is added in vitro, transcription yields mostly full-length probes with lower specific activity. For in situ hybridization experiments, it is recommended to have a mixture of truncated and full-length probes. This is achieved by adding, to the probe synthesis reaction, unlabeled UTP at a concentration ranging from 0 to 5 μM.
7. Radioactive probes should be used within 2–4 days from their synthesis, because of their brief shelf life, owing to rapid degradation mainly due to radiolysis. Therefore, it is recommended to plan several in situ hybridization experiments in the aforementioned time frame. As the radioactive probe ages, the signal/background decreases, resulting in a high background ultimately masking the specific hybridization signal.
8. Hybridization temperature is usually determined by the Tm of the probe, which is directly related to its GC content. The hybridization temperature should be set 15–30 °C below the calculated value of Tm.
9. Stringency washes are designed to destabilize and rinse off any partially hybridized nucleic acids, therefore reducing the general background. **Section 3.3.3, steps 6–11**, describes the "desalting" wash method, which is accomplished by combining the gradual reduction of salt concentrations and high temperatures (50–65 °C). Formamide reduces the thermal stability of double-stranded nucleic acids, notably cRNA–mRNA hybrids. This characteristic implies that, in the presence of formamide, stringency washes can be performed at lower temperatures and, therefore, results in a better preservation of morphology in the tissue sections. The desalting method (**Section 3.3.3, steps 6–11**) can be substituted with a single wash, containing 50% formamide and 2X SSC (with DTT, when using radioactive probes) at a lower rinse temperature (~ 40 °C). This substitution can help in better preservation of cochlear morphology.
10. In non-radioactive in situ hybridization slides, it is recommended to monitor the progress of the chromogenic reaction. In the presence of the enzyme AP NBT/BCIP, substrates

produce a deep blue-purple precipitate. Optimally the reaction should be stopped just when the signal saturates (when the intensity of the signal stops changing).

11. Photographic emulsion has a short shelf life (2–3 months at 4 °C). The background signal increases as the photographic emulsion ages. It is recommended to use it when it is fresh.
12. The photographic emulsion is extremely light sensitive, even to direct red safelight. It is strongly suggested to handle coated experimental slides or the emulsion container either in complete dark or in limited red safelight (adjust the red safelight toward the ceiling or the wall).

Acknowledgments

The author wishes to thanks Dr. Paul A. Fuchs, Dr. Elisabeth Glowatzki, and Dr. Lisa Grant for their advice and their valuable comments. This work was supported by the National Institute of Deafness and other Communication Disorders (NIDCD) Grant R01DC001508 to P.A.F. and A.B.E.

References

1. John, H. A., Birnstiel, M. L., and Jones, K. W. (1969) RNA-DNA hybrids at the cytological level. *Nature* **223**, 582–587.
2. Gall, J. G. and Pardue, M. L. (1969) Formation and detection of RNA-DNA hybrid molecules in cytological preparations. *Proc. Natl. Acad. Sci. USA* **63**, 378–383.
3. Pardue, M. L. and Gall, J. G. (1969) Molecular hybridization of radioactive DNA to the DNA of cytological preparations. *Proc. Natl. Acad. Sci. USA* **64**, 600–604.
4. Wada, E., Wada, K., Boulter, J., Deneris, E., Heinemann, S., Patrick, J., and Swanson, L. W. (1989) Distribution of alpha 2, alpha 3, alpha 4, and beta 2 neuronal nicotinic receptor subunit mRNAs in the central nervous system: a hybridization histochemical study in the rat. *J. Comp. Neurol.* **284**, 314–335.
5. Ryan, A. F., Watts, A. G., and Simmons, D. M. (1991) Preservation of mRNA during *in situ* hybridization in the cochlea. *Hear. Res.* **56**, 148–152.
6. Ryan, A. F., Brumm, D., and Kraft, M. (1991) Occurrence and distribution of non-NMDA glutamate receptor mRNAs in the cochlea. *Neuroreport* **2**, 643–646.
7. Judice, T. N., Nelson, N. C., Beisel, C. L., Delimont, D. C., Fritzsch, B., and Beisel, K. W. (2002) Cochlear whole mount *in situ* hybridization: identification of longitudinal and radial gradients. *Brain Res. Brain Res. Protoc.* **9**, 65–76.
8. Safieddine, S. and Eybalin, M. (1992) Co-expression of NMDA and AMPA/kainate receptor mRNAs in cochlear neurons. *Neuroreport* **3**, 1145–1148.
9. Lécuyer, E., Yoshida, H., Parthasarathy, N., Alm, C., Babak, T., Cerovina, T., Hughes, T. R., Tomancak, P., and Krause, H. M. (2007) Global analysis of mRNA localization reveals a prominent role in organizing cellular architecture and function. *Cell* **131**, 174–187.
10. Bertrand, S., Thisse, B., Tavares, R., Sachs, L., Chaumot, A., Bardet, P. L., Escrivà, H., Duffraisse, M., Marchand, O., Safi, R., Thisse, C., and Laudet, V. (2007) Unexpected novel relational links uncovered by extensive developmental profiling of nuclear receptor expression. *PLoS Genet.* **3**, e188 [Epub ahead of print].
11. Hiel, H., Navaratnam, D. S., Oberholtzer, J. C., and Fuchs, P. A. (2002) Topological and developmental gradients of calbindin expression in the chick's inner ear. *J. Assoc. Res. Otolaryngol.* **3**, 1–15.

Chapter 4

Lineage Analysis of Inner Ear Cells Using Genomic Tags for Clonal Identification

Takunori Satoh and Donna M. Fekete

Abstract

To understand the mechanisms of development of the inner ear, it is important to know the lineal relationships among the different cell types and the migrational boundaries of individual clones within the inner ear. This chapter details the basic methods for performing lineage analysis of the inner ear using replication-defective retroviral vectors in chicken embryos. Protocols are provided for generating avian retroviruses pseudotyped with vesicular stomatitis virus (VSV) envelopes to improve infectivity in early embryos. Moreover, we include the pioneering methods of the Cepko laboratory, whereby a library of DNA tags was developed to allow clonal relationships to be confirmed by PCR amplification and sequencing of the tag in dispersed clonal progeny. By varying the site and time of viral delivery, these methods are appropriate to study cell lineages in other tissues of the developing chicken.

Key words: Lineage analysis, inner ear, retrovirus, pseudotyping, VSV-G.

1. Introduction

The vertebrate inner ear consists of several types of sensory organs; a complicated "labyrinth" of non-sensory epithelial ducts, tubules, and recesses; and the auditory and vestibular ganglia that innervate the sensory organs. Each sensory epithelium contains supporting cells and mechanosensory hair cells. While the non-sensory epithelium has a simple architecture in much of the ear, there are subregions containing specialized differentiated cell types. The sensory and non-sensory epithelial cells become the central cavity of the adult inner ear (the scala media) and, together with the neurons of auditory and vestibular ganglia, are generated from otic placode. The tissues surrounding the scala media,

Bernd Sokolowski (ed.), *Auditory and Vestibular Research: Methods and Protocols, vol. 493*

DOI 10.1007/978-1-59745-523-7_4 Springerprotocols.com

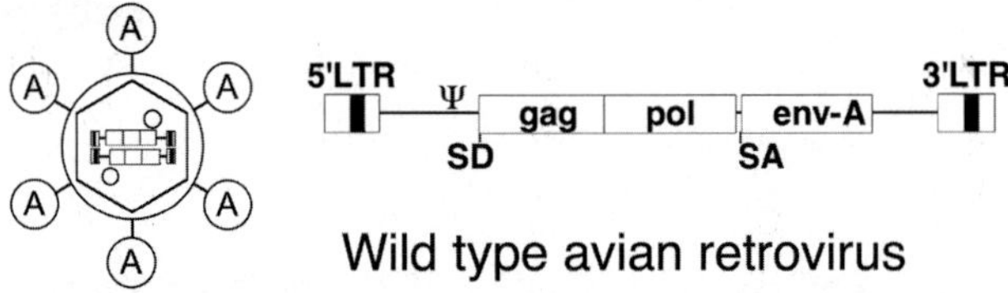

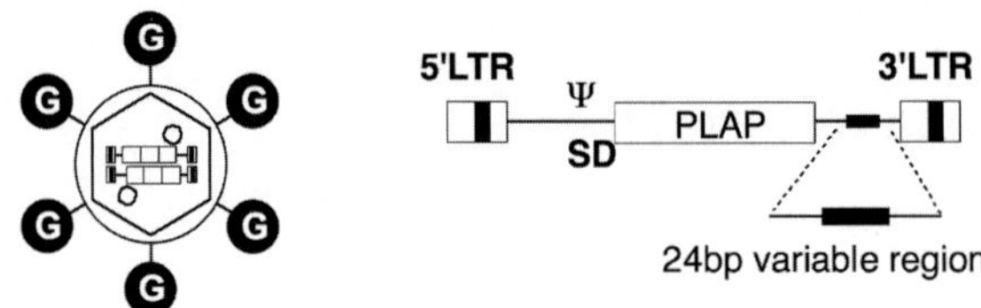

CHAPOL (G) Replication incompetent virus library

B

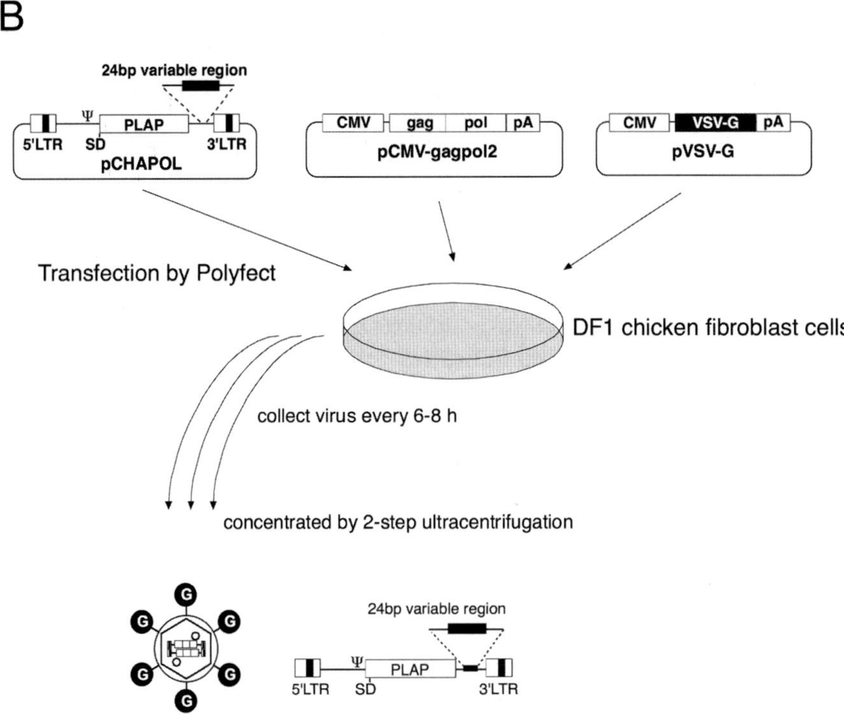

Fig. 4.1. (continued).

which include additional fluid-filled tubes and the bony otic capsule, are of mesenchymal origin. Schwann cells and satellite cells of the eighth cranial nerve and ganglion, respectively, derive from neural crest. Lineage analysis provides information about what kind of cells can be generated from a single progenitor and how far their progeny disperse throughout the tissue, when present in their natural context. Such data provide a baseline for understanding the mechanisms acting to specify cell fates and regulate morphogenesis in the inner ear.

We have explored lineage relationships in the chicken inner ear by marking cells at two different stages: the otic cup and otocyst. Otic cup lineage studies revealed that neurons can be related to cells of the otic epithelium *(1)*. Both otic cup and otic vesicle lineages were consistent in showing limited spatial dispersion of clones; specifically, the vast majority of clones were restricted to a single anatomical subdivision of the sensory periphery or its associated ganglia *(1, 2)*. Finally, by harvesting tissues at late stages of embryonic development, we confirmed that hair cells and supporting cells share a common progenitor, probably as late as their final mitotic division *(3)*. These lineage studies all used variations on retrovirus-mediated gene transfer to mark individual progenitor cells and follow the fate of their progeny.

In contrast to fate mapping, which can be accomplished by labeling clusters of cells, in lineage analysis, it is important to mark single progenitor cells to follow their progeny *(4)*. Replication-incompetent retroviral vectors allow us to study lineage in the inner ear and its associated neural and mesenchymal tissues. By integrating a gene for a histological marker (human placental alkaline phosphatase) into the viral vector, the distributions and morphology of virally transduced cells can be easily traced. PCR amplification of an oligonucleotide library (**Fig. 4.1**) cloned into the viral vectors is used to distinguish the clones generated from separate progenitors *(5)*.

Fig. 4.1. (continued) (**A**) Comparison of replication-competent and replication-incompetent viruses. Schematic drawings of wild-type avian leukemia virus (*top*) and VSV-pseudotyped CHAPOL virus (*bottom*). CHAPOL-G is pseudotyped with the G-envelope of vesicular stomatitis virus (VSV-G). The CHAPOL virus particle is infectious, but replication incompetent, because the mature virus particle does not contain any ALV genes required for replication (*gag, pol*, or *env*). (**B**) Schematic drawing of the production of replication-incompetent CHAPOL-G virus stock. The pCHAPOL plasmid library is co-transfected with helper plasmids into DF1 cells. The helper plasmid, pCMV-gag-pol2, provides genes encoding the protein components derived of avian leukosis virus (ALV), including viral core proteins from the *gag* gene and RNA polymerase from the *pol* gene. The helper plasmid, pVSV-G, provides the gene encoding the G-envelope protein of VSV. Together they create producer cells that package the CHAPOL RNA genome into an infectious, replication-defective virion.

2. Materials

To generate and use (VSV) pseudotyped retrovirus, the facility for cell culture and egg incubation must meet BL2 biosafety. Fixed specimens can be treated as non-infectious. To avoid contamination of PCR products, the work area and instruments to handle PCR products must be separated from those for other steps (*see* **Note 1**).

2.1. Production and Concentration of VSV-G-Pseudotyped Virus

1. DF1 Medium: 10% heat-inactivated fetal calf serum, 2% chicken serum (cat. no. C5405, Sigma-Aldrich, St. Louis, MO), and 1X penicillin/streptomycin (cat. no.15140-122, Invitrogen, Carlsbad, CA) in Dulbecco's Modified Eagle's Medium (DMEM) (cat. no. 51443C, JRH Biosciences, Lenexa, KS). Make 250 mL of DF1 Medium from stock solutions and store at 4 °C for up to 1 month. Stock solutions are frozen aliquots of 25 mL each of heat-inactivated fetal calf serum, 5 mL each of chicken serum, and 2.5 mL each of 100X penicillin/streptomycin. DMEM is stored at 4 °C until used; discard when it reaches the expiration date.
2. 1 *M* N-(2-hydroxyethyl)-piperazine-N′-2-ethanesulfonic acid (HEPES) buffer, pH 8.0: Dissolve 1 *M* HEPES in Milli-Q water, and adjust pH to 8.0 by adding NaOH. Sterilize by autoclaving and store at 4 °C.
3. DF1 Medium plus 10 m*M* HEPES: Add 1% volume of 1 *M* HEPES to DF1 Medium.
4. Solution of trypsin (0.05%) and 1 m*M* ethylenediamine tetraacetic acid (EDTA) in Hanks' buffered saline solution (HBSS) (cat. no. 25300, Invitrogen, Carlsbad, CA). Stored frozen in 5 mL aliquots.
5. Phosphate buffered saline (PBS), without Ca^{2+}/Mg^{2+} (cat. no. 21-040-CV, Mediatech, Herndon, VA). Stored at 4 °C in 25 mL aliquots.
6. DF1 chicken fibroblast cell line *(6)* (cat. no. CRL-12203, American Type Culture Collection (ATCC), Manassas, VA). Cells are maintained in DF1 Medium on 6 cm tissue culture plates by splitting (using trypsin digestion) every 2–3 days at 1:4, 1:6, and 1:8 dilutions when they reach approximately 80% confluence. If left to overgrow, these cells will assume extreme densities (perhaps greater than 200% confluent), and they do not like to be grown at low density. Furthermore, every few months the cells can "crash" for no apparent reason, where they look sick and stop dividing rigorously; at this point it is necessary to thaw a new aliquot from liquid nitrogen storage.

7. Polyfect transfection reagent (cat. no. 301105, QIAGEN, Valencia, CA).
8. Cooled bench-top centrifuge, with rotor for 50 mL disposable Falcon tubes (BD Biosciences, San Jose, CA).
9. Ultracentrifuge with swinging bucket rotor (e.g., Beckman SW28 or equivalent).
10. Disposable polyallomer centrifuge tubes (e.g., 5052, Seton Scientific, Los Gatos, CA). The tubes should be autoclaved and dried.
11. Rocking platform maintained in a cold room or refrigerator at 4 °C.
12. CHAPOL library plasmid DNA *(5)*: Purified by QIAGEN Maxi-Prep (QIAGEN) and dissolved in water. The library must have a high complexity of variable sequences (*see* **Note 2**). 2 μg/10 cm plate will be used. A typical virus prep requires 40 μg to transfect twenty 10 cm plates.
13. Helper plasmid DNA: pCMV-gagpol2 *(1)* or equivalent, purified by QIAGEN Maxi-Prep (QIAGEN). 1 μg/10 cm plate will be used.
14. Helper plasmid DNA: pVSV-G (pMD-G, *(7)*) or equivalent, purified by QIAGEN Maxi-Prep. 1 μg/10 cm plate will be used.
15. 10 cm culture dishes (e.g., Falcon cat. no. 353003).

2.2. Titering Virus

1. 80% confluent DF1 cells from a 6 cm plate are used to seed two 6-well plates.
2. DF1 Medium (*see* **Section 2.1, step 1**).
3. Six-well cluster plates (e.g., Corning cat. no. 3527, Sigma-Aldrich).
4. PBS: Prepare 10X stock with 1.37 *M* NaCl, 27 m*M* KCl, 100 m*M* Na_2HPO_4, and 18 m*M* KH_2PO_4; adjust to pH 7.4 with 1 *N* HCl if necessary; and autoclave before storage at room temperature. Prepare working solution by dilution of one part with nine parts Milli-Q water.
5. 4% Paraformaldehyde in PBS: Using microwave, heat 850 mL of Milli-Q water in glass beaker to approximately 60 °C. Add 40 g of paraformaldehyde and stir on magnetic stirrer until completely dissolved. Adding a few drops 10 *N* NaOH will help crystals go into solution. Add 100 mL of 10X PBS. Bring pH to between 7.2 and 7.4 by adding NaOH or HCl. Adjust the volume to 1,000 mL by adding Milli-Q water. Filter through Whatman no. 4 filter paper (Maidstone, UK). Store at 4 °C for no longer than 1 week, or freeze in 50 mL aliquots and thaw as needed.
6. Tris-HCl, pH9.5: Prepare as 1 *M* Tris-Base in Milli-Q water and adjust to pH 9.5 with HCl. Autoclave and store at room temperature.

7. X-Phos detection buffer: Prepare as 100 m*M* Tris-HCl, pH 9.5, 100 m*M* NaCl and 5 m*M* $MgCl_2$ in Milli-Q water by diluting from sterile stock solutions (1 *M* Tris-Cl, pH 9.5; 1 *M* NaCl; 5*M* $MgCl_2$).
8. 500X BCIP stock solution: 50 mg/mL 5-bromo-4-chloro-3′-indolylphosphate (BCIP/X-Phosphate, cat. no. 1-383-221, Roche, Indiannapolis, IN) stored at −20 °C.
9. 100X NBT stock solution: 100 mg/mL nitroblue tetrazolium (NBT, cat. no. 1-183-213, Roche) stored at −20 °C.
10. 0.45 μm syringe filter.

2.3. Helper Test

1. DF1 Medium (*see* **Section 2.1, step 1**).
2. Six-well cluster plates such as Corning 3527.
3. 40% confluent DF1 cells on a 3 cm plate grown at 37 °C in 5% CO_2.
4. 10 cm cell culture plates (e.g., cat. no. Falcon 353003).
5. 0.05% Trypsin-EDTA in HBSS (Invitrogen).
6. 0.45 μm syringe filter.
7. X-Phos detection buffer, 500X BCIP stock solution, and 100X NBT stock solution (*see* **Section, 2.2, step 9**).

2.4. Injection of Virus

1. Equipment for microinjection: dissection microscope, micromanipulator, pressure injector, pipette puller.
2. Fertilized chicken eggs, tested and confirmed to be free of Rous sarcoma virus (RSV). We prefer SPAFAS eggs (Charles River, Wilmington, MA) (*see* **Note 3**). For lineage analysis in mouse, *see* **Note 4**.
3. Bench-top egg incubator, non-rotating, kept at 37.5–38.5 °C. Keep humidifying pan with reverse-osmosis water containing 0.006% Lysol® (Reckitt Benckiser Group, Berkshire, England).
4. Glass micropipettes (1.5 mm O.D., 1 mm I.D.,) prepared by pulling fiber-filled borosilicate capillary into needles for microinjections and breaking or beveling the tips to about 12–15 μm diameter. Check tip diameter using an ocular micrometer on a compound microscope.
5. Frozen aliquots of CHAPOL(G) virus library. We used 10^6–10^7 per mL of CHAPOL(G) for cup-stage injections.
6. Microloader tip (cat. no. 4910 000.417, Eppendorf North America, Westbury, NY) to be used to load virus solution into the injection pipette.
7. 0.25% fast green in PBS or Milli-Q water, sterilized by filtering through a 0.22 μm syringe filter.
8. 4% paraformaldehyde in PBS (*see* **Section 2.2, step 5**).
9. 10%, 20%, and 30% sucrose solutions by mixing in PBS (*see* **Section 2.2, step 4**). Add sodium azide to 0.05% to prevent bacterial growth. Store at 4 °C for a month.

10. Gelatin for embedding tissue: 7.5% 300 bloom gelatin, 15% sucrose, and 0.05% sodium azide in PBS. In a 500 mL glass bottle, mix 40 mL of 10X PBS, 60 g of sucrose, and 200 mL of Milli-Q water using magnetic stirrer with heater. Add 30 g of 300 bloom gelatin (cat. no. G-2500, Sigma-Aldrich) and warm with gentle stir, until gelatin is completely dissolved. Add 2 mL of 10% sodium azide, bring total volume to 400 mL, and store at 4 °C.
11. Gelatin molds to mount specimen (e.g., 10 mL plastic medicine cup).
12. Liquid nitrogen for freezing tissue.
13. Cryomicrotome (cryostat) such as Leica CM1900 (Leica Microsystems Inc., Bannockburn, IL), set to −20 °C.
14. Superfrost plus slides (Fisher, Pittsburgh, PA) 5 × 75 × 1 mm.

2.5. Staining of Sections

1. Water bath kept at 65 °C. Temperature maintained with an automatic controller.
2. Staining dishes: We use staining dishes that hold 50 slides. Note that serial sections of each embryonic day 10 chicken inner ear can require more than 10 slides.
3. Acrylic humidifying chamber/box for slides for X-Phos reaction.
4. 10 mg/mL Hoechst 33258 fluorescent stain for cellular DNA (cat. no. H1398, Invitrogen) is dissolved in water. Store at 4 °C.
5. X-Phos detection buffer, NBT, BCIP stock solution (*see* **Section 2.2, steps 7–9**).
6. Gelvatol mounting medium: In 240 mL of PBS, add 50 g Gelvatol (Airvol 205 polyvinyl alcohol, Air Products and Chemicals, Inc., Allentown, PA) at 65 °C and stir 1–2 h with heat. Add 130 mL glycerol and stir 24 h with heat. Adjust pH by adding NaOH to 7.2–7.4.
7. Cover-slips, 24 × 60 mm.
8. Liquid blocker super PAP pen (Daido Sangyo, Tokyo, Japan).

2.6. Pick AP-Positive Cells in Sections

1. Thermal cycler.
2. PCR tubes, thin walled, 0.2 mL.
3. PCR Water (such as molecular biology grade water, Ivitrogen).
4. 5X PCR buffer: 500 μL 10X buffer for Taq DNA polymerase, 300 μL 25 m*M* $MgCl_2$ (included in Taq DNA polymerase, cat. no. M1861, Promega, Madison, WI), 200 μL PCR water. If Taq polymerase buffer does not contain Triton-X 100, add 0.5% to 5X buffer.
5. Proteinase K, 20 mg/mL in PCR water. Store in 10 μL aliquots at −20 °C.

6. Whatman no. 1 filter paper.
7. Disposable needle, 30 gauge.
8. Disposable 5 mL syringe.
9. Digital imaging device on compound microscope.
10. Computer on the side of dissection microscope, with imaging software such as Adobe Photoshop, to record the position of picks.

2.7. PCR Amplification of Variable Sequence

1. Taq DNA polymerase (cat. no. M1861, Promega)
2. Oligonucleotide primers, 10 pM/μL *(5, 8)*:
   CLAP0 5′-TGTGGCTGCCTGCACCCCAGGAAAG-3′
   CLAP5 5′-GTGTGCTGTCGAGCCGCCTTCAATG-3′
   CLAP2 5′-GCCACCACCTACAGCCCAGTGG-3′
   CLAP3 5′-GAGAGAGTGCCGCGGTAATGGG-3′
3. 10 mM each deoxy-nucleotide mix (cat. no. D-7295, Sigma-Aldrich).
4. PCR water and 5X PCR buffer (*see* **Section 2.6, step 4**).
5. ExoSAP-IT (cat. no. US78200, Amersham, Piscataway, NJ).
6. Standard equipment for agarose gel electrophoresis and imaging.
7. Tris-acetate-EDTA stock buffer (TAE): Prepare 50X stock with 242 g Tris-Base, 57.1 mL glacial acetic acid, and 100 mL of 0.5 M EDTA, pH 8.0, in 1,000 mL of reverse-osmosis water. For working solution take stock and dilute 1:50 in reverse-osmosis water.
8. 2.5% low-melting point agarose in 1X TAE.

3. Methods

Below we offer methods to generate high-titer, pseudotyped retrovirus vectors and to perform helper tests to confirm that they are replication defective (i.e., free of replication-competent helper viruses) before they are used for lineage studies. Helper viruses are unwanted viruses that provide *gag pol* and *env* genes that could allow otherwise replication-incompetent viruses to replicate in vivo and in vitro, and allow the vector to spread beyond the initially infected lineages. This so-called horizontal transmission leads to errors in clonal assignments. Instead, inheritance of the virus must be only “vertical” from parent to daughter cells. The DF1 cells used for virus production do not carry active avian virus in their genome; so what is the potential source of helper virus? Like most domestic chickens, the DF1 cells probably have inherited inactive “endogenous” viral sequences; these are typically incomplete pieces of viral genomes. As a result, when making viral stocks in DF1 cells, there is a finite chance of having recombination events between chromosomally integrated helper plasmids, proviral vectors, and endogenous viral loci to produce replication-competent virus. Such spontaneously arising "helper" viruses can spread in the cell culture in the course of virus production. So it is essential to check every independently generated virus stock to confirm it is "helper-free." In 10 years of making replication-defective viral stocks with DF1 producer cells, we have never detected helper virus arising de novo.

Also, to avoid creating helper virus after injection into host embryos, the flocks providing the eggs for viral injections must be routinely screened and certified to be free of exogenous avian viruses. While such eggs are considerably more expensive than farm-raised eggs, their "pathogen-free" status makes this an essential expense.

3.1. Production and Concentration of Virus

The G-envelope protein of VSV (VSV-G) can be used in place of avian envelope proteins to create a so-called pseudotyped viral stock (**Fig. 4.1**) that significantly boosts infectivity of early otic progenitors *(1)* (*see* **Note 5**). While the pseudotyped virus is more stable than standard avian viral vectors, G-envelope protein can cause fusion of the cultured packaging cells at pH 6.8 or lower *(9)*. To minimize fusion, the supernatant should be harvested before the media turns orange/yellow. HEPES buffer helps to maintain a high pH. (The low pH can increase the fusion of viral particles, resulting in multiple viruses entering single progenitor cells.) Despite careful attention to the pH of the medium, we find approximately 15% of inner ear clones infected with VSV-G-pseudotyped viruses carry more than one viral sequence. Below we outline a two-step concentration for generating high-titer virus. However, if the titer of virus is high enough, a one-step concentration may be sufficient.

1. Prepare 40%-confluent DF1 cells on about twenty 10 cm plates. Throughout this protocol, DF1 cells will be cultured at 37 °C in 5% CO_2.
2. Transfection mix per 10 cm plate: Combine 300 μl DMEM, 2 μg pCHAPOL, 1 μg pMD-G, 1 μg pCMV-gagpol2, and 25 μl Polyfect. Incubate at room temperature for 10 min. Mix with 6 mL of DF1 Media per 10 cm plate. Scale up when making a large virus prep; we typically add all reagents together for a prep of 20 plates.
3. Wash the plate with warmed DMEM.
4. Add transfection mix onto the washed DF1 cells. Incubate overnight.
5. Wash the plate twice with DMEM warmed to 37 °C. Add 5 mL of DF1 Media containing 10 m*M* HEPES.
6. Harvest the culture medium (100 mL of unconcentrated virus off 20 plates) every 8 h (longer intervals are possible, depending on the investigator preference). At each harvest, add back 5 mL per plate of fresh DF1 Media plus 10 m*M*

HEPES. If the harvested unconcentrated virus is noticeably orange/yellow, add 1% volume of 1 *M* HEPES to neutralize it. Store the unconcentrated virus at −80 °C, if needed. Take a small amount of unconcentrated virus, pass it through 0.45 μm filter, and store at −80 °C for titering. We typically repeat this process for up to 4 days or until the cells are so overgrown they are beginning to die, yielding a total volume of 800 mL.

7. When all of the harvests are completed, freeze unconcentrated virus at −80 °C or continue immediately by adding the last harvest to the earlier harvests that were thawed on ice a few minutes previously. Remove cellular debris from the unconcentrated virus by centrifugation at 720 g for 20 min. Pass the supernatant through 0.45 μm filter. Keep the filtered, unconcentrated virus on ice. We place approximately 33 mL of virus in each of the six round-bottomed centrifuge tubes designed for a swinging bucket rotor (*see* **Section 2.1, step 9**).
8. Centrifuge approximately 200 mL of the filtered unconcentrated virus at 4 °C at 72,000 g for 2 h. After the spin, remove the supernatant by aspiration and refill the same centrifuge tubes with another 200 mL of filtered unconcentrated virus, and repeat the ultracentrifugation. This step can be repeated another 1–3 times, as needed, depending on the total volume harvested over 4 days.
9. Add up to 5 mL per tube of DF1 Media plus 10 m*M* HEPES. Seal the tubes with Parafilm, and resuspend the virus on the rocking platform at 4 °C, overnight. Triturate gently approximately 10 times.
10. Collect all of the resuspended virus (~ 30 mL) into one tube. Concentrate again using the ultracentrifuge at 4 °C at 72,000 g for 2 h.
11. Remove the supernatant by aspiration, add 50 μL of PBS, seal the tube with Parafilm, and resuspend the virus on the rocking platform at 4 °C, overnight. Triturate gently approximately 10 times with an Eppendorf pipette.
12. Collect the concentrated virus into a 1.5 mL microcentrifuge tube. The volume of virus will be about 200 μL. Spin down the debris at 300 g for 2 min, make 20 μL aliquots of the supernatant, freeze in liquid nitrogen, and store at −80 °C.

3.2. Titering Virus

1. The day before adding virus for titering, seed DF1 cells onto a six-well culture plate. This is usually accomplished by taking a 6 cm plate at 70–80% confluence, resuspending the cells by trypsin digestion, and plating them onto two six-well plates (for a 1:6 split). The cells should be about 30% confluent the next day.

2. Make serial dilutions of virus stock in 3 mL DF1 Media. Remove the media from the titer plate and replace with 1 mL each of diluted virus per well. Set up two replicated wells per dilution. It is necessary to calculate the dilutions to attempt to obtain at least one dilution that gives 10–100 colonies per well. In the case of the CHAPOL virus in this protocol, one can expect to obtain unconcentrated CHAPOL virus at about 10^4–10^5/mL and concentrated virus at 10^6–10^7/mL. For example, you might set unconcentrated virus wells at dilutions of 10^{-2}, 10^{-3}, 10^{-4}, and 10^{-5}, and concentrated virus wells at 10^{-4}, 10^{-5}, 10^{-6}, and 10^{-7}. Leave 1–2 wells without virus as uninfected negative controls for the staining procedure.
3. Incubate overnight and replace media with 2 mL of fresh DF1 Media.
4. After 2 days of culture, fix the cells in each titer plate with 4% paraformaldehyde in PBS for 10 min at room temperature.
5. Wash three times with PBS at room temperature being careful to aspirate only on the edge of the wells and add media carefully along the edge.
6. To inactivate endogenous alkaline phosphatase, add preheated PBS into the wells and incubate at 65 °C for 30 min.
7. Wash once with PBS at room temperature.
8. Rinse with working solution of X-Phos detection buffer.
9. Set up working solution of X-Phos reaction: Mix X-Phos detection buffer with 1/100 volume of 50 mg/mL of NBT and 1/500 volume of 100 mg/mL of BCIP. Add 1 mL each of X-Phos reaction mix per well and incubate at 37 °C for 1–2 h.
10. Wash twice with PBS.
11. Count the number of clones. Small clusters of positive cells, ranging from 1 to 16 cells, are considered clones. Because the cells can migrate the equivalent of a few cell diameters, positive cells will be intermingled with negative cells. Use the data from the wells containing 10–100 clones per well.
12. Calculate the efficiency of the virus concentration procedure. This is done by comparing the titers of unconcentrated and concentrated virus. The concentration by titer is usually around 2,000- to 3,000-fold. The concentration by volume is approximately 4,000-fold (from 800 mL to 0.2 mL). Thus, the efficiency of the virus centrifugation procedure should be approximately 50–75%. This information can be used to determine if lower-than-expected titers are a consequence of problems with virus production or virus concentration.

3.3. Helper Test

1. Grow DF1 cells in 35 mm cell culture dish containing DF1 medium.

2. When cells are 30–40% confluent, remove the media, and replace with 2 mL of fresh media containing one vial (at least 5 μL, which is more than 1,000 times the volume one can inject in an otic vesicle) of concentrated virus stock.
3. After overnight incubation, split the cells onto a 10 cm plate, and keep cells growing for more than 2 weeks with appropriate passage, to let potential helper virus spread in the culture.
4. Set up three wells of a six-well plate with DF1 cells, calculated to be at 30% confluent on the day of supernatant transfer. This is the helper test plate. Collect the media (supernatant) from the virus-infected 10 cm plate; keep the plate, refeed the cells with DF1 Media, and return it to the incubator. Pass the supernatant through a 0.45 μm filter (to remove any transferred cells) and transfer 2 mL each to two wells of uninfected DF1 cells on the helper test plate. Keep one well as a negative control for the staining procedure.
5. Incubate overnight and replace the media on the helper test plate with DF1 Media.
6. After 2 days of culture, remove the media, fix both the helper test plate and the passaged virus-infected plate with 4% paraformaldehyde at room temperature for 10 min, and stain for alkaline phosphatase as described in **Section 3.2, steps 5–10**.
7. Confirm that the virus-infected cells are well infected but that there are no positive cells in the helper test plate.

3.4. Injection of Virus

1. Prepare windowed SPAFAS eggs at the desired stage as follows. Eggs should be incubated on their sides at all times; the embryo will develop on the top side. On day 2, mark the top of egg, remove 1.5–2 mL of thin albumen through the pointed end of the egg with an 18-gauge needle, and then seal the hole with Scotch 810 magic tape. Cover the top half of the egg with Scotch 3710 packing tape, and then cut a hole in the top of the shell with fine surgical scissors to visualize the embryo and provide enough working space for virus injections (an oval window approximately 1 × 2 cm should suffice).
2. Thaw an aliquot of concentrated virus stock. Add 1/10 volume of 0.25% fast green. Place on ice.
3. Load the needle from the back with 2–3 μL of virus solution using Microloader tips. Loading from the tip is more difficult because exposing the virus solution to air causes clumps.
4. Inject the virus into the appropriate lumen for the purpose, such as otic cup or vesicle. Reseal the egg with Scotch packing tape and return the egg to the humidified egg incubator.
5. Sacrifice the embryo at the desired stage. For inner ear lineage, we harvest no later than embryonic day 10 to avoid the onset of skull ossification. Fix the head in 4% paraformaldehyde in PBS at 4 °C, for 30–60 min or overnight.

6. Wash the specimen in cold PBS, three times for 30 min each. Infiltrate with 10% and 20% sucrose in PBS for 2 h each at 4 °C, and then 30% sucrose in PBS overnight at 4 °C.
7. Embed the specimen in gelatin embedding solution in appropriately sized plastic containers; we use 25 mL graduated beakers.
8. Trim the gelatin block in the appropriate orientation for sectioning, and place it on a piece of index card. Freeze the specimen on a flat piece of metal (such as head of bolt and nut) chilled with liquid nitrogen. Wrap the specimen tightly in aluminum foil; collect all specimens in a zip-lock plastic bag and store at −80 °C.
9. Using a cryomicrotome (cryostat), cut 15–35 μm serial sections onto Superfrost-Plus slides (Fisher). Air dry at room temperature, and then store the slides at −20 °C.

3.5. Staining of Sections

1. Defrost the sections and encircle each section using a PAP-Pen. Fix the slides in 4% paraformaldehyde in PBS for 15 min at room temperature.
2. Wash the specimen in PBS three times for 3–5 min.
3. To inactivate endogenous alkaline phosphatase, incubate the sections in a slide holder, containing preheated PBS, in a 65 °C water bath for 30 min.
4. Rinse with PBS at room temperature.
5. Rinse with X-Phos detection buffer at room temperature.
6. Mix X-Phos detection buffer with 1 mg/mL of NBT and 0.1 mg/mL of BCIP. Put the mix on the slides in a humidifying chamber. Incubate at room temperature for 2–3 h, or 37 °C for 1–2 h.
7. Wash the slides in PBS three times for 5 min at room temperature.
8. Stain with 1 μg/mL Hoechst (cat. no. 33258) in PBS for 5 min.
9. Wash the slides in PBS three times for 5 min each.
10. Mount in Gelvatol mounting medium by wiping excess PBS off the slide, and lay on a paper towel. Apply approximately 300 μL of Gelvatol on sections. Avoid air bubbles. Gently lay a cover-slip over the sections. Stand the slides on their side and let excess Gelvatol flow away for about 1 h.

3.6. Pick AP-Positive Cells in Sections

1. Analyze the distribution of AP-positive cells in the tissue and determine the samples to be tested by PCR. Note that a mesenchymal clone can spread widely in the chick head. In selected ears with AP-positive cells in an interesting pattern, take publication-quality images of all AP-positive cells in the ear. As a PCR-positive control, take images of AP-positive cells in surrounding tissue, if any, that you can assume will

be unrelated by lineage (e.g., mesenchymal picks can serve as positive controls for epithelial clones).

2. Soak the slides in PBS at 4 °C overnight in a coplin staining jar.
3. Soak the slides in PBS at 37 °C. Remove the cover-slip gently using forceps while still immersed in PBS in a coplin staining jar. Store the slides in PBS at 4 °C.
4. Prepare a 10 μL proteinase K solution in PCR tubes: 7.9 μL of PCR water, 2 μL of 5X PCR buffer, and 0.1 μL of a 20 mg/mL proteinase K stock solution.
5. Rinse a slide in freshly dispensed Milli-Q water.
6. Use a Kimwipe to wipe excess water off the back side of the slide and place the slide on a filter paper. Absorb remaining water around the sections using pieces of filter paper.
7. Open the image file of the section to facilitate finding the correct region. Under a dissection microscope, allow the targeted section (one with a clone to be picked) to half-dry. Identify the target AP-positive cells in the section. If needed, apply more water using a pipette, or remove excess water using a small piece of filter paper.
8. Attach a 30-gauge needle to a disposable syringe.
9. Using the needle, cut the tissue around the targeted AP-positive cell; the "pick" will include some of the surrounding, AP-negative tissue. Change the needle and pick up the piece of tissue containing the AP-positive cell.
10. Immerse the tip of the needle in the proteinase K solution. Label the tube and dispose of the needle. Record the ID, size, and position of the pick on the image file.
11. As a negative control, take a larger piece of surrounding tissue without AP-positive cells. As a positive control, take a piece of AP-positive cells that are lineally independent from the cells of interest.

3.7. PCR Amplification of Variable Sequence

1. Using a thermal cycler, digest the chosen tissue in proteinase K at 65 °C for 3 h; then heat inactivate the proteinase K at 85 °C for 20 min and then at 95 °C for 10 min.
2. Prepare 40 μL per tube of PCR1 reaction mix: 29 μL PCR water, 8 μL of 5X PCR buffer, 1 μL of 10 m*M* of dNTP, 1 μL of primer CLAP0, 1 μL of primer CLAP5, and 0.1 μL of 5 unit/μL Taq DNA polymerase.
3. Add 40 μL each of PCR1 reaction mix into the PCR tube with digested tissue.
4. Run program PCR1: one cycle at 93 °C for 2 min 30 s; 40 cycles at 94 °C for 45 s, 58 °C for 45 s, and 72 °C for 2 min; one cycle at 72 °C for 5 min and hold at 4 °C.
5. Prepare 40 μL per tube of PCR2 reaction mix: 29 μl PCR water, 8 μl 5X PCR buffer, 1 μL of 10 m*M* dNTP, 1 μL of

primer CLAP2, 1 μL of primer CLAP3, and 0.2 μL of 5 units/μL Taq DNA polymerase.

6. Transfer 3 μL each of PCR1 reaction into the PCR2 reaction mix.
7. Run program PCR2: one cycle at 93 °C for 2 min 30 s; 30 cycles at 94 °C for 45 s, 58 °C for 45 s, 72 °C for 2 min; one cycle at 72 °C for 5 min and hold at 4 °C.
8. Run 3 μL each of PCR2 products on a 2.5% agarose gel in 1X TAE. The PCR2 product will be 121 base pairs. PCR amplification will be successful in 30–50% of picks (*see* **Note 6**).
9. Take an image of the gel and store the PCR2 products at −20 °C. Estimate the concentration of the PCR products from the image (*see* **Note 7**).
10. Organize data and determine the samples to be sequenced.
11. Aliquot 3 μL of ExoSAP-IT enzyme mix into a PCR tube. Add 7 μL of selected PCR2 products and mix gently.
12. Using a thermal cycler, run program Preseq: 37 °C for 15 min, 80 °C for 15 min, and hold at 4 °C.
13. Sequence the PCR product using CLAP3 as a primer.
14. Organize sequence data. A pick may contain multiple sequences in the variable region (*see* **Note 8**); these are uninformative. Compare the sequence of each pick to sequences from other ears. Each clone should have a unique sequence in its variable region, assuming that the viruses carrying any single tag are not over-represented within the library or there is no contamination of PCR products (*see* **Note 1**). If the same sequence appears in more than one ear, the cells with this sequence should be eliminated from the lineage study, because it suggests that the sequence is over-represented in the library (*see* **Note 2**).

4. Notes

1. Contamination of PCR products can result in false relationships between independent clones. Aerosol-barrier tips must be used in all pipetting. Furthermore, the area for handling PCR products must be separate from the area used to process specimens (to take "picks"). Finally, it is helpful to include AP-negative picks and negative controls (no pick) in every PCR reaction to ensure the absence of PCR contaminants.
2. The complexity of the CHAPOL library (number of different variable sequence) was estimated to be > 10^5 in the first generation. The probability of detecting two independently infected clones with the same variable sequence in one specimen has been calculated based on the complexity of the library. However, as the library goes through subsequent

amplifications with each additional plasmid preparation, complexity may decrease and some sequences may become overrepresented. To ensure the quality (and complexity) of the library, all sequence data within a plasmid preparation of the library should be compared.

3. Because the active RSV/ALV can help the reproduction of replication-incompetent retroviral vector in vivo and allow the vector to spread beyond the lineage, the host chicken embryo has to be free of active RSV/ALV.
4. For lineage analysis in mouse, there is an available CAP library based on murine leukemia virus *(10)*.
5. In the case of ALV/RSV-based vectors, it is more difficult to obtain high titers when pseudotyping with VSV-G than with native A-envelope. However, VSV-G-pseudotyped virus has a higher infectivity than native A-envelope in early chicken otocysts *(1)* and in the middle ear mesenchyme (D. M. Fekete, unpublished observations). The RSV/ALV vectors with native envelopes of the A or B subgroups are not as stable as the VSV-G subgroup, particularly at room temperature. The titer of the native virus may drop quickly during the course of an injection session. In contrast, the RSV/ALV vectors with VSV-G envelope are stable on ice for several hours. Unlike virus with native B-subgroup envelope, in our hands, polybrene did not increase the infection efficiency (measured by viral titer on DF1 cells) of VSV-G-pseudotyped retroviral vectors.
6. In most of the picks, PCR amplification from a single copy of integrated DNA is required. In our experience, the yields tend to be low. We typically have an efficiency of PCR amplification of 30–50%. That is, more than half the picks may not yield a PCR product and, therefore, they will be uninformative for lineage analysis.
7. ExoSAP-IT enzyme mix degrades unincorporated dNTP and primers, instead of removing them. Therefore, the concentration of the PCR product cannot be determined by a spectrophotometer. For sequencing, the concentration of the PCR products must be estimated by comparing with marker lanes in the gel image.
8. As shown in **Fig. 4.2**, PCR products may contain more than one library sequence. Such clones should not be considered further. Multiple sequences are thought to originate from progenitor cells that accept more than one virion. This could be because the VSV-G envelope is fusigenic. In some extreme cases, the majority of the clones can contain multiple sequences, making the stock unusable for lineage analysis. If this problem is encountered, we suggest supplementing the media with HEPES buffer and harvesting the virus more frequently when making the next virus stock.

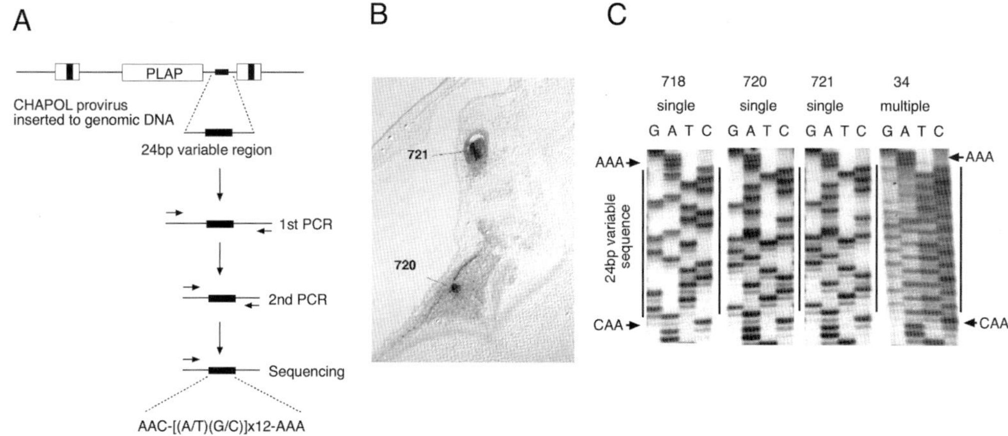

Fig. 4.2. (**A**) Schematic drawing of nested PCR amplification of a variable sequence in the CHAPOL vector. The variable sequence can be amplified from a single insertion of CHAPOL in the genome of host cells. (**B**) An example of a clone found in the middle ear. Pick 720 contains a cell body of a neuron in the geniculate ganglia. Pick 721 contains cells in the sensory epithelium in the paratympanic organ. (**C**) Examples of sequence results of the variable sequence. 5′ is *bottom* and 3′ is *top*. The variable sequences are between base pairs AAA and CAA. Picks 720 and 721 have identical sequences, whereas pick 718 has a different sequence in the variable region. Pick 34 is an example of a pick that yielded multiple sequences.

Acknowledgments

We thank Connie Cepko (Harvard Medical School) for sharing reagents. This work was supported by NIH 5RO1DC2756.

References

1. Satoh, T. and Fekete, D. M. (2005) Clonal analysis of the relationships between mechanosensory cells and the neurons that innervate them in the chicken ear. *Development* **132**, 1687–1697.
2. Lang, H. and Fekete, D. M. (2001) Lineage analysis in the chicken inner ear shows differences in clonal dispersion for epithelial, neuronal, and mesenchymal cells. *Dev. Biol.* **234**, 120–137.
3. Fekete, D. M., Muthukumar, S., and Karagogeos, D. (1998) Hair cells and supporting cells share a common progenitor in the avian inner ear. *J. Neurosci.* **18**, 7811–7821.
4. Kil, S. H. and Collazo, A. (2002) A review of inner ear fate maps and cell lineage studies. *J. Neurobiol.* **53**, 129–142.
5. Golden, J. A., Fields-Berry, S. C., and Cepko, C. L. (1995) Construction and characterization of a highly complex retroviral library for lineage analysis. *Proc. Natl. Acad. Sci. USA* **92**, 5704–5708.
6. Himly, M., Foster, D. N., Bottoli, I., Iacovoni, J. S., and Vogt, P. K. (1998) The DF-1 chicken fibroblast cell line: transformation induced by diverse oncogenes and cell death resulting from infection by avian leukosis viruses. *Virology* **248**, 295–304.
7. Naldini, L., Bloemer, U., Gallay, P., Ory, D., Mulligan, R., Gage, F. H., Verma, I. M., and Trono, D. (1996) In Vivo gene delivery and stable transduction of nondividing cells by a lentiviral vector. *Science* **272**, 263–267.
8. Campbell, J. K. (2002) CHAPOL Lineage Analysis. http://genetics.med.harvard.edu/~cepko/protocol/CHAPOL.pdf
9. Fredericksen, B. L. and Whitt, M. A. (1995) Vesicular stomatitis virus glycoprotein mutations that affect membrane fusion activity and abolish virus infectivity. *J. Virol.* **69**, 1435–1443.
10. McCarthy, M., Turnbull, D. H., Walsh, C. A., and Fishell, G. (2001) Telencephalic neural progenitors appear to be restricted to regional and glial fates before the onset of neurogenesis. *J. Neurosci.* **21**, 6772–6781.

Chapter 5

Genetic Fate-Mapping Approaches: New Means to Explore the Embryonic Origins of the Cochlear Nucleus

Jun Chul Kim and Susan M. Dymecki

Abstract

Greatly impacting the field of neural development are new technologies for generating fate maps in mice and thus for illuminating relationships between embryonic and adult brain structures. Until now, efforts in mammalian models such as the mouse have presented challenges because their in utero development limits the access needed for traditional methods involving tracer injection or cell transplantation. But access is no longer an obstacle. It is now possible to deliver cell lineage tracers via noninvasive genetic, rather than physical, means. The hinge-pin of these new "genetic fate mapping" technologies is a class of molecule called a site-specific recombinase. The most commonly used being Cre and Flp. Through the capacity to produce precise DNA excisions, Cre or Flp can act as an on-switch, capable of transforming a silenced reporter transgene, for example, into a constitutively expressed one. A reporter transgene is, in effect, transformed by the excision event into an indelible cell-lineage tracer, marking ancestor and descendant cells. The actual cell population to be fate mapped is determined by recombinase parameters. Being genetically encoded, Cre or Flp is "delivered" to specific cells in the embryo using transgenics – promoter and enhancer elements from a gene whose expression is restricted to the desired cell type is used to drive recombinase expression. Thus, recombinase delivery is not only noninvasive but also restricted to specific embryonic cells based on their gene expression phenotype, lending molecular precision to the selection of cells for fate mapping.

Resolution in cell type selection has recently been improved further by making lineage tracer activation dependent on two DNA excision events rather than just one. Here, in what is referred to as "intersectional genetic fate mapping," lineage tracer is expressed only in those cells having undergone a Flp-dependent excision as well as a Cre-dependent excision, thus marking the embryonic cells lying at the intersection of two gene (Flp and Cre driver) expression domains. The field of hindbrain development, in particular, has seen great advances through application of these new approaches. For example, genetic fate maps of the cochlear nucleus have yielded surprising information about where in the embryonic hindbrain its constituent neurons arise and journey and what genes are expressed along the way. In this chapter, we detail materials and methods relevant to genetic fate mapping in general and intersectional genetic fate mapping in particular.

Key words: Cre, Flpe, intersectional genetic fate mapping, mice, hindbrain, cochlear nucleus.

Bernd Sokolowski (ed.), *Auditory and Vestibular Research: Methods and Protocols, vol. 493*

DOI 10.1007/978-1-59745-523-7_5 Springerprotocols.com

1. Introduction

Fate maps, by defining relationships between embryonic and adult structures, are one of the most important tools on hand to developmental biologists and stem cell scientists. Yet, generating such maps in mammalian systems, where in utero development restricts access, has proven challenging. This is because most traditional methods involve physical manipulations, such as injection of retroviral *(1–3)*, fluorescent *(4)*, or vital dye lineage tracers *(5)* or the grafting of marked cells *(6)*, all of which are difficult to perform in mammalian embryos in ways that do not interfere with development. This constraint on access to mammalian embryos has been tackled in mice over the last decade through the development of noninvasive, *genetic* approaches for delivering lineage tracers to cells in utero (reviewed in *((7)–9))*. These genetic approaches offer other powerful features as well, for example the capacity for defining the initial embryonic population by gene expression phenotype, thus paving the way toward identifying molecules that may play a role in the specification and differentiation of the identified cell lineages. As well, these genetic methods allow for fate maps to be generated in mutant tissues, making it now possible to go beyond speculation and actually test how certain gene modifications alter the migratory routes or ultimate fates of specific cells. In this chapter, we introduce the general approach of "genetic fate mapping" *(10–12)* along with a newer variation called "intersectional" genetic fate mapping *(9, 13–15)*, the latter offering improved map resolution through the capacity to more precisely define the initial embryonic population under study. As illustration of these approaches, we refer the reader to a recent set of papers *(14, 16)* where these methods have been used to explore the embryonic origins and development of the different neuron types comprising the brainstem cochlear nuclear complex, the entry point for all central auditory processing.

Genetic fate mapping involves using an enzyme, a site-specific recombinase, to activate expression of a reporter molecule, like ß-galactosidase (ß-gal) or green fluorescent protein (GFP), in an indelible, cell-heritable fashion. In effect, recombinase action alters the configuration of a reporter-encoding transgene, turning reporter expression into a lineage tracer (**Fig. 5.1**) *(10–12, 17)*. The site-specific recombinase is typically either Cre recombinase (*c*auses *re*combination of the bacteriophage P1 genome) or Flp recombinase (named for its ability to "flip" a DNA segment in *S. cerevisiae*) (for reviews see *(7–9, 18, 19)*.

Being genetically encoded, the recombinase is "delivered" to specific cells in the embryo by exploiting promoter and enhancer elements from a gene (e.g., hypothetical *gene A* as schematized in **Fig. 5.1**) whose expression is restricted to the desired cell type.

A. Transgene manipulation

B. Recombinase-based genetic fate mapping

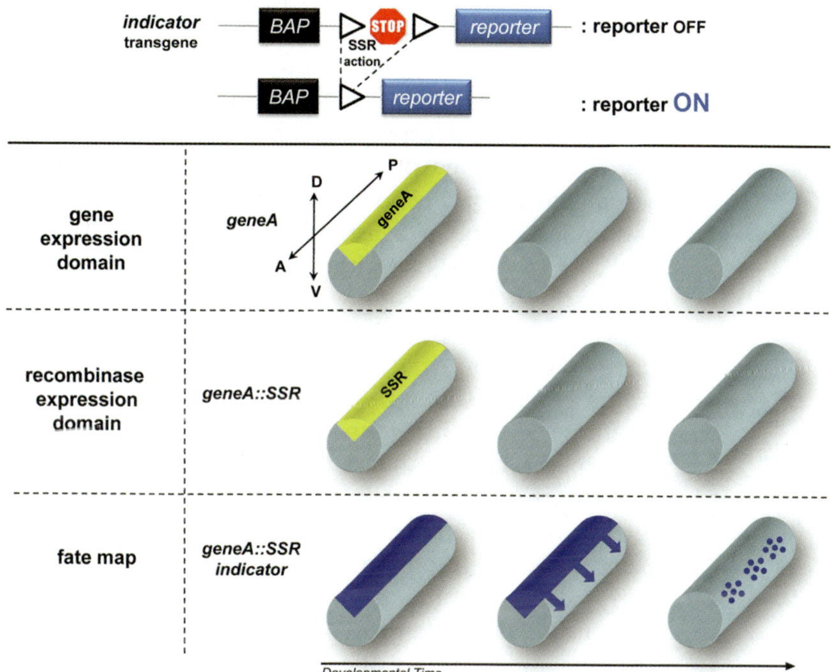

Fig. 5.1. DNA Excisions mediated by site-specific recombinase (SSR) serve as "On–Off" Switches for gene expression and as the basis for genetic fate mapping. (**A**) Structure of a generic SSR-responsive transgene inserted as a single copy into the mouse genome. SSR-mediated recombination between directly repeated SSR recognition sites (*triangles*) results in deletion of intervening transcriptional stop sequences (*red octagonal stop sign*) and consequent expression of a reporter molecule. (**B**) Illustration of how site-specific recombination can be used for genetic fate mapping. The generic SSR-responsive transgene of panel **A** is modified here (*upper panel*) by incorporation of a broadly active promoter (BAP) ideally capable of driving transgene expression in any cell type at any stage in development, such that after a recombination event in a given cell, that cell and all its progeny cells should be marked by reporter expression regardless of subsequent cell differentiation. *Lower panel*, strategy for SSR-based genetic fate mapping with development of the neural tube rendered as a simple cylinder and progressing left to right in each row. *Top row*: hypothetical *gene A* is expressed transiently by progenitor cells located in the dorsal neural tube (*yellow domain*) at an early developmental stage. Middle row: SSR-expressing transgene utilizes enhancer elements from *gene A*. *Bottom row*: When *geneA::SSR* is coupled with an indicator transgene, cells expressing the SSR will activate production of the reporter molecule (e.g., ß-gal). Activation of reporter molecule expression is permanent, and all cells descended from the SSR-expressing (*gene A*–expressing) progenitors will continue expressing the reporter, thereby marking a genetic lineage as it contributes to different brain regions during development. Descendant cells are depicted here as blue circles. (Reproduced from **Ref.** *(9)* with permission from Elsevier Science.)

The cell type–specific gene regulatory elements serve as "drivers" of the system and are incorporated into a transgene along with the coding sequence for the recombinase. In this way the initial embryonic population is defined by gene expression phenotype. Such recombinase-expressing mice can be generated in a variety of ways including conventional transgenic methods, transgenic strategies involving bacterial artificial chromosomes (BACs), or knock-in approaches; choice of method is driven by experimental need and availability of elements (reviewed in *(9)*). Materials and methods for identifying transgenic mice from wild-type littermates by PCR-based genotyping are detailed in **Sections 2.1 and 3.1**.

While the selection of embryonic cells for study involves cell type–restricted expression of Cre or Flp, actual fate mapping of the selected population involves recombinase-mediated rearrangement of a reporter-encoding transgene into a configuration sustaining of constitutive reporter expression (schematized in **Fig. 5.1**). Constitutive reporter expression is achieved upon excision of a disrupting DNA cassette; the latter otherwise blocks transcription of the downstream reporter. Because the target transgene is integrated into the mouse genome, it is cell heritable. Should recombination occur in a progenitor cell, all of its daughter and grand-daughter cells will go on to inherit the target transgene in its recombined form. This heritability feature, coupled with the use of widely active promoter/enhancer sequences to drive reporter expression once recombination has occurred, means that the recombined transgene, in most cases, will be expressed in the progeny as well as parental progenitor cells, thus turning a simple gene activation strategy (**Fig. 5.1A**) into a fate-mapping strategy (**Fig. 5.1B**). If the promoter/enhancer sequences employed in the target transgene are capable of driving reporter expression in any cell type at any stage of development (ideally), then, after a recombination event in a given cell, that cell and all its progeny cells should be marked by the reporter regardless of subsequent cell differentiation. Thus ideally, the target transgene exploits transcriptional regulatory elements with the broadest range of activity; this contrasts with the recombinase transgene that strives for maximal cell type restriction. In practice, the actual range of cell types that can be marked by a given target transgene needs to be determined empirically. For example, an approximation of scope can be gained by analyzing tissue from an animal in which the target transgene has been partnered with a broadly expressed recombinase transgene, such that target transgene activation occurs in most, if not all, cell types, each of which then can be sampled for robustness of reporter expression. Assays for detecting reporter molecule expression with cellular resolution in tissue section are described in **Sections 2.3**

and 3.3. After validating that a given indicator line is suitable for studying lineages in a given tissue, these same reporter assays can be used in the generation of actual fate maps – but now the target transgene is partnered with the desired cell type–selective recombinase driver line. Typically, we refer to the target transgene as an "indicator" transgene or allele, both to distinguish it from more conventional, constitutively driven (non-recombinase dependent) *promoter::reporter* transgenes and to emphasize that the target transgene, through reporter expression, serves to "indicate" or provide a permanent record of all earlier occurring recombination events.

For a given recombinase transgenic, it is critical to establish the extent to which recombinase expression matches the expected endogenous driver gene expression profile – a focus of **Sections 2.2 and 3.2**. For example, it is crucial to determine whether there is any unexpected ectopic recombinase expression, as this would confound subsequent fate-mapping studies by switching on the lineage tracer in unrelated cells that would be erroneously interpreted as part of a given lineage.

Not only is it critical to establish that the recombinase is expressed in a spatiotemporal pattern that matches the pattern of the intended endogenous gene (as mentioned above), but it is also important to establish the extent to which indicator transgene recombination (reporter activation) matches the initial recombinase driver gene expression profile – later they will diverge because the reporter expression is cumulatively and permanently tracking (mapping the fate of) all cells that ever in their history expressed the driver gene, whereas the driver gene expression is transient in a set of embryonic cells, for example. Methods for comparing initial recombinase expression versus reporter activation are detailed in **Sections 2.3 and 3.3**.

Despite the powerful features inherent to genetic fate mapping, many biological questions remain unanswerable because so many individual driver genes mark too broad an extent of cells. For example, in the embryo, individual gene expression domains often restrict along one axis of a tissue or germinal zone but extend along the orthogonal axes (**Fig. 5.2A**). The more expansive dimension will often intersect with multiple other gene-expression domains such that it actually contains multiple uniquely coded molecular subpopulations *(13, 14, 20)*. Resolving these subpopulations and their specific descendant lineages, and thus the relationship between a combination of expressed genes and future cell fate, is not possible using the above described single recombinase-based genetic fate-mapping approach. Toward improving the ability to select cells for fate mapping, we have devised a method called "intersectional" genetic fate mapping where the initial embryonic population to

A. Multiple uniquely coded molecular subdomains may comprise a single gene expression domain

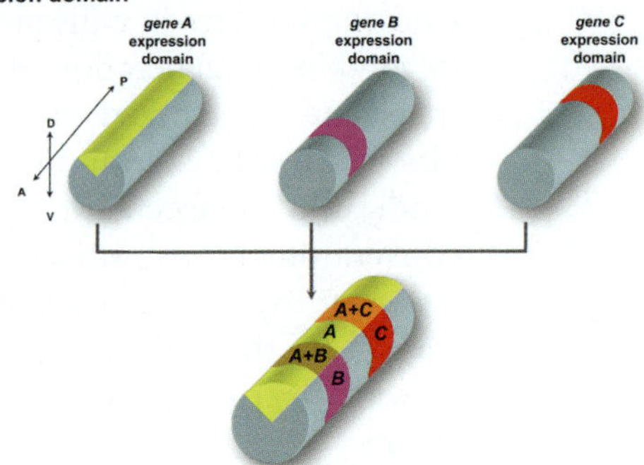

B. Dual recombinase-responsive indicator allele for acessing gene expression subdomains

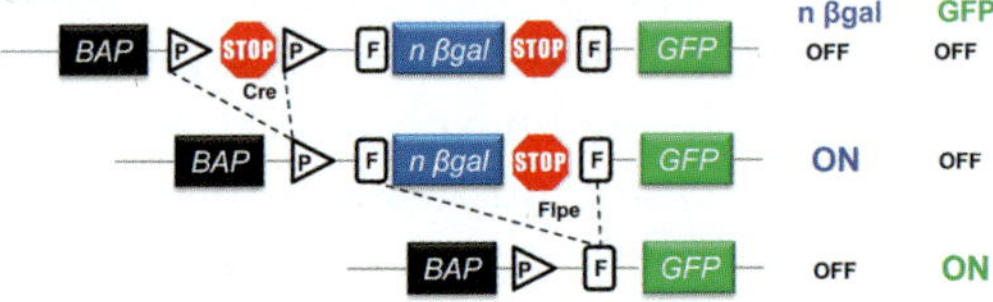

C. Stop-cassette order determines the subtractive population relative to the intersection

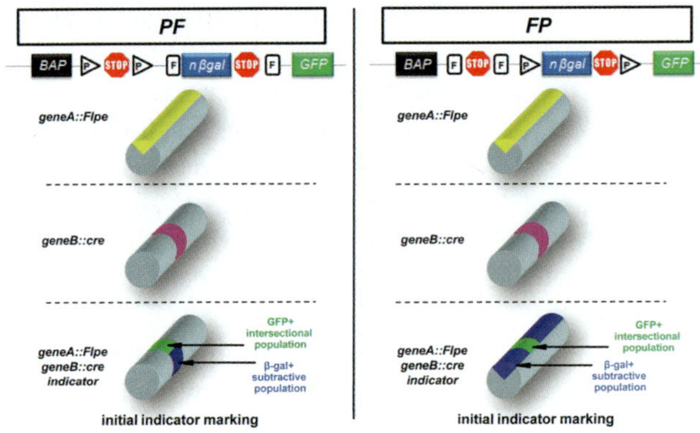

D. Intersectional and subtractive genetic fate mapping

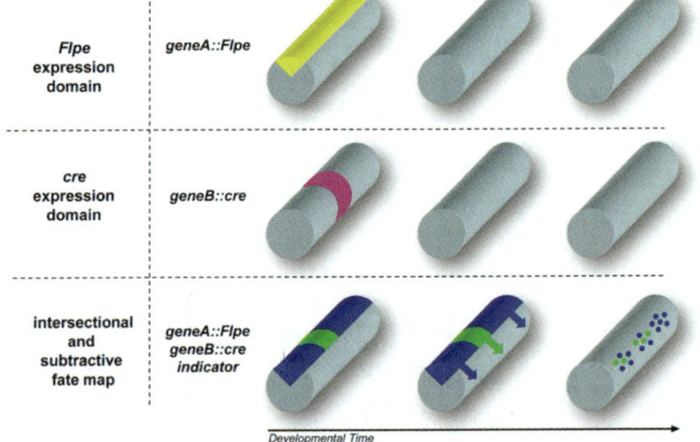

Fig. 5.2. (continued).

be traced is selected based on its expression of two genes rather than one (**Fig. 5.2C,D**).

For intersectional genetic fate mapping, two site-specific recombinases, Cre and Flp, are paired in a dual recombinase-mediated transgene activation paradigm (**Fig. 5.2B**) *(13, 14)*. Lineage-tracer expression, for example GFP, is switched on only in cells that have undergone two genetic events in their history, one mediated by Cre and the other by Flpe (**Fig. 5.2B–D**). Only those cells at the intersection of the two gene-expression domains will activate GFP. In addition to fate mapping intersecting Cre/Flp cell subpopulations, the approach can be used to trace simultaneously the Cre/non-Flp lineages. These lineages are referred to as "subtractive" populations (*see* ß-gal+ in **Fig. 5.2C**, PF configuration) because they are what remain when Cre/Flp-

Fig. 5.2. (continued) Intersectional and subtractive genetic fate-mapping strategy and a prototypical dual recombinase-responsive indicator allele. (**A**) Multiple uniquely coded molecular subdomains may comprise a single gene-expression domain. Shown are schematics of the neural tube (*gray cylinder*), with different gene-expression domains depicted in different colors. The expression domain for hypothetical *gene A* (*yellow*) restricts along the dorsoventral (DV) axis but extends along the anteroposterior (AP) axis; by contrast, the expression domains for *genes B* (*pink*) and *C* (*red*) restrict along the AP axis but extend along the DV axis. Thus, the *gene A* expression domain (*yellow*) is subdivided into three molecularly distinct subdomains: one in which *genes A* and *B* are coexpressed (*tan domain*), another in which *genes A* and *C* are coexpressed (*orange domain*), and finally, that territory (*yellow*) marked by *gene A* expression, but not *B* or *C*. Similarly, both the *gene B* and *C* expression domains are each subdivided. (**B**) Structure of a prototypical dual recombinase (Cre and Flpe)–responsive indicator allele. In contrast to a single recombinase-responsive indicator allele (**Fig. 5.1B**), a dual recombinase-responsive indicator allele has two stop cassettes, one flanked by directly oriented loxP sites (*triangles*) and the other by FRT sites (vertically oriented *rectangles*). Cre-mediated stop cassette removal results in expression of nβ-gal, while the remaining FRT-flanked stop cassette prevents GFP expression. Following removal of both stop cassettes, requiring Cre- and Flpe-mediated excisions, GFP expression is turned on and nβ-gal expression off. (**C**) Illustration of intersectional and subtractive populations and the latter dependency on stop-cassette order. In the "PF" configured allele, the loxP-flanked stop cassette precedes the FRT-flanked cassette (*left panel*), while the reciprocal order characterizes the "FP" configuration (*right panel*). Shown are schematics of the neural tube (*gray cylinder*), with the expression domain for hypothetical *gene A* and *Flpe* recombinase (*yellow*) restricting along the DV axis but extending along the AP axis (*top row*); in contrast, the expression domain for *gene B* (*pink*) restricts along the AP axis but extends along the DV axis (*middle row*). When *geneA::Flpe* and *geneB::cre* are coupled with a PF dual recombinase-responsive indicator allele (*bottom row, left*), cells expressing cre and Flpe activate production of GFP (*green domain*, intersectional population) while cells expressing only cre activate production of nβ-gal (*blue domain*, subtractive population). When *geneA::Flpe* and *geneB::cre* are coupled with an FP configured allele (*bottom row, right*), cells expressing cre and Flpe still activate production of GFP in the same intersectional population (*green domain*), but now cells expressing only Flpe (rather than cre) activate production of nβ-gal (*blue domain*, new subtractive population). (**D**) Illustration of the selective fate mapping achievable using an intersectional and subtractive approach. *Top row: gene A* drives transient Flpe expression in progenitor cells located in the dorsal neural tube (*yellow domain*) at an early developmental stage. *Middle row: gene B* drives transient cre expression in progenitor cells located at a particular AP level of the neural tube at an early developmental stage (*pink domain*). *Bottom row*: when *geneA::Flpe* and *geneB::cre* are coupled with a dual recombinase-responsive indicator allele (FP configuration), cells expressing Flpe and cre activate production of GFP, while cells expressing only Flpe activate production of nβ-gal. Activation of reporter molecule expression is permanent, and all cells descended from Flpe-expressing or Flpe- and cre-expressing progenitors will continue expressing the *blue* or *green* marker, respectively. Descendant cells from the intersectional domain are denoted by *green circles*, those from the subtractive population by *blue circles*. (Reproduced from **Ref.** *(9)* with permission from Elsevier Science.)

intersecting cells are subtracted from the Cre-only expressing domain *(14)*. Alternative dual recombinase-responsive indicator transgenes exist *(7, 13, 15)* in which the subtractive population is the reverse Flpe/non-Cre population (**Fig. 5.2C**, FP configuration). Materials and methods specific to intersectional and substractive approaches, where two reporter molecules are detected simultaneously, are presented in **Sections 2.3 and 3.3**.

Finally, it should be noted that there are three variants of Flp recombinase that have been employed in mice – enhanced Flp (Flpe), Flp-wt, and low-activity Flp (FlpL) (reviewed in *(9)*). The set collectively spans greater than a 10-fold range in activity in mice, with Flpe functioning in mice with similar efficacy as Cre *(12–14, 17, 21)*.

2. Materials

2.1. PCR Genotyping Animals Carrying Recombinase and/or Indicator Transgenes Using DNA Isolated from Tail Biopsy or Embryo Yolk Sac

2.1.1. DNA Preparation from Tail Biopsy or Embryo Yolk Sac

1. 50 m*M* NaOH: Dilute from 10 *N* NaOH stock solution with distilled water. Store at room temperature.
2. 1 *M* Tris-HCl, pH 8.0: Dissolve 121.1 g of Tris in 900 mL of distilled water. Add concentrated HCl to adjust to pH 8.0 and add distilled water up to 1 L.

2.1.2. PCR-Genotyping

1. 10X PCR buffer: 100 m*M* Tris-HCl, 500 m*M* KCl, 15 m*M* $MgCl_2$, pH 8.3.
2. 50X dNTP: 10 m*M* dATP, dTTP, dGTP, and dCTP in distilled water. Aliquot and store at −20 °C.
3. PCR primers, examples: Store at −20 °C as concentrated stocks (e.g., 200 μ*M*). The primers for genotyping *cre, Flpe, GFP*, and *lacZ* are as follows (*see* **Note 1**).
   a. *cre* (forward primer): 5′-GGCATGGTGCAAGTTGAAT AACC-3′
   b. *cre* (reverse primer): 5′-GGCTAAGTGCCTTCTCTAC AC-3′
   c. *Flpe* (forward primer): 5′-GCATCTGGGAGATCACTG AG-3′
   d. *Flpe* (reverse primer): 5′-CCCATTCCA TGCGGGG TATCG-3′
   e. *GFP* (forward primer): 5′-TACGGCAAGCTGACCCTG AAGTTC-3′

f. *GFP* (reverse primer): 5′-AAGTCGATGCCCTTCAGC TCGATG-3′
g. *lacZ* (forward primer): 5′-CACGAGCATCATCCTCT GCATG-3′
h. *lacZ* (reverse primer): 5′-CAGCGACTGATCCACCC AGTCC-3′

2.2. Profiling Recombinase mRNA Expression by In Situ Hybridization

2.2.1. Preparing Tissue from Prenatal Mice for In Situ Hybridization

1. Phosphate-buffered saline (PBS): 137 m*M* NaCl, 2.6 m*M* KCl, 10 m*M* Na_2HPO_4, 1.8 m*M* KH_2PO_4, pH 7.4. Autoclave and store at room temperature.
2. Diethyl pyrocarbonate (DEPC)-PBS: Add 1 mL of DEPC to 1 L of PBS, stir overnight in fume hood, autoclave, and store at room temperature.
3. 4% paraformaldehyde (PFA) fix solution: Prepare fresh by diluting 16% PFA/PBS stock solution with PBS. 16% PFA stock solution: In a fume hood, dissolve 160 g PFA in 800 mL of distilled water stirring at 60 °C. Add 1 *N* NaOH drop wise until solution clears. Add 100 mL of 10X PBS. Cool and adjust pH to 7.4. Add distilled water to 1 L and filter. Aliquot into 50 mL tubes and store at −20 °C
4. 30% sucrose/PBS solution: 30% sucrose (w/v) in PBS. Filter (0.2 um) – sterilize and store at 4 °C.
5. Embedding media: OCT (VWR Scientific, West Chester, PA) or TBS (Triangle Biomedical Sciences, Inc., Durham, NC).
6. Superfrost® Plus microscope slides (cat. no. 483II-703, VWR).
7. Equipment: embedding molds (cat. no. 12-550-15, Polysciences, Niles, IL), freezer-safe slide boxes (cat. no. 634200, Central Carolina Products), and a cryostat.

2.2.2. Preparing Tissue from Postnatal Mice for In Situ Hybridization

1. As per **Section 2.2.1, steps1–6**.

2.2.3. Synthesis of Digoxigenin-Labeled Riboprobes for In Situ Detection of mRNA – Endogenous or Recombinase Transcripts

1. DEPC water: Add 1 mL of DEPC to 1 L of distilled water, stir overnight in the fume hood, autoclave, and store at room temperature.
2. 10X transcription buffer (Roche Applied Science, Indianapolis, IN): 400 m*M* Tris-HCl, pH 8.0, 60 m*M* $MgCl_2$, 100 m*M* dithioerythritol, 20 m*M* spermidine, and 100 m*M* NaCl.

3. 10X NTP labeling mixture (Roche): 10 m*M* ATP, 10 m*M* CTP, 10 m*M* GTP, 6.5 m*M* UTP, and 3.5 m*M* digoxigenin-UTP in Tris-HCl, pH 7.5.
4. RNase inhibitor (Roche).
5. T3, T7, or SP6 RNA polymerase (Roche).
6. DNase I (RNase-free) (Roche).
7. 0.5 *M* EDTA: Add 186.1 g of EDTA (disodium, dihydrate) to 800 mL of distilled water and adjust pH to 8.0 by adding NaOH pellets (~ 20 g). Autoclave and store at room temperature.
8. TE: 50 m*M* Tris-HCl, 1 m*M* EDTA, pH 8.0. Autoclave and store at room temperature.
9. Digoxigenin-labeled RNA standards (Roche).

2.2.4. mRNA Detection on Tissue Sections by In Situ Hybridization

1. DEPC–PBS: Add 1 mL of DEPC to 1 L of PBS, stir overnight in fume hood, autoclave, and store at room temperature.
2. Proteinase K stock solution (10 mg/mL): Add 0.1 g of proteinase K to 10 mL of TE. Store in aliquots at −20°C.
3. Glycine/DEPC–PBS solution: 2 mg/mL glycine in PBS–DEPC. Prepare just before use.
4. 0.25% acetic anhydride in 0.1 *M* triethanolamine: Add 625 μL of acetic anhydride in 250 mL of 0.1 *M* triethanolamine.
5. Prehybridization/hybridization solution: 50% formamide, 5X SSC, 50 g/mL yeast tRNA, 1% SDS, and 50 g/mL heparin in DEPC water. Store at −20 °C.
6. Siliconized cover-slip: Treat cover-slips as follows. Dip 15 times in 3% silicon in chloroform (v/v). Repeat with fresh 3% silicon in chloroform (v/v). Dip 15 times in 100% ethanol. Repeat two times with fresh 100% ethanol. Air dry under hood.
7. 5X SSC solution: Dilute from 20X stock solution with DEPC-treated water. 20X SSC stock solution: 3 *M* NaCl, 0.3 *M* sodium citrate in DEPC water. Adjust to pH 4.5 with 0.6 *M* citric acid.
8. Wash buffer I: 50% formamide, 5X SSC, 1% SDS in DEPC water.
9. Wash buffer II: 0.5 *M* NaCl, 10 m*M* Tris-HCl, pH 7.5, 0.1% polyoxyethylene-sorbitan monolaurate (TWEEN-20) in DEPC water.
10. Wash buffer III: 50% formamide, 2X SSC in DEPC water.
11. 1X TBST: Dilute from 10X stock solution with distilled water. 10X TBST: 1.4 *M* NaCl, 27 m*M* KCl, 250 m*M*

Tris-HCl pH 7.5, and 10% TWEEN-20. Store at room temperature.

12. Heat-inactivated sheep serum (Invitrogen, Carlsbad, CA).
13. Anti-digoxigenin antibody conjugated to alkaline phosphatase (cat. no. 1 093 274, Roche).
14. NTMT: 100 m*M* NaCl, 100 m*M* Tris-HCl, pH 9.5, 50 m*M* $MgCl_2$, 1% TWEEN-20.
15. 4-Nitro-blue tetrazolium chloride (NBT) (cat. no.1383213, Roche).
16. X-phoshate/5-bromo-4-chloro-3-indolyl-phosphate (BCIP) (cat. no.1383221, Roche).
17. Alkaline phosphatase staining solution: Add 4. 5 μl of NBT and 3. 5 μL of BCIP to 1 mL of NTMT. Prepare immediately before use.
18. Mounting medium: Aqua/PolyMount (Polysciences, Inc).
19. Equipment: 24-slide slide holders with handles (Tissue Tek), plastic containers for slide holders (cat. no. 25608-904, VWR), humidified chamber for riboprobe hybridization (e.g., plastic slide box; cat. no. 423843, Becton Dickinson). For steps involving blocking, antibody incubation, and alkaline phosphatase staining, we use plastic dishes (cat. no. 73520-774, VWR) in which slides rest on a raised track, where the latter is configured from 5 mL plastic pipettes glued length wise on the bottom of the dish.

2.2.5. Comparing the Spatiotemporal Profile of Endogenous Driver Gene mRNA and Recombinase mRNA by In Situ Hybridization on Adjacent Tissue Sections

1. As per **Sections 2.2.1–2.2.4**.

2.2.6. Comparing the Spatiotemporal Profile of Recombinase Expression to that of Actual Recombinase Activity by Detecting Recombinase mRNA versus Lineage Tracer on Adjacent Tissue Sections

1. As per **Sections 2.2.1–2.2.4, and 2.3.1–2.3.3**.

2.3. Identifying Progenitor Cells and Their Descendants by Enzymatic or Immunofluorescence Detection of the Lineage Tracer Molecule – ß-gal or GFP

2.3.1. Tissue Preparation from Prenatal Mice for X-gal Detection of ß-gal

1. PBS: 137 m*M* NaCl, 2.6 m*M* KCl, 10 m*M* Na_2HPO_4, 1.8 m*M* KH_2PO_4, pH 7.4. Autoclave and store at room temperature.
2. 4% PFA fix solution: Prepare fresh by diluting 16% PFA/PBS stock solution with PBS. 16% PFA stock solution: In a fume hood, dissolve 160 g PFA in 800 mL of distilled water stirring at 60 °C. Add 1 N NaOH drop wise until solution clears. Add 100 mL of 10X PBS. Cool and pH to 7.4. Add distilled water to 1 L and filter. Aliquot into 50 mL tubes and store at −20 °C.
3. 30% sucrose/PBS solution: 30% sucrose (w/v) in PBS. Filter sterilize (0. 2 μm) and store at 4 °C.
4. Embedding media: OCT (cat. no. 100498-158, VWR). TBS (cat. no. H-TFM Triangle Biomedical Sciences).
5. Superfrost® Plus microscope slides (cat. no. 483II-703, VWR scientific).
6. Aqua/PolyMount mounting medium (Polysciences, Inc.)
7. Equipment: *see* **Section 2.2.1, step 7**.

2.3.2. Tissue Preparation from Postnatal Mice for X-gal Detection of ß-gal

1. *See* **Section 2.3.1**.

2.3.3. Detecting β-gal Activity on Tissue Section by X-gal Histochemistry

1. PBS: 137 m*M* NaCl, 2.6 m*M* KCl, 10 m*M* Na_2HPO_4, 1.8 m*M* KH_2PO_4, pH 7.4. Autoclave and store at room temperature.
2. X-gal staining solution: 5 m*M* $K_3[Fe(CN)_6]$, 5 m*M* $K_4[Fe(CN)_6]$, 1 mg/mL X-gal, 0.01% sodium deoxycholate, 0.02% Nonidet P-40 (NP-40), 2 m*M* $MgCl_2$ in PBS.
3. Aqua/PolyMount mounting medium.
4. Equipment: glass coplin jars (cat. no. 25457-006, VWR).

2.3.4. Detecting Lineage Tracer Molecule (e.g. ß-gal or GFP) by Immunofluorescence

1. PBS: 137 m*M* NaCl, 2.6 m*M* KCl, 10 m*M* Na_2HPO_4, 1.8 m*M* KH_2PO_4, pH 7.4. Autoclave and store at room temperature.
2. PBS-T: 0.05% Triton X100 in PBS.
3. DAPI staining solution: 300 n*M* 4′,6-Diamidino-2-phenyindole (DAPI) in PBS.
4. Aqua/PolyMount mounting medium.
5. Equipment: *see* **Section 2.2.4, step 19**.

2.3.5. Co-detection of Subtractive and Intersectional Lineage Tracer Molecules

1. *See* **Section 2.3.4**.

3. Methods

3.1. PCR Genotyping Animals Carrying Recombinase and/or Indicator Transgenes Using DNA Isolated from Tail Biopsy or Embryo Yolk Sac

3.1.1. DNA Preparation from Tail Biopsy or Yolk Sac

1. Place a yolk sac or a 0.25 cm tail biopsy into a microcentrifuge tube.
2. Add 200 μL of 50 m*M* NaOH and incubate at 94 °C for 20 min.
3. Add 50 μL of 1 *M* Tris-HCl, pH 8.0, and vortex briefly.
4. Centrifuge at 10,000 g for 10 min and save supernatants as tail or yolk sac DNA material for PCR genotyping (*see* **Note 2**).

3.1.2. PCR Genotyping

1. To assemble 25 μL of PCR genotyping reaction, mix the following reagents: 18. 5 μL of distilled water, 2. 5 μL of 10X PCR buffer containing 15 m*M* $MgCl_2$, 1 μL of forward primer (20 pmole), 1 μL of reverse primer (20 pmole), 0. 5 μL of 50X dNTP, 0. 5 μL of Taq polymerase (5 u/μL), and 1 μL of tail or yolk sac DNA.
2. Run the following PCR program: 10 min at 94 °C, 30 cycles of 1 min at 94 °C, 1 min at 55 °C, 1 min at 72 °C, and 10 min at 72 °C (*see* **Note 3**).
3. Load the entire reaction on a 1–2% agarose gel, containing 0. 5 μg/mL ethidium bromide.

3.2. Profiling Recombinase Expression by In Situ Hybridization

3.2.1. Preparing Tissue from Prenatal Mice for In Situ Hybridization

1. Dissect embryos in chilled DEPC–PBS (*see* **Note 4**) and remove extra-embryonic membranes. Rinse with cold DEPC–PBS to remove any residual blood.
2. Replace DEPC–PBS with 4% PFA fix solution and gently rock at 4 °C for 4–6 h.
3. Wash three times with PBS, for 5 min each, to remove residual fixative.
4. To preserve tissue architecture, soak tissues in 30% sucrose/PBS solution, rocking gently at 4 °C overnight or until tissue sinks.
5. Carefully roll tissues in an aliquot of embedding media (e.g., OCT) to remove residual solution before embedding.
6. Submerge and position tissue in an embedding mold-containing media. Gently push out air bubbles with forceps,

and immediately freeze by floating mold in a dry ice/ethanol bath until embedding media turn from clear to white.

7. Carefully remove mold from dry ice/ethanol bath and wipe off excess ethanol. Wrap mold tightly in aluminum foil and place in an airtight plastic bag. Store at −80 °C until use.
8. Cryosection tissue and mount on charged slides (e.g., Superfrost® Plus microscope slides). Air dry slides for 1 h. Store sections in sealed box at −80 °C.

3.2.2. Preparing Tissue from Postnatal Mice for In Situ Hybridization

1. Perfuse animal transcardially with ice-cold PBS solution followed by ice-cold 4% PFA fix solution.
2. Dissect tissue of interest and incubate in 4% PFA fix solution, rocking gently at 4 °C overnight.
3. Rinse with ice-cold PBS at 4 °C to remove residual fixative.
4. Follow **steps 4–8** of **Section 3.2.1**.

3.2.3. Synthesis of Digoxigenin-labeled Riboprobes for In Situ Detection of mRNA – Endogenous or Recombinase Transcripts

1. To prepare DNA template (1 μg/μL) for in vitro transcription, cut plasmid DNA containing cDNA of interest (e.g., *Cre* or *Flpe*) downstream of T3, T7, or SP6 promoter by appropriate restriction enzyme (*see* **Note 5**). Plasmid DNA should contain a promoter and cDNA sequences in an orientation that will generate anti-sense RNA during the in vitro transcription reaction. After restriction digestion, purify linearized DNA using preferred methods.
2. To generate approximately 10 μg of a given riboprobe, assemble the following reagents: 2 μL of 10X transcription buffer, 2 μL of 10X NTP labeling mixture, 1 μL of linearized plasmid (1 μg/μL), 1 μL of RNase inhibitor (40 U/μL), and 1 μL of T3, T7, or SP6 RNA polymerase (20 U/μL); add DEPC water up to 20 μL. Incubate at 37 °C for 2 h.
3. To digest template DNA, add 1 μL of DNase I (RNase free, 10u/μL) and incubate for 25 min at 37 °C.
4. Add 2 μL of 0. 5 *M* EDTA to stop the polymerase reaction.
5. Add 100 μL of TE, 10 μL of 4 *M* LiCl, and 300 μL of ice-cold 100% ethanol. Mix well and incubate at −20 °C for 2 h to overnight.
6. Spin at 10,000 g at 4 °C for 15 min in a microcentrifuge. Discard supernatant while ensuring that the riboprobe pellet remains in the tube. Wash pellet with ice-cold 70% ethanol and air dry for 5 min.
7. Resuspend pellet in 100 μL of TE to achieve a final concentration of approximately 0. 1 μg/μL (*see* **Note 6**).
8. To more precisely estimate riboprobe concentration, run 1 μL of riboprobe on a 1% agarose gel, containing 0. 5 μg/mL ethidium bromide besides an RNA standard of known concentration (*see* **Note 7**). Compare band intensities to estimate the amount of riboprobe synthesized.

3.2.4. RNA Detection on Tissue Sections by In Situ Hybridization

1. Remove tissue slides from −80 °C and allow them to come to room temperature.
2. Transfer slides into DEPC–PBS. Incubate for 5 min at room temperature.
3. Transfer slides into 1 μg/mL proteinase K in DEPC–PBS (prepared from freshly thawed 10 mg/mL proteinase K stock). Incubate for 5 min at room temperature (*see* **Note 8**).
4. To quench proteinase K activity, transfer slides into glycine/DEPC–PBS solution. Incubate for 5 min at room temperature.
5. Transfer slides into fresh DEPC–PBS. Incubate for 5 min at room temperature.
6. Transfer slides into 0.25% acetic anhydride in 0. 1 *M* triethanolamine. The acetic anhydride should be added to the triethanolamine solution just before use. Incubate for 10 min at room temperature.
7. Transfer slides into fresh DEPC–PBS. Incubate for 5 min at room temperature.
8. Prehybridization: Transfer slides into prewarmed (65 °C) prehybridization solution (*see* **Note 9**). Incubate for 15 min at 65 °C.
9. Hybridization: Place aliquot of probe (0. 1 μg per slide) in microcentrifuge tube and put in 85 °C heat block for 3 min. Transfer to ice immediately. Add hybridization buffer (100 μL per slide) directly to the tube containing riboprobe and maintain at 65 °C until needed. Apply 100 μL of hybridization buffer, containing 0. 5–1 μg/mL digoxigenin-labeled riboprobe, to each slide and cover with siliconized cover-slip. Hybridize in sealed, humid chamber overnight at 65 °C (*see* **Note 10**). This can be achieved by placing slides horizontally in a plastic slide box that is covered at the bottom with Kim Wipes soaked in 5X SSC.
10. Transfer slides into prewarmed (65 °C) 5X SSC. Incubate for 5 min at room temperature and dip until cover-slips fall off (*see* **Note 11**).
11. Transfer slides into prewarmed (65 °C) wash buffer I. Incubate for 15 min at 65 °C. Repeat twice using fresh prewarmed (65 °C) wash buffer I.
12. Transfer slides into fresh prewarmed (65 °C) wash buffer I:wash buffer II (1:1 ratio). Incubate for 10 min at 65 °C.
13. Transfer slides into prewarmed (65 °C) wash buffer II. Incubate for 5 min at 65 °C. Repeat twice using fresh prewarmed (65 °C) wash buffer II.
14. Transfer slides into RNase A containing wash buffer II (RNase A 25 mg/mL). Incubate for 30 min at 37 °C.
15. Transfer slides into prewarmed (65 °C) wash buffer II. Incubate for 5 min at 65 °C.

16. Transfer slides into TBS-T. Incubate for 10 min at room temperature. Repeat twice using fresh TBS-T.
17. Apply 0.5 mL of TBS-T containing 10% heat-inactivated lamb serum to each slide. Incubate in humid chamber for 30 min at room temperature.
18. Apply 0.5 mL of antibody (alkaline phosphatase–conjugated anti-digoxigenin antibody) diluted at 1:5,000 in TBS-T to each slide. Incubate in humid chamber for 2 h at room temperature or overnight at 4 °C.
19. Transfer slides into TBS-T. Incubate for 5 min at room temperature.
20. Transfer slides into fresh TBS-T. Incubate for 15 min at room temperature. Repeat four times using fresh TBS-T.
21. Transfer slides into NTMT. Incubate for 5 min at room temperature. Repeat twice using fresh NTMT.
22. Apply 1 mL of freshly prepared alkaline phosphatase staining solution to each slide. Incubate in humid chamber in the dark at room temperature until desired color development is achieved (a few hours to a few days depending on the abundance of the mRNA). Change alkaline phosphatase staining solution every day.
23. Stop reaction by transferring slides into 20 m*M* EDTA in PBS.
24. Rinse slides in PBS for 10 min at room temperature.
25. Remove excess PBS, apply mounting media, and carefully place a glass cover-slip on the slide.
26. Air dry slides until mounting medium is solidified. Slides should be stored in the dark to prevent additional color development.

3.2.5. Comparing the Spatiotemporal Profile of Endogenous Driver Gene mRNA and Recombinase mRNA by In Situ Hybridization on Adjacent Tissue Sections

As indicated in the "Introduction," it is critical to establish the extent to which recombinase expression mirrors spatiotemporally the driver gene expression profile. This is because any unexpected ectopic recombinase expression could lead to erroneous inclusion of cells into the fate map that are, in fact, lineally unrelated. To assess this, we recommend performing *in situ* hybridizations on adjacent tissue sections prepared from a developmental series of staged transgenic embryos (e.g., embryos collected at 24–48 h intervals until birth) as well as a range of postnatal animals (*see* **Sections 3.2.1 and 3.2.2**). Adjacent sections are prepared and probed for either recombinase or driver gene mRNA by *in situ* hybridization (*see* **Section 3.2.4**), followed by high magnification imaging to compare expression domains.

1. Identify animals carrying recombinase allele by genotyping as described in **Section 3.1**.
2. Prepare pre- or postnatal tissues as described in **Sections 3.2.1 and 3.2.2**.

3. Cryosection tissues to collect serial sections.
4. Perform *in situ* hybridization as described in **Section 3.2.4** with two sets of adjacent sections (one set using riboprobe that recognizes the recombinase and the other riboprobe to the driver gene) (*see* **Note 12**).

3.2.6. Comparing the Spatiotemporal Profile of Recombinase Expression to That of Actual Recombinase Activity by Detecting Recombinase mRNA Versus Lineage Tracer on Adjacent Tissue Sections

As indicated in the "Introduction," it is important to establish the extent to which indicator transgene recombination (actual reporter activation) matches the initial recombinase driver gene expression profile (mRNA and/or protein) – later they will diverge because the reporter expression is cumulatively and permanently tracking all cells that ever in their history expressed the driver gene, whereas the driver gene expression is transient in a set of embryonic cells, for example. In other words, it is important to determine whether all or only a part of an initial gene expression domain is being fate mapped. This becomes of particular concern if the driver gene exhibits a gradient in its expression; it is then possible that the lowest expressors in the gradient may be missed because adequate levels of recombinase are not achieved (reviewed in *(9)*). Thus, in double transgenic (recombinase; indicator) mice it is important to check whether cells positive for recombinase mRNA are also positive for recombinase activity, as reflected in reporter molecule expression. Sets of serial sections are obtained and analyzed for recombinase mRNA on one set and reporter molecule on the other (e.g., X-gal detection for ß-gal activity).

1. Genotype as described in **Section 3.1**, to identify double transgenic animals carrying both the recombinase and indicator alleles.
2. Prepare embryonic tissue as described in **Section 3.2.1**.
3. Cryosection and collect adjacent tissue sections into serial sets.
4. Perform *in situ* hybridization on one set of sections as described in **Section 3.2.4,** using riboprobe that recognizes recombinase mRNA, and perform X-gal staining with the other set of adjacent sections as described in **Section 3.3.3** (*see* **Note 13**).

3.3. Identifying Progenitor Cells and Their Descendants by Enzymatic or Immunofluorescence Detection of the Lineage Tracer Molecule – ß-gal or GFP

For intersectional genetic fate mapping, both a *Cre-* and a *Flpe*-encoding transgene are combined with a dual recombinase responsive indicator allele of choice (PF or FP variations, *see* **Fig. 5.2C**), thus generating triple transgenic animals (harboring *cre, Flpe*, and indicator alleles). Breeding strategies to combine these alleles varies depending on the reproductive rate of each strain. A typical breeding scheme, for example, might involve first generating *cre; Flpe* double transgenics, which are then crossed to indicator homozygotes. Triple transgenic progeny are identified by PCR genotyping (*see* **Section 3.1**), followed by tissue harvest, preparation (*see* **Sections 3.3.1 and 3.3.2**), and analyses to locate marked descendant cells by fluorescence or enzymatic detection of indicator molecules (*see* **Sections 3.3.3–3.3.5**).

3.3.1. Tissue Preparation from Prenatal Mice for X-gal Detection of ß-gal

1. Dissect embryos in ice-cold PBS, removing extra-embryonic membranes.
2. Wash with cold PBS to remove any residual blood.
3. Replace PBS with 2% PFA fix solution and gently rock at 4 °C for 4–6 h (*see* **Note 14**).
4. Wash three times with PBS for 5 min each to remove residual fixative.
5. To preserve tissue architecture, soak tissues in 30% sucrose/PBS solution, rocking gently at 4 °C overnight or until tissue sinks.
6. Carefully roll tissues in an aliquot of embedding media (e.g., OCT) to remove residual solution before embedding.
7. Submerge and position tissue in an embedding mold-containing embedding media. Gently push out air bubbles with forceps and immediately freeze, by floating mold in a dry ice/ethanol bath until embedding media turn from clear to white.
8. Carefully remove mold from dry ice/ethanol bath and wipe off excess ethanol. Wrap mold tightly in aluminum foil and place in an airtight plastic bag. Store at −80 °C until use.
9. Cryosection tissue and mount on charged slides (e.g., Superfrost® Plus microscope slides). Air dry slides for 1 h and store sections in a sealed box at −80 °C.

3.3.2. Tissue Preparation from Postnatal Mice for X-gal Detection of ß-gal

1. Perfuse animal transcardially with ice-cold PBS followed by ice-cold 4% PFA fix solution.
2. Dissect tissue of interest and incubate in 4% PFA fix solution, rocking gently at 4 °C overnight (*see* **Note 14**).
3. Rinse with ice-cold PBS at 4 °C to remove residual fixative.
4. Follow **steps 5–8 of Section 3.3.1**.

3.3.3. Detecting β-gal Activity on Tissue Section by X-gal Histochemistry

1. Collect and process tissue as described in **Sections 3.3.1 and 3.3.2**.
2. Remove the tissue slides from −80 °C and allow them to come to room temperature.
3. Transfer slides into PBS. Incubate for 5 min at room temperature. Repeat with fresh PBS.
4. Transfer slides into glass coplin jar containing X-gal staining solution and incubate at 37 °C in the dark until color develops to the desired extent (2–48 h) (*see* **Note 15**).
5. Transfer slides into PBS. Incubate for 5 min at room temperature. Repeat with fresh PBS.
6. Remove excess PBS, apply mounting media, and place a glass cover-slip carefully.
7. Air dry slides until mounting medium is solidified.

3.3.4. Detecting Lineage Tracer Molecule (e.g., ß-gal or GFP) by Immunofluorescence

1. Collect and process tissue as described in **Sections 3.3.1 and 3.3.2**.
2. Remove the tissue slides from −80 °C and allow them to come to room temperature.
3. Transfer slides into PBS. Incubate for 5 min at room temperature.
4. Transfer slides into PBS-T (*see* **Note 16**). Incubate for 5 min at room temperature.
5. Blocking: Apply 5% normal serum (use serum of species that secondary antibody is raised in) in PBS-T and incubate in humid chamber at room temperature for 1 h.
6. Transfer slides into PBS-T. Incubate for 5 min at room temperature.
7. Apply primary antibody (e.g., anti-GFP or β-gal antibody) diluted in PBS-T to each slide and incubate in humid chamber at 4 °C overnight (*see* **Note 17**).
8. Transfer slides into PBS-T. Incubate for 5 min at room temperature. Repeat twice with fresh PBS-T.
9. Apply fluorophore-conjugated secondary antibody diluted in PBS-T to each slide and incubate in humid chamber at room temperature for 1 h.
10. Transfer slides into PBS. Incubate for 5 min at room temperature. Repeat twice with fresh PBS.
11. If necessary, stain with DAPI by incubating slides in DAPI staining solution for 5 min at room temperature (*see* **Note 18**).
12. Transfer slides into PBS. Incubate for 5 min at room temperature.
13. Remove excess PBS, apply mounting media, and place a glass cover-slip carefully.
14. Air dry slides until mounting medium is solidified. Store slides at 4 °C in the dark.

3.3.5. Co-detection of Subtractive and Intersectional Lineage Tracer Molecules

1. Detect the first lineage tracer molecule (either subtractive or intersectional) as described in **steps 1–10 of Section 3.3.4**.
2. Continue with detecting the second lineage tracer molecule by following **steps 5–14 of Section 3.3.4** (*see* **Note 19**).

4. Notes

1. Three variants of *Flp* have been employed in mice; *Flp* wild type, *Flp*-L (low-activity Flp), and Flpe-enhanced *Flp*. The Flpe genotyping primers presented in **Section 2.1.2** can be used for genotyping all three variants.
2. Tail/yolk sac DNA samples can be stored at 4 °C for long term.

3. Annealing temperature varies depending on primer pair.
4. DEPC is suspected to be a carcinogen and should be handled with care.
5. To minimize the contamination of vector sequence, use a restriction enzyme that cuts adjacent to the cDNA sequences on the side distal to the promoter. As long as the restriction enzyme does not cut between promoter and cDNA sequences, the restriction site need not be unique.
6. If the labeled probe is not used immediately, store the probe solution at −80 °C. Avoid repeated freezing and thawing of the probe.
7. To avoid RNase contamination, all electrophoresis apparatus should be cleaned thoroughly with DEPC-treated water and treated with reagents containing RNase inhibitor. We found the commercially available reagent such as RNaseZap (cat. no. 9780-9784, Ambion, Austin TX) works efficiently.
8. Optimal concentration of proteinase K and incubation time should be empirically decided depending on thickness of tissue.
9. Prehybridization solution can be made ahead of time and stored at −20 °C.
10. If working with multiple probes, change gloves between each probe to avoid contamination. Avoid placing multiple slides with different probes in the same box.
11. Do not pull out cover-slips, since it will damage the tissue section.
12. Expression of recombinase and driver gene can also be detected in the same section simultaneously using double *in situ* hybridization methods – although robust expression is typically required as this assay sensitivity may be less than standard single probe detection protocols *(22)*.
13. In addition to assaying for *cre* mRNA, it is also possible to detect Cre protein using commercially available antibodies *(23)*. Currently, antibodies to detect Flp protein in tissue section are lacking.
14. Over-fixation can affect β-gal activity. Optimal fixation time may be empirically determined.
15. X-gal staining solution can be reused several times if stored at 4 °C and protected from light.
16. Detection of membrane-bound antigens may be affected by Triton X treatment.
17. We found it is very important to determine the optimal titers of primary antibody by titration. For example, if the manufacturer suggests a dilution of 1:500, it is useful to test a range such as 1:500, 1:1,000, 1:2,000, 1:4,000, 1:8,000, and 1:16,000.
18. DAPI staining solution can be stored as 10X stock at 4 °C.
19. Two different antigens can be detected simultaneously by adding two different primary antibodies in the same solution.

However, we found that in some cases, more sensitive detection can be achieved by detecting multiple antigens sequentially, as described in **Section 3.3.5**.

References

1. Cepko, C. L., Austin, C. P., Walsh, C., Ryder, E. F., Halliday, A., and Fields-Berry, S. (1990) Studies of cortical development using retrovirus vectors. *Cold Spring Harb. Symp. Quant. Biol.* 55, 265–278.
2. Galileo, D. S., Gray, G. E., Owens, G. C., Majors, J., and Sanes, J. R. (1990) Neurons and glia arise from a common progenitor in chicken optic tectum: demonstration with two retroviruses and cell type-specific antibodies. *Proc. Natl. Acad. Sci. USA* 87, 458–462.
3. Walsh, C. and Cepko, C. L. (1988) Clonally related cortical cells show several migration patterns. *Science* 241, 1342–1345.
4. Wetts, R. F. and Fraser, S. E. (1988) Multipotent precursors can give rise to all major cell types of the frog retina. *Science* 239, 1142–1145.
5. Keller, R. E. (1975) Vital dye mapping of the gastrula and neuraula of *Xenopus laevis.* I. Prospective areas and morphogenetic movements of the superficial layer. *Dev. Biol.* 42, 222–241.
6. Le Douarin, N. (1982) *The Neural Crest*, Cambridge University Press, Cambridge.
7. Branda, C. and Dymecki, S. M. (2004) Talking about a revolution: The impact of site-specific recombinases on genetic analyses in mice. *Develop. Cell* 6, 7–28.
8. Joyner, A. and Zervas, M. (2006) Genetic inducible fate mapping in mouse: Establishing genetic lineages and defining genetic neuroanatomy in the nervous system. *Dev. Dynamics* 235, 2376–2385.
9. Dymecki, S. M. and Kim, J. C. (2007) Molecular neuroanatomy's "Three Gs": A Primer. *Neuron* 54, 17–34.
10. Dymecki, S. M. and Tomasiewicz, H. (1998) Using Flp-recombinase to characterize expansion of Wnt1-expressing neural progenitors in the mouse. *Dev. Biol.* 201, 57–65.
11. Zinyk, D., Mercer, E. H., Harris, E., Anderson, D. J., and Joyner, A. L. (1998) Fate mapping of the mouse midbrain-hindbrain constriction using a site-specific recombination system. *Curr. Biol.* 8, 665–668.
12. Rodriguez, C. I. and Dymecki, S. M. (2000) Origin of the precerebellar system. *Neuron* 27, 475–486.
13. Awatramani, R., Soriano, P., Rodriguez, C., Mai, J. J., and Dymecki, S. M. (2003) Cryptic boundaries in roof plate and choroid plexus identified by intersectional gene activation. *Nat. Genet.* 35, 70–75.
14. Farago, A., Awatramani, R., and Dymecki, S. (2006) Assembly of the brainstem cochlear complex is revealed by intersectional and subtractive genetic fate maps. *Neuron* 50, 205–218.
15. Jensen, P., Farago, A. F., Awatramani, R. B., Scott, M. M., Deneris, E. S., Dymecki, S. M. (2008) Redefining the serotonergic system by genetic lineage. *Nat Neurosci.* April; *11(4)*, 417–419.
16. Wang, V. Y., Rose, M. F., and Zoghbi, H. Y. (2005) *Math1* expression redefines the rhombic lip derivatives and reveals novel lineages within the brainstem and cerebellum. *Neuron* 48, 31–43.
17. Landsberg, R. L., Awatramani, R. B., Hunter, N. L., Farago, A. F., DiPietrantonio, H. J., Rodriguez, C. I., and Dymecki, S. M. (2005) Hindbrain rhombic lip is comprised of discrete progenitor cell populations allocated by Pax6. *Neuron* 48, 933–947.
18. Stark, W. M., Boocock, M. R., and Sherratt, D. J. (1992) Catalysis by site-specific recombinases. *Trends Genet.* 8, 432–439.
19. Dymecki, S. M. (2000) Site-specific recombination in cells and mice, in *Gene Targeting: A Practical Approach* (Joyner, A. L., ed.) second ed., Oxford University Press, Oxford, pp. 37–99.
20. Lumsden, A. and Krumlauf, R. (1996) Patterning the vertebrate neuraxis. *Science* 274, 1109–1115.
21. Rodriguez, C. I., Buchholz, F., Galloway, J., Sequerra, R., Kasper, J., Ayala, R., Stewart, A. F., and Dymecki, S. M. (2000) High-efficiency deleter mice show that FLPe is an alternative to Cre-loxP. *Nat. Genet.* 25, 139–140.
22. Kosman, D., Mizutani, C. M., Lemons, D., Cox, W. G., McGinnis, W., and Bier, E. (2004) Multiplex detection of RNA expression in Drosophila embryos. *Science* 305, 846.
23. Tsien, J. Z., Chen, D. F., Gerber, D., Tom, C., Mercer, E. H., Anderson, D. J., Mayford, M., Kandel, E. R., and Tonegawa, S. (1996) Subregion- and cell type-restricted gene knockout in mouse brain. *Cell* 87, 1317–1326.

Chapter 6

The Practical Use of Cre and *loxP* Technologies in Mouse Auditory Research

Yiling Yu and Jian Zuo

Abstract

Gene manipulation, specifically in the hair cells of the inner ear during development and adulthood in mice, is crucial for understanding the physiology of hearing and the pathology of deafness in humans. Recent advances have demonstrated that gene expression can be manipulated in developing mouse hair cells in a spatially and temporally controlled manner. The Cre–*loxP* system has been widely used for such purposes. Many laboratories, including ours, have developed and characterized transgenic mouse lines that express or induce Cre activity specifically in inner ear hair cells. These *Cre* lines have been used with high efficiency to inactivate several genes such as *Rb* in hair cells. Here we discuss the use of these *Cre* lines in inner ear research with emphasis on practical issues for researchers who are not familiar with these particular techniques but are interested in using these Cre mice and floxed mice to inactivate genes of their interest specifically in inner ear hair cells. We provide detailed protocols for the use of these techniques and reagents. These considerations and protocols can be easily applied to other cell types in the inner ear and other parts of the auditory pathways. Because the NIH Knockout Mouse Project (KOMP) and the European Conditional Mouse Mutant Program (EUCOMM) have initiated plans to create conditional (floxed) knockout strains for every gene in the mouse genome and because numerous *Cre*-expressing mouse lines have already been created in various systems, including the nervous system, it is our hope that many hearing researchers will benefit from the detailed protocols and practical considerations described in this review.

Key words: Cre, *loxP*, transgenic mice, knockout mice, bacterial artificial chromosome, hair cell, auditory.

1. Introduction

Mouse models have been widely recognized as a crucial component for delineating the pathogenesis of human hearing disorders and elucidating the functions of important inner ear genes. Studies of knockout mice with null alleles or germline

Bernd Sokolowski (ed.), *Auditory and Vestibular Research: Methods and Protocols, vol. 493*

DOI 10.1007/978-1-59745-523-7_6 Springerprotocols.com

inactivation of a few of these genes have been fruitful *(1)*; however, this approach has serious limitations. For instance, a germline mutation of *Myosin 1c (Myo1c)*, which is expressed in many tissues and cell types, results in embryonic death *(2, 3)*. In fact, among more than 1,500 knockout alleles, about 15% cause embryonic death *(4)*. Furthermore, some hair cell genes affect not only the cell fate and differentiation of hair cells during development, but also the maintenance of their mature function. For these genes, germline inactivation would not distinguish their pleiotropic functions. This approach also precludes the generation of accurate mouse models of human hearing disorders.

To overcome these spatial and temporal limitations, it is essential to examine the function of a gene of interest not only in a specific cell type, but also at a given time in the life of the animal. Three general approaches have been used successfully for this purpose, all of which depend on the ability to modify the genomic DNA sequences (site-directed genome modification): virus-induced, transposon-based, and recombinase-mediated DNA integration systems *(1, 5)*. The virus-induced and transposon-based approaches are generally not uniform across cell types of interest and are not site specific, whereas Cre and Flp recombinase-mediated DNA integration has been successfully used in conditional gene expression in various tissues and cell types *(6)*. While both systems work in vivo, here we focus on the most commonly used Cre–*loxP* system.

1.1. Cre Recombinase and loxP Sites

Cre recombinase is originally a 38-kDa product of the *Cre* gene of bacteriophage P1. It targets *loxP*, a 34-bp site present in the P1 genome but absent in the mouse genome. It efficiently catalyzes the insertion, excision, or inversion of DNA between two *loxP* sites (**Fig. 6.1A**). Cre is widely used as an important tool in mice for the excision of DNA flanked by 2 *loxP* sites ("floxed" DNA; **Fig. 6.1B**) *(7)*. Cre can be expressed as a transgene under the control of a tissue- or cell-type-specific promoter. These lines can then be crossed with lines containing a floxed gene of interest to generate conditional gene knockouts. Alternatively, a floxed "stop" DNA segment can be added upstream from a transgene to prevent its expression. The floxed "stop" can be removed to overexpress the downstream gene of interest only in cells expressing the *Cre* gene (**Fig. 3** in *(8)*; **Fig. 1** in *(1)*).

1.2. Inducible Gene Expression

To express genes in a temporally controlled manner, several systems based on Cre–*loxP* have been developed and used successfully that utilize interferon, muristerone, and tetracycline *(9–14)*. One of the most successful systems uses a fusion protein comprising Cre and a mutated ligand-binding domain of a steroid hormone receptor such as the estrogen receptor (ER) (**Fig. 6.1A**).

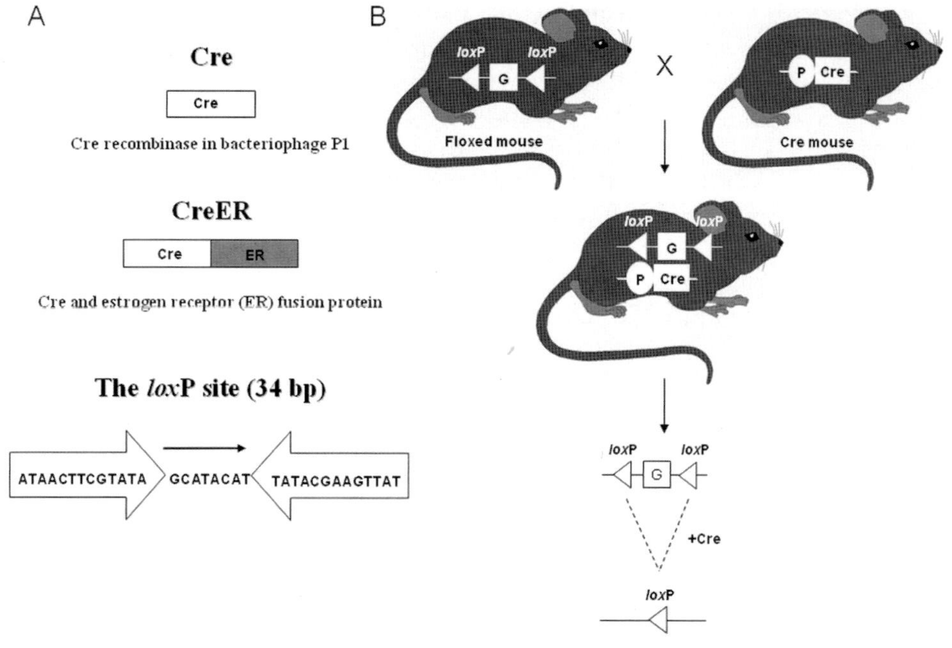

Fig. 6.1. (**A**) Cre, Cre-ER, and the loxP site. (**B**) Mouse breeding scheme to inactivate gene (G) in Cre-expressing cells. Two loxP sites (*triangles*) flank the gene (G); Cre expression is driven by a tissue-specific promoter (P). Gene (G) is deleted only in Cre-expressing cells.

Cre-ER can bind to the exogenous ligand tamoxifen but not to the corresponding endogenous ligand, estrogen. Only upon binding to tamoxifen can Cre-ER be translocated into the nuclei, where Cre excises the floxed gene of interest (*see* **Fig. 4** in *(8)*; **Fig. 2** in *(6)*). In a given tissue, all cell types that express Cre-ER can be induced to excise the floxed DNA *(15)*. To overcome potential tamoxifen-induced toxicity and undesired side effects, the fusion protein Cre-ER^{T2} was developed. The binding affinity of tamoxifen to Cre-ER^{T2} is 4- to 10-fold higher than to Cre-ER. Thus, a much lower dose of tamoxifen (0.1 mg/mouse/day) is sufficient to induce Cre-mediated excision of the floxed DNA *(16–18)*.

1.3. Creation of Floxed Alleles of Every Gene in the Mouse Genome

The NIH KOMP (http://www.genome.gov/17515708) and the EUCOMM have initiated a plan to create conditional (floxed) knockout strains for every gene in the mouse genome *(19)*. To take full advantage of these invaluable resources, it would be ideal to synchronize the generation of inner ear–specific *Cre* lines with specific designs of the large-scale conditional targeting strategies. Therefore, studies of gene functions in developing and adult inner ear cells will motivate auditory researchers to generate *Cre* or inducible *Cre* lines specific for these cells. In fact, a recent publication described the creation and characterization of 14 *Cre* lines, including one Cre-ER line specifically for the nervous system, by using bacterial artificial chromosome (BAC) transgenics *(20)*. Moreover, several large grants have been awarded to create such Cre "drivers" that include inner ear cell types (e.g., NIH 5U01MH078833). Much progress in creating many lines of useful Cre mouse strains for auditory research is expected in the foreseeable future.

1.4. Inner Ear- and Hair Cell–Specific Cre and Inducible Cre Lines

Inner ear–specific Cre lines *(1, 21–25)*, hair cell–specific Cre lines *(26,27)*, and an inducible Cre line for neonatal hair cells *(28)* have been developed and characterized by many researchers. Specifically, our laboratory has developed three hair cell *Cre* lines: the first line expresses *Cre* using a prestin enhancer element driving the *Cre* transgene in IHCs starting at postnatal day 14 (P14) *(26)*, the second expresses *Cre* using a modified prestin-BAC transgene in both IHCs and OHCs starting at P5 *(27)*, and the third expresses Cre-ER using an Atoh1 (*Math1*) enhancer driving the *Cre*-ER transgene in both IHCs and OHCs at P0–P2 *(28)*. In addition, several inner ear–specific *Cre* lines have been reported by others (**Table 3** in *(1)*; *(6)*). These lines have been useful in knocking out several genes, including *Rb*, specifically in hair cells with high efficiency (data not shown).

However, several features of these lines make studies of gene functions in mature hair cells difficult. First, none of these lines expresses *Cre* exclusively in hair cells. The IHC-*Cre* line also expresses Cre in many other tissues and in all cochlear spiral ganglia; the OHC/IHC-prestin BAC-*Cre* line expresses Cre in about 10% of spiral ganglia and several other tissues, such as testes; and in the Atoh1 enhancer-driven *Cre-ER* line, cerebellar granule cells also express Cre-ER. Second, Cre is not inducible in adult hair cells in these lines. In our *Atoh1–Cre-ER* line, the transgene is downregulated at P6 and Cre-ER cannot be induced at P7 or in adulthood (*28*). Therefore, there is a significant need to develop inducible Cre expression specifically in adult hair cells. Other cell types (i.e., supporting cells in the organ of Corti and marginal cells in the stria) in the inner ear can be targeted with similar strategies and efficiency.

1.5. Practical Considerations for the Use of Cre and loxP Technologies in Auditory Research in Mice

1.5.1. Cre Toxicity

A general concern recently expressed by the scientific community is that overexpression of Cre might have toxic effects to cells *(29)*. To overcome this, one would simply need to analyze various controls (i.e., *Cre*-positive and floxed-negative or wild-type littermates). Based on studies of *Cre* lines in the inner ear, there seems to be no such toxicity. However, because the Cre expression level varies between lines, investigators need to make efforts to address this issue in each case before publication. Nonetheless, neither tamoxifen treatment nor expression of Cre or Cre-ER in hair cells alone had any detectable effect on hearing sensitivity in mice, as measured by auditory brainstem response thresholds, at any ages we tested *(26–28)*.

1.5.2. Reporter Lines for Cre Activity

It is crucial to use the appropriate *Cre* reporter mouse lines to visualize the Cre activity patterns. The most commonly used line is *ROSA26R (30)*. In this line, a *LacZ* reporter is inserted in the *ROSA26* gene locus that is presumably ubiquitously expressed in every cell of the mouse. Only when Cre excises/removes the *loxP*-flanked stop codon will the *LacZ* reporter be expressed; thus, the pattern of X-gal staining can indicate Cre activity.

Alternatively to *ROSA26R*, several other *Cre* reporter lines have been developed and used. The commonly used reporter line Z/EG expresses *LacZ* throughout embryonic development and adult stages; Cre deletes the *LacZ* gene and activates expression of the second reporter, enhanced green fluorescent protein (EGFP) *(31)*. This double-reporter Z/EG line has been used successfully for early embryonic to adult cell lineage studies. Because EGFP is used rather than *LacZ*, Cre activity can be monitored in live samples, and live cells with Cre activity can be isolated using fluorescence-activated cell sorting (FACS). However, it remains debatable which reporter line (*ROSA26R* or Z/EG) has higher sensitivity for Cre activity.

It is unclear whether the *ROSA26* locus is indeed expressed ubiquitously in all cells of the mouse, especially in the adult mouse; therefore, there might be inconsistency between Cre activity and X-gal staining patterns. In other words, an X-gal-negative cell could still express Cre activity by itself or in its progenitors, yielding a false-negative result.

One important notion of the Cre-*loxP* system is that the reporter (*LacZ*) pattern does not always reflect that of the Cre expression at that particular time of analysis; it visualizes the developmental history of Cre expression in the cell lineage (**Fig. 6.2**). In other words, the detection of X-gal-positive offspring cells in this *LacZ* reporter line indicates that either these cells are currently Cre positive, or their progenitor was Cre positive, or both. One perfect example is the ectopic expression of the reporter EGFP observed in the adult brain of *Th BAC-Cre+:Rosa26 EGFP+* mice. The endogenous *Th* is not expressed at this adult stage but is expressed during development in these ectopic brain areas *(20)*. Similarly, our *prestin-Cre* lines that displayed IHC or IHC/OHC Cre activity patterns might also be attributed to such developmental ectopic expression, in addition to many other possible reasons *(26,27)*. Moreover, it is possible that the Pou4f3-Cre activity pattern observed in early embryonic presensory hair cell and supporting cell precursors was also due to the ectopic expression of Pou4f3 in the transgenic line *(25)*. Nevertheless, these ectopic Cre activity patterns have been useful for elucidating gene

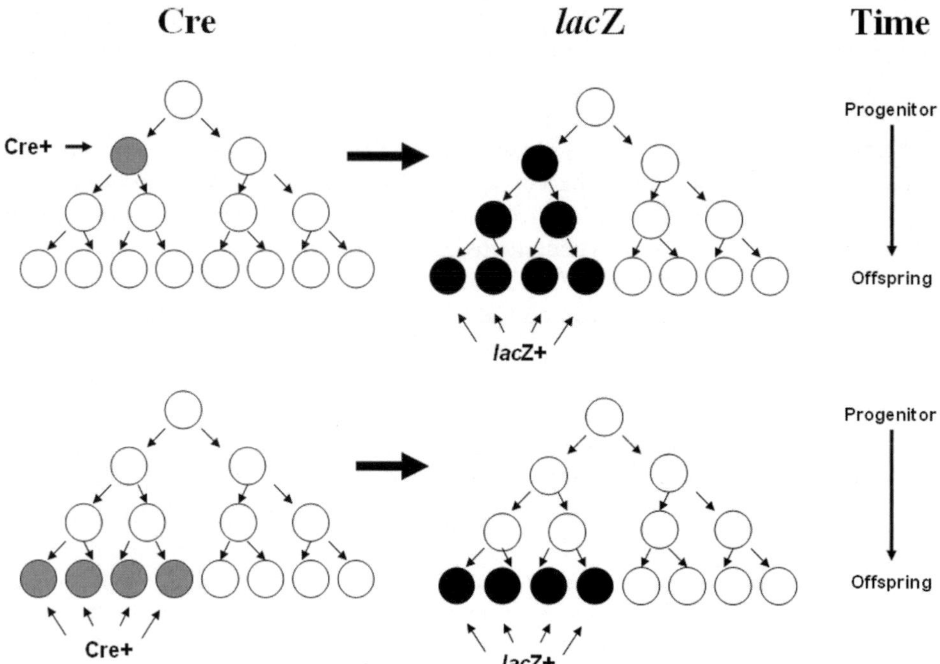

Fig. 6.2. Cre-mediated cell lineage studies during development. The same pattern of lacZ-positive offspring cells (*filled circles*) is detected when Cre is expressed only in progenitor cells (*top left, filled circle*) or only in offspring cells (*bottom left, filled circles*).

functions in these studies as long as the Cre activity patterns are carefully described.

1.5.3. Cell Lineage Tracing Using Cre Lines

Because *Cre*-mediated reporter expression labels both progenitor cells and their offspring, one would expect that all X-gal-positive cells could be derived from the same progenitors, thus providing evidence for cell lineage during development (**Fig. 6.2**). This is, in fact, frequently used by developmental biologists. Recently, this *Cre*-mediated lineage tracing has been used to study auditory brainstem pathways during development *(32)*.

1.5.4. Mosaic Patterns of Cre Activity

It is sometimes necessary to create a mosaic pattern of Cre activity. For example, one can generate a mosaic collection of *Cre*-positive and -negative HCs in a single cochlea, so that some HCs would have the gene of interest inactivated among many HCs that are wild type like. In this system, one can determine whether the effects of gene inactivation are cell autonomous (independent of neighboring cells). One can also investigate whether a random mix of HCs with variable levels of the gene of interest would cause variable defects in the physiology of the mouse.

One such example is our mouse model for the study of Rb in HC regeneration. First, in addition to completely eliminating HCs specifically by injecting tamoxifen at P0–P1, we can generate a mosaic of surviving and dying HCs by simply delaying the tamoxifen injection for 2 days. Because of the downregulation of Atoh1–Cre-ER in a cochlear longitudinal gradient, we expect to have Cre activity in some HCs but not in all HCs and more Cre-positive cells in the apex than in the base. Thus, individual surviving HCs can be studied in apical turns to determine the cell autonomous mode of Rb-induced HC loss.

Another potential use for such mosaic patterns of Cre activity is to create a knockout of *prestin* in a mosaic pattern so that prestin can be manipulated to be expressed at various levels (from 0 to 100%). In these mosaic prestin models, one can assess quantitative contributions of prestin to cochlear amplification in vivo.

1.5.5. BAC-Cre Transgenic Lines

Conventional transgenic lines use the minimum promoter regions (< 10 kb) as promoters to drive transgene expression. Because of the large insert of > 100 kb, BAC transgenic lines use entire endogenous promoter regions as promoters. Therefore, BAC transgenic lines should duplicate the endogenous expression patterns of the genes of interest. Because of the recent development of several BAC-Cre lines in the nervous system *(20)*, it is important for us to understand some of the limitations of these lines before using them.

1.5.5.1. Integration Sites

BAC integration sites sometimes still make differences in terms of Cre expression patterns among different mouse lines derived from the same constructs. Therefore, it might be useful to first identify multiple transgenic founder lines that produce Cre expression patterns either matching that of the endogenous gene or consistent with each other.

1.5.5.2. Rearrangements

It is well known that BACs frequently rearrange when integrated into the host genome. So do the BAC transgenes, particularly in successive generations of offspring of these transgenic mice. Such rearrangements could have effects on Cre expression. Cre expression/activity may not mimic the endogenous pattern in the founder lines, or Cre activity patterns become progressively more unstable in successive generations. Therefore, it should not be surprising that the Cre activity pattern in mice of subsequent generations is not identical to what was originally observed.

1.5.5.3. Copy Number

To excise a *loxP* site, Cre expression must reach a minimum level (threshold). Presumably, as few as four Cre molecules per cell are enough for Cre to effectively excise floxed DNA *(7)*. When Cre expression is lower than the threshold, floxed DNA may not be excised successfully. Furthermore, the thresholds may vary in different cells of the same tissue. Thus, the copy numbers of the BAC transgenes are important. However, because of the large promoter inserts, BAC transgenic mice normally have only 1–2 copies per line (unlike 10–100 copies per line in conventional transgenic mice). Such low copy numbers of BAC transgenes ultimately yield faithful expression of Cre at the expense of high expression levels. These low BAC copy numbers may also restrict expression of Cre to only a subset of cells expressing the endogenous gene. For example, in the *Chat* BAC-Cre line (GM60) crossed with ROSA26-EGFP, Cre-induced EGFP expression is restricted to a subset of cells (motor neurons in the brainstem and spinal cord) *(20)*.

1.5.6. Strain Background and Age-Related Hearing Loss

Previous transgenic mice and knockout mice, which were generated in FVB/NJ or 129S7/C57BL/6J mixed strains, did not display significant hearing or vestibular phenotypes before 3 months of age *(1)*. Therefore, these strains were suitable for auditory studies in younger mice. However, because of *Ahl* loci in these strains *(33–35)*, if aging studies are to be done, it would be necessary to further transfer these strains into the CBA/CaJ strain, the "gold strain" for auditory research. CBA/CaJ mice display the least variation in their hearing phenotypes during aging *(1)*. We have previously transferred our *prestin*-knockout strain into the CBA/CaJ strain successfully (**Fig. 6.3**), and the physiological defects in the *prestin*-CBA background remain similar to those in

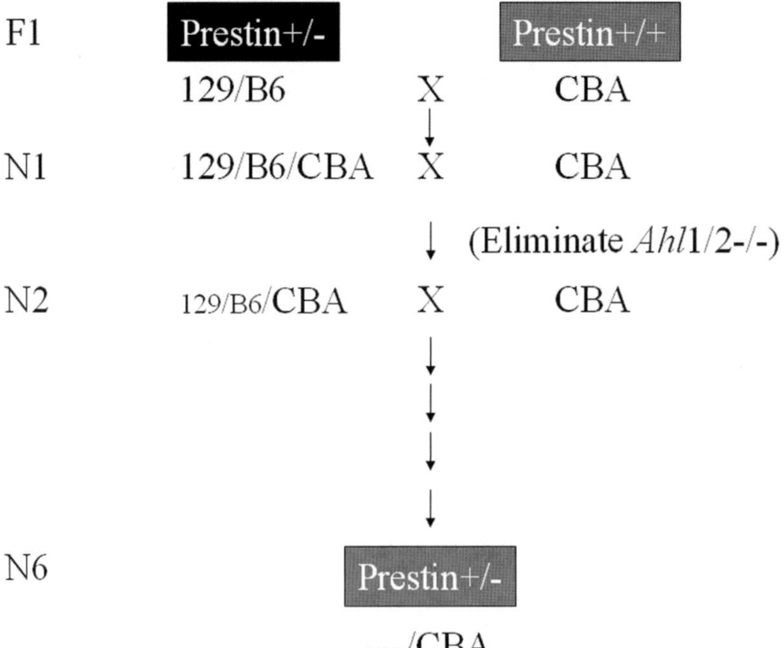

Fig. 6.3. Transfer of prestin +/− mice in original 129/B6 mixed strain background into CBA/CaJ. Prestin +/− mice in the original 129/B6 mixed strain (*black*) can be crossed with wild-type prestin +/+ mice in CBA/CaJ (*gray*). The offspring will be further screened by PCR for the two major known age-related hearing loss loci (*Ahl1* and *Ahl2*) in chromosomes 10 and 5 to ensure that only N1 offspring prestin +/− mice without *Ahl1-2* mutant loci can be further bred. N2–N6 successive backcrosses would in theory result in significant reduction of 129/B6 (*smaller letters*) and increase CBA content (*larger letters*). The N6 offspring should be mostly in CBA/CaJ background (*gray*) and prone to *Ahl* mutant effects.

the C57B6/129S7 mixed background but with delayed onset of cell death *(36)*.

2. Materials

2.1. Genotype

1. Lysis buffer: 100 m*M* Tris-HCl, pH 8.0, 5 m*M* ethylenediaminetetraacetic acid (EDTA), 0.2% SDS, 200 m*M* NaCl.
2. Proteinase K stock solution (20 mg/mL) in double distilled (dd) H_2O and stored at 4 °C.
3. Proteinase K working solution: Add 6 μl of 20 mg/mL proteinase K stock solution to 600 μL of lysis buffer for the use of one mouse tail sample.
4. Phenol:chloroform:isoamyl alcohol: Mix at a ratio of 25:24:1.
5. dd H_2O: water for molecular biology reagent (Sigma-Aldrich, St Louis, MO)
6. PuRe Taq Ready-To-Go PCR beads (GE Healthcare Illustra, UK).
7. Primers typically used for genotyping by polymerase chain reaction (PCR) to detect the *Cre* transgene are: forward, 5′-ACG ACC AAG TGA CAG CAA TG-3′; reverse, 5′-CCA TCG CTC GAC CAG TTT AG-3′. Primers for the Rosa26 reporter line are designed as follows: forward, 5′-CTA CAG GAA GGC CAG ACG CGA-3′; reverse, 5′-ACG TAG TGT GAC GCG ATC GGC-3′. Prepare 33 μ*M* primer in ddH_2O for Cre-forward and reverse primers and 4 μ*M* for ROSA26-forward and -reverse primers.
8. 1X TAE: 40 m*M* Tris-acetate, 1 m*M* EDTA.
9. 1% agarose gel: Melt 1.2 g agarose (molecular biology grade) in 120 mL of 1X TAE buffer using a microwave oven. When it cools down to ~ 50 °C, add 6 μL of 10 mg/mL ethidium bromide. Ethidium bromide is added to the gel at a concentration of 0. 5 μg/mL. The gel is poured into a gel tray sealed by buffer chamber and the comb is put into the gel. The gel is allowed to dry for 30 min or until hard. Buffer of 1X TAE is added into the buffer chamber until the gel is covered.

2.2. Induction of Cre Activity in Cre-ER Mice

1. Tamoxifen working solution: Prepare the solution in a 50 mL centrifuge tube by diluting 100 mg of tamoxifen (Sigma-Aldrich) in 20 mL of corn oil (Sigma-Aldrich) to obtain a 5 mg/mL tamoxifen solution. Shake the solution with a vortex mixer and incubate at 37 °C (*see* **Note 1**). Use a 30 mL syringe and a 0. 45 μm syringe filter to filter sterilize the solution into a 50 mL centrifuge tube where it is stored at 4 °C for a maximum of 7 days.

2.3. LacZ Staining

1. X-gal staining solution: Prepare a fresh solution of 1% sodium deoxycholate by mixing 100 mg of deoxycholic acid in 10 mL of PBS. To prepare the X-gal solution for one cochlea, prepare a 300 μL solution by taking 15 μl of X-gal solution and 285 μL of iron buffer from a β-Gal Staining Kit (Roche Biochemicals, Indiananapolis, IN) and mixing thoroughly for 10 min using a vortexer. Add 0. 6 μL of 10% NP-40 (AG Scientific, Inc., San Diego, CA; Protein Solubilzer 40) and 3 μL of 1% sodium deoxycholate to the solution. The staining solution should be prepared freshly before each use.
2. PBS (0.01 *M*): 138 m*M* NaCl, 2.7 m*M* KCl, pH 7.4 (Sigma-Aldrich).
3. Paraformaldehyde (Sigma-Aldrich): Prepare a 20% (w/v) stock solution by mixing the paraformaldehyde powder in 0.01 *M* PBS and store at −20 °C until use. For a working solution, dilute with 0.01 *M* PBS to a desired concentration (2% or 4%) before use.

3. Methods

3.1. Genotype for Cre Mice and ROSA26 Reporter Line Mice

3.1.1. Purification of Genomic DNA

1. Cut about 1 mm of mouse tail tissue at the tail tip using a scissor and place it in a 1.5 mL centrifuge tube.
2. Pipette 600 μL of the proteinase K working solution into the tube and incubate at 50 °C from 2 h to overnight.
3. After incubation add 600 μL of the phenol:chloroform: isoamyl alcohol solution (25:24:1) to each sample, shake vigorously by hand, and centrifuge at 15,000 g for 5 min.
4. Transfer 500 μL of the aqueous phase into a new 1.5 mL centrifuge tube (*see* **Note 2**), add 500 μL of chloroform to the contents, and mix by shaking vigorously by hand. Centrifuge at 15,000 g for 5 min.
5. Transfer the supernatant to another new 1.5 mL centrifuge tube (*see* **Note 3**), add 1 mL of 100% EtOH, and mix again by shaking vigorously by hand. Centrifuge at 15,000 g for 10 min.
6. Discard the supernatant and add 500 μL of 70% EtOH to rinse the pellet. Centrifuge at 15,000 g for 5 min.
7. Discard the supernatant, keep the DNA pellet in the tube, and air dry by opening the top of tube for 3 min. Add 50 μL of ddH_2O to dissolve the pellet (*see* **Note 4**).

3.1.2. Polymerase Chain Reaction

1. Prepare reaction components according to the following recipe (**Table 6.1**). Usually, we use 1 μl of 1:10 diluted, purified DNA sample for PCR.

Table 6.1
Reaction components for PCR

Components	Per reaction
DNA sample	1 μL
Primer 1 (forward)	1 μL
Primer 2 (reverse)	1 μL
ddH_2O	22 μL

2. Pipette the components into PCR tubes and keep them on ice. Add one ready-to-go PCR bead (GE Healthcare) to each reaction tube.
3. Run in a thermocycler using the following conditions for *Cre*: 94 °C, 2 min; 30 cycles (94 °C, 30 sec; 56 °C, 30 sec; 72 °C, 30 sec) 72 °C, 5 min; 4 °C. The conditions for *ROSA26* are as follows: 95 °C, 4 min; 30 cycles (94 °C, 30 sec; 59 °C, 30 sec; 72 °C, 30 sec) 72 °C, 8 min; 4 °C.

3.1.3. Analyze the PCR Product

1. Electrophorese 10 μL from each reaction on a 1% agarose gel stained with ethidium bromide using 120 mV for 30 min.
2. The expected PCR products are ~ 400 bp for *Cre* and ~ 300 bp for *ROSA26* (*see* **Note 5**).

3.2. Induction of Cre Activity in Cre-ER Mice

There are several alternatives for administering tamoxifen to initiate *Cre* excision: intraperitoneal (IP) injections, gavage, and as a supplement in chow *(37)*.

3.2.1. Intraperitoneal Injections

To start *Cre* excision at embryonic or postnatal stages, the pregnant females at mid-gestation or new born pups are given IP injections of tamoxifen. According to some published reports on using tissue-specific Cre-ER, tamoxifen (2 mg per 30 g of body weight) can be injected into pregnant females at E10.5 and embryos can be harvested at E13.5 or E14.5 *(38)*. Tamoxifen doses applied in each lab should be optimized.

The protocol used in our lab for Cre activity induction on P0 and P1 is as follows:

1. The tamoxifen solution is warmed to 37 °C and then injected (3 mg per 40 g of body weight; *see* **Note 6**) into mouse pups IP, using a 30-gauge needle attached to an insulin syringe (Becton Dickinson, Franklin Lakes, NJ).
2. For new born pups, inject IP through the bottom of the left leg to enter the pelvis (*see* **Note** 7) *(28)*.

3.2.2. Other Tamoxifen Administration Choices

Some studies have shown that oral tamoxifen administration is an acceptable alternative method of gene induction. One protocol used for the *KspCad–Cre-ER*^{T2} line is to dissolve 600 mg of tamoxifen (Sigma) in 900 μl of ethanol and suspend in 5.1 mL of sunflower oil. This 100 mg/mL tamoxifen solution is sonicated for 2 min (two times for 1 min each using 6-sec pulses) and then incubated at 55 °C until thawed completely and administered to mice (50 μL) by a feeding needle at 55 °C *(39)*.

An oral administration protocol can be used because IP injection and gavage supplement are likely to put considerable stress on mice *(37)*. Add 0.36 g/kg of tamoxifen citrate salt and sucrose at a final concentration of 5% to a soy-free, low-phytoestrogen diet (Ssniffl M-Z Phytoestrogenarm, Ssniff Spezialdiäten GmbH, Soest, Germany). The tamoxifen-containing chow is given ad libitum for 3–5 weeks. This induction by a medium dose of oral tamoxifen proved as efficient as IP injections of 0.5 mg tamoxifen daily for five consecutive days in a heart-specific $Txnrd2^{fl/-,Tg[MerCreMer]}$ knockout mouse line *(37)*.

3.3. LacZ Staining in Cochleae

1. Excise cochleae using conventional dissection tools.
2. Fix the cochleae in 2% paraformaldehyde in 0.01 *M* PBS for 6 h at 4 °C.
3. Rinse the cochleae one time with 0.01 *M* PBS at room temperature.
4. Make a small puncture in the apical part of the cochlea using the tip of forceps.
5. Inject the X-gal staining solution into the round and oval windows and then place the whole cochlea in the rest of the staining solution.
6. Stain the cochleae for up to 16 h at 37 °C with X-gal (*see* **Note 8**).
7. Rinse the cochleae with 0.01 *M* PBS at room temperature.
8. Postfix the cochleae with 4% paraformaldehyde in 0.01 *M* PBS for another 6–10 h at 4 °C.
9. Decalcify the cochleae by placing in 120 m*M* EDTA for 1–3 days at 4 °C (*see* **Note 9**).
10. To make frozen sections, move the decalcified cochleae into a 30% sucrose solution overnight at 4 °C.
11. Embed the cochleae in tissue freezing medium (Triangle Biomedical Sciences, Durham, NC).
12. Store the cochleae at −80 °C until sections are cut on a cryostat.

3.4. EGFP Visualization

Utilize three different systems for EGFP visualization depending on the sample size *(31)*.

1. A GFsP-5 light source (BLS-Ltd., Hungary) served to visualize all live animals.
2. A stereo-microscope with a MAA-02 universal light source (BLS-Ltd., Hungary) served to visualize embryos and wholemount organs.

3. A microscope equipped with epifluorescence lighting and an FITC filter set served to visualize tissue sections and whole mounts (*see* **Note 10**).

3.5. Transfer of 129/B6 Strain Background into CBA/CaJ in Successive Backcrosses

3.5.1. An Example

We first cross *prestin* knockout mice with CBA/CaJ to obtain F1 offspring; then we cross F1 *prestin* heterozygotes with CBA/CaJ wildtype mice to obtain N2 prestin heterozygous offspring. We further use PCR genotyping methods to identify the N2 offspring that contain the *Ahl1* and *Ahl2* genes from CBA/CaJ "good" background, which are then used for obtaining the subsequent N3-N6 offspring. For CBA/CaJ genotyping, two markers D10Mit80 (4 cM) and D10Mit12 (56 cM) and two other markers D5Mit346 (1 cM) and D5Mit168 (78 cM) are closely linked to *Ahl1* and *Ahl2* loci on chromosomes 10 (~ 30 cM) and 5 (~ 44 cM), respectively (http://www.informatics.jax.org). These polymorphic markers were chosen because they flank the *Ahl1* and *Ahl2* loci and can distinguish 129S and C57BL6/J mice from CBA/CaJ mice. For a specific mouse, the sizes of PCR products from the markers are different between CBA/CaJ and 129S or C57BL6/J. At the N2 generation, if both markers displayed the CBA/CaJ characteristic sizes, we would conclude that this mouse most likely contains the CBA/CaJ "good" allele at these *Ahl* loci, thus displaying no *Ahl* phenotypes. One N2 offspring *prestin* knockout heterozygote that contains only CBA/CaJ background at the two *Ahl* loci was chosen to breed further for N3-N6 generations. At the N6 generation, mice have CBA/CaJ content in greater than 95% of the genome.

3.5.2. PCR Primers

1. D10Mit12: forward//5′-ATG TCC AAA ACA CCA GCC AG-3′ and reverse///5′-GGA AGT GAT GGA GCT CTG TT-3′;
2. D10Mit80: forward//5′-CAA AAA AAA CCC TGA TTC TAC CA-3′ and reverse///5′-GTG TGC ATA TGG CAG TAA CTT TG-3′;
3. D5Mit346: forward//5′-TCA AAC TCC TCT AAT ATG GAA GTG C-3′ and reverse///5′-CTG TCT CAT TAA TCC ATG GAT CC-3′;
4. D5Mit168: forward//5′-CAG GTG ACA GTT GTT CTC TTC C-3′ and reverse///5′-CAT GCA TGA ACA CAC ATC ACA-3′.

The thermocycler conditions are as follows: 93 °C, 2 min; 40 cycles (93 °C, 30 sec; 54 °C, 30 sec; 72 °C, 1 min) 72 °C, 5 min; 4 °C. The PCR products are visualized in either 1.5–2% agarose or 5–10% acrylamide gel, and the expected sizes normally range from 100 to 250 bps.

4. Notes

1. It is important to keep shaking when making the tamoxifen solution at 37 °C until it is dissolved.
2. There will be three phases in the tube after centrifuging. The top phase is an aqueous phase containing nucleic acids, which should be taken out and used for the next step. The bottom phase contains phenol-chloroform and isoamyl alcohol. Wedged between the two is a thin phase, which contains protein. Try to avoid pipetting any protein from this phase.
3. Nucleic acid is in the supernatant and chloroform in the bottom.
4. The obtained genomic DNA can be stored at 4 °C until use for PCR.
5. In some cases, β-gal immunostaining needs to be used to confirm the X-gal staining pattern *(27)*.
6. Usually a P0 pup's body weight is around 1.2–1.5 g, and a pregnant female's is 30–40 g. However, tamoxifen doses applied in each lab should be optimized.
7. Multiple injections in the same mouse should be separated by 24 h due to cytotoxicity. The *Cre* excision pattern in the Z/AP and Z/EG reporter lines was seen 2 days after tamoxifen administration as single cells scattered throughout the embryo, but 1 day later, as cell patches *(40)*.
8. The incubation time should vary depending on the expression level of β-gal.
9. For mice older than P10, cochleae should be decalcified. After decalcification, the cochleae can be subjected to whole mount or section immunohistochemistry.
10. If EGFP fluorescence is not strong enough to be visualized directly, antibodies to EGFP (rabbit anti-EGFP) can be used at a dilution of 1:1,000 for visualization *(20)*.

Acknowledgments

This work was supported in part by grants from the National Institutes of Health (DC06471, DC05168, DC008800, CA21765) and the American Lebanese Syrian Associated Charities of St. Jude Children's Research Hospital. J. Zuo is a recipient of a Hartwell Individual Biomedical Research Award.

References

1. Gao, J., Wu, X. and Zuo, J. (2004) Targeting hearing genes in mice. *Brain Res. Mol. Brain Res.* **132**, 192–207.
2. Holt, J. R., Gillespie, S. K., Provance, D. W., Shah, K., Shokat, K. M., Corey, D. P., Mercer, J. A., and Gillespie, P. G. (2002) Achemical-

genetic strategy implicates myosin-1c in adaptation by hair cells. *Cell* **108**, 371–381.
3. Stauffer, E. A., Scarborough, J. D., Hirono, M., Miller, E. D., Shah, K., Mercer, J. A., Holt, J. R., and Gillespie, P. G. (2005) Fast adaptation in vestibular hair cells requires myosin-1c activity. *Neuron* **47**, 541–553.
4. Zambrowicz, B. P., Abuin, A., Ramirez-Solis, R., Richter, L. J., Piggott, J., BeltrandelRio, H., et al. (2003) Wnk1 kinase deficiency lowers blood pressure in mice: a gene-trap screen to identify potential targets for therapeutic intervention. *Proc. Natl. Acad. Sci. USA* **100**, 14109–14114.
5. Coates, C. J., Kaminski, J. M., Summers, J. B., Segal, D. J., Miller, A. D., and Kolb, A. F. (2005) Site-directed genome modification: derivatives of DNA-modifying enzymes as targeting tools. *Trends Biotechnol.* **23**, 407–419.
6. Tian, Y., James, S., Zuo, J., Fritzsch, B., and Beisel, K. W. (2006) Conditional and inducible gene recombineering in the mouse inner ear. *Brain Res* **1091**, 243–254.
7. Nagy, A. (2000) Cre recombinase: the universal reagent for genome tailoring. *Genesis* **26**, 99–109.
8. Zuo, J. (2002) Transgenic and gene targeting studies of hair cell function in mouse inner ear. *J. Neurobiol.* **53**, 286–305.
9. Kuhn, R., Schwenk, F., Aguet, M., and Rajewsky, K. (1995) Inducible gene targeting in mice. *Science* **269**, 1427–1429.
10. No, D., Yao, T. P., and Evans, R. M. (1996) Ecdysone-inducible gene expression in mammalian cells and transgenic mice. *Proc. Natl. Acad. Sci. USA* **93**, 3346–3351.
11. Shimizu, E., Tang, Y. P., Rampon, C., and Tsien, J. Z. (2000) NMDA receptor-dependent synaptic reinforcement as a crucial process for memory consolidation [In Process Citation]. *Science* **290**, 1170–1174.
12. Utomo, A. R., Nikitin, A. Y., and Lee, W. H. (1999) Temporal, spatial, and cell type-specific control of Cre-mediated DNA recombination in transgenic mice. *Nat. Biotechnol.* **17**, 1091–1096.
13. Bond, C. T., Herson, P. S., Strassmaier, T., Hammond, R., Stackman, R., Maylie, J., and Adelman, J. P. (2004) Small conductance Ca2+-activated K+channel knock-out mice reveal the identity of calcium-dependent afterhyperpolarization currents. *J. Neurosci.* **24**, 5301–5306.
14. Hammond, R. S., Bond, C. T., Strassmaier, T., Ngo-Anh, T. J., Adelman, J. P., Maylie, J., and Stackman, R. W. (2006) Small-conductance Ca2+-activated K+ channel type 2 (SK2) modulates hippocampal learning, memory, and synaptic plasticity. *J. Neurosci.* **26**, 1844–1853.
15. Brocard, J., Warot, X., Wendling, O., Messaddeq, N., Vonesch, J. L., Chambon, P., and Metzger, D. (1997) Spatio-temporally controlled site-specific somatic mutagenesis in the mouse. *Proc. Natl. Acad. Sci. USA* **94**, 14559–14563.
16. Danielian, P. S., Muccino, D., Rowitch, D. H., Michael, S. K., and McMahon, A. P. (1998) Modification of gene activity in mouse embryos in utero by a tamoxifen-inducible form of Cre recombinase. *Curr. Biol.* **8**, 1323–1326.
17. Indra, A. K., Warot, X., Brocard, J., Bornert, J. M., Xiao, J. H., Chambon, P., and Metzger, D. (1999) Temporally-controlled site-specific mutagenesis in the basal layer of the epidermis: comparison of the recombinase activity of the tamoxifen-inducible Cre-ER(T) and Cre-ER(T2) recombinases. *Nucleic Acids Res.* **27**, 4324–4327.
18. Li, M., Indra, A. K., Warot, X., Brocard, J., Messaddeq, N., Kato, S., Metzger, D., and Chambon, P. (2000) Skin abnormalities generated by temporally controlled RXRalpha mutations in mouse epidermis. *Nature* **407**, 633–636.
19. Auwerx, J., Avner, P., Baldock, R., Ballabio, A., Balling, R., Barbacid, M., et al. (2004) The European dimension for the mouse genome mutagenesis program. *Nat. Genet.* **36**, 925–927.
20. Gong, S., Doughty, M., Harbaugh, C. R., Cummins, A., Hatten, M. E., Heintz, N., and Gerfen, C. R. (2007) Targeting Cre recombinase to specific neuron populations with bacterial artificial chromosome constructs. *J. Neurosci.* **27**, 9817–9823.
21. Hebert, J. M. and McConnell, S. K. (2000) Targeting of cre to the Foxg1 (BF-1) locus mediates loxP recombination in the telencephalon and other developing head structures. *Dev. Biol.* **222**, 296–306.
22. Matei, V., Pauley, S., Kaing, S., Rowitch, D., Beisel, K. W., Morris, K., Feng, F., Jones, K., Lee, J., and Fritzsch, B. (2005) Smaller inner ear sensory epithelia in Neurog 1 null mice are related to earlier hair cell cycle exit. *Dev. Dyn.* **234**, 633–650.
23. Ohyama, T. and Groves, A. K. (2004) Generation of Pax2-Cre mice by modification of a Pax2 bacterial artificial chromosome. *Genesis* **38**, 195–199.
24. Sage, C., Huang, M., Karimi, K., Gutierrez, G., Vollrath, M. A., Zhang, D. S., Garcia-Anoveros, J., Hinds, P. W., Corwin, J. T., Corey, D. P., and Chen, Z. Y. (2005)

Proliferation of functional hair cells in vivo in the absence of the retinoblastoma protein. *Science* **307**, 1114–1118.
25. Sage, C., Huang, M., Vollrath, M. A., Brown, M. C., Hinds, P. W., Corey, D. P., Vetter, D. E., and Chen, Z. Y. (2006) Essential role of retinoblastoma protein in mammalian hair cell development and hearing. *Proc. Natl. Acad. Sci. USA* **103**, 7345–7350.
26. Li, M., Tian, Y., Fritzsch, B., Gao, J., Wu, X., and Zuo, J. (2004) Inner hair cell Cre-expressing transgenic mouse. *Genesis* **39**, 173–177.
27. Tian, Y., Li, M., Fritzsch, B., and Zuo, J. (2004) Creation of a transgenic mouse for hair-cell gene targeting by using a modified bacterial artificial chromosome containing Prestin. *Dev. Dyn.* **231**, 199–203.
28. Chow, L. M., Tian, Y., Weber, T., Corbett, M., Zuo, J., and Baker, S. J. (2006) Inducible Cre recombinase activity in mouse cerebellar granule cell precursors and inner ear hair cells. *Dev. Dyn.* **235**, 2991–2998.
29. Editorial (2007) Toxic alert. *Nature* **449**, 378.
30. Soriano, P. (1999) Generalized lacZ expression with the ROSA26 Cre reporter strain [letter]. *Nat. Genet.* **21**, 70–71.
31. Novak, A., Guo, C., Yang, W., Nagy, A., and Lobe, C. G. (2000) Z/EG, a double reporter mouse line that expresses enhanced green fluorescent protein upon Cre-mediated excision. *Genesis* **28**, 147–155.
32. Goodrich, L. V. (2007) A genetic dissection of auditory circuit assembly and function. *ARO MidWinter Meeting*, 241.
33. Johnson, K. R. and Zheng, Q. Y. (2002) Ahl2, a second locus affecting age-related hearing loss in mice. *Genomics* **80**, 461–464.
34. Ohlemiller, K. K. (2004) Age-related hearing loss: the status of Schuknecht's typology. *Curr. Opin. Otolaryngol. Head Neck Surg.* **12**, 439–443.
35. Kujawa, S. G. and Liberman, M. C. (2006) Acceleration of age-related hearing loss by early noise exposure: evidence of a misspent youth. *J. Neurosci.* **26**, 2115–2123.
36. Huynh, K., Edge, R., Gao, J., Zuo, J., Dallos, P., and Cheatham, M. A. (2006) Anatomy and physiology of prestin knockout mice backcrossed to the CBA/CaJ strain. *ARO Midwinter Meeting*, Abstr. 379.
37. Kiermayer, C., Conrad, M., Schneider, M., Schmidt, J., and Brielmeier, M. (2007) Optimization of spatiotemporal gene inactivation in mouse heart by oral application of tamoxifen citrate. *Genesis* **45**, 11–16.
38. Hong, S. B., Furihata, M., Baba, M., Zbar, B., and Schmidt, L. S. (2006) Vascular defects and liver damage by the acute inactivation of the VHL gene during mouse embryogenesis. *Lab Invest.* **86**, 664–675.
39. Lantinga-van Leeuwen, I. S., Leonhard, W. N., van de Wal, A., Breuning, M. H., Verbeek, S., de Heer, E., and Peters, D. J. (2006) Transgenic mice expressing tamoxifen-inducible Cre for somatic gene modification in renal epithelial cells. *Genesis* **44**, 225–232.
40. Guo, C., Yang, W., and Lobe, C. G. (2002) A Cre recombinase transgene with mosaic, widespread tamoxifen-inducible action. *Genesis* **32**, 8–18.

Chapter 7

Helios® Gene Gun–Mediated Transfection of the Inner Ear Sensory Epithelium

Inna A. Belyantseva

Abstract

Helios® Gene Gun–mediated transfection is a biolistic method for mechanical delivery of exogenous DNA into cells *in vitro* or *in vivo*. The technique is based on bombardment of a targeted cellular surface by micron- or submicron-sized DNA-coated gold particles that are accelerated by a pressure pulse of compressed helium gas. The main advantage of Helios® Gene Gun–mediated transfections is that it functions well on various cell types, including terminally differentiated cells that are difficult to transfect, such as neurons or inner ear sensory hair cells, and cells in internal cellular layers, such as neurons in organotypic brain slices. The successful delivery of mRNA, siRNA, or DNA of practically any size can be achieved using biolistic transfection. This chapter provides a detailed description and critical evaluation of the methodology used to transfect cDNA expression constructs, including green fluorescent protein (GFP) tagged full-length cDNAs of myosin XVa, whirlin, and β-actin, into cultured inner ear sensory epithelia using the Bio-Rad Helios® Gene Gun.

Key words: Biolistic, transfection, Gene Gun, culture, ear, hair cell, stereocilia, myosin, whirlin, actin, immunofluorescence, GFP.

1. Introduction

Mammalian inner ear sensory epithelia of hearing and balance end-organs (organ of Corti and vestibular epithelia, respectively) are tiny delicate tissues embedded in a bony labyrinth of the temporal bone and are not readily accessible for non-traumatic, non-surgical interventions. Using explanted (cultured) rodent inner ear neurosensory epithelium, one can study the development and function of these sensory cells *(1, 2)* as well as investigate the effects of epitope-tagged cDNA constructs that are transfected into particular cell types of the inner ear of mice and

Bernd Sokolowski (ed.), *Auditory and Vestibular Research: Methods and Protocols, vol. 493*

DOI 10.1007/978-1-59745-523-7_7 Springerprotocols.com

rats *(3–6)*. Moreover, using inner ear tissues from deaf mutant mice, researchers can evaluate the effects of exogenous wild-type cDNA on the phenotype of auditory and vestibular hair cells as well as other cell types from a malfunctioning inner ear *(6, 7)*. However, inner ear hair cells cannot be transfected using conventional transfection techniques such as lipofection. In wild-type animals, auditory hair cells are terminally differentiated cells; they exit the cell cycle during embryonic development and lose their ability to divide. Lipofection is a technique designed to deliver genetic material into a cell by means of liposomes, phospholipid-based vesicles that merge with the cell membrane and release their contents into the cell *(8)*. Lipofection can be used to transfect many cell lines but does not work well with terminally differentiated cells. Another method of delivering exogenous DNA uses viruses *(9–11)*. The drawback for the most commonly used virus vectors is that the maximum DNA insert size is limited *(9)*, and there are concerns over the toxicity and immunogenicity of viral DNA delivery systems *(12)*. An alternative is electroporation-mediated transfection, which is based on the application of an electric field pulse that creates transient aqueous pathways in lipid bilayer membranes, allowing polar molecules to pass *(13)*. Electroporation causes a brief increase in membrane permeability after which the membrane quickly reseals. This method is effective with nearly all cell types and species and is used in many applications, including *in vivo* transfection of embryonic mouse brain *(14)* and *in vitro* transfection of immature hair cells from embryonic inner ear explants *(15)*. One disadvantage of this method is the exposure of targeted and non-targeted cells in complex tissues to a potentially damaging current. Excessive cell damage and death were a long-standing concern *(16)* until electroporation devices were improved *(17)*. However, electroporation so far has not been used successfully to transfect structurally developed, nearly mature hair cells from the auditory epithelium of postnatal mice.

Transfection mediated by calcium phosphate precipitation also has a low efficiency for terminally differentiated cells. However, it was shown recently that co-precipitating adenoviral vectors with calcium phosphate increased gene expression and transduction efficiency in mouse dendritic cells, primarily owing to receptor-independent viral uptake. This approach combines the efficiency of adenoviral-mediated endosomal escape and nuclear trafficking with the receptor independence of non-viral gene delivery *(18)*. Improved transfection efficiency was observed also when combining viral or non-viral vectors with paramagnetic nanoparticles and targeted gene delivery by application of a magnetic field. This method of transfection is called magnetofection *(12, 19, 20)*. While magnetofection does not necessarily improve the overall performance of any given standard gene transfer method *in vitro*, its potential lies in rapid and efficient transfection at low vector doses and the possibility of remotely controlled vector targeting *in vivo* *(21)*. A microinjection method of delivering exogenous DNA into a cell, although precise, is labor intensive. Recently, a fully automated robotic system for microinjection was developed and used in zebrafish embryos *(22)*. Another method that might allow for targeted transfection of cells is an ultrashort (femtosecond), high-intensity, near-infrared laser that can make a tiny, localized transient perforation in the membrane through which plasmid DNA can enter the cell *(23)*. However, this method has not yet been adapted to transfect inner ear sensory epithelia.

The method of DNA delivery using sub-micron-sized particles (microcarriers) that are accelerated to high velocity was developed in the late 1980s by Sanford, Johnston, and colleagues *(24–26)*. This biolistic method was designed to circumvent difficulties in transfecting plant cells whose cell walls present a physical barrier for simple diffusion and/or internalization of transfected material and was subsequently shown to be applicable to mammalian cells *(26, 27)*. In the early 1990s, it was used to deliver exogenous DNA to tissue in a live mouse *(28, 29)*. Since then biolistic devices were modified for particular applications and used *in vitro* to transfect cultured cells and tissues, from yeast to mouse brain slices *(25, 27, 30–32)*, and *in vivo* for intradermal vaccination of human and animals using DNA and mRNA vaccines *(33, 34)*. In the Bio-Rad handheld Helios® Gene Gun delivery system (Bio-Rad Laboratories, Inc., Hercules, CA), DNA-coated gold particles (bullets) are accelerated to high speed by pressurized helium and are able to overcome physical barriers such as the stratum corneum in the epidermis *(35)* or the actin-rich cuticular plate of inner ear hair cells. This method is suitable for the delivery of mRNA, siRNA, or cDNA to terminally differentiated cells that are difficult to transfect such as neurons, inner ear sensory cells, or cells from internal cellular layers *(33, 36, 37)*. Gene Gun-mediated transfection works well even with postnatal inner ear sensory epithelial explants, can be used to co-transfect two or more different plasmids on the same bullet, and is suitable for delivery of large cDNAs.

Using the Helios® Gene Gun, we successfully transfected hair cells with cDNA expression constructs of GFP-tagged full-length myosin Ic, myosin VI, myosin VIIa, myosin XVa, whirlin, espin, and γ- and β-actin (*4, 6, 7,* and unpublished data). Some of our results with cDNA expression constructs of GFP-tagged myosin XVa, whirlin, and β-actin will be used in this chapter to illustrate the Gene Gun transfection method. Our data show that Helios® Gene Gun–mediated transfection in combination with fluorescence immunostaining and genetic and phenotype analyses of mouse models of human deafness is a valuable tool to elucidate functions of these genes and their encoded proteins.

Inner ear sensory epithelia are populated by many cell types with apical surfaces that have different physical properties. Directly underneath the apical plasma membrane of the sensory hair cells of the organ of Corti is a dense actin meshwork referred to as the cuticular plate. The cuticular plate helps hair cells withstand disturbances due to acoustic stimulations. It also provides support for the rows of stereocilia, which are rigid microvilli-like projections on the apical surface of a hair cell.

Stereocilia are susceptible to damage due to the applied pulse of helium pressure as well as to gold particle bombardment. Meanwhile, the dense cuticular plate is an obstacle to the introduction of gold particles into sensory hair cells. These factors require careful consideration of the many parameters and settings needed for using the Gene Gun to transfect cDNA into sensory hair cells. Variables include the distance between the cartridge with bullets and the targeted tissue, the angle at which bullets strike the cells, the helium pressure applied to propel the bullets toward the tissue, the thickness of the residual liquid layer that covers the tissue during bombardment, the density of bombarding gold particles over the surface area of targeted cells, the purity and concentration of DNA, and the general quality of the cartridges and bullets, which is discussed in Section 3 of this chapter (*see* **Note 1**). This chapter describes in detail the experimental protocol, including preparation of the organotypic cultures of the sensory epithelia of the inner ear from postnatal mice and rats, coating microcarriers with plasmid DNA, cartridge preparation, and bombarding tissues with DNA-coated gold particles accelerated by a pulse of helium gas pressure (*see* **Note 2**).

2. Materials

2.1. Preparation of Inner Ear Sensory Epithelial Explants

1. Experimental animals. Mouse or rat pups of postnatal days 0–4 (*see* **Note 3**).
2. Dissection tools and microscope (*see* **Note 4**).
3. Sterile 60 × 15 mm polystyrene tissue culture dishes (Becton Dickinson and Co., Franklin Lakes, NJ).
4. Leibowitz's L-15 medium without phenol red (Invitrogen, Carlsbad, CA). Store at 4 °C.
5. Sterile MatTek glass bottom Petri dishes (cat. no. P-50G-0-14-F, MatTek Corp, Ashland, MA) (*see* **Note 5** and **Fig. 7.1**).
6. 2.18 mg/mL Cell-Tak cell and tissue adhesive (BD Biosciences, San Jose, CA). Store at 4 °C.
7. Tissue culture grade water (Invitrogen).
8. Dulbecco' Modified Eagle's Medium (DMEM) with high glucose content (4.5 g/L) and 25 m*M* HEPES buffer

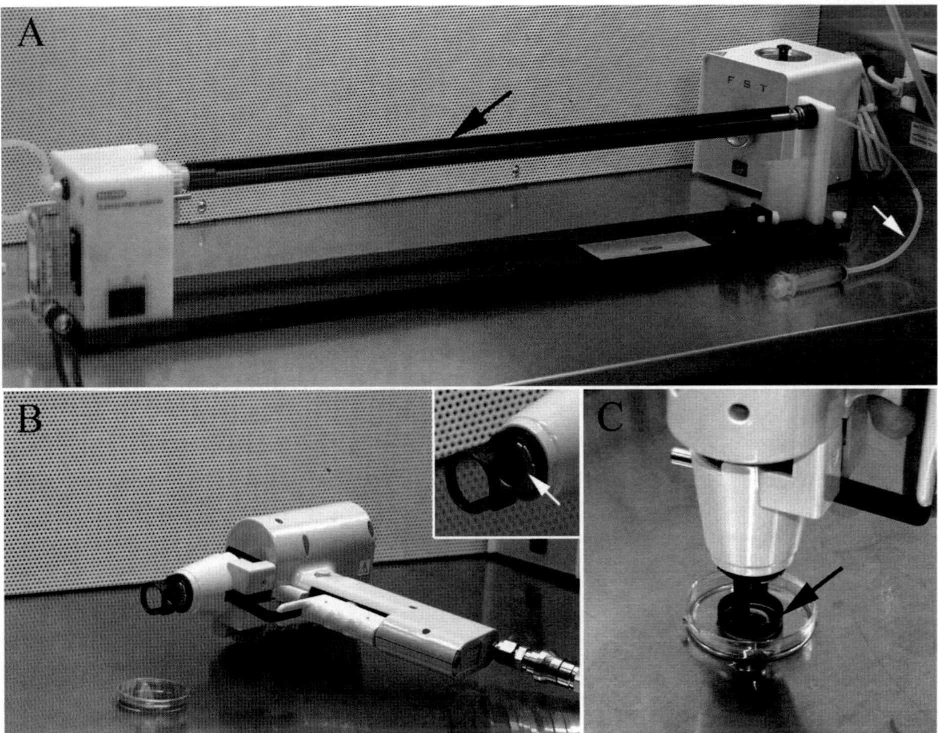

Fig. 7.1. **Bio-Rad Helios® Gene Gun and Tubing Prep Station.** (**A**) Tubing Prep Station with Tefzel tubing inserted into the tubing support cylinder (*black arrow*). The right end (~15 cm) of the Tefzel tubing is sticking out and is connected to the 10 cc syringe with adaptor tubing (white arrow). (**B**) An assembled Gene Gun with a diffusion screen inserted into the barrel. The insert shows a close view of a barrel with a diffusion screen (*white arrow*). Next to the Gene Gun, there is a MatTek glass bottom Petri dish, containing the attached sensory epithelium explant in DMEM. (**C**) Correct placement of the Gene Gun while transfecting inner ear sensory epithelium cultured in a MatTek Petri dish. The plastic ring at the end of the barrel (*black arrow*) is positioned so that the explant appears in the center of the ring. DMEM was removed in preparation for firing.

(Invitrogen) supplemented with 7% (v/v) fetal bovine serum (FBS, Invitrogen). Store at 4 °C (*see* **Note 6**).

9. Sterile microdissecting curette, 12.7 cm, size 3, 2.5 mm (Biomedical Research Instruments, Rockville, MD) (*see* **Note 7**).
10. Tissue culture incubator set at 37 °C and 5% CO_2 (*see* **Note 8**).

2.2. Preparation of Bullets with DNA-Covered Gold Microcarriers

1. 50 μg of plasmid DNA at 1 mg/mL (*see* **Note 9**). Store at −20 °C.
2. Fresh (unopened) bottle of 100% ethyl alcohol. Store in a cabinet for flammable alcohol reagents at room temperature (*see* **Note 10**).
3. 1 *M* $CaCl_2$: Dilute in the DNase, Rnase-free molecular biology grade water from 2*M* $CaCl_2$ molecular biology grade stock solution. Prepared stock solutions can be purchased

from several vendors (e.g., Quality Biological, Inc., Gaithersburg, MD).

4. 1 μm gold microcarriers or tungsten microcarriers (Bio-Rad) (*see* **Note 11** , *(38)*).
5. 20 mg/mL polyvinylpyrrolidone (PVP, Bio-Rad): Weigh out 20 mg of crystallized PVP, add 1 mL of 100% ethanol, and vortex. PVP becomes fully dissolved within 5–10 min at room temperature. Store at 4 °C and use within 1 month (*see* **Note 12**).
6. 0.05 *M* spermidine (cat. no. S0266, Sigma-Aldrich Inc. St. Louis, MO) stock solution: Dilute the content of one ampule (1 g) of spermidine in 13.6 mL of DNase, RNase-free molecular biology grade water to get a 0.5 *M* stock solution. Store this solution as single-use aliquots at −20 °C for 1 month. For a working solution to use in bullet preparation, thaw one aliquot of stock solution, take 5 μL, and add 45 μL of DNase, RNase-free molecular biology grade water to obtain a final concentration of 0.05 *M*. Use the same day (*see* **Note 13**).
7. Two sterile 15 mL conical tubes and sterile 1.5 mL centrifuge tubes.
8. Ultrasonic cleaner (waterbath sonicator) (e.g., Model 50D, VWR International, West Chester, PA) (*see* **Note 14**).
9. Tubing Prep Station (**Fig. 7.1A**) (cat. no. 1652418, Bio-Rad). Clean by wiping with 70% (v/v) ethanol before each use.
10. Nitrogen gas tank, grade 4.8 or higher, and nitrogen regulator (cat. no. 1652425, Bio-Rad). Also, see the Bio-Rad Helios® Gene Gun System instruction manual for nitrogen gas requirements.
11. Tefzel tubing (cat. no. 165-2441, Bio-Rad).
12. Tubing cutter and disposable blades (cat. no. 165-2422 and 165-2423, Bio-Rad).
13. 10 cc syringe with ~12–15 cm of syringe adaptor tubing (**Fig. 7.1A**, white arrow) (Bio-Rad).
14. 20 mL disposable scintillation vials with caps (cat. no. 7451020 Kimble Glass Inc., Vineland, NJ) and desiccating capsules of drycap dehydrators type 11 (Ted Pella, Inc., Redding, CA).

2.3. Helios® Gene Gun Transfection Procedure

1. Helium gas tank grade 4.5 (99.995%) or higher should be used together with a helium pressure regulator (cat. no. 165-2413, Bio-Rad).
2. Helios Gene Gun System, 100/120 V (**Fig. 7.1B**) (cat no. 1652431, Bio-Rad).
3. A diffusion screen (**Fig. 7.1B**, white arrow in the insert) (cat no. 165-2475, Bio-Rad) can be reused with the same DNA preparation (*see* **Note 15**).

4. Inner ear sensory epithelial explants attached to the bottom of a glass bottom MatTek Petri dish (prepared as described in "Methods" section).

2.4. Counterstaining, Immunostaining, and Imaging of Transfected Samples

1. 1X phosphate buffered saline (PBS) without Ca^{2+} and Mg^{2+}: 1.06 m*M* KH_2PO_4, 155.17 m*M* NaCl, 2.97 m*M* Na_2HPO_4 (*see* **Note 16**). Store at 4 °C.
2. 4% (v/v) paraformaldehyde fixative: Dilute 16% paraformaldehyde (Electron Microscopy Sciences, Hatfield, PA) 1:4 with 1X PBS (1X PBS is without Ca^{2+} and Mg^{2+} if not specifically mentioned otherwise). Store at 4 °C.
3. 0.5% (v/v) Triton X-100: Dilute 0.25 mL of 100% Triton X-100 (ACROS Organics, New Jersey, USA) in 50 mL of 1X PBS (*see* **Note 17**).
4. Blocking solution: Dilute 0.2 g of bovine serum albumin fraction V, protease-free (Roche Diagnostics, Indianapolis, IN) and 0.5 mL goat serum (Invitrogen) in 10 mL of 1X PBS. Keep refrigerated and use within 48 h. Every time before use, filter the desired volume of blocking solution using a syringe-driven MF membrane filter unit (25 mm in diameter and 22 μm pore size; Millipore Corporation, Bedford, MA) for sterilization of aqueous solutions.
5. Primary antibody to recognize endogenous native or fluorescently tagged newly synthesized protein. For example, to recognize endogenous whirlin or GFP-tagged whirlin, we use polyclonal rabbit anti-whirlin antibody, diluted 1:400 in blocking solution (**Fig. 7.2**, *(6)*). Store at −80 °C.
6. Secondary antibody conjugated to a fluorophore. For example, to bind to polyclonal rabbit anti-whirlin primary antibody (**step 5 of Section 2.4**), we use Alexa 643-conjugated goat anti-rabbit secondary antibody (Invitrogen). Store at 4 °C. Dilute 1:500 in blocking solution at time of use.
7. Rhodamine-phalloidin (Invitrogen). Dilute 1:100 in 1X PBS or blocking solution before use (*see* **Note 18**).
8. A short (146 mm) glass Pasteur pipette (Ted Pella, Inc., Redding, CA) to transfer inner ear sensory epithelial explant from MatTek petri dish to a glass slide to mount. To make a tip opening of the Pasteur pipette wider to accommodate the explant, cut the narrow part of the pipette with glass cutter tool.
9. Anti-fade kit (Invitrogen). Store at −20 °C. Use according to manufacturer's instructions (*see* **Note 19**).
10. LSM510 confocal microscope (Carl Zeiss Inc., Göttingen, Germany) equipped with a 100X, 1.4 numerical aperture objective or another confocal microscope suitable for fluorescence imaging.

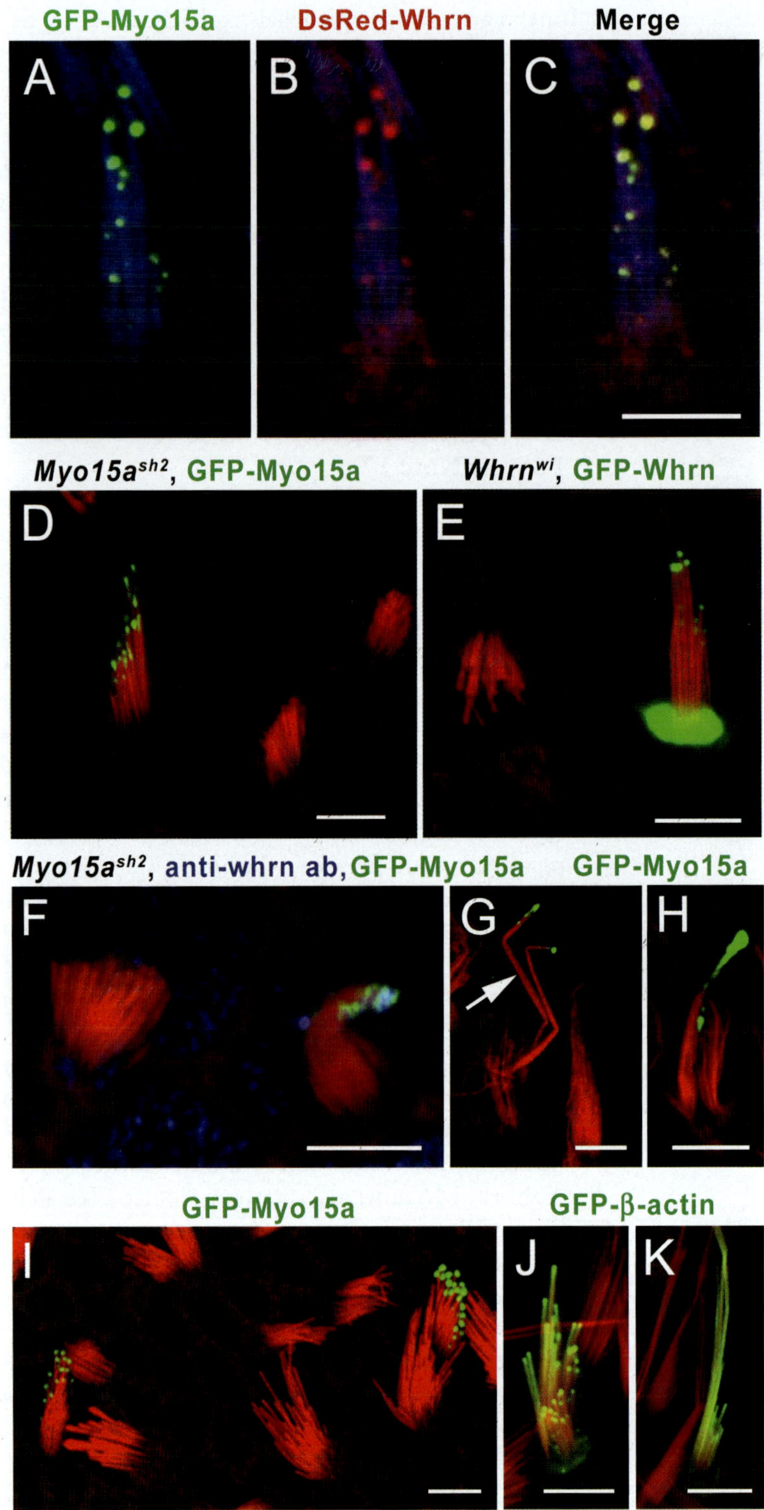

Fig. 7.2. **Gene Gun-transfected vestibular hair cells from organotypic cultures of wild type and mutant mouse inner ear sensory epithelia**. (**A–C**) Simultaneous transfections of GFP-Myo15a and DsRed-Whrn into the vestibular

3. Methods

3.1. Preparation of Inner Ear Sensory Epithelial Explants

1. Prepare MatTek Petri dishes by coating the entire glass bottom of the dish with Cell-Tak diluted 1:6 (v/v) in tissue culture grade water. Let Cell-Tak dry. Immediately before transferring the tissue into the Cell-Tak-covered MatTek Petri dish, wash the dish once briefly with DMEM without FBS (*see* **Note 20**).
2. Dissect sensory epithelia from postnatal day 0–4 (P0–P4) mouse or rat inner ear in a 60-mm sterile cell culture dish with the tissue submerged in L-15 medium. In the case of the organ of Corti, remove spiral ligament and the stria vascularis, remove the tectorial membrane, dissociate the organ of Corti from the modiolus, and then cut the entire organ of Corti into the desired number of pieces. Microdissection of the vestibular sensory epithelia should include removal of utricular and saccular otoconia using a 26-gauge needle.
3. Transfer one or two pieces of the epithelia into the MatTek Petri dish with 2 mL of DMEM supplemented with 7% (v/v) FBS using a microdissecting curette (*see* **Note 7**). Submerge all pieces and gently push them against the surface of the dish coated with Cell-Tak to attach them to the substrate (*see* **Note 21**). Immediately place the dish with attached organotypic culture in an incubator at 37 °C and 5% CO_2. Let the tissue adhere to the dish undisturbed overnight.

Fig. 7.2. (continued) hair cell of a wild type mouse. GFP-myosin XVa (*left*) and DsRed-whirlin (*middle*) accumulate at the tips of stereocilia in direct proportion to each other and to the length of stereocilia. The merged image (*right*) shows overlapping localization of GFP-myosin XVa and DsRed-whirlin. Cytoskeletal actin is visualized using phalloidin 633 (*blue*). (**D**) Restoration of the staircase shape of a stereocilia bundle of a homozygous $Myo15a^{sh2}$ vestibular hair cell 67 h after GFP-Myo15a transfection. Note the short length of stereocilia of non-transfected neighboring $Myo15a^{sh2}$ hair cells. (**E**) Restoration of the staircase shape of the stereocilia bundle in a homozygous $Whrn^{wi}$ vestibular hair cell 48 h after transfection with GFP-Whrn. (**F**) Exogenous GFP-myosin-XVa (*green*) recruits endogenous whirlin stained with anti-whirlin HL5136 antibody (*blue*) to stereocilia tips of a $Myo15a^{sh2}$ in transfected vestibular hair cell. Note that there is no anti-whirlin immunoreactivity in the stereocilia of neighboring non-transfected hair cells (*left*). Images in **A-F** are reproduced from Belyantseva et al., 2005 *(6)*. (**G**) Rat vestibular hair cell stereocilia bundle (transfected with GFP-Myo15a) degenerates as a result of excessive helium pressure and particle bombardment. Two giant, over-elongated, and deformed stereocilia (*arrow*) were observed 40 h post-transfection. The stereocilia bundle of non-transfected neighboring hair cell remains intact. (**H**) Degeneration of a stereocilia bundle transfected with GFP-myosin XVa (GFP-Myo15a) in a mouse vestibular hair cell 26 h post-transfection. There is an enormous accumulation of GFP-myosin XVa at the tips of fused stereocilia. (**I**) Accumulation of variable amounts of GFP-myosin XVa at the tips of stereocilia of two transfected mouse vestibular hair cells from the same sensory epithelium explant 45 h post-transfection. Images is reproduced from Belyantseva et al., 2003 *(4)*. (**J, K**) Different patterns of GFP-β-actin distribution in stereocilia of simultaneously transfected mouse vestibular hair cells, from the same explant 48 h post-transfection. (**J**) GFP-β-actin is mostly at the tips of stereocilia. (**K**) GFP-β-actin is distributed along the length of stereocilia. Cytoskeletal actin is visualized by rhodamine-phalloidin (*red*) in panels **D–K**. Sensory explants were harvested at P2–P5 and transfected the next day. Scale bars: 5 μm.

3.2. Preparation of Bullets with DNA-Covered Gold Microcarriers

1. Weigh out 25 mg of gold microcarrier into a 1.5 mL centrifuge tube.
2. Add 100 μL of 0.05 *M* spermidine (*see* **Notes 9, 13**). Vortex for 10–15 sec, and then sonicate the tube for 30 sec by dipping the tube half-way into the water bath of the sonicator.
3. Add 50 μg of plasmid DNA (50 μL of 1 mg/mL). Vortex briefly (~ 5 sec) to ensure even distribution of DNA in gold suspension (*see* **Note 9**).
4. Add 100 μL of 1 *M* $CaCl_2$ one drop at a time to the tube with DNA. Vortex briefly (~ 3 sec) after each drop. If more than 100 μL of DNA is used, match this volume with the same amount of $CaCl_2$. Incubate the tube at room temperature for 10 min.
5. Cut ~76 cm of Tefzel tubing using scissors. Trim both ends using the Bio-Rad Tubing cutter. Position the Tubing Prep Station so that the nitrogen gas meter and the "ON-OFF" switch for rotation are facing toward you (**Fig. 7.1A**). Insert the tubing into the Tubing Prep Station, leaving about 10 cm of the tube sticking out on the right hand side (**Fig. 7.1A**).
6. Turn "on" the nitrogen gas to 0.3–0.4 L/min and flush the tubing for 10–15 min.
7. Dilute PVP to 50 μg/mL using 100% ethanol. (Add 10 μL of 20 mg/mL of PVP to 4 mL of ethanol in 15 mL conical tube.) (*see* **Note 12**).
8. Microfuge the gold with DNA at 1,000 g for 2 min at room temperature to pellet gold particles. Aspirate excess supernatant using a 1 mL pipette tip, leaving about 20 μL. Resuspend the gold pellet in this residual volume by gently tapping on the lower part of the tube.
9. Wash three times in 1 mL of 100% ethanol. Pellet gold by centrifugation at 1,000–2,000 g at room temperature for 10 sec, aspirate supernatant as in **step 8 of Section 3.2**, resuspend gold, and add 1 mL of 100% ethanol. After the last wash, remove most of the ethanol (*see* **Note 10**).
10. Add 200 μL of the 50 μg/mL PVP in ethanol made in **step 7 of Section 3.2**. Pipette up and down to break up clumps. Transfer the contents of the tube to a 15 mL conical tube. Add another 200 μL of 50 μg/mL PVP in ethanol to the centrifuge tube, repeat the pipetting, and transfer to the same 15 mL tube until all the gold particles are transferred. Bring the final volume to 3 mL with fresh 50 μg/mL PVP in ethanol. Vortex briefly for about 15 sec to ensure an even distribution of gold particles in the suspension. Close the tube and keep inverting it by hand to prevent the gold from clumping.
11. Turn off the nitrogen gas on the Tubing Prep Station. Insert the right end of the Tefzel tubing into the adaptor tubing

(*see* **step 13 of Section 2.2**) attached to an empty 10 cc syringe. Remove the tubing from the apparatus. Remove the cap of the 15 mL tube containing 3 mL of the gold particle suspension (*see* **step 10 of Section 3.2**) and immediately place the left end of the Tefzel tubing at the bottom of this tube. Pull the plunger of the syringe and quickly and consistently draw the gold suspension into the tubing. When the entire volume of gold particle suspension is within the Tefzel tubing, continue drawing the suspension into the tubing to empty ~2–3 cm of the left end segment of the tubing. Make sure that the gold particle suspension is distributed along the Tefzel tubing evenly, without air bubbles, and is not drawn into the adaptor tubing (**Fig. 7.1A**, white arrow). Immediately bring the gold-filled tubing to a horizontal position and slide it, with syringe attached, into the tubing support cylinder (**Fig. 7.1A**, black arrow) of the Tubing Prep Station until the tubing passes through the O-ring.

12. Let the tubing sit undisturbed for 3 min in the tubing apparatus. The gold will settle to the lower side of horizontally positioned tubing.
13. Use the syringe with adaptor tubing (*see* **step 11 of Section 3.2**) attached to the right end of the tube with gold suspension to slowly and consistently pull the supernatant (ethanol) out of the tubing over the course of ~40–45 sec. After all of the ethanol is transferred into the syringe and the connecting tubing, disconnect the syringe with adaptor tubing from the Tefzel tube with gold particles.
14. Turn the rotation switch to the "ON" position on the Tubing Prep Station and rotate the tube with gold particles for 20–30 sec, allowing the gold to uniformly smear on the inside surface of the tube.
15. On the Tubing Prep Station, slowly open the valve on the flowmeter regulating the nitrogen gas to 0.35–0.4 L/min, while rotating the tubing with gold particles for 5 min.
16. Stop rotating the tubing. Turn off the nitrogen gas and remove tubing from the apparatus. Trim the ends of the tubing, which usually are not evenly coated with gold. Cut the tubing into 1.27 cm sections with the Bio-Rad Tubing cutter. These pieces are now the cartridges with bullets.
17. Store these prepared cartridges with bullets in tightly closed scintillation vials (*see* **step 14 of Section 2.2**) with one capsule of drycap dehydrators in each vial (*see* **step 14 of Section 2.2**). The coated gold particles stored at 4 °C are usable for about 1 year (*see* **Note 22**).

3.3. Helios® Gene Gun Transfection Procedure

1. Load bullets with the desired plasmid DNA into the cartridge holder. Leave slot no. 1 empty. Place the cartridge holder into the Gene Gun.

2. Connect the Gene Gun to the helium gas tank via the helium regulator. Open the gas valve and set the pressure to 758.42 kPa (110 psi). Using empty slot no. 1 for firing, discharge the Helios Gene Gun 2–3 times. Be sure that the helium pressure remains stable and does not drop during these trial shots.
3. Switch to slot position no. 2. Insert the diffusion screen into the Helios® Gene Gun barrel as shown in **Fig. 7.1B** (*see* **Note 15**).
4. Remove a dish with the inner ear sensory explants from the incubator and place it in a sterile laminar flow hood.
5. Aspirate culture medium (as much as possible) from the inner ear organotypic culture attached at the bottom of the MatTek Petri dish (**Fig. 7.1B,C**). Immediately position the plastic ring, located at the end of the Helios® Gene Gun barrel (**Fig. 7.1C**, black arrow), at the bottom of the Petri dish so that the targeted tissue is located in the center of the ring, and the Gene Gun barrel is perpendicular to the dish bottom. Discharge the Gene Gun (**Fig. 7.1C**; *see* **Note 23**).
6. Immediately add 2 mL of fresh DMEM medium containing 7% FBS to the dish. Without delay, place the dish in the incubator for the desired number of hours/days.
7. Repeat steps 1 through 5 for the other dishes with organotypic culture using fresh bullet cartridges in the consecutive slots of the same cartridge holder.

3.4. Immunostaining and Imaging of Transfected Samples

1. Wash cultures two times in cold 1X PBS (*see* **Note 16**).
2. Fix in 4% paraformaldehyde for 30 min to 1 h at room temperature or overnight at 4 °C.
3. Wash four times with 1X PBS for 5 min.
4. Permeabilize in 0.5% Triton X-100 for 10–15 min (*see* **Note 17**).
5. Wash four times with 1X PBS for 5 min.
6. Incubate in blocking solution (2% BSA and 5% goat serum in 1X PBS) for 30 min.
7. Incubate in primary antibody diluted in blocking solution for 1–2 h at room temperature or overnight at 4 °C (*see* **Note 18**).
8. Wash four times with 1X PBS for 5 min.
9. Incubate simultaneously for 20 min at room temperature in secondary antibody diluted in blocking solution (*see* **step 6 of Section 2.4**) and phalloidin conjugated to a particular fluorophore (Invitrogen) and diluted 1:100 in blocking solution (*see* **step 7 of Section 2.4**) (*see* **Note 18**).
10. Wash four times in 1X PBS.
11. Using a 26-gauge needle, lift the inner ear explant off the glass bottom of the MatTek Petri dish. Transfer the explant to a glass slide by carefully drawing it into a short (146 mm)

glass Pasteur pipette (*see* **Section 2.4, step 8**). Position the sensory epithelium on the glass slide with stereocilia facing up; remove the surrounding liquid as much as possible before adding anti-fade mounting medium. Immediately apply a drop of mounting medium. Mount the tissue using the ProLong Antifade kit (Invitrogen) according to the manufacturer's instructions (*see* **Note 24**).

12. Keep slides protected from light in a slide box overnight to let the mounting media solidify (*see* **Note 25**). Acquire images on the next day using a confocal microscope equipped with a 100X, 1.4 numerical aperture objective.

3.5. Biolistic Transfection of Inner Ear Sensory Epithelium: Interpretation of Results

The advantages and disadvantages of the described method can be illustrated by analyzing the data obtained from Gene Gun–mediated transfections of GFP-myosin XVa and/or DsRed- or GFP-whirlin into hair cells of inner ear sensory epithelial explants obtained from wild-type and mutant mice (*6*). In wild-type hair cells, the unconventional motor protein myosin XVa and the PDZ domain-containing protein whirlin localize together at the tips of stereocilia *(4, 6, 7, 39)*. We used a Helios® Gene Gun to co-transfect two expression vectors. Gold particles were coated with two different cDNAs at a 1:1 ratio of molecules. GFP-myosin XVa and DsRed-whirlin on the same gold bullet were co-transfected into wild-type hair cells. Both corresponding epitope-tagged proteins were localized to the tips of stereocilia *(6, 7)*. These data complemented observations of the endogenous myosin XVa and whirlin at the tips of stereocilia as revealed by immunofluorescence and, most importantly, demonstrated that GFP and DsRed epitope tags do not interfere with proper targeting or function of these two proteins (**Fig. 7.2A–C**). Moreover, over-expression of GFP-myosin XVa in stereocilia of wild-type hair cells causes distention of stereocilia tips due to an accumulation of an excessive amount of GFP-myosin XVa. No over-elongation of stereocilia of wild-type hair cells due to over-expression of myosin XVa was observed *(4, 7)*.

Myosin XVa mutant ($Myo15a^{sh2}$) and whirlin mutant ($Whrn^{wi}$) strains of deaf mice have hair cells with abnormally short stereocilia bundles that fail to elongate to a normal length due to mutations of these genes *(4, 6, 40–42)*. Gene Gun–mediated transfections of wild-type GFP-myosin XVa and wild-type GFP-whirlin into the hair cells of sensory epithelial explants from the corresponding mutant mice resulted in the restoration of a normal length of stereocilia bundles (**Fig. 7.2D–E**). Moreover, using Gene Gun–mediated transfections of domain-deletion constructs of myosin XVa and whirlin cDNAs into $Myo15a^{sh2}$ and $Whrn^{wi}$ hair cells, we found that these two proteins interact *in vitro* through the C-terminal PDZ ligand of myosin XVa and the third PDZ domain of whirlin. This interaction allows myosin XVa to deliver whirlin to the tips of stereocilia. Using an anti-whirlin specific antibody in combination with Gene Gun transfection, we found that exogenous wild-type GFP-myosin XVa "reawakens" the elongation process in the abnormally short *Myo15a*^{sh2} hair cell stereocilia by recruiting endogenous whirlin to stereocilia tips (**Fig. 7.2F**).

3.5.1. Advantages

In lipofection, viral mediated transfection, electroporation, and calcium phosphate precipitation a transfected cell may receive a very small amount of fluorescently tagged cDNA expression construct. In this case, the amount of synthesized fluorescently tagged protein might be below the threshold of detection by fluorescence microscopy, and this cell might be indistinguishable from and mistaken for untransfected control cells with background fluorescence. Nevertheless, a low level of the epitope-tagged protein may subtly alter the phenotype of the cell. One of the advantages of Gene Gun transfection is that you can usually see a gold particle inside the body of a transfected cell while imaging, thereby distinguishing a transfected cell from adjacent cells that lack such a particle and may serve as an untransfected control. Thus, true control cells can be identified and distinguished from transfected cells. Moreover, in our experiments (*see* **Section 3.5**), stereocilia undergo elongation only when wild-type GFP-tagged myosin XVa or whirlin is transfected into hair cells of *Myo15a*^{sh2} and *Whrn*^{wi}, respectively. In contrast, stereocilia bundles of non-transfected hair cells from the same explant remain short, because they are deficient for functional myosin XVa (**Fig. 7.2D,F**) or whirlin (**Fig. 7.2E**). Thus, Gene Gun transfections provide the opportunity to quantitatively measure the induced elongation of stereocilia, by comparing the lengths of restored stereocilia of transfected hair cells to the lengths of short hair bundles of non-transfected neighboring control cells. These types of experiments, with the Helios® Gene Gun, can reveal the importance of specific proteins to key developmental events, as demonstrated by the importance of myosin XVa and whirlin to the differential elongation of stereocilia during hair bundle morphogenesis *(6)*.

3.5.2. Disadvantages

Helios® Gene Gun–mediated transfections of cultured inner ear sensory epithelia have limitations. First, hair cells survive at most for 2 weeks in culture once a mouse organ of Corti is explanted at postnatal day 0 through day 4 (P0–P4). Second, stereocilia are sensitive to mechanical disturbances and are easily damaged by a pulse of helium gas pressure, triggering hair cell degeneration. An early sign of degeneration is an abnormal hair bundle shape (**Fig. 7.2G–H**). Stereocilia show unrestrained elongation and/or fusion and some stereocilia proteins localize abnormally, such as GFP-tagged proteins introduced via cDNA constructs (**Fig. 7.2G–H**). Thus, care should be exercised in interpreting the results if the hair bundle of a transfected cell looks abnormal, especially in explants from mutant mice. For example, an over-elongated or disorganized stereocilia bundle of a transfected *Myo15a*^{sh2} hair cell may likely result from degeneration rather than over-expression of GFP-myosin XVa. Moreover, individual stereocilia may elongate abnormally within a short stereocilia bundle, even in non-transfected *Myo15a*^{sh2} hair cells that lack functional myosin XVa (*see* cover image of *6, 7*). Therefore, the presence of abnormal hair bundles with over-elongated stereocilia should not be interpreted as a result of myosin XVa-GFP over-expression in transfected wild-type hair cells (e.g., **Fig. 7.2G–H**). Rather, in both transfected and non-transfected *Myo15a*^{sh2} hair cells, over-elongation of stereocilia can be interpreted as a consequence of degeneration caused by mechanical damage. In general, hair cell degeneration may be a secondary effect of a variety of environmental or genetic insults to the sensory epithelium, including mechanical damage due to helium gas pressure or noise, gold particle bombardment, culture conditions for a prolonged period of time, and/or a gene mutation.

Third, dissimilar amounts of plasmid DNA on the gold microcarriers may affect data interpretation. While only one gold particle penetrates a hair cell for most transfections, transfected hair cells may show different levels of GFP-tagged protein expression at a specific point in time. **Fig. 7.2I** shows different levels of myosin GFP-XVa accumulation at the stereocilia tips of hair cells from the same explant 45 h post-transfection. A second example is GFP-β-actin, which first appears at the tips of stereocilia, then supposedly incorporates into actin filaments and treadmills toward the apical surface of the hair cell at a particular rate *(3, 5)*. Hair cells from the same explant simultaneously transfected with full-length GFP-β-actin may show dissimilar amounts and different distribution patterns of this protein within stereocilia. These differences in GFP-β-actin distribution in stereocilia, after Gene Gun–mediated transfection, are similar to differences described for viral infection/delivery of GFP-β-actin to hair cells *(11)*. **Figure 7.2J–K** shows Gene Gun–transfected hair cells from the same explant. Approximately 72 h post-transfection, GFP-β-actin is present primarily at the tips of the stereocilia in one hair cell (**Fig. 7.2J**), while GFP-β-actin highlights most of the stereocilia length of another hair cell (**Fig. 7.2K**). The time of appearance and the amount of GFP-tagged protein visualized in a transfected cell are probably related to the amount of DNA transfected into a cell. Therefore, it is important to repeat the above-described experiments (independent determinations each with replicas) to document the range of variation in the kinetics of appearance and localization of the tagged protein. In the absence of such data, one should view any interpretations of time-sensitive expression of GFP-tagged proteins with some skepticism. Rigorous evaluation and accurate interpretation of the data from Gene Gun–mediated transfections may require, for example, immunofluorescence to confirm correct localization of GFP-tagged proteins and genetic analyses of the phenotype of relevant mutant mice.

4. Notes

1. Other Gene Gun models, including Accell®, have been developed by Aurogen, Inc., a Bio-Rad collaborator (*see* also Helios® Gene Gun System Instruction Manual from Bio-Rad, which is available online at http://www.bio-rad.com/LifeScience/pdf/Bulletin_9541.pdf). Cell penetration, gene expression, and other parameters vary with the model of the Gene Gun. Therefore, users must be careful to optimize the operating parameters for their particular model. O'Brien and Lummis *(37)* developed a modified barrel for the Bio-Rad handheld Helios® Gene Gun, which reportedly improves the penetration of gold particles into cultured brain slices and allows the use of lower gas pressures without the loss of transfection efficiency. This modified Gene Gun barrel is available from Modolistics (see http://www2.mrc-lmb.cam.ac.uk/personal/job/index.html). However, there are no reported data on the use of this modification in transfections of inner ear sensory epithelial explants. Meanwhile, we have optimized transfection conditions, taking into account all of the above-mentioned variables for our application, using the original Bio-Rad Helios® handheld Gene Gun.
2. You can find information about the assembly, operation, maintenance, spare parts, and general optimization of particle delivery in the Helios® Gene Gun System Instruction Manual from Bio-Rad. Also, there is a helpful troubleshooting section.
3. All experimental animals should be handled according to the protocols of the Instituional Animal Care and Use Committee.
4. All microdissections of the inner ear sensory epithelia should be carried out under sterile conditions using autoclaved instruments. Preferably, a clean dissection microscope should be placed in a laminar flow hood.
5. Glass-bottomed Petri dishes of different diameters can also be used and purchased from other companies (e.g., World Precision Instruments, Inc., Sarasota, FL or Electron Microscopy Sciences).
6. DMEM/F12 media supplemented with 7% (v/v) FBS can also be used.

7. This microdissecting curette will allow you to transfer pieces of sensory epithelia submerged in a limited volume of L-15 media. Pieces of the sensory epithelia can also be transferred using a glass Pasteur pipette (*see* **step 8 of Section 2.4**) attached to a pipette holder (e.g., cat. no. 378980000, A. Daigger & Co., Vernon Hills, IL), which allows you to control the release of liquid from the pipette tip. To transfer your specimen using this pipette, aspirate in the specimen with some L-15 media, let the specimen settle down toward the opening of the pipette (you can help this along by tapping gently on the glass pipette), and then touch the surface of the liquid (DMEM), where you want to transfer your specimen, with the tip of the glass pipette. The specimen will be released into the dish containing DMEM with minimum contamination by L-15 media.
8. It is important to frequently check the water level in the incubator tray to maintain an appropriate humidity level.
9. pAcGFP1-Actin vector is available from BD Biosciences (cat. no. 632453). To use a different vector with your cDNA of interest, you can purchase a Quantum Prep Plasmid Miniprep Kit (100 preps, Bio-Rad, cat. no. 7326100) or Qiagen's QIAfilter plasmid midi kits (Qiagen, Valencia, CA, cat. no. 12243, *see* also *4, 6*) to prepare plasmid DNA of a high purity suitable for Helios® Gene Gun transfection. To use more than 100 μL of a less concentrated plasmid DNA preparation, match the volume of spermidine, but try to avoid volumes larger than 150 μL. More concentrated plasmid DNA (greater then 1.0 mg/mL) may cause gold particles to cluster. It is also crucial to use purified plasmid DNA since impure plasmid DNA may result in poor transfection and/or gold particle clustering. After purification, DNA should be diluted to 1 mg/mL in molecular biology grade water (*see* also *31, 37*).
10. It is very important that the ethanol is free of water (200 proof). A fresh bottle of 100% ethanol should be opened on the day of use in bullet preparation procedures.
11. The size of the microcarrier should be optimized for the particular application, cell types, etc. It is possible to use tungsten particles instead of gold. They are less expensive, but be aware that tungsten can oxidize and may be toxic to the cells *(38)*.
12. Old PVP may cause uneven gold coating of the Tefzel tubing, inefficient release of gold particles during the shot, lower tissue penetration, and reduced transfection efficiency. Do not keep diluted PVP for more than 1 month. The concentration of PVP should be optimized for each particular instrument and application. Crystallized PVP is hygroscopic. Store it at room temperature in a tightly closed desiccated vial.

13. Spermidine solution should be sterile filtered using an 0.22 μm pore filter, if sterile solution is necessary. Spermidine deaminates with time; solutions should be stored frozen. Old spermidine may cause poor precipitation of DNA onto the gold particles and subsequently reduce the transfection efficiency. Do not keep spermidine for more than 1 month even at −20 °C.
14. O'Brien and Lummis recommend omitting the sonication step *(31, 37)*. However, sonication seems to be efficient in mixing spermidine with gold particles and keeping gold particles in suspension. In the protocol described in this chapter, sonication is omitted only in steps when DNA is added to gold particles to avoid the possible destructive effect of sonication on DNA.
15. The use of a Bio-Rad diffusion screen reduces the damage to the inner ear sensory epithelia and especially to the hair cell stereocilia bundles by reducing the density of the gold particles in the center of the shot during bombardment. After repeatedly firing the Gene Gun, gold particles build up on the center of the diffusion screen, which may become gold colored. Use a dedicated diffusion screen for each plasmid DNA used for firing to avoid cross-contamination. A dedicated diffusion screen exclusively used for a particular plasmid DNA can be reused many times and does not require frequent autoclaving. However, it is necessary to clean diffusion screens before using them to fire bullets coated with a different plasmid DNA. Diffusion screens can be cleaned by soaking in 70% or 100% ethanol and by sterilizing them in an autoclave that uses only distilled water.
16. You can use 1X PBS containing Ca^{2+} and Mg^{2+} if the presence of these ions is required.
17. 100% Triton X-100 is a viscous solution. It is useful to prepare a 1% stock solution by adding 500 μL of 100% Triton X-100 to 49.5 mL of 1X PBS. Before introducing 100% Triton X-100 into a 1 mL pipette tip, cut off about 1 cm of the tip to make a wider opening. This will help to draw a viscous solution in and out of the pipette tip.
18. Rhodamine-phalloidin and other phalloidin conjugates can be used to visualize filamentous actin. Phalloidin 633 can be used to highlight the actin cytoskeleton when cells are co-transfected using two cDNA plasmids tagged with GFP and dsRed. In this case, no primary and secondary antibodies are required.
19. If component B is still too viscous at room temperature, warm it to 37 °C for 30 min before use.
20. A higher concentration of Cell-Tak may improve the adhesion of the sensory epithelia but it can be toxic to hair cells. While diluting Cell-Tak in a 1.5 mL centrifuge tube, make sure you

use it immediately as Cell-Tak quickly adheres to the walls of the tube. Also, you can prepare your dishes using rat tail collagen, type I (Upstate, Lake Placid, NY) *(1, 2)*. Alternatively, you may choose to attach your sample directly to a glass surface not covered with any substrate. In this case, use DMEM without serum during the attachment period.

21. If you have problems with tissue attachment, try DMEM without serum for several hours or overnight. The next morning change the media to DMEM with 7% (v/v) FBS. Also, the freshly dissected tissues seem to adhere to Cell-Tak better than tissues kept in L-15 media for more than ~10 min after microdissection. DMEM/F12 medium (Invitrogen) with 7% (v/v) FBS seems to be better for rat inner ear organotypic culture.
22. Alternatively, to store cartridges you can use tightly closed 15 mL conical tubes, each with one capsule of dehydrator as described in **step 14 of Section 2.2**. Some preparations of plasmid DNA coated onto gold particles were used successfully for transfections after more than two years of storage with proper desiccation.
23. To prevent culture contamination, wipe the plastic ring of the end of a barrel (**Fig. 7.1C**, black arrow) with 70% ethanol after each shot. It is advisable to wear ear protection (earmuffs, cat. no. 56219268, VWR, or earplugs, cat. no. 56610680) when firing the Gene Gun.
24. It is important to remove the 1X PBS from the slide as much as possible without over-drying the sample before adding a drop of Antifade solution. Residual liquid around the sample will interfere with anti-fade properties of the mounting media. In general, samples should not be allowed to dry out at any time during the immunostaining and mounting procedures.
25. You can use clear nail polish to seal the perimeter edge of a coverslip onto a slide. (This step is not necessary if you intend to keep your slide for less than one week). Let it dry before observing the slide under a confocal microscope. For long-term storage up to a few months, store slides at 4 °C with several capsules of desiccant (Ted Pella, Inc.) in the slide box.

Acknowledgment

I thank Thomas Friedman for support and encouragement as well as for critical reading of this chapter, Jonathan Gale (University of London, UK) from whom I learned the organ of Corti explant technique, Erich Boger for preparing myosin XVa and whirlin cDNA expression constructs, critical reading of the chapter and helpful discussions, Andrew Griffith, Valentina Labay,

Polina Belyantseva and Shin-Ichiro Kitajiri for critical reading of the manuscript. Supported by funds from the NIDCD Intramural Program (1 Z01 DC 00035-11) to Thomas B. Friedman.

References

1. Russell, I. J., Richardson, G. P., and Cody, A. R. (1986) Mechanosensitivity of mammalian auditory hair cells in vitro. *Nature* **321**, 517–519.
2. Russell, I. J. and Richardson, G. P. (1987) The morphology and physiology of hair cells in organotypic cultures of the mouse cochlea. *Hear. Res.* **31**, 9–24.
3. Schneider, M. E., Belyantseva, I. A., Azevedo, R. B., and Kachar, B. (2002) Rapid renewal of auditory hair bundles. *Nature* **418**, 837–838.
4. Belyantseva, I. A., Boger, E. T., and Friedman, T. B. (2003) Myosin XVa localizes to the tips of inner ear sensory cell stereocilia and is essential for staircase formation of the hair bundle. *Proc. Natl Acad. Sci. USA* **100**, 13958–13963.
5. Rzadzinska, A. K., Schneider, M. E., Davies, C., Riordan, G. P., and Kachar, B. (2004) An actin molecular treadmill and myosins maintain stereocilia functional architecture and self-renewal. *J. Cell Biol.* **164**, 887–897.
6. Belyantseva, I. A., Boger, E. T., Naz, S., Frolenkov, G. I., Sellers, J. R., Ahmed, Z. M., et al. (2005) Myosin-XVa is required for tip localization of whirlin and differential elongation of hair-cell stereocilia. *Nat. Cell Biol.* **7**, 148–156.
7. Boger, E. T., Frolenkov, G. I., Friedman, T. B., and Belyantseva, I. A. (2008) Myosin XVa, in *Myosins: A Superfamily of Molecular Motors* (Coluccio, L. M., ed.), Springer, The Netherlands, pp. 441–467.
8. Felgner, P. L., Gadek, T. R., Holm, M., Roman, R., Chan, H. W., Wenz, M., et al. (1987) Lipofection: a highly efficient, lipid-mediated DNA-transfection procedure. *Proc. Natl. Acad. Sci. USA* **84**, 7413–7417.
9. Parker Ponder, K. (2001) Vectors of gene therapy, in *An Introduction to Molecular Medicine and Gene Therapy* (Kresina, T. F., ed.), A John Wiley and Sons, Inc., New York, pp. 77–112.
10. Stone, I. M., Lurie, D. I., Kelley, M. W., and Poulsen, D. J. (2005) Adeno-associated virus-mediated gene transfer to hair cells and support cells of the murine cochlea. *Mol. Ther.* **11**, 843–848.
11. Di Pasquale, G., Rzadzinska, A., Schneider, M. E., Bossis, I., Chiorini, J. A., and Kachar, B. (2005) A novel bovine virus efficiently transduces inner ear neuroepithelial cells. *Mol. Ther.* **11**, 849–855.
12. Bhattarai, S. R., Kim, S. Y., Jang, K. Y., Lee, K. C., Yi, H. K., Lee, D. Y., et al. (2008) Laboratory formulated magnetic nanoparticles for enhancement of viral gene expression in suspension cell line. *J. Virol. Methods.* **147**, 213–218.
13. Favard, C., Dean, D. S., and Rols, M. P. (2007) Electrotransfer as a non viral method of gene delivery. *Curr. Gene Ther.* 7, 67–77.
14. Saito, T. and Nakatsuji, N. (2001) Efficient gene transfer into the embryonic mouse brain using in vivo electroporation. *Dev. Biol.* **240**, 237–246.
15. Jones, J. M., Montcouquiol, M., Dabdoub, A., Woods, C., and Kelley, M. W. (2006) Inhibitors of differentiation and DNA binding (Ids) regulate Math1 and hair cell formation during the development of the organ of Corti. *J. Neurosci.* **26**, 550–558.
16. Weaver, J. C. (1995) Electroporation theory: concepts and mechanisms. *Methods Mol. Biol.* **47**, 1–26.
17. Atkins, R. L., Wang, D., and Burke, R. D. (2000) Localized electroporation: a method for targeting expression of genes in avian embryos. *Biotechniques* **28**, 94–96, 98, 100.
18. Seiler, M. P., Gottschalk, S., Cerullo, V., Ratnayake, M., Mane, V. P., Clarke, C., et al. (2007) Dendritic cell function after gene transfer with adenovirus-calcium phosphate co-precipitates. *Mol. Ther.* **15**, 386–392.
19. Scherer, F., Anton, M., Schillinger, U., Henke, J., Bergemann, C., Krüger, A., et al. (2002) Magnetofection: enhancing and targeting gene delivery by magnetic force in vitro and in vivo. *Gene Ther.* **9**, 102–109.
20. Mykhaylyk, O., Antequera, Y. S., Vlaskou, D., and Plank, C. (2007) Generation of magnetic nonviral gene transfer agents and magnetofection in vitro. *Nat. Protoc.* **2**, 2391–2411.
21. Plank, C., Schillinger, U., Scherer, F., Bergemann, C., Rémy, J. S., Krötz, F., et al. (2003) The magnetofection method: using magnetic force to enhance gene delivery. *Biol. Chem.* **384**, 737–747.
22. Wang, W., Liu, X., Gelinas, D., Ciruna, B., and Sun, Y. (2007) A fully automated robotic system for microinjection of zebrafish embryos. *PLoS ONE* **2**, e862.
23. Tirlapur, U. K. and König, K. (2002) Targeted transfection by femtosecond laser. *Nature* **418**, 290–291.

24. Klein, T. M., Wolf, E. D., Wu, R., and Sanford, J. C. (1987) High-velocity microprojectiles for delivering nucleic acids into living cells. *Nature* **327**, 70–73.
25. Klein, T. M., Fromm, M., Weissinger, A., Tomes, D., Schaaf, S., Sletten, M., et al. (1988) Transfer of foreign genes into intact maize cells with high-velocity microprojectiles. *Proc. Natl. Acad. Sci. USA* **85**, 4305–4309.
26. Zelenin, A. V., Titomirov, A. V., and Kolesnikov, V. A. (1989) Genetic transformation of mouse cultured cells with the help of high-velocity mechanical DNA injection. *FEBS Lett.* **244**, 65–67.
27. Johnston, S. A. (1990) Biolistic transformation: microbes to mice. *Nature* **346**, 776–777.
28. Yang, N. S., Burkholder, J., Roberts, B., Martinell, B., and McCabe, D. (1990) In vivo and in vitro gene transfer to mammalian somatic cells by particle bombardment. *Proc. Natl Acad. Sci. USA* **87**, 9568–9572.
29. Williams, R. S., Johnston, S. A., Riedy, M., DeVit, M. J., McElligott, S. G., and Sanford, J. C. (1991) Introduction of foreign genes into tissues of living mice by DNA-coated microprojectiles. *Proc. Natl. Acad. Sci. USA* **88**, 2726–2730.
30. Sanford, J. C., Smith, F. D., and Russell, J. A. (1993) Optimizing the biolistic process for different biological applications. *Methods Enzymol.* **217**, 483–509.
31. O'Brien, J. A. and Lummis, S. C. (2002) An improved method of preparing microcarriers for biolistic transfection. *Brain Res. Protoc.* **10**, 12–15.
32. Thomas, J. L., Bardou, J., L'hoste, S., Mauchamp, B., and Chavancy, G. (2001) A helium burst biolistic device adapted to penetrate fragile insect tissues. *J. Insect Sci.* **1**, 9.
33. Kim, T. W., Lee, J. H., He, L., Boyd, D. A., Hardwick, J. M., Hung, C. F., et al. (2005) Modification of professional antigen-presenting cells with small interfering RNA in vivo to enhance cancer vaccine potency. *Cancer Res.* **65**, 309–316.
34. Pascolo, S. (2006) Vaccination with messenger RNA. *Methods Mol. Med.* **127**, 23–40. [Review].
35. Yang, C. H., Shen, S. C., Lee, J. C., Wu, P. C., Hsueh, S. F., Lu, C. Y., et al. (2004) Seeing the gene therapy: application of gene gun technique to transfect and decolour pigmented rat skin with human agouti signalling protein cDNA. *Gene Ther.* **11**, 1033–1039.
36. Shefi, O., Simonnet, C., Baker, M. W., Glass, J. R., Macagno, E. R., and Groisman, A. (2006) Microtargeted gene silencing and ectopic expression in live embryos using biolistic delivery with a pneumatic capillary gun. *J. Neurosci.* **26**, 6119–6123.
37. O'Brien, J. A. and Lummis, S. C. (2006) Biolistic transfection of neuronal cultures using a hand-held gene gun. *Nat. Protoc.* **1**, 977–981.
38. Russell, J. A., Roy, M. K., and Sanford, J. C. (1992) Physical trauma and tungsten toxicity reduce the efficiency of biolistic transformation. *Plant Physiol.* **98**, 1050–1056.
39. Delprat, B., Michel, V., Goodyear, R., Yamasaki, Y., Michalski, N., El-Amraoui, A., et al. (2005) Myosin XVa and whirlin, two deafness gene products required for hair bundle growth, are located at the stereocilia tips and interact directly. *Hum. Mol. Genet.* **14**, 401–410.
40. Probst, F. J., Fridell, R. A., Raphael, Y., Saunders, T. L., Wang, A., Liang, Y., et al. (1998) Correction of deafness in shaker-2 mice by an unconventional myosin in a BAC transgene. *Science* **280**, 1444–1447.
41. Holme, R. H., Kiernan, B. W., Brown, S. D., and Steel, K. P. (2002) Elongation of hair cell stereocilia is defective in the mouse mutant whirler. *J. Comp. Neurol.* **450**, 94–102.
42. Mburu, P., Mustapha, M., Varela, A., Weil, D., El-Amraoui, A., Holme, R. H., et al. (2003) Defects in whirlin, a PDZ domain molecule involved in stereocilia elongation, cause deafness in the whirler mouse and families with DFNB31. *Nature Genet.* **34**, 421–428.

Chapter 8

Electroporation-Mediated Gene Transfer to the Developing Mouse Inner Ear

John V. Brigande, Samuel P. Gubbels, David W. Woessner, Jonathan J. Jungwirth, and Catherine S. Bresee

Abstract

The mammalian inner ear forms from a thickened patch of head ectoderm called the otic placode. The placodal ectoderm invaginates to form a cup whose edges cinch together to establish a fluid-filled sac called the otic vesicle or otocyst. The progenitor cells lining the otocyst lumen will give rise to sensory and non-sensory cells of the inner ear. These formative stages of inner ear development are initiated during the first week of postimplantation embryonic development in the mouse. The inaccessibility of the inner ear in utero has hampered efforts to gain insight into the molecular mechanisms regulating essential developmental processes. An experimental embryological method to misexpress genes in the developing mammalian inner ear is presented. Expression plasmid encoding a gene of interest is microinjected through the uterine wall into the lumen of the otocyst and electroporated into otic epithelial progenitor cells. Downstream analysis of the transfected embryonic or postnatal inner ear is then conducted to gain insight into gene function.

Key words: Inner ear, transuterine microinjection, beveled microinjection pipette, in vivo electroporation, square wave pulse train, mouse experimental embryology, mouse survival surgery, in utero gene transfer.

1. Introduction

The first developmental biologists were experimental embryologists who physically manipulated embryos to describe gross mechanisms of development *(1)*. Dye injection, carbon particle transfer, extirpation, and tissue transplantation were the methods of choice *(2)*. Indeed, our first insights into neural plate induction *(3)*, optic nerve regeneration *(4)*, vertebrate limb formation *(5)*, and neural crest migration *(6, 7)* were gleaned

Bernd Sokolowski (ed.), *Auditory and Vestibular Research: Methods and Protocols, vol. 493*

DOI 10.1007/978-1-59745-523-7_8 Springerprotocols.com

through experimental embryological studies in lower vertebrates. Modern molecular embryologists who use the mouse as a model system are constrained by the inaccessibility of the embryo in vivo. In the case of the embryonic inner ear, accessibility is further complicated because the otic epithelium that will give rise to the vestibular and auditory sensory structures develops within a membranous labyrinth cloistered in the head mesenchyme *(8)*. The elucidation of experimental embryological techniques that permit manipulation of gene expression in the developing mouse inner ear may advance our mechanistic understanding of how the inner ear forms.

A solution to the challenge of gene transfer to the mammalian inner ear in utero is conceptually straightforward: gain access to the embryo, atraumatically introduce a reagent capable of transducing a bioactive signal in otic epithelial progenitor cells, and maintain a healthy pregnancy postoperatively. An experimental embryological approach to misexpress a gene in the developing mammalian inner ear is presented in three procedural stages: (1) sodium pentobarbital-based anesthesia for mouse survival surgery, (2) ventral laparotomy to expose the uterine horns, and (3) transuterine microinjection and electroporation of expression plasmid to transfect otic epithelial progenitors.

2. Materials

2.1. General Anesthesia for Mouse Ventral Laparotomy

1. 50 mg/mL Nembutal (pentobarbital sodium solution, United States Pharmacopeia[USP]; this is a controlled substance requiring a prescription or license to obtain and store).
2. Magnesium sulfate heptahydrate ($MgSO_4 \cdot 7H_2O$) dissolved in water at 65 mg/mL. Sterile filter and store in 1 mL aliquots at −20 °C (*see* **Note 1**).
3. Propylene glycol, USP. Sterile filter and store in 1 mL aliquots at room temperature.
4. Ethanol, absolute, 200 proof for molecular biology (Sigma-Aldrich, St. Louis, MO). Aliquot in 1.5 mL sterile tubes and store at room temperature.
5. Becton Dickinson (BD) Allergy Syringe Tray 0.5 mL, 27G, 3/8″ Testing (cat. no. 305536, Becton Dickinson, Franklin Lakes, NJ).
6. Sterile ophthalmic ointment (non-prescription item available from any drug store).
7. T/Pump (Gaymar Industries, Inc, Orchard Park, NY) connected to a Hallowell EMC (Pittsfield, MA) Heated Hard Pad.
8. Oster (Shelton, CT) Grooming Shears with a #40 (fine) blade.

2.2. Ventral Laparotomy

1. Surgical instruments (Fine Science Tools, Foster City, CA): needle driver (cat. no. 12502-12), ball-tipped scissors (cat. no. 14109-09), and ring forceps (cat. no. 11106-09) (*see* **Note 2**).
2. Suture: 6-0 (0.7 metric) Polysorb braided lactomer 9-1, 30″ (75 cm) violet, CV-11, taper (cat. no. GL-889, Syneture, United States Surgical, Norwalk, CT).
3. Sterile disposable supplies: surgical drape, cotton balls, and cotton-tipped applicators.
4. Disinfection solutions: 70% ethanol and 10% povidone iodine (10% Betadine®, Purdue Pharma, L. P., Stamford, CT) solution.
5. Lactated Ringer's Injection USP (Baxter 2B2323, Deerfield, IL). This is a sterile, 500 mL intravenous (IV) bag of lactated Ringer's solution whose precise electrolyte composition is described under the USA National Drug Code 0338-0117.
6. Non-di(2-ethylhexyl) phthalate (DEHP) IV Fat Emulsion Administration Set (cat. no. 2C1145, Baxter). This is a sterile IV tube set used to aseptically dispense lactated Ringer's solution from the IV bag.
7. Dedicated mouse survival surgical area (*see* **Note 3**).

2.3. Transuterine Microinjection and In Vivo Electroporation

1. Leica stereofluorescence dissecting microscope (MZ10F with PLAN 0.8X long working distance objective, 10X eyepieces, and GFP2 filter set).
2. Borosilicate glass capillaries with filament (cat. no. GC150F-10, Harvard Apparatus, Holliston, MA).
3. CUY21 Electroporator with foot pedal external trigger (Protech International Inc., San Antonio, TX).
4. Tweezers-style electrode with 5 mm diameter platinum disks (cat. no. CUY650P5, Protech International Inc.).
5. Micropipette puller: Flaming/Brown Model P-97 (Sutter Instrument Co., Novato, CA).
6. Micropipette beveler: K.T. Brown Type with 104C abrasive plate (Sutter Instrument Co.).
7. PicoSpritzer III (Parker Hannifin Corporation, General Valve Operation, Fairfield, NJ) with 60 psi regulator and foot pedal external trigger. Source gas is compressed nitrogen (>99% purity).
8. M33 roller bearing micromanipulator (Stoelting Co., Wood Dale, IL).
9. Articulated arm on a magnetic base (Stoelting Co.).
10. Accublock digital dry bath (LabNet International, Woodbridge, NJ) with aluminum block to hold five sterile, 50 mL conical centrifuge tubes.
11. Phosphate buffered saline: 137 mM NaCl, 2.7 mM KCl, 9.9 mM Na_2HPO_4, 2 mM KH_2PO_4, pH 7.2.

12. Fast green, crystalline (Sigma-Aldrich).
13. Timed-pregnant mouse (dam) whose embryos are at embryonic day 11.5 (E11.5). Noontime on the day a vaginal plug is detected is considered E0.5 (i.e., E11.5 is 11 days after the day the plug is detected).

3. Methods

3.1. General Anesthesia for Mouse Ventral Laparotomy

1. Assemble a fresh working solution of the anesthetic mixture each day from stock solutions: Mix 320 μL of the 65 mg/mL magnesium sulfate stock with 100 μL of absolute ethanol, 400 μL of propylene glycol, and 180 μL of 50 mg/mL pentobarbital sodium solution. Vortex the solution and briefly spin down.
2. Weigh the dam to the nearest 0.1 g.
3. Administer 7.2 μL of anesthetic mixture per gram body weight by intraperitoneal injection with the BD allergy syringe. For example, a 25.0 g mouse receives 7.2 μL of the anesthetic mixture *per gram body weight* for a total of 0.18 mL (*see* **Note 4**).
4. Place the injected mouse in a warmed cage for 4–6 min.
5. Assess responses to tail/toe pinches and the intactness of the ocular reflex. Proceed only after the mouse is unresponsive to these noxious stimuli and the reflex is absent.
6. Apply a thin layer of sterile ophthalmic ointment over the corneas of each eye.
7. Proceed with ventral laparotomy.

3.2. Ventral Laparotomy on the Anesthetized Dam

1. Shave the fur from the suprapubic region to just beneath the rib cage with fine (#40 blade) shears (*see* **Note 5**).
2. Disinfect the shaved skin by alternating ethanol and povidone iodine:
   a. 70% ethanol: gentle swipes from rostral to caudal with ethanol-dampened surgical cotton ball (*see* **Note 6**).
   b. 10% povidone iodine: gentle swipes from rostral to caudal with an iodine-saturated cotton-tipped applicator.
   c. 70% ethanol: Repeat as indicated above.
   d. Place the mouse in a supine position on a sterile drape and allow the skin to air dry.
3. Make a 2–3 mm ventral midline cut through the skin *only* with scissors.
4. Validate the location of the *linea alba*, an avascular connective tissue band extending from the xiphoid process past the navel. The small skin incision provides a viewing portal to search for the *linea alba*. Gently slide the incised skin from

side to side to identify the precise location of the *linea alba* if it is not visible beneath the initial incision.

5. Extend the initial midline incision by cutting the skin that overlies the *linea alba*. This incision may involve a slight course correction to align it with the underlying *linea alba*. The total length of the skin incision is 12–15 mm.
6. Gently grasp the connective tissue overlying the abdominal wall with blunt forceps and lift it away from the abdominal organs. This maneuver will pull the abdominal wall away from the underlying abdominal viscera.
7. Make a ventral midline scissor nick through the abdomen on the *linea alba* at the level of the navel. Since the *linea alba* is avascular, there should be no bleeding post-incision (*see* **Note 7**).
8. Insert the ball-tipped scissor through the abdominal slit and incise the *linea alba* first rostrally and then caudally. The total abdominal incision length is 10–14 mm. The ball tip prevents accidental nicking of any abdominal visera. If the bowel is nicked, euthanize the dam. Do not attempt a surgical repair.
9. Immediately irrigate the abdomen with sterile, 37 °C lactated Ringer's solution. Ringer's solution is kept at a set temperature in a sterile, 50 mL conical centrifuge tube resting in the tapered, aluminum heating block. Dispense Ringer's solution with a sterile Pasteur pipette (*see* **Note 8**).
10. Gently displace the intestines and inguinal fat that obstruct access to the right uterine horn with blunt forceps.
11. Gently lift the right horn of the uterus through the abdominal incision with ring forceps and rest it on the abdomen.
12. Validate that the entire uterine horn is externalized by identification of the ovary/oviduct rostrally and the cervical region of the uterus caudally. The oviduct appears as a coiled, muscular tube, and the ovary looks like a small granular white cluster of tissue embedded in fat. The cervical region of the uterus is the pronounced, Y-shaped bifurcation point of the two horns.
13. Perform the transuterine microinjection and in vivo electroporation procedures.

3.3. Transuterine Microinjection and In Vivo Electroporation

1. Load a 1.5 mm outer diameter by 0.86 mm inner diameter borosilicate glass capillary pipette into the Sutter P-97 micropipette puller outfitted with a 3 mm box filament that is 3 mm wide. *Conduct the ramp test as indicated in the P-97 instruction manual.* Program the instrument as follows: pressure = 200, heat = ramp value plus 3 units, pull = 0, velocity = 46, and time = 110. The capillary pipette produced is then manually broken with forceps to approximately 14 μm outer diameter and beveled at 20 degrees on the BV-10

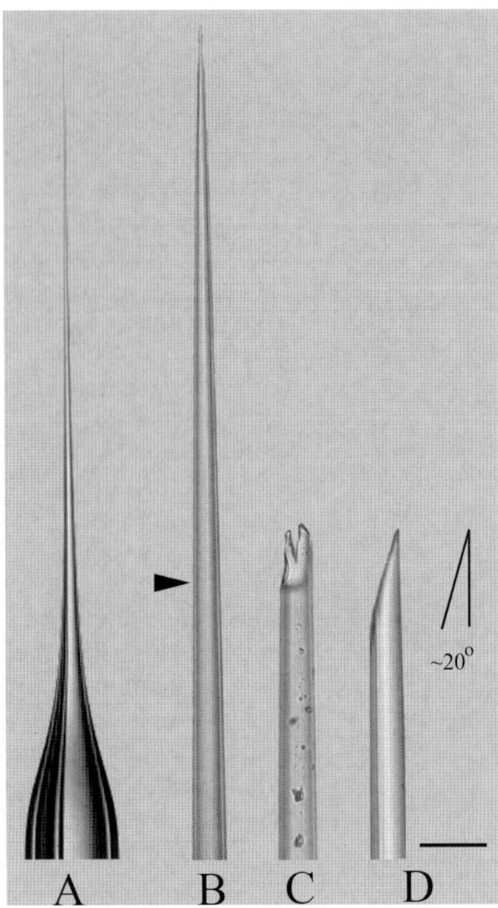

Fig. 8.1. Fabrication of a transuterine microinjection pipette. A pipette was pulled with the P-97 micropipette puller using the pressure:heat:pull:velocity:time settings indicated in **Section 3.3, step 1**. (**A**) The outer diameter of the unpulled shaft is 1.5 mm and the length of the tapered part is ~12 mm. The tip of this pipette was imaged in three successive stages of preparation (**B–D**). (**B**) The approximate location of the manual breakpoint at the tip of the pipette is indicated by the arrowhead. The pipette tip was broken by pinching the glass with Biologie #5 forceps (Fine Science Tools). (**C**) The broken pipette presents a bifurcated, coarse tip with particulate debris on the outside of the glass. (**D**) The pipette tip was beveled with the BV-10 beveler to replace the jagged leading edge and establish a 20° bevel. The final outer diameter of the injection pipette is ~20 μm (target range is 18–24 μm). Scale bar = 50 μm, applies to **B–D**.

micropipette beveler fitted with the Sutter 104C (gold) abrasive plate (**Fig. 8.1**) (*see* **Note 9**).

2. Backfill the beveled capillary pipette with 3 μg/μL of expression plasmid in sterile phosphate buffered saline (*see* **Note 10**).
3. Attach the pipette to its holder, and secure the shaft firmly to avoid ejection of the pipette from the holder during a pressure injection cycle (*see* **Note 11**).

4. Set the PicoSpritzer to 20 ms injection duration and 15 psi injection pressure. Test the ejection of plasmid by injecting into a drop of sterile lactated Ringer's solution. Confirm that all air bubbles are purged from the plasmid solution within the pipette (*see* **Note 12**).
5. Transillumination of the uterus enables visual identification of the gross features of the E11.5 mouse embryo. An implantation site housing the embryo is indicated by a bulge in the uterus. Gently grasp the uterine horn with latex-gloved thumb and forefinger, and press the output end of the fiber optic light guide against the freshly irrigated uterine wall. Center the light on a uterine bulge that houses an embryo. Gently knead the uterus between the thumb and the forefinger to reposition the embryo, so that the left side of the embryo is parallel to the surface of the dissecting scope objective (the embryo is oriented as though it is lying on its right side; *see* **Fig. 8.2B**). Search for the hindbrain, which looks like an angular white notch in the caudal-most region of the cephalic neural tube. Ventral to the hindbrain and dorsal to the branchial arches, identify a blood vessel pattern that looks like the uprights of an American football goalpost: a thick base with rostral and caudal branches that are finer (**Fig. 8.2**). The otocyst is superficially set within the periotic

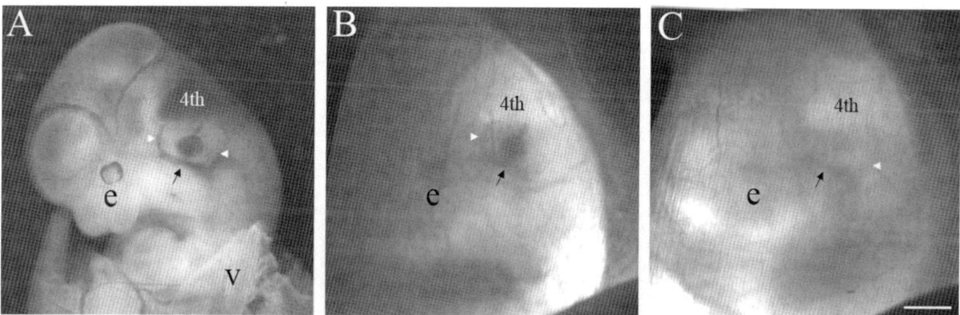

Fig. 8.2. Transuterine microinjection into the embryonic day 11.5 mouse otocyst. (**A**) An aqueous solution (0.1%) of fast green dye was microinjected into the left otocyst through the uterine wall at E11.5. The embryo was dissected free of the uterus and the extraembryonic membranes and its left side was imaged. The otocyst is located superficially in the lateral head mesenchyme, dorsal and slightly medial to the main trunk of the primary head vein (*black arrow*) and approximately midway between its rostral and caudal branches (*white arrowheads*). The trunk of the primary head vein forms the base of an American football (or rugby) goalpost, and its rostral and caudal branches form the uprights. The otocyst displays a clamshell-shaped vestibule and tapered dorsal endolymphatic duct. *Abbreviations*: e, left eye; 4th, nascent 4th ventricle; v, visceral yolk sac. (**B**) The same embryo imaged in panel **A** immediately after transuterine microinjection of dye into the left otocyst. The uterus and extraembryonic membranes are intact. The 4th ventricle, eye, and rostral branch (*white arrowhead*) of the primary head vein (*black arrow*) are visible. The caudal branch of the primary head vein is not apparent. (**C**) The left side of the E11.5 mouse embryo prior to transuterine microinjection into the otocyst. The transillumination scheme in this image highlights the caudal branch (*white arrowhead*) of the primary head vein (*black arrow*), but the rostral branch is not clearly visible. Since the otocyst cannot be visually identified at E11.5, its location is inferred by interpretation of the anatomical landmarks provided by the primary head vein vasculature. Scale bar = 1 mm and applies to all panels.

mesenchyme midway between the uprights, though it is not possible to see the fluid-filled otocyst directly. The location of the otocyst is therefore interpolated using the flanking vasculature as an instructive anatomical landmark.

6. Advance the beveled microinjection pipette through the uterine wall along a trajectory that will place the tip of the pipette in the head mesenchyme between the uprights. Pulse the PicoSpritzer to eject some fast green-tinged plasmid solution as an aid to define pipette tip position. If the pipette tip is located too lateral with respect to the otocyst, dye will collect in either the exocoelomic or the amniotic cavities. If the pipette is advanced through the otocyst, dye will collect in the medial periotic mesenchyme or neural tube. When the otocyst is properly targeted, the vestibule and endolymphatic duct fill with dye and the borders of the otocyst become clearly outlined (**Fig. 8.2A,B**). The number of injection pulses required to fill the otocyst is typically 10–15 with an 18–24 μm outer diameter micropipette at 15 psi injection pressure and 20 ms injection duration per pulse. Release finger pressure on the uterus and then remove the pipette in one quick reverse stroke of the micromanipulator (*see* **Note 13**).
7. Immediately irrigate the uterus and the tweezer-style electrode paddles with pre-warmed lactated Ringer's solution. Position the negative electrode on the surface of the uterus that is adjacent to the lateral wall of the injected, left otocyst (**Fig. 8.3A**). Position the positive electrode on the surface of the uterus that is adjacent to the lateral wall of the right, uninjected otocyst (**Fig. 8.3B**). Gently compress the uterus

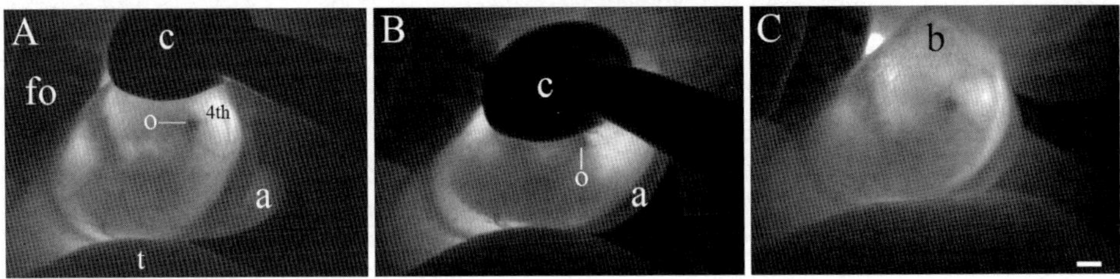

Fig. 8.3. In vivo electroporation of the embryonic day 11.5 mouse otocyst. (**A**) The uterus was transilluminated with light from a fiber optic cable whose output end (fo) was directly in contact with the irrigated uterus. The left otocyst (o) of the E11.5 mouse embryo was injected with fast green solution, and its gross morphology is discernable beneath the nascent 4th ventricle (4th) in the caudal hindbrain. The insulated surface of the cathode (c) and the reflective, platinum surface of the anode (a) were grossly positioned around the uterus to flank the embryo. (**B**) Gentle compression of the uterus with the tweezer-style electrodes forced a counter-clockwise rotation of the embryo, placing the injected otocyst toward the center of the 5 mm cathode–anode field. The goal is to drive negatively charged DNA into ventral progenitors within the otic vesicle. (**C**) Bubbles (b) on the surface of the uterus after execution of the 5-pulse train appear as graininess in the image. Light pressure from the thumb (t) reorients the embryo within the uterus to facilitate placement of the electrodes (panel **A**). Both the fiber optic light head and the thumb are removed from contact with the uterus prior to triggering the square wave pulse train. Scale bar = 1 mm and applies to all panels.

to securely hold and center the otocyst between the electrode paddles (**Fig. 8.2B**) and trigger the electroporator by actuating the foot pedal. When the 5-pulse train is completed, release the tweezer-style electrode grip and immediately irrigate the uterus with pre-warmed, lactated Ringer's solution (*see* **Note 14**).

8. Inject and electroporate 2–3 embryos per uterine horn for a total of 4–6 injected embryos per dam. Document which embryos were injected (right horn, embryo 1 is the embryo closest to the right ovary), the quality of the otocyst injections (a weak fill fails to demonstrate the endolymphatic duct and a strong fill does), and the current transferred to each embryo during the electroporation (*see* **Note 15**).
9. Liberally irrigate the uterus and any other externalized abdominal viscera and return them to their native positions within the abdominal cavity. Irrigate the abdominal cavity with 4–8 mL of lactated Ringer's solution, allowing the excess to spill out of the incision site. Approximately 2 mL of lactated Ringer's solution should remain in the abdominal cavity to facilitate rehydration (*see* **Note 16**).
10. Carefully place the dam on a fresh, dry sterile drape and begin the two-stage closure with 6–0 resorbable suture. A running stitch consisting of alternating a conventional throw and a locking stitch is ideal for both the abdominal wall and the skin closures.
11. Remove excess moisture from the dam's fur by gently blotting with sterile, absorbent paper towel. Clear any bedding away from the bottom of the heated recovery cage; tuck the dam into a folded, sterile paper towel; and set her down on the floor to maximize heat transfer. The dam's eyes should be fully shielded from direct light (*see* **Note 17**).
12. Conduct postoperative monitoring of the dam every 30 min until conscious by assessing the surgical site for bleeding or exudate, the quality of respiration (unlabored breathing), and the time of ambulation. The dam should take initial, unsteady steps within 1.5–2 h and begin grooming to remove the ophthalmic ointment from her eyes. Record these observations in the mouse survival surgery log book. Construct a calendar to record the dam's identification number, the bioactive reagent electroporated, and the date of embryo or pup harvest.
13. Re-evaluate the dam the morning after surgery and assess the quality of respirations, the surgical site, and evidence of eating (moist feces) and drinking (pick the dam up by the tail and look for stress-induced urination). Return the dam to the main mouse colony in the same cage (*see* **Note 18**).
14. Harvest electroporated inner ears at a relevant downstream embryonic or postnatal time point and analyze the phenotype. **Figure 8.4** presents a representative analysis of an

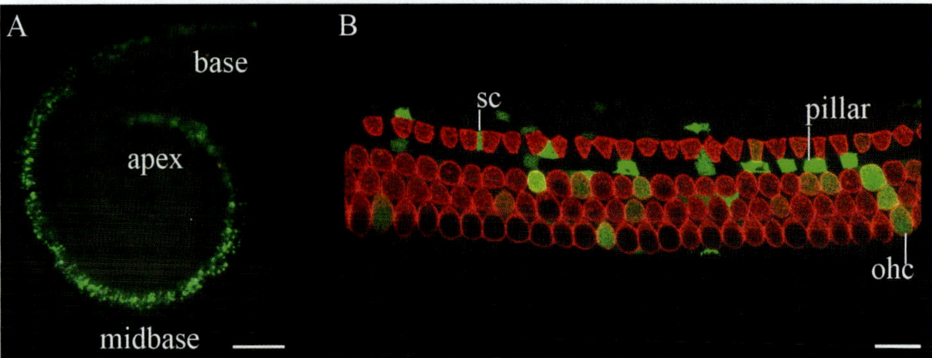

Fig. 8.4. Electroporation-mediated transfer of an expression plasmid encoding green fluorescent protein transfects progenitors that give rise to the organ of Corti. An expression plasmid encoding green fluorescent protein (GFP) driven by the human elongation factor 1-α promoter (EF1-α) was electroporated into the E11.5 mouse otocyst. The inner ear was harvested 6 days later at E17.5 and fixed in 4% paraformaldehyde in PBS for 12 h. The cartilaginous otic capsule and the cochlear lateral wall were removed, and the whole mount preparation was imaged in panel **A**. GFP expression is detectable in the base, midbase, and proximal apex of the cochlea. (**B**) Laser confocal microscopy of a representative 100 μm section of the EF1-α/GFP-transfected cochlea immuostained with an antibody against myosin 7a (*red*) to identify the single row of inner hair cells and three rows of outer hair cells. Several supporting cells (sc), pillar cells (pillar), and outer hair cells (ohc) express GFP. These data indicate that progenitors giving rise to supporting cells and hair cells of the organ of Corti were transfected and expression of the transgene was maintained in differentiated cell types of the maturing cochlea. Scale bar in **A** = 100 μm and in **B** = 10 μm.

E17.5 inner ear that was transfected at E11.5 with a construct encoding enhanced green fluorescent protein.

4. Notes

1. All stock solutions are prepared in sterile, pyrogen-free water with a resistivity of 18.2 MΩ-cm unless otherwise indicated. The concentration of magnesium sulfate administered does not induce paralysis but merely relaxes uterine tone. This enables the embryos to be repositioned for microinjection by gentle uterine palpation with the thumb and index finger.
2. Instruments suitable for mouse ventral laparotomy should be of surgical grade stainless steel since the rigors of repeated sterilization will dull cutting surfaces and corrode hinge points of lower quality scissors and needle drivers. Surgical-grade instruments are expensive, though the veterinary staff of your home institution may be able to purchase these items at reduced cost.
3. Animal care and use guidelines of the home institution should be consulted regarding mouse survival surgery requirements. The use of a horizontal laminar flow hood is encouraged to prevent exposure of the dam to potentially infectious agents. Scrupulous attention to asepsis in the surgical arena is

essential. A dedicated mouse survival surgery suite or station is highly recommended.

4. Intraperitoneal injection in a robustly pregnant mouse requires practice, since it is fairly easy to misplace the needle in abdominal viscera that are displaced by enlarged uterine horns. The BD allergy syringe has an intradermal bevel designed for presentation of antigen beneath the human skin. This syringe is ideal for *intraperitoneal* injections of pregnant mice because the needle is short; the bevel pierces the skin atraumatically, and virtually all of the anesthetic mixture is transferred to the abdominal cavity (i.e., the syringe has an insignificant volume of dead space). The dam should lose consciousness within 3–5 min after a successful intraperitoneal injection of the sodium pentobarbital anesthesia mixture. The intraperitoneal injection is unsuccessful if the dam remains ambulatory after 5–7 min. If this occurs, do not redose the dam: move on to another dam. Redosing with sodium pentobarbital correlates with reduced embryonic survival.
5. Complete fur removal facilitates disinfection and promotes the rapid healing of the skin incision. At least 30% of individuals who work with mice will eventually develop an allergy to them. Shave the fur in a chemical fume hood to keep the free-fur and dander contained.
6. Surgical cotton balls are less likely to shed fibers and contaminate the surgical field compared with cosmetic cotton balls. Store the cotton balls in a closed container of 70% ethanol. Aggressively squeeze the residual ethanol out of the cotton ball before contact with the abdominal integument. A thin film of ethanol should coat the skin to achieve disinfection. Care should be exercised to avoid excessive superficial ethanol exposure, because excessive evaporative cooling can induce physiological stress.
7. The presence of active blood flow from the incised abdominal wall indicates that the incision extended from the *linea alba* to the adjacent musculature. Identify the precise location of the bleeding by stereomicroscopic inspection, and apply direct pressure with forceps for 1–2 min to stop the hemorrhage. Do not proceed with the laparotomy until all bleeding vessels are managed. The observation of vascular compromise is fortunately rare during *linea alba* incision, and when it is encountered, direct pressure stops the bleeding.
8. The compromise of the abdominal cavity necessitates aggressive irrigation of the abdominal organs. The small and large intestines, and the inguinal fat pads in particular, will suffer from desiccation that will affect the progression of the pregnancy. Irrigation with sterile, pre-warmed (37 °C) lactated Ringer's solution is therefore critical. The heat block temperature is set a few degrees above 37 °C to account for cooling

of the lactated Ringer's solution during pipette transfer from the culture tube to the abdomen.

9. Carefully fabricated micropipettes are critical to the success of the transuterine microinjection procedure. An appropriately crafted pipette enables accurate, atraumatic injection into the otocyst. The recommended glass is thick walled with a filament to provide durability during targeting and to facilitate backfilling with plasmid. Pipettes should be broken under a suitable microscope with the aid of a reticule that enables estimation of outer diameter after the break. Beveling is normally conducted in an aqueous environment with a surfactant added. For in vivo applications, bevel the pipettes in water alone to avoid the possibility of introducing chemical contaminants to the embryo in utero. However, this method will most certainly reduce the lifespan of the 104C abrasive plate, an acceptable tradeoff for enhanced embryonic survival. The *P-97 Pipette Cookbook* (2006, revision C, Sutter Instrument Co.) is an excellent discourse on the theory underlying micropipette fabrication. Refer to the *Pipette Cookbook* for a comprehensive discussion of all issues related to microinjection pipette fabrication with the P-97 puller and beveling with the BV-10 beveler.
10. Expression plasmids are isolated and purified with the Qiagen HiSpeed Plasmid Maxi Kit (cat. no. 12662, Qiagen, Valencia, CA) according to manufacturer's instructions. The plasmid is sterile filtered prior to standard ethanol precipitation. The plasmid is resuspended in sterile phosphate buffered saline, diluted to 3 μg/μL final concentration, and stored at −20 °C in 8 μL aliquots. Prior to use, crystalline fast green is added to the thawed plasmid solution, which is then triturated 50 times, spun at 10,000 g for 15 s, and backfilled into the beveled microinjection pipette.
11. The PicoSpritzer contains high-performance pneumatics that reliably deliver pressurized nitrogen to the injection pipette. The pipette is secured to the pipette holder with a screw-type fitting that compresses a gasket in contact with the pipette shaft. Wear of the pipette holder gasket can result in a weak seal that permits gas leakage and inefficient ejection of plasmid DNA. Replace the gasket prophylactically every 3 months or sooner if performance issues arise. Moreover, an inexperienced operator may be unfamiliar with the requisite torque needed to secure the pipette firmly in the holder. In this case, the pipette can be ejected from the holder with considerable force. As a general precaution, the operator's fingers and hands should never be placed in the putative trajectory path of the pipette.
12. Pull micropipettes in advance and store them unbroken in a plastic, covered Petri dish by lengthwise insertion into a thin

band of modeling clay. Break and bevel three pipettes prior to administering anesthesia to the dam. Pipettes broken and beveled the day before may become clogged as the residual water within the pipette evaporates and deposits glass fines inside the pipette lumen.

13. Transuterine microinjection into the E11.5 mouse otocyst is challenging at several levels. Identification of gross embryonic anatomy by transillumination requires practice. Start with finding the beating heart, the pigmented epithelium of the eye, the limb buds, and brain vesicles. Then find the hindbrain notch and the goalpost vasculature. Experiment with various lighting intensities and orientations since individuals vary tremendously in what "looks good" to them. It is certainly best to use the minimum intensity of illumination necessary. Once the target area cloistering the otocyst is identified, microinjection should be conducted with an "outside-in" progression. Advance through the uterus, and pulse to view the tracer dye as an indication of approximate pipette tip position. Advance under micrometer control and pulse again to track pipette position, working from lateral to medial or outside-in with respect to the embryo. It may be possible to see a slight "toggle" when the pipette tip contacts the lateral side of the embryo's head. Typically, 3–5 pulses executed in rapid succession are required to fill the otocyst with enough dye to discern that indeed the pipette is properly positioned. Thereafter, fill the otocyst fully so that the endolymphatic duct and the vestibule proper swell with injected DNA solution. Freshly irrigate the uterus and abdominal viscera immediately before and after the targeting/microinjection procedures, to ensure that these tissues do not desiccate. Replace the sterile drape that is beneath the mouse when it becomes saturated with lactated Ringer's solution.

14. The 5 mm electrode paddles must be coupled to the uterus with lactated Ringer's solution, an efficient charge carrier. Poor hydration of the uterus will reduce current delivered to the embryo and focally heat the uterine vasculature. Bubbles will be generated on the surface of the hydrated uterus during the electroporation cycle, and their presence is a dynamic indication of current flow (**Fig. 8.3C**). Compressing the uterus with too much force will rupture the amnion and/or the visceral yolk sac resulting in embryonic lethality. The CUY21 will report the current transferred during the last pulse of the 5-pulse train. Typically, current transfer of 50–100 mA per pulse is sufficient to transfect the otic epithelium.

15. Develop a mouse survival surgery log book with the guidance of your institutional animal care and use committee and veterinary staff. This resource serves as a permanent record of preoperative, operative, and postoperative animal care and should

be included in your animal care and use protocol. Subtle observations during targeting, injection, and electroporation are recorded for each embryo manipulated, and these data are correlated with transfection efficiency after tissue harvest. Follow postoperative recovery for at least 24 h. Ventral laparotomy and transuterine microinjection are extremely well tolerated. Minor, temporary vaginal bleeding can occur and likely results from focal compromise of embryonic or uterine vasculature during an injection. When present, active blood flow ceases while the dam is still under anesthesia in her recovery cage. Euthanize the dam if blood flow does not cease during this time. Perform a necropsy to determine the source of the bleeding and try to correlate the outcome with a causal procedural misstep. To avoid microinjection-induced bleeding, concentrate on fabrication of beveled pipettes of the appropriate diameter and geometry.

16. Mice experiencing ventral laparotomy will tend to drink and eat less immediately after surgery. Filling the abdominal cavity with lactated Ringer's solution prior to suturing is therefore vitally important to ensure normal fluid homeostasis. Postoperative supplemental fluid administration is not necessary, following ventral laparotomy, provided that sufficient care is taken to load the abdominal cavity with Ringer's solution prior to closing. Food placed on the bottom of the cage facilitates access and obviates the need for stretching to reach food in an overhead hopper. Dams recovering from ventral laparotomy benefit from the community effect provided by co-housing. Up to four post-surgical dams may be co-housed through the 24 h postoperative recovery period. House a post-surgical dam alone rather than pairing her with a mouse that has not undergone surgery.
17. It is essential to provide 10–12 h of accessory heating for the dam immediately after surgery to reduce physiological stress and promote healing. The entire recovery cage is efficiently warmed with a Gaymar T/Pump that recirculates water through a Hallowell hard pad. Use a funnel to replace evaporated water in the T/Pump fluid reservoir, since water spilled during refill will track directly to sensitive electrical components in the pump and cause a short circuit that is expensive to repair.
18. Housing mice overnight in any location other than the animal facility will require justification in the animal care protocol.

Acknowledgments

We thank Tomomi Fukuchi-Shimogori and Elizabeth Grove for demonstrating transuterine microinjection techniques and Donna M. Fekete for inspired postdoctoral mentorship (to JB). Funding from the National Institute on Deafness and Other Communication Disorders and the McKnight Fund for Neuroscience is gratefully appreciated.

References

1. Schoenwolf, G. C. (2001) Cutting, pasting and painting: experimental embryology and neural development. *Nat. Rev. Neurosci.* 2, 763–771.
2. Hamburger, V. (1990) The rise of experimental neuroembryology: a personal reassessment. *Int. J. Devl. Neurosci.* 8, 121–131.
3. Spemann, H. and Mangold, H. (2001) Induction of embryonic primordia by implantation of organizers from a different species. 1923. *Int. J. Dev. Biol.* 45, 13–38.
4. Sperry, R. (1945) Restoration of vision after crossing of optic nerves and after contralateral transplantation of eye. *J. Neurophysiol.* 8, 15–28.
5. Hamburger, V. (1939) The development and innervation of transplanted limb primordia of chick embryos. *J. Exp. Zool.* 80, 347–389.
6. Le Douarin, N. M. and Teillet, M. A. (1974) Experimental analysis of the migration and differentiation of neuroblasts of the autonomic nervous system and of neurectodermal mesenchymal derivatives, using a biological cell marking technique. *Dev. Biol.* 41, 162–184.
7. Le Douarin, N. (1973) A biological cell labeling technique and its use in experimental embryology. *Dev. Biol.* 30, 217–222.
8. Kelley, M. W. (2006) Regulation of cell fate in the sensory epithelia of the inner ear. *Nat. Rev. Neurosci.* 7, 837–849.

Chapter 9

Isolation of Sphere-Forming Stem Cells from the Mouse Inner Ear

Kazuo Oshima, Pascal Senn, and Stefan Heller

Abstract

The mammalian inner ear has very limited ability to regenerate lost sensory hair cells. This deficiency becomes apparent when hair cell loss leads to hearing loss as a result of either ototoxic insult or the aging process. Coincidently, with this inability to regenerate lost hair cells, the adult cochlea does not appear to harbor cells with a proliferative capacity that could serve as progenitor cells for lost cells. In contrast, adult mammalian vestibular sensory epithelia display a limited ability for hair cell regeneration, and sphere-forming cells with stem cell features can be isolated from the adult murine vestibular system. The neonatal inner ear, however, does harbor sphere-forming stem cells residing in cochlear and vestibular tissues. Here, we provide protocols to isolate sphere-forming stem cells from neonatal vestibular and cochlear sensory epithelia as well as from the spiral ganglion. We further describe procedures for sphere propagation, cell differentiation, and characterization of inner ear cell types derived from spheres. Sphere-forming stem cells from the mouse inner ear are an important tool for the development of cellular replacement strategies of damaged inner ears and are a bona fide progenitor cell source for transplantation studies.

Key words: Cochlea, vestibular, utricle, spiral ganglion, hair cell, regeneration, neurosphere, stem cell, progenitor cell.

1. Introduction

The ability to form floating clonal colonies or "spheres" is not only a hallmark of certain stem and progenitor cell populations, but is also a useful feature for isolation of these cells from complex cell mixtures. Several laboratories have shown that sphere-forming cells reside in the neonatal and even in the adult inner ear *(1–11)*. It has also been demonstrated that some of the inner ear–derived sphere-forming cells have the ability to self-renew, which

Bernd Sokolowski (ed.), *Auditory and Vestibular Research: Methods and Protocols, vol. 493*

DOI 10.1007/978-1-59745-523-7_9 Springerprotocols.com

is a characteristic feature of stem cells *(2, 7)*. Self-renewal has been reported for sphere-forming cells from the neonatal spiral ganglion, the organ of Corti (OC), and vestibular sensory epithelia, as well as from the adult utricular sensory epithelium. In this chapter, we refer to these sphere-forming and self-renewing cells as inner ear stem cells. Previous results show that different tissues of the neonatal inner ear harbor distinct populations of stem cells, each one displaying specific features. Spiral ganglion–derived spheres, for example, give rise to neurons and glial cell types and other unidentified cells after withdrawal of growth factors and attachment to a substrate *(7, 8, 12)*. Only occasionally have we found cells positive for hair cell markers in cell populations differentiated from spiral ganglion–derived spheres. In contrast, we frequently observed the generation of hair cell marker expressing cells from spheres isolated from the OC or vestibular sensory epithelia. These spheres also readily gave rise to neurons and glial cell types, albeit with lower frequency when compared with spiral ganglion–derived spheres.

Here we provide detailed protocols for the isolation of sphere-forming cells from various parts of the neonatal inner ear. We also describe how to propagate spheres and how to initiate spontaneous differentiation of inner ear cell types. Spheres generated from inner ear stem cells can be used not only for in vitro studies but also for transplantation experiments, for example, in explorative studies aimed at development of cellular therapies.

2. Materials

2.1. Cell Culture

2.1.1. Tissue Dissection and Cell Culture Equipment

1. Dedicated room or work area (*see* **Note 1**).
2. Biosafety cabinet or laminar flow hood.
3. Humidified incubator at 37 °C with 5% CO_2/95% air atmosphere.
4. Dissection microscope (e.g., Zeiss Stemi 2000-C or equivalent, Carl Zeiss MicroImaging, Thornwood, NY) with a light source (Zeiss KL1500 or equivalent).
5. Surgical forceps (#5 and #55, Roboz, Gaithersburg, MD) and micro-dissecting scissors.
6. Pipetman® (20-, 200-, and 1,000 μL).
7. 15- and 50-mL sterile conical tubes.
8. 5-, 10-, and 25-mL sterile plastic pipettes.
9. Pipet-Aid (Drummond Scientific, Broomall, PA) or other automatic pipettor.

10. Pipette tips (20–300 μL work well for tissue trituration, cat. no. 022491245, Eppendorf of North America, Westbury, NY).
11. Petri dishes (or cell suspension culture dishes) (we successfully used the following two brands: BD Falcon 35 mm Petri dish, BD Biosciences, San Jose, CA; Greiner 6-well suspension culture plate, Greiner Bio-One, Monroe, NC).
12. Cell strainer (70 μm, sterile).

2.1.2. Media, Supplements, Solutions, and Other Supplies

1. Dulbecco's phosphate-buffered saline (DPBS): 2.67 m*M* KCl, 1.47 m*M* KH_2PO_4, 137.93 m*M* NaCl, and 8.06 m*M* Na_2HPO_4, pH 7.3.
2. Hanks' balanced salt solution (HBSS): 1.26 m*M* $CaCl_2$, 0.493 m*M* $MgCl_2$, 0.407 m*M* $MgSO_7$, 5.33 m*M* KCl, 0.441 KH_2PO4, 4.17 m*M* $NaHCO_3$, 137.93 m*M* NaCl, 0.338 m*M* Na_2HPO_4, and 5.56 m*M* D-glucose, pH 7.3.
3. 0.25% Trypsin/EDTA solution (Invitrogen or similar product).
4. Trypsin inhibitor/DNaseI cocktail: Transfer 100 mg of trypsin inhibitor and 100 mg of DNaseI into a 50-mL conical tube. Add 20 mL of PBS and mix by shaking. Sterilize the solution using a syringe equipped with a 0.2 μm sterile syringe filter. Aliquot for single use (50 μL and 100 μL) and store at −20 °C (*see* **Note 2**).
5. Dulbecco's Modified Eagle Medium/F12 mixed 1:1 (DMEM/F12).
6. N-2 cell culture supplement, 100X (Invitrogen or similar product).
7. B-27 cell culture supplement, 50X (Invitrogen or similar product).
8. Ampicillin sodium salt. Make a 1,000X stock solution in Milli-Q water (50 mg/mL), filter sterilize, and store in 500-μL aliquots at −20 °C. Working concentration is 50 μg/mL.
9. 10% (*w/v*) bovine serum albumin (BSA), Fraction V solution: Dissolve 1 g of BSA in 10 mL of PBS, filter sterilize, and store in small aliquots at −20 °C. Prepare a 0.1% BSA solution (*w/v*) by diluting the 10% stock solution 1:100 in PBS.
10. Recombinant human epidermal growth factor (EGF): To generate a 5,000X stock solution (100 μg/mL), reconstitute the content of a 200-μg vial in 2 mL sterile PBS containing 0.1% BSA and store in 40 μL aliquots at −20 °C.
11. Recombinant human fibroblast growth factor-basic (bFGF). To generate a 2,500X stock solution (25 μg/mL), reconstitute the content of a 25-μg vial in 1 mL sterile PBS containing 0.1% BSA and store in 80-μL aliquots at −20 °C.

12. Recombinant mouse insulin-like growth factor-I (IGF-1): To generate a 2,000X stock solution (100 μg/mL), reconstitute the content of a 50-μg vial in 0.5 mL sterile PBS containing 0.1% BSA and store in 100-μL aliquots at −20 °C.
13. 2,000X Heparan sulfate proteoglycan solution (100 μg/mL): Dilute the content of a 100-μg vial with 750 μL sterile PBS containing 0.1% BSA and store in 100-μL aliquots at −20 °C.
14. Differentiation medium: Mix 500 mL of DMEM/F12 + 5 mL of N-2 supplement +10 mL of B-27 supplement +500 μL of ampicillin stock solution.
15. Sphere culture medium: Mix 200 mL of differentiation medium +40 μL of EGF stock solution +80 μL of bFGF stock solution +100 μL of IGF-1 stock solution +100 μL of heparan sulfate stock solution. The final concentrations of EGF, bFGF, IGF-1, and heparan sulfate are 20, 10, 50, and 50 ng/mL, respectively.

2.2. Passaging of Spheres

2.2.1. For Spheres from the Utricle and Organ of Corti

1. 0.25% Trypsin/EDTA.
2. Trypsin inhibitor/DNaseI cocktail (*see* **Section 2.1.2, step 4**).
3. Eppendorf pipette tips (*see* **Section 2.1.1, step 10**).

2.2.2. For Spiral Ganglion Spheres

1. Accumax (cat. no. AM105, Innovative Cell Technologies Inc., San Diego; CA).
2. Fire-polished Pasteur pipettes (*see* **Note 3**).

2.3. Differentiation

1. Gelatin solution: Dilute 2% stock gelatin solution (from bovine skin, i.e., cat. no. G1393, Sigma-Aldrich, or similar product) in sterile water or DPBS to make a 0.2% working solution that can be stored for up to 4 weeks at 4 °C.
2. Poly-L-ornithine solution (50-mL bottle, 0.1 mg/mL, Mw 30,000–70,000).
3. Fibronectin from bovine plasma (1-mL bottle, 1 mg/mL). Aliquot into 10 μL stock solution aliquots and store at −20 °C.
4. 35 mm four-well plates (Greiner Bio-One).
5. Eight-well Lab-Tek II chamber slides (cat. no. 154534, Nunc, Rochester, NY).
6. Six-well tissue culture plates.

2.4. Immunocytochemistry

1. Paraformaldehyde solution, 16% (10 × 10 mL ampoules, cat. no. 15710, Electron Microscopy Sciences, Hatfield, PA, or similar product): Dilute 1:4 in DPBS to generate a 4% paraformaldehyde solution, which can be stored at 4 °C for up to two additional days.

2. Triton X-100 (250-mL bottle)
3. BSA, Fraction V.
4. Goat serum (100-mL bottle): Make 10-mL aliquots and heat inactivate for 30 min at 56 °C, store frozen at −20 °C.
5. PBS/BSA/Triton (PBT1): 0.1% Triton X-100 (*v/v*), 1% BSA (*w/v*), and 5% (*v/v*) heat-inactivated goat serum in PBS. Prepare 500 mL and store at −20 °C in 25-mL aliquots.
6. PBT2: 0.1% Triton X-100 (*v/v*) and 0.1% BSA (*w/v*) in PBS. Prepare 500 mL and store at −20 °C in 25-mL aliquots.
7. DakoCytomation fluorescent mounting medium (15-mL bottle, cat. no. S3023, Dako, Carpinteria, CA).
8. 4′,6-diamidino-2-phenylindole (DAPI) (1 mg vial): Prepare a stock of 2 mg/mL (1,000X stock solution) in water and store frozen and protected from light in small aliquots. To generate a working solution, dilute the stock solution 1:1,000 with PBS.
9. Cover glass (24 × 60 mm, No. 1).
10. Round cover glass (10 mm diameter, No. 1).

2.5. RT-PCR

1. RNeasy Mini Kit (Mini kit (50), cat. no. 74104, Qiagen, Valencia, CA).
2. RNaseZap (250-mL bottle, cat. no. 9780, Ambion, Foster City, CA, or similar product).
3. 2-Mercaptoethanol (ß-ME), pure liquid, 14.3 *M* (100-mL bottle).
4. Buffer RLT-ß-ME: Buffer RLT is supplied with the RNeasy Mini kit. Add 40 μL of ß-ME to 4 mL of buffer RLT (*see* **Note 4**).
5. 20-gauge needle (Box of 100).
6. 1-mL syringe (Box of 100).
7. Ribonuclease (Rnase)-free water (500-mL bottle).
8. 200-proof ethanol.
9. 70% ethanol: Mix from 200 proof ethanol and RNase-free water.
10. UV spectrophotometer.
11. Oligo$(dT)_{12-18}$ (25 μg vial, 500 μg/mL).
12. dNTP mix, 10 m*M* (100 μL vial).
13. SuperScript™ II reverse transcriptase, including 5X first-strand buffer and 0.1 *M* DTT solution (2,000 units, cat. no. 18064-022, Invitrogen).
14. RNaseOUT™ (5,000 units, 40 units/μL, cat. no. 10777-019, Invitrogen).
15. Microcentrifuge (e.g., Eppendorf 5417C or similar apparatus).
16. Thermal cycler (e.g., GeneAmp PCR system 9700, Applied Biosystems, or similar apparatus).
17. Taq DNA Polymerase (500 units).

18. Taq DNA Polymerase 10X standard reaction buffer with $MgCl_2$.
19. Dimethyl sulfoxide (DMSO) (100-mL bottle).
20. Nuclease-free microcentrifuge tubes (1.5 mL and 0.5 mL).
21. PCR reaction tubes, thin wall, 0.2 mL.

3. Methods

3.1. Isolation of Sphere-Forming Stem Cells from the Neonatal Mouse Inner Ear

3.1.1. Preparation for Mouse Inner Ear Dissection

1. Postnatal day 1 (P1) mouse.
2. Fill sterile 35 mm Petri dishes with 2.5 mL of ice-cold HBSS.
3. Put 50 μL drops of DPBS into each well of a six-well suspension culture plate placed onto a cooled metal plate on ice.
4. Sterilize forceps (#5 and #55) and surgical scissors with 70% ethanol.
5. Pre-warm Trypsin/EDTA at 37 °C.
6. Thaw trypsin inhibitor/DNaseI cocktail; keep it at room temperature.
7. Pre-warm the sphere culture medium at 37 °C.

3.1.2. Dissection of Mouse Inner Ear

1. Follow your institution's guidelines for euthanization of neonatal mice, and decapitate the euthanized P1 mouse. Thoroughly spray the decapitated head with 70% ethanol to sterilize its surface.
2. Using a pair of surgical scissors cut the head into two halves by making an incision at the midline. Remove the brain and brainstem to expose the temporal bone (**Fig. 9.1A**). Dissect out the temporal bones with scissors and transfer them into 2.5 mL of ice-cold HBSS in a 35 mm Petri dish.
3. Set the Petri dish under a dissection microscope, and remove as much of the surrounding connective and muscle tissue from the temporal bone. The utricle is dissected as shown in **Fig. 9.1**. Poke a small hole in the cartilage at the position marked with an asterisk (**Fig. 9.1B,C**). Carefully free the utricle after cutting the connections with the neighboring ampullae and nerve fibers, and clean from surrounding tissue using fine forceps (*see* **Note 5**). Transfer the utricle into fresh ice-cold HBSS, and remove the pigmented membrane covering the utricle (this membrane is not pigmented in albino mice). The sticky otoconia layer can be removed with fine forceps by gently touching its surface and lifting from the sensory epithelium (**Fig. 9.1D,E**). To separate the sensory epithelium from the underlying non-sensory tissue (**Fig. 9.1F**), incubate the utricles in DMEM/F-12 with 0.5 mg/mL of thermolysin for 45 min at 37 °C in a cell culture incubator. The sensory epithelium can be removed after

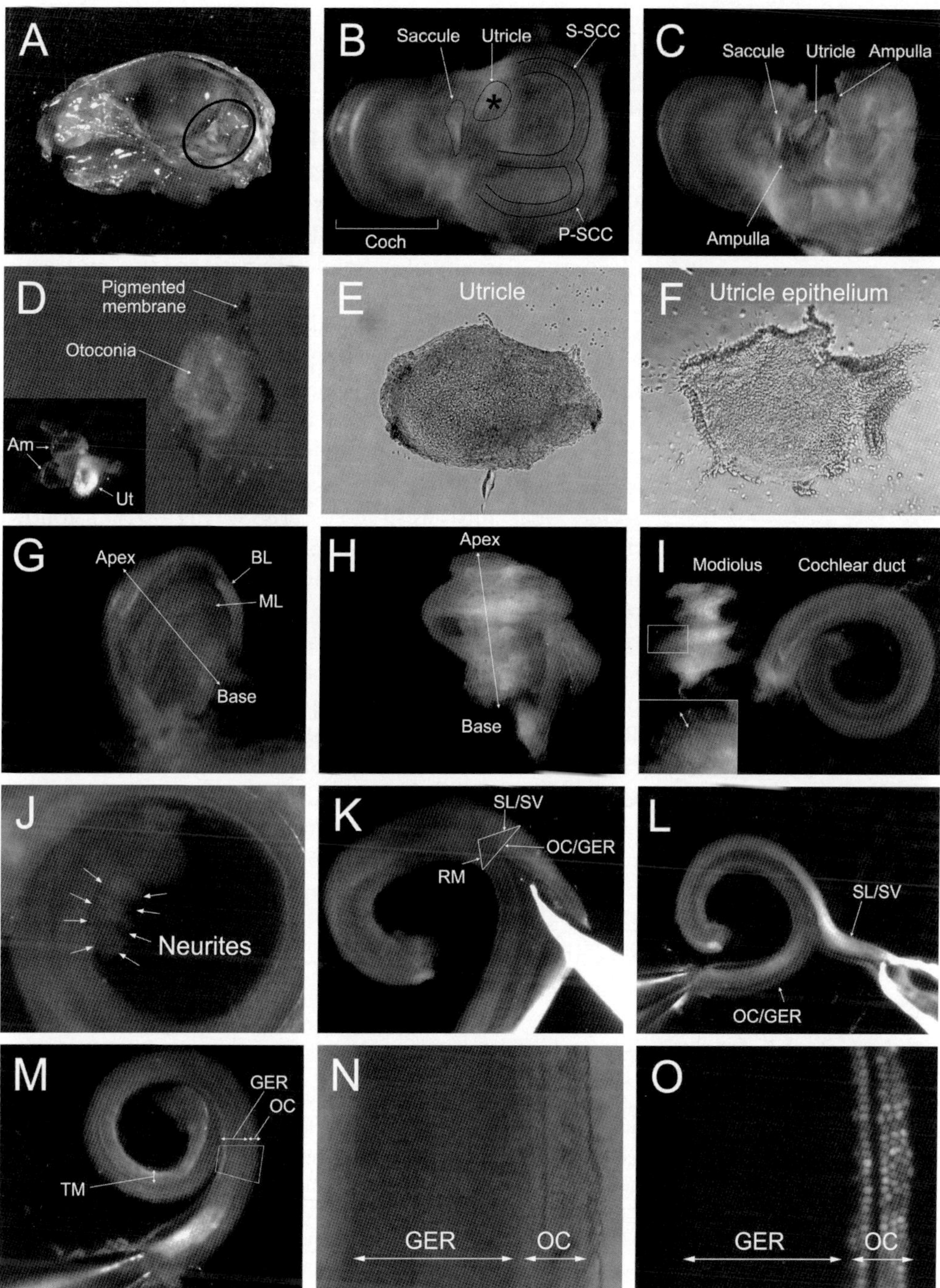

Fig. 9.1. Dissection of utricle, cochlear duct, and modiolus. (**A**) Right side of a bisected skull of a 1-day-old (P1) mouse after removal of the brain. Circled area shows the petrous (temporal) bone containing the inner ear organs to be dissected. (**B**) The inner ear after removal of the bulla and surrounding tissue. The asterisk (*) indicates the cartilage overlying the utricle. Coch = cochlear part, S-SCC = superior semicircular canal, and P-SCC = posterior semicircular canal. (**C**) The utricle is exposed by fenestration of the overlying cartilaginous plate. (**D**) Dissected utricle. The pigmented membrane is pulled aside and mostly removed to expose the epithelium covered with white otoconia. The inset picture shows the utricle (Ut) with two ampullae (Am), which belong to the superior SCC and horizontal SCC. (**E**) The utricle after complete

a short (30 sec) rinse with ice-cold DMEM/F-12 containing 5% fetal bovine serum to stop the tryptic digestion (*see* **Note 6**).

4. For dissection of the cochlear organs, remove the otic bulla to visualize the otic capsule. Poke a small hole in the basal part, and detach the membranous labyrinth from the otic capsule by inserting a forceps' tip between the capsule and the membranous labyrinth (**Fig. 9.1G**). Remove the cartilaginous otic capsule completely with forceps and expose the membranous labyrinth of the cochlea (**Fig. 9.1H**). Grab the most basal part of the cochlear duct (containing the scala media), and peel it away starting at the base around the modiolus until you reach the apex. This will result in a preparation of the cochlear duct that contains the OC, spiral ligament (with the stria vascularis), and Reissner's membrane (*see* **Note 7**). Transfer the dissected duct into fresh cold HBSS (**Fig. 9.1I**) followed by separating the OC and greater epithelial ridge (GER) from Reissner's membrane (**Fig. 9.1K**) and the spiral ligament (with the stria vascularis) (**Fig. 9.1L**). The OC/GER region and Reissner's membrane/spiral ligament appear like two parallel ribbons adjacent to each other that can be pulled apart (**Fig. 9.1L**). Separate the OC and GER from Reissner's membrane (**Fig. 9.1K**) and the spiral ligament (with the stria vascularis) (**Fig. 9.1L**). The OC and the GER (**Fig. 9.1M–O**) (*see* **Note 8**) will be used for isolation of cochlear sphere-forming stem cells. The residual modiolus (**Fig. 9.1I**) contains the spiral ganglion and can be further cleaned by cutting off the basal part. Prepare extra Petri dishes with HBSS on ice in advance to ensure that there is an ample supply of cold HBSS and sterile dishes.

Fig. 9.1. (continued) removal of the pigmented membrane and otoconia. (**F**) The utricle epithelium obtained from the same utricle as in *E* after thermolysin treatment. (**G**) The bony labyrinth (BL) of the cochlea is partially removed to expose the membranous labyrinth (ML) of the cochlea. (**H**) Side view of the whole membranous labyrinth of the cochlea. (**I**) The cochlear duct is peeled off from the modiolus. The tiny protrusions (arrow in the inset) around the modiolus are the neuronal fibers that innervated the hair cells. Note that there should be no neuronal tissue (neurites or cellular material) visibly attached to the duct. (**J**) Example of an unacceptable cochlear duct preparation. The arrows labeled "Neurites" indicate contaminating neuronal tissue attached to the duct. (**K**) Reissner's membrane is cut with forceps from the most basal to the apical part to open the cochlear duct. The triangle illustrates the cross-section of the cochlear duct. RM = Reissner's membrane, SL/SV = spiral ligament with stria vascularis, and OC/GER = the organ of Corti (OC) with the greater epithelial ridge (GER). (**L**) The spiral ligament and stria vascularis (SL/SV) are removed from the OC by carefully tearing the tissues apart. (**M**) View of the OC with the GER and tectorial membrane (TM). (**N**) A higher magnified image of the squared area in (**M**). OC indicates the organ of Corti and GER labels the greater epithelial ridge. (**O**) An epifluorescent image of (**N**) in which the nuclei of cochlear hair cells can be identified by their bright nuclear fluorescence. This nuclear green fluorescence of hair cells is a distinct feature of the Math1-nGFP transgenic mouse *(16)* used in the dissection shown.

3.1.3. Tissue Dissociation and Cell Culture

Dissected tissues are dissociated by enzymatic digestion and mechanical trituration. Greiner six-well cell suspension culture plates are suitable for production of floating colonies because their non-stick surface does not allow adherent cell growth.

1. Using forceps, transfer the dissected organs into individual drops of 50 μL DPBS in a six-well suspension culture plate (*see* **Note 9**).
2. Add 50 μL of pre-warmed 0.25% Trypsin/EDTA to each of the drops, and incubate at 37 °C for 5 min in a CO_2 incubator.
3. Add 50 μL of trypsin inhibitor/DNaseI cocktail to each of the drops.
4. Add 50 μL of sphere culture medium (this step is to increase the volume to 200 μL).
5. Triturate the organs by pipetting up and down 30–40 times without generating bubbles (*see* **Note 10**).
6. After one round of trituration, inspect the cells with an inverted microscope to assess the dissociation. The best result is attained if there are many single cells with some residual cell clumps of 2–5 cells. Do not attempt to dissociate these remaining clumps completely because you will lose many of the already dissociated single cells. If you do not have achieved good dissociation, triturate again and monitor the process microscopically (*see* **Note 11**).
7. Transfer 200 μL of the cell suspension into a 70 μm cell strainer placed into a fresh well of a six-well suspension plate. Take 900 μL of sphere culture medium with a 1,000 μL pipette, wash out the remaining cells on the first plate with the medium, and transfer this medium through the cell strainer. Repeat with a second batch of 900 μL. The resulting final volume of the cell suspension in the new six-well plate is 2 mL (200 + 900 + 900 μL).
8. Culture the cell suspension in a CO_2 incubator for 5–7 days. Many spheres will form in these conditions (**Fig. 9.2**) (*see* **Note 12**).

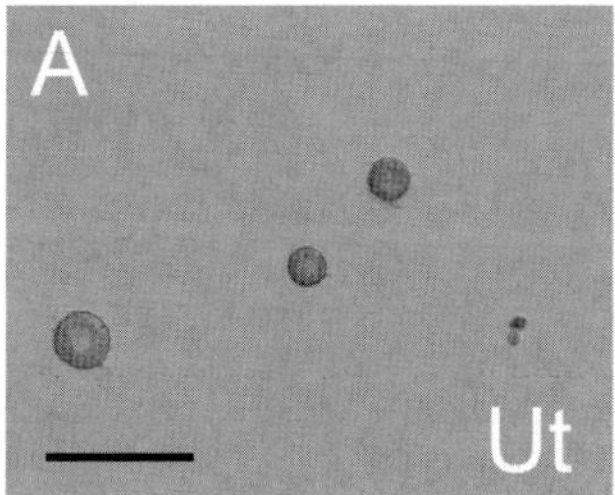

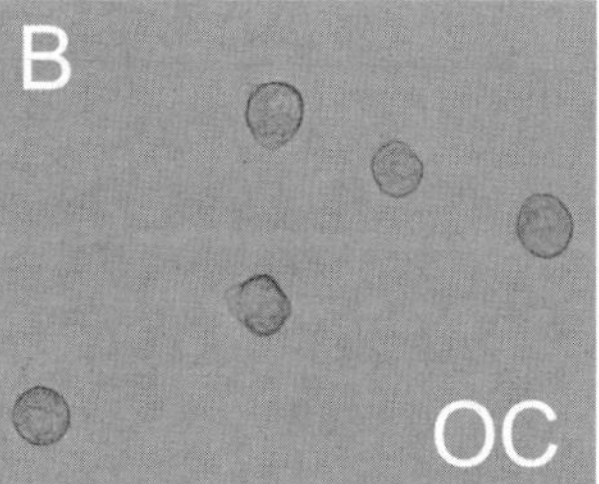

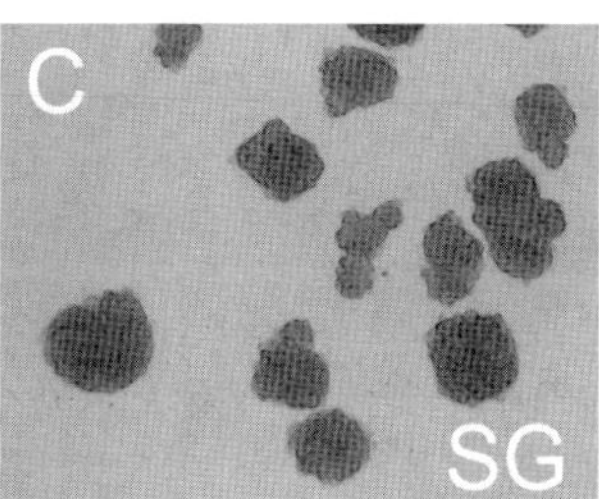

Fig. 9.2. Morphology of spheres isolated from various inner ear tissues. Spheres (**A**) from the utricle (Ut), (**B**) from the cochlear epithelium (OC), and (**C**) from the spiral ganglion (SG). Scale bar = 200 μm.

3.2. Passaging of Spheres

3.2.1. Passaging Spheres from the Utricle and Organ of Corti

1. Use an inverted microscope placed into a tissue culture hood to identify spheres (**Fig. 9.2**). Pick batches of 50 spheres from the suspension cultures with a P-200 Pipetman®. Place each batch of 50 spheres into a 50 μL drop in a six-well suspension plate.
2. Add 50 μL of pre-warmed 0.25% Trypsin/EDTA into the drops, and incubate at 37 °C for 5 min in a cell culture incubator.
3. Add 50 μL of trypsin inhibitor/DNaseI cocktail to each of the drops.
4. Add 50 μL of sphere culture medium, and gently triturate the spheres by pipetting up and down 40–50 times using Eppendorf pipette tips (20–300 μL) without generating bubbles.
5. Assess the dissociation with a microscope. The best results are attained if there are many single cells with some residual cell clumps of 2–10 cells. Do not attempt to dissociate the spheres completely. If cells are not dissociated, additional trituration is needed followed by microscopic assessment. In some instances, large spheres may be more easily dissociated mechanically with sharpened tungsten needles (*see* **Note 13**).
6. Add 1,800 mL of cell culture medium and carefully redistribute the cells in the suspension by gently pipetting up and down.
7. For expansion, split equally and fill each well with medium for a total volume of 2 mL.
8. Culture the cell suspension in a CO_2 incubator for 5–7 days until next-generation spheres have grown (**Fig. 9.2**).

3.2.2. Passaging Spiral Ganglion Spheres

1. Transfer the complete spiral ganglion suspension culture with all spheres that formed after 5–7 days into a 15-mL conical tube. The best way to do this is to hold the plate at an angle and transfer 1 mL of the suspension at a time with a P-1000 Pipetman®. Wash the plate with additional 2 mL of culture medium while still holding it slanted. In the conical tube, you will now have a total volume of 4 mL of diluted suspension culture.
2. Add 4 mL of Accumax solution to the conical tube (1:1 dilution) and mix by gently shaking the tube.
3. Place the conical tube for 5 min into a 37 °C water bath.
4. Centrifuge the tube at 200 g for 5 min.
5. Aspirate the supernatant and leave the pellet at the bottom of the tube in ~ 50 μL of medium.
6. Transfer the remaining medium with the pellet into a fresh suspension culture six-well plate. Wash the tube with 150 μL

of sphere culture medium and add it to the 50 μL drop into a six-well plate.

7. Gently triturate the cells with a fire-polished glass pipette 10–20 times (*see* **Note 3**). Check the status of dissociation under the microscope. If there are too many clumps, continue to triturate. Do not generate bubbles. After trituration, most of the cells will be dissociated.
8. Dilute the suspension with 1,800 μL of sphere culture medium and mix gently.
9. For expansion, split 1:2 or 1:3 and fill each well with sphere culture medium for a total volume of 2 mL.
10. Incubate the culture at 37° and 5% CO_2 for 5–7 days until next-generation spheres have grown.

3.3. Differentiation

1. Utricle and OC-derived spheres attach to poly-L-ornithine and fibronectin-coated surfaces. Prepare poly-L-ornithine-coated 35 mm four-well plates (or eight-well Lab-Tek II chamber slides) in advance by pipetting enough poly-L-ornithine solution into each well to completely cover the surface. Let the plates sit over night at room temperature and wash twice with sterile water. Aspirate the water and let the plates dry in a biosafety cabinet or hood. The dry poly-L-ornithine-coated plates can be stored for several months in a tightly sealed container at 4 °C.
2. Fibronectin coating is done 2–4 h before spheres are attached for differentiation. Prepare a fibronectin working solution of 5 μg/mL by diluting the stock solution 1:200. Cover the poly-L-ornithine-coated surfaces with fibronectin working solution and let the plates sit for 2–4 h at room temperature. Remove the fibronectin solution and fill the plates with 100 μL (200 μL for eight-well Lab-Tek II chamber slide) of the differentiation medium.
3. Spiral ganglion–derived spheres are plated onto gelatin-coated surfaces. For gelatin coating of 35 mm four-well plates (or eight-well Lab-Tek II chamber slide), add 0.2% gelatin solution into the plates to completely cover their surface and incubate for 30 min at room temperature. Remove the gelatin solution and fill the plates with 100 μL (200 μL for eight-well Lab-Tek II chamber slide) of the differentiation medium.
4. Use an inverted microscope placed into a tissue culture hood to pick batches of 30–100 spheres from the suspension cultures with a P-20 Pipetman®. Attempt to capture a complete batch of spheres at once in 20 μL or less to reduce the amount of the sphere culture medium carried over into the differentiation culture.
5. Transfer the spheres into the medium on four-well plates (or eight-well Lab-Tek II chamber slide).

6. Exchange the whole medium the next day after inspecting the attachment of the spheres, which should form small islands of cells (*see* **Note 14**). Exchange 80% of the medium every 3–4 days. Culture the cells for 10–14 days to observe differentiated cells. (For spiral ganglion-derived spheres, *see* **Note 15.**)

3.4. Screening for Phenotypic Marker Expression

3.4.1. Immunofluorescence

1. After a differentiation period of at least 10–14 days in four-well plates or eight-well Lab-Tek II chamber slides, carefully aspirate the medium. Wash the cells in each well once with 100 μL of DPBS, add 100 μL of 4% paraformaldehyde solution, and fix the cells for 10 min at room temperature (*see* **Note 16**). Use 200 μL of each solution for eight-well Lab-Tek II chamber slides.
2. Wash the fixed cells twice with 100 μL DPBS and then incubate in PBT1 solution for 15–30 min at room temperature. PBT1 solution permeabilizes the cells' membranes and blocks unspecific antibody binding sites.
3. Incubate cells with the primary antibody diluted in PBT1 for 2 h at room temperature or overnight at 4 °C (*see* **Note 17** and **Table 9.1**).
4. Wash the cells twice for 5 min each with PBT1.
5. Wash the cells once for 5 min with PBT2.

Table 9.1
Selected Antibodies

Antigen	Details	Supplier/reference
Myosin VIIa	Guinea pig polyclonal	Oshima et al., 2007 *(7)*
Parvalbumin 3	Rabbit polyclonal	Heller et al., 2002 *(14)*
Espin	Rabbit polyclonal	Li et al., 2004 *(15)*
Phalloidin	A toxin from the death cap (Amanita phalloides) that binds to F-actin	Molecular probes; A12379 (FITC), A22283 (TRITC)
Math 1	Mouse monoclonal, IgG1	Developmental Studies Hybridoma Bank, Iowa City, IA; cat. no. Math1-s.
TuJ	Mouse monoclonal, IgG2a	Covance; cat. no. MMS-435P.
Nestin	Mouse monoclonal, clone Rat-401 (recognizes mouse, does not cross-react with human), IgG1	Developmental Studies Hybridoma Bank; cat. no. Rat-401.

6. Incubate cells with the fluorophore-conjugated secondary antibody in PBT2 for 2 h at room temperature. Shield from light (*see* **Note 18**).
7. Wash twice for 5 min each with PBT2.
8. Incubate the cells for 15 min with DAPI in PBS (2 μg/mL).
9. Wash twice for 5 min each with PBS.
10. Mount with DakoCytomation mounting medium. Cover with 10-mm round cover-slip (for four-well plates) or 24 × 60 mm cover glass (for Lab-Tek II chamber slides), and carefully aspirate excess mounting medium with a pipette tip connected to a vacuum source with liquid trap.
11. Analyze with a fluorescence microscope (Zeiss AxioImager or similar) (**Fig. 9.3**).

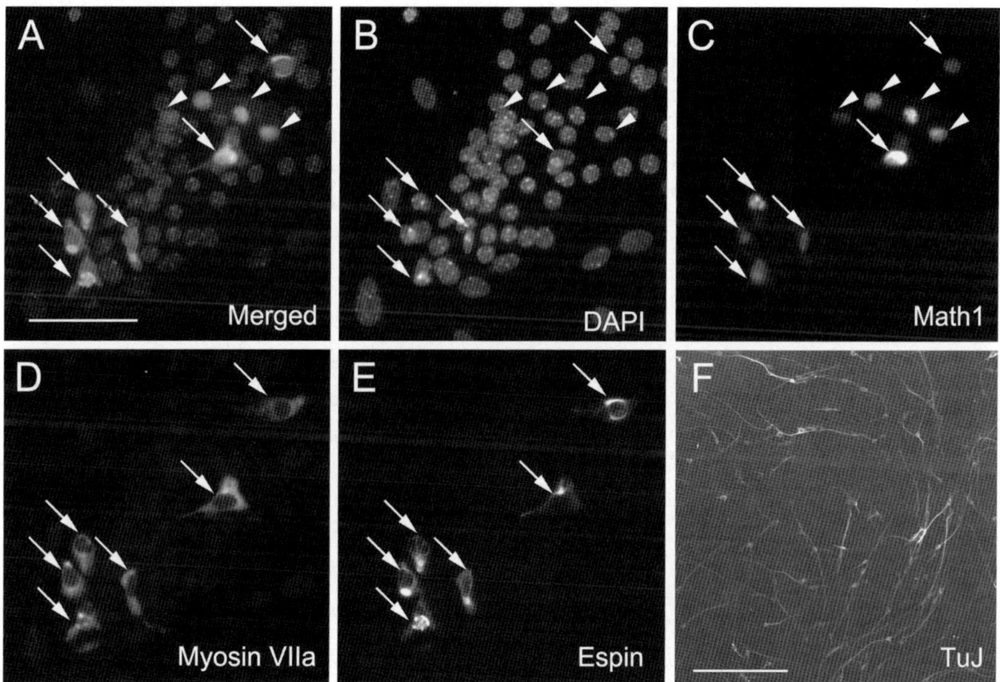

Fig. 9.3. Differentiation of spheres derived from the utricle and spiral ganglion. (**A–E**) After a 2-week differentiation period, hair cell–like cells appear in some of the cells derived from utricle-derived spheres. (**A**) Merged image of (**B–E**). (**B**) All cell nuclei are visualized with DAPI. (**C**) Cells expressing nuclear green fluorescence as seen by bright spots that overlap with the nuclear staining shown in (**B**) are indicative of *Atoh1* promoter activity in the Math1/nGFP mouse used for sphere preparation *(16)* (Math1). (**D**) Myosin VIIA immunoreactivity, and (**E**) immunostaining for espin expression. Arrows in (**A–E**) point to triple hair cell marker–positive cells expressing nGFP (Math-1), myosin VIIA, and espin. Arrowheads point to cells that are only positive for nGFP (Math-1) and negative for the other two markers, suggesting that these cells have not yet upregulated myosin VIIA and espin. (**F**) Cells that differentiated from spiral ganglion-derived spheres. After the 2-week differentiation period, (TuJ)-positive neuronal marker is detectable in cells with distinct neural morphology. Scale bar = 50 μm in (**A–E**) and 500 μm in (**F**).

3.4.2. Comparative (Semi-Quantitative) RT-PCR

3.4.2.1. Total RNA Preparation

Total RNA is isolated from inner ear tissue samples, undifferentiated spheres, and differentiated sphere-derived cells using silica-gel-based membrane spin columns (Qiagen RNeasy Mini kit) (*see* **Note 19**). For the RT-PCR assay, at least 300–1,000 spheres should be analyzed directly or cultured for differentiation in six-well tissue culture plates (as described in **steps 1–6, Section 3.3**), since sufficient amount of RNA cannot be obtained in our hands from smaller samples.

1. For undifferentiated spheres: Collect undifferentiated spheres by centrifugation. Transfer 2 mL of the sphere suspension from a six-well plate into a 15-mL conical tube, wash the well with 2 mL of the sphere culture medium and transfer it to the same conical tube for a total of 4 mL. Spin for 5 min at 200 g and carefully aspirate the supernatant. Add 4 mL of buffer RLT-β-ME to the cell pellet and mix by pipetting up and down.
2. For differentiated cells: Carefully aspirate medium from differentiated cells grown on a six-well tissue culture plate and wash cells once with 10 mL PBS. Add 600 μL of buffer RLT-β-ME to the cell culture plate for direct lysis of the cells. Collect the lysate, transfer it into a microcentrifuge tube, and mix by pipetting up and down.
3. Homogenize by passing the lysates 15–20 times through a 20-gauge needle fitted to a sterile, plastic 1-mL syringe. Avoid bubbles.
4. Add 600 μL of 70% ethanol to the homogenized lysates and mix well by pipetting.
5. Apply up to 700 μL of the sample to an RNeasy Mini column placed in a 2-mL centrifuge tube and spin for 15 sec at 8,000 g. Discard the flow-through and apply the remaining 500-μL sample; spin again for 15 sec at 8,000 g and discard the flow-through.
6. Wash the column by applying 700 μL of buffer RW1, spin for 15 sec at 8,000 g, and discard the flow-through.
7. Add 500 μL of buffer RPE to the column, spin for 15 sec at 8,000 g, and discard the flow-through. Repeat this step once.
8. Transfer the column to a fresh 15-mL centrifuge tube and spin for 5 min at 8,000 g to dry the column.
9. Elute into a fresh 1.5-mL collection tube by adding 40 μL of RNase-free water directly onto the silica-gel membrane of the column, let it sit for 1 min, and then spin for 1 min at 8,000 g. Keep the RNA solution on ice if used immediately for reverse transcription or store frozen at −80 °C for future use.

10. Measure the absorption at 260 nm with a spectrophotometer and calculate the RNA concentration assuming absorption of 1 at 260 nm equals 40 ng/μL.

3.4.2.2. Reverse Transcription

Screening for up- or downregulation of mRNAs by comparative RT-PCR works only when equal amounts of total RNA are used from the beginning. Good results have been achieved when starting with 5–10 μg aliquots of total RNA with equal concentrations isolated from progenitor cells or differentiated cells. cDNA generated by reverse transcription can be used directly for PCR amplification or can be stored at −20 °C for future use.

1. Mix the following components in a nuclease-free microcentrifuge tube: 1 μL of oligo(dT)$_{12-18}$, 1 μL of dNTP mix, 4 μL of RNA solution (obtained from **step 9**, **Section 3.4.2.1**), and 6 μL of RNase-free water (*see* **Note 20**).
2. Heat at 65 °C for 5 min, quickly chill on ice, and spin briefly to collect the liquid at the bottom of the tube.
3. Place the tube on ice; add 4 μL of 5X First-Strand Buffer, 1 μL of 0.1 *M* DTT solution, and 1 μL of RNaseOUT; mix; and spin briefly in a microcentrifuge to collect the liquid at the bottom of the tube.
4. Incubate at 42 °C for 2 min and then add 1 μL (200 U) of SuperScript II reverse transcriptase to the mixture.
5. Incubate at 42 °C for 50 min.
6. Inactivate the enzyme by incubation at 70 °C for 15 min.
7. The resulting cDNA can be immediately used as a template for PCR; otherwise store frozen at −20 °C.

3.4.2.3. PCR

Oligonucleotide primers for PCR should be carefully selected to discriminate between cDNA and genomic DNA, when possible. Using individual primers specific for different exons is a simple way to achieve this; amplification from genomic DNA with these primers will create a larger reaction product that often will not be efficiently amplified. **Table 9.2** lists several primers that have been successfully used to compare marker gene expression of inner ear tissue, selected inner ear progenitor cells, and differentiated inner ear cell types *(2, 7, 8, 12, 13)*. The following protocol is for one comparative (semi-quantitative) PCR using three samples, such as cDNA from inner ear tissue, selected progenitors, and differentiated cells.

1. Thaw cDNA aliquots for the samples to be tested; keep the tubes on ice.
2. Prepare master mix for Gapdh (MM1) in a sterile 1.5-mL reaction tube on ice: 107.5 μL of sterile Milli-Q water, 17.5 μL of 10X Taq reaction buffer, 17.5 μL of DMSO, 7.0 μL of 10 m*M* dNTP mix, 7.0 μL of *Gapdh* forward primer, and 7.0 μL of *Gapdh* reverse primer (*see* **Note 21**).

Table 9.2
PCR Primer Pairs

Marker	Forward Primer (5′-3′)	Reverse Primer (5′-3′)	Product length
Gapdh	AACGGGAAGCCCATCACC	CAGCCTTGGCAGCACCAG	442 bp
Otx2	CCATGACCTATACTCAGGCTTCAGG	GAAGCTCCATATCCCTGGGTGGAAAG	211 bp
Nestin	GCCGAGCTGGAGCGCGAGTTAGAG	GCAAGGGGGAAGAGAAGGATGTCG	694 bp
Pax2	CCAAAGTGGTGGACAAGATTGCC	GGATAGGAAGGACGCTCAAAGAC	544 bp
BMP4	TGGTAACCGAATGCTGATGGTCG	GTCCAGTAGTCGTGTGATGAGGTG	598 bp
BMP7	TGGGCTTCTGAGGAGGGCTGGTTG	TGGCGTGGTTGGTGGCGTTCAT	484 bp
Jagged-1	CAGAATGACGCCTCCTGTCG	TGCAGCTGTCAATCACTTCG	361 bp
p27^{Kip1}	CTGGAGCGGATGGACGCCAGAC	CGTCTGCTCCACAGTGCCAGC	525 bp
Math1	AGATCTACATCAACGCTCTGTC	ACTGGCCTCATCAGAGTCACTG	449 bp
Myosin VIIA	CTCCCTCTACATCGCTCTGTTCG	AAGCACCTGCTCCTGCTCGTCCACG	628 bp
Espin	CAGCCTGAGTCACCGCAGCCTC	TGACCTGTCGCTGCCAGGGCGCG	475 bp
Brn3.1	GCCATGCGCCGAGTTTGTC	ATGGCGCCTAGATGATGC	368 bp
Musashi1	ACCTACGCCAGCCGGAGTTACAC	CTGGGGCGCTCCTGCTACCTC	444 bp
Neurofilament M	GCACATCACGGTAGAGCGCAAAG	TCGTGCGCGCACTGGAATGCG	450 bp
Peripherin	GTGAGCGTAGAGAGCCAGCAGG	TCGAAGCTCTTCCTCCAGCCGT	474 bp
GluR2	TAAAATGTGGACTTATATGAGGAGTG	CTCTCGATGCCATATACGTTGTAAC	573 bp
GluR3	GAAAATGTGGTCTTACATGAAATCCG	TGAGTGTTGGTGGCAGGAGCA	525 bp
GluR4	ATGAGGATTATTTGCAGGCAG	TCAATGAAGGTCTTAGCTGAAG	415 bp

(continued)

Table 9.2 (continued)

GFAP	CCTCCGCCAAGCCAAACACGAA	ACCATCCCGCATCTCCACAGTC	433 bp
BDNF	CGCAAACATGTCTATGAGGGTTC	TAGTAAGGGCCCGAACATACGAT	302 bp
NT3	TAGAACCTCACCACGGAGGAAAC	AGGCACACACACAGGAAGTGTCT	359 bp
TrkB	GTACTGAGCCTTCTCCAGGCATC	CGTCAGGATCAGGTCAGACAAGT	305 bp
TrkC	TACTACAGGGTGGGAGGACACAC	TTTAGGGCAGACTCTGGGTCTCT	225 bp
$p75^{NTR}$	CCGATGCTCCTATGGCTACTACC	CTATGAGGTCTCGCTCTGGAGGT	353 bp

3. Mix by vortexing, spin at 8,000 g for 1 sec to collect the liquid to the bottom of the tube, and keep on ice.
4. Prepare master mixes for each specific primer pair (e.g., MM2, MM3, etc.) in a sterile 1.5-mL reaction tube on ice: 107.5 μL of sterile Milli-Q water, 17.5 μL of 10X Taq reaction buffer, 17.5 μL of DMSO, 7.0 μL of 10 m*M* dNTP mix, 7.0 μL of specific forward primer, and 7.0 μL of specific reverse primer (*see* **Note 21**).
5. Mix by vortexing, spin at 8,000 g for 1 sec to collect the liquid at the bottom of the tube, and keep on ice.
6. Distribute three 47.5-μL aliquots of MM1 and three 47.5-μL aliquots each of MM2, MM3, and so on into thin-wall PCR reaction tubes. Keep the tubes on ice.
7. Add 2.5 μL of each specific cDNA (*see* **Section 3.4.2.2, step 7**) to one MM1 aliquot and to each one of the MM2, MM3, and so on aliquot. For example, 2.5 μL of tissue cDNA is added to MM1 and 2.5 μL of progenitor cell cDNA is added to MM2 and so forth. Keep the tubes on ice.
8. Place the tubes directly into the preheated thermal cycler (idling at 94 °C) and execute the following program (*see* **Note 22**): denaturation at 94 °C for 60 sec; then *n* cycles of the following: denaturation at 94 °C for 30 sec, annealing at 58 °C for 30 sec, and extension at 72 °C for 60 sec.
   The number of cycles (*n*) has to be predetermined in pilot experiments. With the instrumentation described here and the primers listed in **Table 9.2**, good results were obtained when using 22 cycles for *Otx2*; 25 cycles for *Gapdh*; 30 cycles for *myosin VIIA*, *espin*, and *Brn3.1*; and 32 cycles for all other primer pairs (*see* **Note 23**).
9. Place samples on ice until all amplification reactions are completed.
10. Separate 15-μL aliquots of the reaction products on 2% agarose gels and compare intensity of the bands.

4. Notes

1. All cell culture is done in a dedicated room, separated from the main laboratory by a closed door. Traffic in and out of the cell culture room has to be minimized. All supplies and instruments have to be dedicated only for cell culture use and should never be carried into the main laboratory and used for other experiments. To avoid contamination, all surfaces are wiped before and after use with 70% ethanol. Nothing inside the room should be touched without wearing gloves. Sterile technique and common sense are usually effective means

to avoid loss of cell lines due to contamination. It is recommended that sterile plasticware be used instead of glassware.

2. 50 μL of trypsin inhibitor is sufficient to completely deactivate 50 μL of 0.25% trypsin/EDTA solution. DnaseI helps to reduce viscosity caused by DNA released from damaged cells during trypsinization and trituration. We have successfully used inhibitor and DNaseI from Worthington, Lakewood, NJ (cat. nos. LS003570 and LS002139).
3. Attach a rubber bulb at the wide end of an autoclaved Pasteur pipette. Rotate the pipette tip in a flame (Bunsen burner) to fire polish it for a few seconds. From time to time, squeeze the rubber bulb to check for increasing resistance of airflow through the polished pipette tip. Airflow should be slightly but noticeably restricted. The edges of the pipettes should be nice and rounded. Freshly prepared fire polished pipettes are sterile and can be used directly after a 2-min cooling period.
4. ß-ME is toxic; dispense in a fume hood and wear protective clothing.
5. Alternatively, leave the ampullae in contact with the utricle and take out the utricle with 2 ampullae (**Fig. 9.1D** inset). Grabbing one of the ampullae prevents the utricle from being damaged by forceps.
6. It is not absolutely necessary to use pure sensory epithelium (**Fig. 9.1F**) for sphere generation. Spheres usually form from cells present in the sensory epithelia and not from the underlying non-sensory tissue *(2)*.
7. Make sure that there is no residue of neuronal fibers attached to the cochlear duct (**Fig. 9.1J**). Clean separation of the duct and the modiolus is also important for spiral ganglion preparation.
8. The OC/GER region may contain adjacent mesenchymal tissue and some lesser epithelial ridge (LER) cells. Sphere-forming cells from the OC/GER region are likely to contain the proliferative cell populations previously reported to reside in the GER and LER *(4, 11)*.
9. Attempt to reduce the volume of HBSS, which contains calcium and magnesium ions, carried over into the DPBS drop since those ions will disturb the enzymatic activity of trypsin by obscuring the peptide bonds on which trypsin acts. A small amount of the ions will be chelated by the EDTA that is part of the trypsin solution. If this is not possible, rinse the organs briefly in DPBS, which is calcium and magnesium free.
10. An Eppendorf pipette tip (cat. no. 022491245, 20–300 μL, Eppendorf) appears to be well suited for this specific kind of trituration. Fire-polished and silanized Pasteur pipettes can be used as a good substitute.

11. Avoid bubbles because you will lose the majority of sphere-forming cells. Setting the dial of a 200-μL Pipetman® at 150–180 μL will help.
12. If you want to ensure clonal sphere formation, you need to plate the cells at a much lower density (10 cells/cm^2). The most stringent generation of clonal spheres can only be achieved with flow-cytometric placement of single cells into individual wells of 96-well non-stick plates. Keep in mind that 0.1–0.2% or less of the cells have the capacity for sphere formation.
13. Tungsten needles can be sharpened by electrolytical erosion in a beaker of 1 *M* NaOH or KOH. The needle is connected to the positive terminal of a 9 V battery using a crocodile clip. A copper or carbon cathode attached to the negative terminal of the battery is immersed in a small beaker containing 1 *M* NaOH (or KOH). Slowly move the needle up and down in the electrolyte. Resharpening the needle can be done by repeating the same process.
14. Most of the spheres will be attached within 10 h.
15. Leukemia inhibitory factor (LIF) promotes neuronal differentiation from spiral ganglion–derived spheres. If you want to make use of this ability, plate the spheres and culture them in differentiation medium containing 1 ng/mL of LIF (recombinant rat LIF, cat. no. LIF3005 Chemicon, or similar product) for 10–14 days. The number of neurons is significantly increased up to sevenfold in a dose-dependent manner by the addition of LIF, reaching a maximum at 1 ng/mL *(12)*.
16. Throughout the procedure, do not allow the cells to dry. If you immunostain more than a few wells, consider working in staggered batches of four wells.
17. Dilute primary antibody according to the supplier's suggestion. If you incubate overnight, place the plates into a larger container and add a wet piece of tissue. Close the container tightly. This humidified chamber will help prevent the wells from drying.
18. Dilute secondary antibody according to the supplier's recommendation.
19. Because of the high stability and high efficacy of Rnases, it is necessary to create an RNase-free environment by wiping the working area with RNaseZap. In addition, it is necessary to wear gloves while handling reagents and samples and to use RNase-free sterile, disposable plasticware for all experimental steps. For details on the abbreviated protocol described in this section, refer to the RNeasy kit protocol booklet (provided with the Qiagen kit).
20. If the two RNA samples are not equally concentrated, use different volumes and reduce the amount of RNase-free water

accordingly. We have used as little as 1 ng total RNA for successful RT-PCR experiments.

21. Custom synthesized oligonucleotide primers are usually shipped lyophilized. Resuspend primers in sterile Milli-Q water at a concentration of 100 p*M*.
22. It is important to establish specific PCR conditions (cycling parameters) for each of the different products so that they are optimized to generate products at the linear portion of the product accumulation curve. These parameters depend, in part, on the thermal cycler used in the experiment. It is recommended that a series of pilot experiments be conducted for each primer pair with all samples that will be compared (e.g., cDNA from tissue, selected progenitors, and differentiated cells). Cycling parameters need to be selected based on the sample that produces the highest amount of amplification product.
23. It is convenient to program the thermal cycler to cool the reactions at 4 °C once the amplification is done. Because of the different cycle numbers of the different reactions, close monitoring of the reaction is required to ensure that samples are removed from the thermal cycler at the appropriate time points. The thermal cycler has to be programmed to run the program with the most cycles. Reactions that undergo less cycles have to be removed at appropriate time points at the end of the 72 °C extension period.

Acknowledgments

The authors would like to thank the members of their research group for critically reading this manuscript. This work was supported by a McKnight Endowment Fund for Neuroscience Brain Disorders Award and grant DC006167 from the National Institutes of Health.

References

1. Malgrange, B., Belachew, S., Thiry, M., et al. (2002) Proliferative generation of mammalian auditory hair cells in culture. *Mech. Dev.* **112**, 79–88.
2. Li, H., Liu, H., and Heller, S. (2003) Pluripotent stem cells from the adult mouse inner ear. *Nat. Med.* **9**, 1293–1299.
3. Rask-Andersen, H., Bostrom, M., Gerdin, B., et al. (2005) Regeneration of human auditory nerve. In vitro/in video demonstration of neural progenitor cells in adult human and guinea pig spiral ganglion. *Hear Res.* **203**, 180–191.
4. Zhai, S., Shi, L., Wang, B. E., et al. (2005) Isolation and culture of hair cell progenitors from postnatal rat cochleae. *J. Neurobiol.* **65**, 282–293.
5. Wang, Z., Jiang, H., Yan, Y., et al. (2006) Characterization of proliferating cells from newborn mouse cochleae. *Neuroreport* **17**, 767–771.
6. Lou, X., Zhang, Y., and Yuan, C. (2007) Multipotent stem cells from the young rat inner ear. *Neurosci. Lett.* **216**, 28–33.
7. Oshima, K., Grimm, C. M., Corrales, C. E., et al. (2007) Differential distribution of stem

cells in the auditory and vestibular organs of the inner ear. *J. Assoc. Res. Otolaryngol.* **8**, 18–31.
8. Senn, P., Oshima, K., Teo, D., Grimm, C., and Heller, S. (2007) Robust postmortem survival of murine vestibular and cochlear stem cells. *J. Assoc. Res. Otolaryngol.* **8**, 194–204.
9. Savary, E., Hugnot, J. P., Chassigneux, Y., et al. (2007) Distinct population of hair cell progenitors can be isolated from the postnatal mouse cochlea using side population analysis. *Stem Cells* **25**, 332–339.
10. Yerukhimovich, M. V., Bai, L., Chen, D. H., Miller, R. H., and Alagramam, K. N. (2007) Identification and characterization of mouse cochlear stem cells. *Dev. Neurosci.* **29**, 251–260.
11. Zhang, Y., Zhai, S. Q., Shou, J., et al. (2007) Isolation, growth and differentiation of hair cell progenitors from the newborn rat cochlear greater epithelial ridge. *J. Neurosci. Methods* **164**, 271–279.
12. Oshima, K., Teo, D. T., Senn, P., Starlinger, V., and Heller, S. (2007) LIF promotes neurogenesis and maintains neural precursors in cell populations derived from spiral ganglion stem cells. *BMC Dev. Biol.* 7, 112.
13. Li, H., Roblin, G., Liu, H., and Heller, S. (2003) Generation of hair cells by stepwise differentiation of embryonic stem cells. *Proc. Natl. Acad. Sci. U. S. A.* **100**, 13495–13500.
14. Heller, S., Bell, A. M., Denis, C. S., Choe, Y., and Hudspeth, A. J. (2002) Parvalbumin 3 is an abundant Ca2+ buffer in hair cells. *J. Assoc. Res. Otolaryngol.* **3**, 488–498.
15. Li, H., Liu, H., Balt, S., Mann, S., Corrales, C. E., and Heller, S. (2004) Correlation of expression of the actin filament-bundling protein espin with stereociliary bundle formation in the developing inner ear. *J. Comp. Neurol.* **468**, 125–134.
16. Lumpkin, E. A., Collisson, T., Parab, P., et al. (2003) Math1-driven GFP expression in the developing nervous system of transgenic mice. *Gene Expr. Patterns* **3**, 389–395.

Chapter 10

Molecular Biology of Vestibular Schwannomas

Long-Sheng Chang and D. Bradley Welling

Abstract

Recent advances in molecular biology have led to a better understanding of the etiology of vestibular schwannomas. The underlying purpose of vestibular schwannoma research is the development of new treatment options; however, such options have not yet been established. A fundamental understanding of the underlying molecular events leading to tumor formation began when mutations in the *neurofibromatosis type 2 (NF2)* tumor suppressor gene were identified in vestibular schwannomas. The clinical characteristics of vestibular schwannomas and neurofibromatosis type 2 (NF2) syndromes have both been related to alterations in the *NF2* gene. Genetic screening for NF2 is now available. When utilized with clinical screening, such as magnetic resonance imaging (MRI), conventional audiometry, and auditory brainstem response (ABR), the early detection of NF2 can be made, which consequently makes a significant difference in the ability to successfully treat vestibular schwannomas. Additionally, the signaling pathways affected by merlin, the product of the *NF2* gene, are becoming better understood. *Nf2*-transgenic and knockout mice as well as vestibular schwannoma xenograft models are now ready for novel therapeutic testing. Hopefully, better treatment options will be forthcoming soon.

Key words: Vestibular schwannoma, Neurofibromatosis type 2 (NF2), *Neurofibromatosis 2 (NF2)* gene, merlin, magnetic resonance imaging (MRI), xenograft, Schwann cells, mutation, severe combined immunodeficiency (SCID) mice.

1. Introduction

Human vestibular schwannomas are Schwann cell tumors originating from the superior or inferior vestibular branches of cranial nerve VIII *(1, 2)*. Although histologically benign, vestibular schwannomas can compress the brainstem leading to hydrocephalus, herniation, and death. Most commonly, however, they are associated with hearing loss, tinnitus, imbalance, facial paralysis, and facial paresthesias. Vestibular schwannomas may occur either as sporadic unilateral tumors or as bilateral tumors.

Bernd Sokolowski (ed.), *Auditory and Vestibular Research: Methods and Protocols, vol. 493*

DOI 10.1007/978-1-59745-523-7_10 Springerprotocols.com

Unilateral tumors are most commonly seen and constitute 95% of all vestibular schwannomas. The development of bilateral vestibular schwannomas is the hallmark of neurofibromatosis type 2 (NF2; OMIM **#101000**), an autosomal dominant disease that is highly penetrant. Another group of unilateral, sporadic schwannomas that need great attention is the cystic schwannoma because these tumors are more aggressive and often invade the surrounding cranial nerves, splaying them throughout the tumors.

Molecular genetic analysis reveals that mutations inactivating both alleles of the *neurofibromatosis type 2 (NF2)* tumor suppressor gene are responsible for the development of vestibular schwannomas *(3, 4)*. Patients with NF2 inherit one mutated copy of the *NF2* gene through autosomal dominant transmission. Inactivation of the second allele results in complete loss of *NF2* tumor suppressor function. In addition, mutations in the *NF2* gene have been identified in unilateral vestibular schwannomas including cystic schwannomas *(5–8)*. Located on chromosome 22q12, the *NF2* gene encodes a protein named merlin (*m*oesin-*e*zrin-*r*adixin-*l*ike prote*in*) *(4)* or schwannomin, a word derived from schwannoma, the most prevalent tumor seen in *NF2 (3)*. Merlin has been shown to play an important role in cell motility, cell adhesion, and cell proliferation *(9, 10)*. Because of the autosomal dominant transmission in *NF2* and the fact that early intervention is very important in clinical outcome, genetic testing for *NF2* mutation and clinical screening in at-risk patients has been advocated *(2)*. Early detection in *NF2* makes a significant difference in the ability to successfully treat vestibular schwannomas.

The gold standard for evaluating vestibular schwannomas in situ is T1 (longitudinal proton relaxation time)–weighted magnetic resonance imaging (MRI) with gadolinium enhancement *(11, 12)*. MRI also distinguishes clearly between the solid and cystic vestibular schwannomas. Cystic regions of the tumors are hyper-intense on T2-weighted images, and the cysts do not enhance with gadolinium administration *(1)*. The non-cystic component of the cystic tumors enhances with gadolinium in a manner similar to the unilateral and NF2-associated schwannomas. Currently, surgical resection and radiation are the choices of treatment for vestibular schwannomas. Although the effectiveness of these treatments is generally good, treatment-related morbidity continues to be problematic. With the recent discoveries on deregulated signaling pathways in vestibular schwannomas *(7, 13, 14)* and the availability of animal models including *NF2* transgenic and knockout mice *(15, 16)* as well as schwannoma xenografts in SCID mice *(17)*, novel therapeutics targeting these signaling pathways are ready for testing. Ultimately, effective drug therapies will be found, offering alternatives to the current treatment options.

2. Materials

2.1. Tissues and DNA Isolation

1. Solution of ammonium (10%) is prepared by diluting concentrated ammonium with dH_2O.
2. To purify genomic DNA, use the QIAamp® DNA Micro kit (Qiagen, Valencia, CA).
3. Dulbecco's Modified Eagle Medium (DMEM) with high glucose (Invitrogen, Carlsbad, CA) supplemented with 10% fetal bovine serum (FBS).
4. A 1-kb Plus DNA Ladder (Invitrogen) is commonly used as DNA molecular-weight markers. Other DNA-size markers can also be used.
5. For purification of DNA from cell culture and tissue, use the Gentra Puregene® DNA isolation kit (Qiagen).
6. The quality of DNA, RNA, proteins, and cells can be analyzed quickly using a micro-fluidics-based analyzer such as the Agilent 2100 Bioanalyzer (Agilent Technologies, Santa Clara, CA). An additional advantage of using a Bioanalyzer is the minimal sample consumption. Otherwise, extracted DNA can be checked by conventional electrophoresis using a 0.7% agarose gel, if the DNA quantity is not a concern (*see* **Section 3.2, step 5**).
7. The NanoDrop® ND-1000 spectrophotometer (NanoDrop Technologies, Wilmington, DE) is used to quantify nucleic acids, proteins, and other compounds. The measurement requires only 1 μL of sample and covers a wide range of DNA concentrations. For most samples, dilution is not required. As an alternative, the Molecular Probe PicoGreen dsDNA Assay kit (Invitrogen) is used to quantify DNA with small yield (*see* **Section 3.2, step 7**).

2.2. Mutation Analysis

1. The *ExTaq*™ DNA polymerase kit, containing *ExTaq*™ DNA polymerase, 10X *ExTaq*™ buffer with 20 m*M* $MgCl_2$, and 2.5 m*M* of each deoxynucleotide trisphosphate, is used for PCR applications (Takara Bio, Madison, WI).
2. The BigDye Terminator Cycle Sequencing FS Ready Reaction Kit is used for gene sequencing (Applied Biosystems, Foster City, CA).

2.3. Cell Culture

1. DMEM and DMEM/F-12 (1:1) culture medium (Invitrogen) supplemented with 10% FBS.
2. Penicillin and streptomycin 100X stock solution (*see* **Note 5**): 5,000 units/mL penicillin G sodium and 5,000 μg/mL streptomycin sulfate in 0.85% saline.

3. Poly-L-lysine or poly-D-lysine hydrobromide, gamma-irradiated powder (molecular weight 70–150 kDa; Sigma-Aldrich, St. Louis, MO): dissolved in 100 m*M* Tris-HCL, pH 7.5, 0.9% w/v NaCl as a 10X stock solution (0.5 mg/ml). Similarly, laminin (Sigma-Aldrich) is prepared as a 100X stock solution (0.5 mg/mL) (*see* **Note 6**).
4. Both 10X Tris-buffered saline and 10X phosphate-buffered saline (PBS) are available from Invitrogen.
5. The dissociation solution is freshly prepared by dissolving collagenase type I (Invitrogen) and dispase (Invitrogen) in DMEM to a final concentration of 0.125% and 1.25 U/mL, respectively, and is sterilized by filtration through a 0. 2-μm syringe filter unit.
6. Lyophilized recombinant human neuregulin-β1/heregulin-β1 containing an epidermal growth factor domain (rhuHRG-β1; R&D Systems, Minneapolis, MN) is reconstituted with PBS according to the manufacturer's instructions. To avoid freeze-thawing, we store rhuHRG-β1 as 10-μl aliquots at −80 °C. The rhuHRG-β1 solution together with forskolin, insulin, transferrin, and Na-selenite (all from Sigma-Aldrich) are freshly added to the FBS-containing medium (*see* **Section 3.4, step 3**).

2.4. Mice and Imaging

1. The Fox Chase severe combined immunodeficiency (SCID) mice can be purchased from Harlan (Indiananapolis, IN; *see* **Section 3.5, step 1**).
2. A stock solution (1 g/mL) of Avertin® (2,2,2-tribromoethanol; Sigma-Aldrich) is prepared by dissolving tribromoethanol in 2-methyl-2-butanol (amylene hydrate, or tertiary amyl alcohol; Sigma-Aldrich). The preparation requires slight heating to about 40 °C and vigorous stirring using a magnetic stir bar. To prepare a working solution (20 mg/mL), the stock solution is diluted 50X in PBS, mixed with a Vortex for 2 min, and filtered through a 0. 2-μm syringe filter unit. Avertin is light sensitive; both the stock and working solutions should be kept in dark container wrapped with aluminum foil and stored under refrigerated conditions (4 °C). The recommended dosage of Avertin® for mice ranges from ~0. 25 to 0.6 mg/g body weight by intraperitoneal (IP) injection (*see* **Section 3.5, steps 4** and **5**).
3. Gadodiamide (Omniscan™; GE Healthcare; Piscataway, NJ) is frequently used in MRI to visualize lesions with abnormal vascularity including malignancies. Gadodiamide is freshly diluted with injection-grade saline to 10 m*M* prior to use in mice.

3. Methods

Human tissue specimens are of vital importance to translational research. Similar to other studies utilizing patient tissue samples, surgically resected vestibular schwannomas and adjacent vestibular nerves are valuable for understanding pathobiology of the disease, improving diagnosis, testing for novel therapeutics, and potentially, identifying biomarkers for treatment outcomes *(1, 2)*. Great care must be taken when procuring the tissue samples *(18)*. Pathological evaluation by a certified neuropathologist is always obtained to ensure proper diagnosis of tissue origin and the quality of tumor specimen. Routinely, tissues are stored frozen or embedded in paraffin blocks. Fresh tissues can also be brought to the research laboratory for preparing primary cell cultures for molecular study or implanting into immunodeficient mice as a xenograft model *(17)*. We discuss below a few methods utilizing vestibular schwannoma and vestibular nerve tissues for molecular and translational research.

3.1. Tissue Procurement

1. Prior to the start of research, an approval of a Human Subjects Protocol for the use of surgically removed vestibular schwannoma and adjacent vestibular nerve specimens must be obtained from the Institutional Review Board. Vestibular schwannoma is resected with informed patient consent and utilized per the Human Subjects Protocol for tissue procurement.
2. Tissues should be properly labeled. A portion of the tumor is snap frozen with liquid nitrogen in the operating room, recorded in the pathology laboratory, and stored at −80 °C (*see* **Note 1**). A piece of the remainder of the tumor is placed in DMEM for cell culture (*see* **Section 3.4, steps 1–2**), and the rest is fixed in 10% buffered formalin (Richard-Allan Scientific; Kalamazoo, MI) and embedded in paraffin. Paraffin-embedded tissue sections of 5 μm are prepared and processed for hematoxylin-eosin staining. A neuropathologist evaluates the stained slides and histologically confirms the tumor type. Also, tissue sections can be processed for immunostaining with antibodies recognizing Schwann cell–specific markers such as S100 and myelin basic protein *(17)*.
3. For primary cell culture or xenograft study, we recommend that tissues be placed into a sterile tube containing DMEM immediately following surgical resection. We note that the cells in the tissue remain viable in the medium for up to 24 h; however, the sooner they are brought to the research laboratory for initiating primary cell cultures, the better the yield of viable cells.

3.2. Preparation of Genomic DNA from Paraffin-Embedded Blocks, Tissue Sections Fixed on Slides, or Frozen Tissues

1. Genomic DNA can be extracted from tissue detached from a slide or from a paraffin-embedded tissue block using the QIAamp® DNA Micro kit (Qiagen) according to the manufacturer's instruction (*see* **Note 2**).
2. When slides containing tissue sections are used, the tissue can be removed by immersing the slide in 10% ammonia solution.
3. Paraffin-embedded tissue blocks are cut into small pieces and placed in a 1.5-mL microcentrifuge tube prior to lysis and DNA extraction using the QIAamp® DNA Micro kit.
4. Because the yield of genomic DNA extracted from a slide or from a small tissue block is expected to be small, we add carrier RNA (1 μg/μL) to the sample, which is supplied in the QIAamp® DNA Micro kit, to allow better recovery during DNA extraction.
5. The quality of extracted DNA should be examined by agarose gel electrophoresis and ethidium bromide staining. Routinely, we run genomic DNA side by side with the 1-kb Plus DNA Ladder (Invitrogen) on a 0.7% agarose gel. As an alternative method, extracted DNA can be analyzed with a Bioanalyzer (Agilent Technologies). Extracted DNA of good quality should predominantly migrate to the high molecular weight (> 10 kb) region; however, DNA migrating to slightly lower molecular weight regions (1–10 kb) may still be useful for mutation analysis by polymerase chain reaction (PCR; **Section 3.3**). Extracted DNA that predominantly migrates below 1 kb is usually not suitable for mutation analysis.
6. When frozen tissues are used for DNA extraction, we employ the salting-out protein precipitation method using the Gentra Puregene® DNA isolation kit (Qiagen). The tissue is homogenized in the lysis buffer (provided in the kit) using a Dounce homogenizer or a disposable pellet pestle with a microtube (Kontes, Vineland, NJ), followed by digestion with proteinase K (200 μg/mL; Roche Biochemicals, Indiananapolis, IN) at 56 °C overnight. The next day, lysates are treated with RNase and protein contaminants are removed by salt precipitation according to the manufacturer's instruction.
7. The yield of DNA can be quantified using a conventional or a NanoDrop® ND-1000 spectrophotometer (NanoDrop Technologies). For small yields of DNA, the concentration can be measured using the Molecular Probe PicoGreen dsDNA Assay kit (Invitrogen).

3.3. Mutation Analysis of the NF2 Gene

1. Detection of *NF2* gene mutations is achieved by direct sequencing after PCR amplification of genomic DNA isolated from vestibular schwannomas or peripheral blood mononuclear cells *(5, 6)*. The human *NF2* gene contains 17 exons *(3, 4, 19, 20)*.

Primers used for amplifying each *NF2* exon are as follows:
Exon-1F, 5′-AGGCCTGTGCAGCAACTC-3′;
Exon-1R, 5′-GAGAACCTCTCGAGCTTCCAC-3′;
Exon-2F, 5′-GAGAGTTGAGAGTGCAGAG-3′;
Exon-2R, 5′-TCAGCCCCACCAGTTTCATC-3′;
Exon-3F, 5′-GCTTCTTTGAAGGTAGCACA-3′;
Exon-3R, 5′-GGTCAACTCTGAGGCCAACT-3′;
Exon-4F, 5′-CCTCACTTCCCCTCACAGAG-3′;
Exon-4R, 5′-CCCATGACCCAAATTAACGC-3′;
Exon-5F, 5′-ATCTTTAGAATCTCAATCGC-3′;
Exon-5R, 5′-AGCTTTCTTTTAGACCACAT-3′;
Exon-6F, 5′-CATGTGTAGGTTTTTTATTTTGC-3′;
Exon-6R, 5′-GCCCATAAAGGAATGTAAACC-3′;
Exon-7F, 5′-CAGTGTCTTCCGTTCTCC-3′;
Exon-7R, 5′-AGCTCAGAGAGGTTTCAA-3′;
Exon-8F, 5′-CCACAGAATAAAAAGGGCAC-3′;
Exon-8R, 5′-GATCTGCTGGACCCATCTGC-3′;
Exon-9F, 5′-GTTCTGCTTCATTCTTCC-3′;
Exon-9R, 5′-GTAATGAAAACCAGGATC-3′;
Exon-10F, 5′-CCTTTTAGTCTGCTTCTG-3′;
Exon-10R, 5′-TCAGTTAAAACAAGGTTG-3′;
Exon-11F, 5′-TCGAGCCCTGTGATTCAATG-3′;
Exon-11R, 5′-AAGTCCCCAAGTAGCCTCCT-3′;
Exon-12F, 5′-CCCACTTCAGCTAAGAGCAC-3′;
Exon-12R, 5′-CTCCTCGCCAGTCTGGTG-3′;
Exon-13F, 5′-GGTGTCTTTTCCTGCTACCT-3′;
Exon-13R, 5′-GGGAGGAAAGAGAACATCAC-3′;
Exon-14F, 5′-TGTGCCATTGCCTCTGTG-3′;
Exon-14R, 5′-AGGGCACAGGGGGCTACA-3′;
Exon-15F, 5′-TCTCACTGTCTGCCCAAG-3′;
Exon-15R, 5′-GATCAGCAAAATACAAGAAA-3′;
Exon-16F, 5′-CTCTCAGCTTCTTCTCTGCT-3′;
Exon-16R, 5′-CCAGCCAGCTCCTATGGATG-3′;
Exon-17F, 5′-GGCATTGTTGATATCACAGGG-3′; and
Exon-17R, 5′-GGCAGCACCATCACCACATA-3′.

2. The PCR is initiated by pipetting ~0.1–1 μg DNA into a 50-μL volume containing 1X *ExTaq*™ buffer with 2 m*M* $MgCl_2$, 25 pmol of each pair of primers, 5 nmol of each deoxynucleotide triphosphate, and 1.25 U of *ExTaq*™ DNA polymerase in a PCR tube. Prior to PCR, the DNA is preheated to 94 °C for 4 min. Amplification is carried out for 40 cycles with the cycle condition of denaturation at 94 °C for 1 min, annealing step at 58 °C for 1 min, and elongation step at 72 °C for 1 min (*see* **Note 2**).
3. A 5-μL aliquot of each PCR product is run on a 2% agarose gel to ensure the success of PCR. The sizes of the expected PCR products are 261 bp for exon 1, 240 bp for exon 2,

271 bp for exon 3, 187 bp for exon 4, 173 bp for exon 5, 161 bp for exon 6, 123 bp for exon 7, 179 bp for exon 8, 122 bp for exon 9, 166 bp for exon 10, 212 bp for exon 11, 284 bp for exon 12, 227 bp for exon 13, 254 bp for exon 14, 245 bp for exon 15, 148 bp for exon 16, and 178 bp for exon 17.

4. Sequencing of the PCR product is carried out using an Applied Biosystems DNA sequencer (Perkin-Elmer Cetus; Waltham, MA) with a PCR primer and a BigDye Terminator Cycle Sequencing FS Ready Reaction kit (Applied Biosystems) (*see* **Notes 3** and **4**).

3.4. Preparation of Primary Schwann Cells and Vestibular Schwannoma Cells

1. Freshly removed vestibular schwannoma and vestibular nerve tissues are placed in DMEM and should be brought to the research laboratory as soon as possible. To prepare primary Schwann cell cultures, the nerve tissue is cut into small pieces and incubated at 37 °C in DMEM supplemented with 10% FBS, 50 U/mL penicillin, and 0.05 mg/mL streptomycin for *in vitro* Wallerian degeneration for 7 days *(21, 22)*. This pre-treatment medium is replaced every 2 days (*see* **Note 5**).
2. To culture Schwann cells or schwannoma cells, dish surfaces should be coated with 0.05 mg/mL of poly-L-lysine and 5 μg/mL of laminin. A stock solution of poly-L-lysine is diluted 10X with ddH_2O and kept on ice. An aliquot of laminin is taken out from −80 °C, thawed on ice, and added to the poly-L-lysine solution to a final concentration of 5 μg/mL. The poly-L-lysine/laminin mixture (3 mL per 100 mm dish) is added to cover the dish surface for 1 h at room temperature. Coated dishes are rinsed with PBS and stored at 4 °C until use (*see* **Note 6**).
3. To dissociate vestibular schwannoma or pre-degenerated nerve tissues, the dissociation solution, containing DMEM, 0.125% collagenase, and 1.25 U/mL dispase, is freshly prepared. The tissues are placed into a sterile 50-mL tube containing 5 mL of the dissociation solution and incubated for 3 h at 37 °C with gentle mixing (60 RPM) in a New Brunswick Scientific (Edison, NJ) series 25-shaker incubator. The digestion is stopped by adding an equal volume of DMEM plus 10% FBS. To completely dissociate tissue into single cells, the suspension is triturated three times through a narrow-bored 9-in. Pasteur pipette. Cells are collected by centrifugation for 5 min at 300 g; resuspended in DMEM/F-12 (1:1) supplemented with 10% FBS, 10 ng/mL rhuHRG-β1, 2 μM forskolin, 0.5 μg/mL insulin, 0.1 mg/mL transferrin, 0.5 ng/mL Na-selenite; and plated in poly-L-lysine and laminin-coated dishes.

4. After incubation at 37 °C overnight, non-adherent cells are removed and the culture medium is changed every 3 days until the cell density reaches ~70% confluence.
5. To enrich Schwann cells or schwannoma cells, the simple cold jet technique is used *(23–25)*. Briefly, the culture medium is removed and 2 mL of ice-cold PBS is added slowly to the dish, followed by immediate aspiration. Schwann cells or schwannoma cells can be detached from the fibroblast substrate by flushing the cells with a stream of 5 mL ice-cold culture medium and pipetting the medium on and off several times. The suspension of floating cells is re-plated on a new coated dish and grown as described above (*see* **Note 7**).
6. **Figure 10.1** shows the typical morphology of primary vestibular schwannoma cells after 24 h in culture. Vestibular schwannoma cells display long and multiple projections in poly-lysine/laminin-coated dishes.

3.5. Vestibular Schwannoma Xenografts in Severe Combined Immunodeficiency Mice

1. SCID mice carry an autosomal recessive mutation on the *Prkdc* (protein kinase, DNA activated, catalytic polypeptide) gene on chromosome 16 *(26)*. The *Prkdc* gene encodes an enzyme involved in the rearrangement of the immunoglobulin and T cell receptor genes. The mutation disrupts the differentiation of both B and T lymphocyte progenitor cells; consequently, SCID mice show a severe combined immunodeficiency affecting both B and T cells. In addition to studying the basic biology of the immune system, SCID mice are extensively used as hosts for normal and malignant tissue transplants and for testing the safety

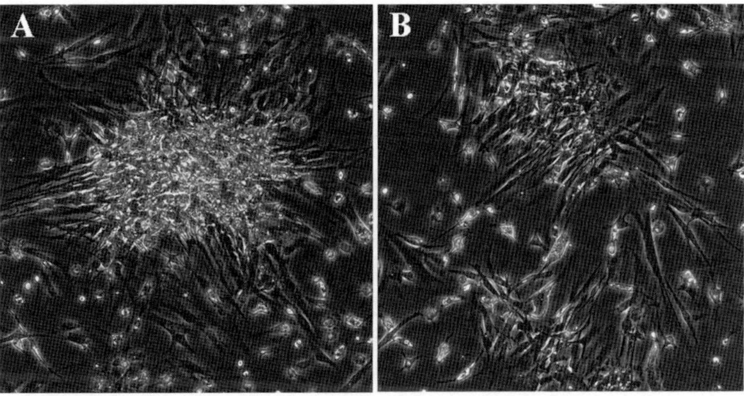

Fig. 10.1. Phase contrast photomicrograph of primary cultures of vestibular schwannoma cells. Tumor tissues were dissociated with collagenase and dispase. The cell suspension was plated in a poly-L-lysine and laminin-coated dish (**A**) or was triturated through a narrow-bored Pasteur pipette to enrich single-cell population before plating (**B**). Note that vestibular schwannoma cells display long and multiple projections.

of new therapeutic agents in immunocompromised individuals. By using SCID mice, we have established a quantifiable vestibular schwannoma xenograft model that utilizes MRI to measure tumor volumes *(17)* as described in **Section 3.5, steps 5–7**.

2. Prior to performing the xenograft experiment, an animal use and care protocol must be prepared and approved by the Institutional Animal Care and Use Committee.
3. Under sterile conditions, freshly procured vestibular schwannoma tissues are cut into sizes of ~2–5 mm in diameter and mixed in Matrigel® (BD Biosciences, San Jose, CA).
4. SCID mice are anesthetized by IP injection of Avertin® (250 mg/kg body weight). Under anesthesia, the left flank of the mouse is shaved and prepped using aseptic techniques. The thigh is selected for ease of implantation and the ability to grossly observe tumor growth *(17)*. Additionally, previous studies indicated that proximity to a peripheral nerve might affect growth of vestibular schwannomas *(27–29)*. An incision is made along the long axis of the proximal thigh. Soft tissues are dissected bluntly in order to identify the biceps femoris muscle and the sciatic nerve. A piece of Matrigel®-coated vestibular schwannoma specimen is implanted *en bloc* near the nerve, and the skin is closed using a single layer of interrupted suture. The contralateral leg is not dissected and is used as a control for comparison. Implanted mice are revived on a warming blanket until recovery and are watched daily for tumor growth.
5. High-field small animal MRI is used to visualize and quantify vestibular schwannoma xenografts in mice *(17)*. We recommend performing the first MRI scans 1 month after surgery to ensure that the animals have healed. Follow-up MRIs are obtained at every 1- or 2-month intervals for up to a year post-procedure. For MRI, mice are anesthetized with Avertin® (250 mg/kg body weight), immobilized on an animal holder, and placed prone in a 4.7 T/cm MRI system with a 120-mm inner diameter gradient coil (max. 400 mT/m), a 72-mm inner diameter proton volume radiofrequency (RF) coil for transmission, and a 4-cm surface receive coil. For T1 (longitudinal relaxation time)–weighted axial and coronal images, a spin echo sequence with repetition time (TR) of 550–600 ms and echo time (TE) of 10.5 ms is used. For T2 (transverse relaxation time)–weighted images, a Rapid Acquisition with Refocusing Echoes (RARE) sequence with TR of 2500–2600 ms, an effective TE of 47–54 ms, and a RARE factor of 4 is employed. In-plane resolution is 156 μm on the axial and 195 μm on the coronal images, and the slice thickness is 0.8 mm with a 0.2-mm gap between slices. Scan time is 5–6 min per scan.

6. Contrast-enhanced T1 axial and coronal images are acquired after a bolus injection of 0.1 mL of 10 mM gadodiamine (Omniscan^{TM}). The contrast agent is injected through a tail vein catheter using a thin (outer diameter of 0.06 mm) polyethylene tubing that extends outside the magnet and allows quick delivery of the contrast agent without changing the position of the mouse inside the magnet.
7. Multi-planar tumor volumes are determined from T1- and T2-weighted images. For this measurement, tumor areas are manually traced on axial and coronal T1 and T2 images. Post-contrast images are also used when available. Tumor volumes are calculated by adding the traced areas from all slices depicting the tumor and multiplying by the distance between slice (i.e., 0.8-mm slice thickness + 0.2-mm gap = 1 mm). All volume measurements are referenced to the first MRIs taken 1 month after implantation (*see* **Note 8**).
8. Vestibular schwannoma xenografts can be dissected from implanted mice and used in immuno- and histopathological analysis *(7, 17)*. The analysis will confirm that the mass seen on the MRI is, in fact, schwannoma tissue by phenotype. In addition to typical vestibular schwannoma characteristics, tumor cells show strong immunoreactivity to S100, p75^{NGFR}, and myelin basic protein (MBP) antigens *(17)*.

4. Notes

1. It is essential that surgically removed tissues be sent to the laboratory as soon as possible. Proper sample collection and storage are also critical for subsequent molecular analysis *(18)*. Methods and time of fixation affect the integrity of archived DNA. Prolonged fixation in formalin rarely yields useful DNA and should be avoided *(30)*.
2. Because the PCR technique is highly sensitive, when extracting genomic DNA and performing PCR reaction, great caution must be taken to avoid cross-contamination among samples. Aerosol-barrier pipet tips should be used at all steps of the procedure.
3. Mutation analysis of the *NF2* gene can also be carried out by heteroduplex or single-strand conformational polymorphism (SSCP) analysis followed by DNA sequencing *(5, 6)*. Methods to detect large mutations or loss of heterozygosity can be found elsewhere *(19, 31–33)*.
4. Additional valuable information about diagnosis of *NF2* can be found at the Children's Tumor Foundation website https://www.ctf.org/about-nf/diagnosis-of-nf2/. Genetic testing service for *NF2* are available in several

laboratories, including Massachusetts General Hospital Neurogenetics DNA Diagnosis Laboratory (http://www.massgeneral.org/neuroDNAlab/neuroDNA_neurofibro.htm), Clinical Laboratories at the Department of Human Genetics, The University of Alabama, Birmingham (http://www.genetics.uab.edu/medgenomics/MedGenomicsTesting-Services.htm), and National Genetics Reference Laboratory, Manchester, UK (http://www.ngrl.org.uk/ Manchester/).

5. To reduce contamination during preparation of primary Schwann cells and vestibular schwannoma cells, antibiotics such as penicillin and streptomycin can be used throughout the experiment. When plating out primary adult Schwann cells, optimal cell density is desired to enhance the plating efficiency and proliferation rate of the culture *(25, 34)*. We recommend maintaining Schwann cell cultures at optimal cell densities with 1:2 or 1:3 dilutions during each cell passage.
6. It is essential to use dishes coated with extracellular matrix (ECM), including poly-lysine and laminin for culturing Schwann cells and schwannoma cells. The ECM enhances cell attachment, spreading, and growth of these cells (**Fig. 10.1**). Either poly-L-lysine or poly-D-lysine is used. Alternatatively, poly-L-ornithine can be used *(25)*. Although the poly-lysine solution is stable at 4 °C, we prefer storing it frozen at −20 °C or below. Laminin stock should be aliquoted and stored at −80 °C. Dishes coated with poly-lysine and laminin are stable for more than 3 months when stored at 4 °C.
7. With the use of the cold jet technique *(23–25)*, the purity of Schwann cells or schwannoma cell cultures can reach more than 90%. An alternative technique for enriching Schwann cells is to employ magnetic cell sorting using an antibody that recognizes the protein found on the surface of Schwann cells, such as the low-affinity nerve growth factor receptor $p75^{NGFR}$ *(35)*. This technique also yields Schwann cell cultures with greater than 90% purity.
8. We have shown that the majority of vestibular schwannoma xenografts persisted but did not show significant growth over time *(17)*. Most tumor volumes were either unchanged or reduced over the 1-year study period. However, even without growth, the xenograft was detectable by MRI scans 1 year after implantation. Also, we detected an increase in tumor volume in two of 15 VS xenografts over time. The tumors retained their original microscopic and immunohistochemical characteristics after prolonged implantation (*see* **Methods 3.5** and **Fig. 10.2**).

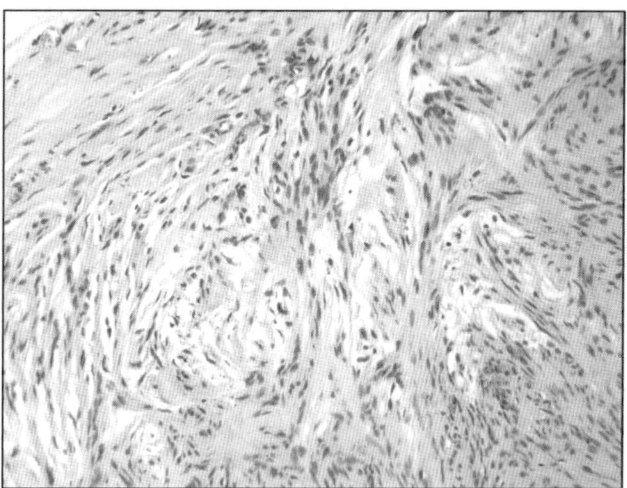

Fig. 10.2. Histological examination of a vestibular schwannoma xenograft with significant tumor growth 6 months after implantation shows that the tumor mass consisted of spindle cells with alternating compact areas of elongated cells. Both the nuclear palisading (Antoni A pattern) and less cellular, loosely textured Antoni B areas could be seen.

Acknowledgments

The authors would like to thank Elena M. Akhmametyeva, Tina Lee, and Mark Packer for discussions and suggestions. This work was supported by grants from the Department of Defense Neurofibromatosis Research Program and National Institute of Deafness and Communicative Disorders.

References

1. Neff, B. A., Welling, D. B., Akhmametyeva, E., and Chang, L. S. (2006) The molecular biology of vestibular schwannomas: dissecting the pathogenic process at the molecular level. *Otol. Neurotol.* **27**, 197–208.
2. Welling, D. B., Packer, M. D., and Chang, L.-S. (2007) Molecular studies of vestibular schwannomas: a review. *Curr. Opin. Otolaryngol. Head Neck Surg.* **15**, 341–346.
3. Rouleau, G. A., Merel, P., Lutchman, M., et al. (1993) Alteration in a new gene encoding a putative membrane-organising protein causes neurofibromatosis type 2. *Nature* **363**, 515–521.
4. Trofatter, J. A., MacCollin, M. M., Rutter, J. L., et al. 1993. A novel Moesin-, Exrin-, Radixin-like gene is a candidate for the neurofibromatosis 2 tumor-suppressor. *Cell* **72**, 791–800.
5. Jacoby, L. B., MacCollin, M., Louis, D. N., Mohney, T., et al. 1994. Exon scanning for mutation of the NF2 gene in schwannomas. *Hum. Mol. Genet.* **3**, 413–419.
6. Welling, D. B., Guida, M., Goll, F., Pearl, D. K., et al. (1996) Mutational spectrum in the neurofibromatosis type 2 gene in sporadic and familial schwannomas. *Human Genet.* **98**, 189–193.
7. Welling, D. B., Lasak, J. M., Akhmametyeva, E., Ghaheri, B., and Chang, L.-S. (2002) cDNA microarray analysis of vestibular schwannomas. *Otol. Neurotol.* **23**, 736–748.
8. Baser M. E. (2006) Contributors to the International NF2 Mutation Database. 2006. The distribution of constitutional and somatic mutations in the neurofibromatosis 2 gene. *Human Mutat.* **27**, 297–306.

9. McClatchey, A. I. and Giovannini, M. 2005. Membrane organization and tumorigenesis – the NF2 tumor suppressor, Merlin. *Genes Dev.* **19**, 2265–2277.
10. Okada, T., You, L., and Giancotti, F. G. (2007) Shedding light on Merlin's wizardry. *Trends Cell Biol.* **17**, 222–229.
11. Welling, D. B., Glasscock, M. E., Woods, C. I., and Jackson, C. G. (1990) Acoustic neuroma: A cost-effective approach. *Otolaryngol. HN Surg.* **103**, 364–370.
12. Schmalbrock, P., Chakeres, D. W., Monroe, J. W., Saraswat, A., Miles, B. A., and Welling, D. B. (1999) Assessment of internal auditory canal tumors: a comparison of contrast-enhanced T1-weighted and steady-state T2-weighted gradient-echo MR imaging. *Am. J. Neuroradiol.* **20**, 1207–1213.
13. Lasak, J. M., Welling, D. B., Akhmametyeva, E. M., Salloum, M., and Chang, L.-S. (2002) Retinoblastoma-cyclin-dependent kinase pathway deregulation in vestibular schwannomas. *Laryngoscope* **112**, 1555–1561.
14. Jacob, A., Lee, T. X., Neff, B. A., Miller, S., Welling, B., and Chang, L. S. (2008) Phosphatidylinositol 3-Kinase/AKT pathway activation in human vestibular schwannomas. *Otol. Neurotol.* 29: 58–68.
15. Giovannini, M., Robanus-Maandag, E., Niwa-Kawakita, M., van der Valk, M., et al. (1999) Schwann cell hyperplasia and tumors in transgenic mice expressing a naturally occurring mutant NF2 protein. *Genes Dev.* **13**, 978–986.
16. Giovannini, M., Robanus-Maandag, E., van der Valk, M., Niwa-Kawakita, M., et al. (2000) Conditional biallelic Nf2 mutation in the mouse promotes manifestations of human neurofibromatosis type 2. *Genes Dev.* **14**, 1617–1630.
17. Chang, L. S., Jacob, A., Lorenz, M., Rock, J., et al. (2006) Growth of benign and malignant schwannoma xenografts in severe combined immunodeficiency mice. *Laryngoscope* **116**, 2018–2026.
18. Check, E. (2007) Cancer atlas maps out sample worries. *Nature* **447**, 1036–1037.
19. Zucman-Rossi, J., Legoix, P., Der Sarkissian, H., Cheret, G., Sor, F., Bernardi, A., Cazes, L., Giraud, S., Ollagnon, E., Lenoir, G., and Thomas G. (1998) NF2 gene in neurofibromatosis type 2 patients. *Hum. Mol. Genet.* **7**, 2095–2101.
20. Chang, L.-S., Akhmametyeva, E. M., Wu, Y., Zhu, L., and Welling, D. B. (2002) Multiple transcription initiation sites, alternative splicing, and differential polyadenylation contribute to the complexity of human neurofibromatosis 2 transcripts. *Genomics* **79**, 63–76.
21. Morrissey, T. K., Kleitman, N., and Bunge, R. P. (1991). Isolation and functional characterization of Schwann cells derived from adult peripheral nerve. *J. Neurosci.* **11**, 2433–2442.
22. Casella, G. T., Bunge, R. P., and Wood, P. M. (1996) Improved method for harvesting human Schwann cells from mature peripheral nerve and expansion in vitro. *Glia* **17**, 327–338.
23. Mauritz, C., Grothe, C., and Haastert, K. (2004) Comparative study of cell culture and purification methods to obtain highly enriched cultures of proliferating adult rat Schwann cells. *J. Neurosci. Res.* 77, 453–461.
24. Haastert, K., Mauritz, C., Matthies, C., and Grothe, C. (2006) Autologous adult human Schwann cells genetically modified to provide alternative cellular transplants in peripheral nerve regeneration. *J. Neurosurg.* **104**, 778–786.
25. Haastert, K., Mauritz, C., Chaturvedi, S., and Grothe, C. (2007) Human and rat adult Schwann cell cultures: fast and efficient enrichment and highly effective non-viral transfection protocol. *Nat. Protoc.* **2**, 99–104.
26. Shultz, L. D., Ishikawa, F., and Greiner, D. L. (2007) Humanized mice in translational biomedical research. *Nat. Rev. Immunol.* **7**, 118–130.
27. Lee, J. K., Sobel, R. A., Chiocca, E. A., Kim, T. S., and Martuza, R. L. (1992) Growth of human acoustic neuromas, neurofibromas and schwannomas in the subrenal capsule and sciatic nerve of the nude mouse. *J. Neurooncol.* **14**, 101–112.
28. Charabi, S., Rygaard, J., Klinken, L., Tos, M., and Thomsen, J. (1994) Subcutaneous growth of human acoustic schwannomas in athymic nude mice. *Acta Otolaryngol.* **114**, 399–405.
29. Stidham, K. R. and Roberson, J. B., Jr. (1997) Human vestibular schwannoma growth in the nude mouse: evaluation of a modified subcutaneous implantation model. *Am. J. Otol.* **18**, 622–626.
30. Ferrer, I., Armstrong, J., Capellari, S., Parchi, P., et al. (2007) Effects of formalin fixation, paraffin embedding, and time of storage on DNA preservation in brain tissue: a BrainNet Europe study. *Brain Pathol.* **17**, 297–303.
31. Szijan, I., Rochefort, D., Bruder, C., Surace, E., et al. (2003) NF2 Tumor suppressor gene: A comprehensive and efficient detection of somatic mutations by denaturing HPLC and microarray-CGH. *NeuroMol. Med.* **3**, 41–52.
32. Wallace, A. J., Watson, C. J., Oward, E., Evans, D. G. R., and Elles, R. G. (2004) Mutation scanning of the *NF2* gene: An improved service based on Meta-PCR/sequencing, dosage analysis, and loss

of heterozygosity analysis. *Genet. Testing* **8**, 368–380.

33. Kluwe, L., Nygren, A. O. H., Errami, A., Heinrich, B., Matthies, C., Tatagiba, M., and Mautner, V. (2005) Screening for Large Mutations of the NF2 Gene. *Genes Chrom. Cancer* **42**, 384–391.
34. Casella, G. T., Wieser, R., Bunge, R. P., Margitich, I. S., Katz, J., Olson, L., and Wood, P. M. (2000) Density dependent regulation of human Schwann cell proliferation. *Glia* **30**, 165–177.
35. Manent, J., Oguievetskaia, K., Bayer, J., Ratner, N., and Giovannini, M. (2003) Magnetic cell sorting for enriching Schwann cells from adult mouse peripheral nerves. *J. Neurosci. Methods* **123**, 167–173.

Chapter 11

Multilocus Sequence Typing and Pulsed Field Gel Electrophoresis of Otitis Media Causing Pathogens

Jonathan C. Thomas and Melinda M. Pettigrew

Abstract

Streptococcus pneumoniae, Haemophilus influenzae, and *Moraxella catarrhalis* are the three leading bacteria species associated with otitis media. Defining the molecular epidemiology of bacteria known to cause otitis media is of great importance, in both clinical and research settings. PFGE and MLST provide data for the characterization of isolates' genetic relatedness, yet they differ in the types of studies for which they are most useful. Consequently, knowledge of both techniques is important for laboratories intending to study the molecular epidemiology of otitis media–associated bacterial pathogens.

Key words: Molecular epidemiology, multilocus sequence typing, pulsed field gel electrophoresis, eBURST, *Streptococcus pneumoniae, Haemophilus influenzae, Moraxella catarrhalis*, otitis media.

1. Introduction

The ability to establish the genetic relatedness of bacteria is important in the study of species known to cause otitis media, including *Streptococcus pneumoniae, Haemophilus influenzae* and *Moraxella catarrhalis.* This is crucial in determining the population structure and transmission patterns of each species, as well as for tracking the spread of antibiotic-resistant clones or those particularly adapted to disease. The two most commonly used techniques used to distinguish between isolates of otitis media–associated pathogens are pulsed field gel electrophoresis (PFGE) and multilocus sequence typing (MLST).

MLST involves the sequencing of internal fragments of, on average, seven housekeeping genes distributed around the bacterial chromosome. The first MLST scheme was proposed

Bernd Sokolowski (ed.), *Auditory and Vestibular Research: Methods and Protocols, vol. 493*

DOI 10.1007/978-1-59745-523-7_11 Springerprotocols.com

for the bacterium *Neisseria meningitidis* by Maiden et al. *(1)*, while a scheme for *S. pneumoniae* soon followed *(2)*. MLST schemes for *H. influenzae (3)* and *M. catarrhalis* (available at http://web.mpiib-berlin.mpg.de/mlst/dbs/Mcatarrhalis) have also been developed. MLST has since become the most frequently used tool for genotypically characterizing a number of bacterial pathogens *(4)*.

Once established, the sequence of each locus is queried against a central, curated database in which each allele is assigned an allelic number. In the event of a sequence not being present in the central database, it can be added by the database's curator. Once an allelic number has been assigned to each locus, an allelic profile will be determined and a sequence type (ST) allocated. The relationship between STs is ascertained by the number of alleles shared between strains.

PFGE involves the comparison of patterns of bacterial genomic DNA digested with a rare cutting restriction enzyme, such as *SmaI (5)*. Genomic DNA is embedded in agarose plugs to prevent shearing of large DNA molecules. The embedded DNA is then digested with a restriction enzyme that cuts the genomic DNA into a variable number of fragments, depending on the polymorphisms within potential cleavage sites throughout the genome. The digested DNA is then subjected to pulsed field electrophoresis on an agarose gel. Standard gel electrophoresis can separate fragments of DNA up to 30–50 kb in size. Fragments larger than 30–50 kb in size are not effectively separated, because they do not tend to experience a difference in mobility *(6)*. In contrast, PFGE is capable of resolving large DNA molecules of up to 5 Mb in size. Rather than the continuous field used in a standard electrophoresis process, the orientation of the field is repeatedly changed, or pulsed, thus causing the separation of larger fragments of DNA. The more often the direction of the field is altered, the greater the separation between fragments and hence the greater resolution of subtypes *(7)*.

PFGE is often regarded as the gold standard in epidemiology due to its ability to discriminate between very closely related isolates. The technique may be used to compare middle ear and nasopharyngeal isolates from a single patient *(8)*, or to distinguish between isolates obtained from children in the same daycare facility *(9)*. PFGE has been shown to be the most discriminatory method for the comparison of nontypeable *H. influenzae* isolates *(10)*. However, there are distinct advantages and drawbacks to both MLST and PFGE. MLST offers discrete, unambiguous data that may be easily shared and compared between laboratories across the globe, in contrast to the banding patterns of PFGE that often limit such data to the laboratory in which it was generated. However, PFGE is far cheaper to perform than MLST and provides more discriminatory results, although the cost of sequencing is continually decreasing. Both methods are hampered by the time taken to perform the protocols each requires *(11)*.

MLST and PFGE protocols exist for *S. pneumoniae*, *H. influenzae*, and *M. catarrhalis*, but for the purposes of this chapter the methodology will concentrate on *S. pneumoniae (2, 3, 9)*. The methodology for *S. pneumoniae* can be adapted for the other two bacteria species. The main differences in protocol involve (1) the media used to grow *H. influenzae* and *M. catarrhalis* for DNA extraction or the preparation of PFGE plugs, (2) the specific housekeeping genes that are amplified for MLST, and (3) the specific PCR conditions for MLST.

2. Materials

2.1. Crude DNA Extraction for PCR Amplification of MLST Loci

1. Trypticase soy agar (TSA) plates, supplemented with 5% (v/v) sheep's blood.
2. TE buffer: 10 m*M* Tris-HCl, pH 7.5, and 1 m*M* EDTA, pH 7.5. Store at room temperature.
3. 48-well PCR plate and adhesive PCR film.

2.2. PCR Amplification of MLST Housekeeping Loci

1. 48-well PCR plate and adhesive PCR film.
2. PCR Master Mix.
3. Seven pairs of PCR primers (**Table 11.1**), designed by Enright and Spratt *(2)* to amplify internal fragments of the seven housekeeping genes *aroE, gdh, gki, recP, spi, xpt*, and *ddl*, although the primers described here have been adapted to include tails that correlate to the M13F and M13R sequencing primers, as described by Pettigrew et al. *(12)* (*see* **Note 1**).
4. 10X Tris-Borate-EDTA (TBE) buffer, pH 8.3: 0.9 *M* Tris-base, 0.9 M boric acid, 30 m*M* EDTA.
5. Ethidium bromide (10 mg/mL) dissolved in distilled water. Store in a darkened environment.
6. 1 kb DNA ladder.

2.3. Purification of PCR Amplicons and DNA Sequence Reactions

1. Polyethylene glycol (PEG) solution: 20% PEG, 2.5 *M* NaCl mixture (w/v). Store at room temperature.
2. 70% and 95% ethanol (v/v), diluted with distilled water. Store at 4 °C.
3. 3 *M* sodium acetate, pH 5.2.
4. 15 mL tubes.
5. Saran wrap.

2.4. DNA Sequencing Reaction

1. 48-well PCR plate and adhesive PCR film.
2. BigDye (version 3; PE Applied Biosystems, UK) fluorescent terminators. The BigDye version required may differ depending on the setup of individual sequencing facilities.
3. Sequencing primers M13F and M13R (**Table 11.1**).

Table 11.1
Primers utilized in the multilocus sequence typing of *S. pneumoniae*

Locus	Primer	Sequence (5′-3′)
aroE	aroM13F	TGTAAAACGACGGCCAGTGCCTTTGAGGCGACAGC
	aroM13R	AGGAAACAGCTATGACCATTGCAGTTCARAAACATWTTCTAA
gdh	gdhM13F	TGTAAAACGACGGCCAGTATGGACAAACCAGCNAGYTT
	gdhM13R	AGGAAACAGCTATGACCATGCTTGAGGTCCCATRCTNCC
gki	gkiM13F	TGTAAAACGACGGCCAGTGGCATTGGAATGGGATCACC
	gkiM13R	AGGAAACAGCTATGACCATTCTCCCGCAGCTGACAC
recP	recPM13F	TGTAAAACGACGGCCAGTGCCAACTCAGGTCATCCAGG
	recPM13R	AGGAAACAGCTATGACCATTGCAACCGTAGCATTGTAAC
spi	spiM13F	TGTAAAACGACGGCCAGTTTATTCCTCCTGATTCTGTC
	spiM13R	AGGAAACAGCTATGACCATGTGATTGGCCAGAAGCGGAA
xpt	xptM13F	TGTAAAACGACGGCCAGTTTATTAGAAGAGCGCATCCT
	xptM13R	AGGAAACAGCTATGACCATAGATCTGCCTCCTTAAATAC
ddl	ddlM13F	TGTAAAACGACGGCCAGTTGCYCAAGTTCCTTATGTGG
	ddlM13R	AGGAAACAGCTATGACCATCACTGGGTRAAACCWGGCAT
M13	M13F	TGTAAAACGACGGCCAGT
	M13R	AGGAAAGACGTATGACCAT

*Underlined sections represent M13 gene specific primer sequences (*see* **Note 1**).

2.5. PFGE

1. PET IV buffer for cell suspension: 1 *M* NaCl, 10 m*M* Tris, pH 7.5. Store at room temperature.
2. InCert agarose (Cambrex, Bio Science Rockland Inc., ME, USA).
3. EC lysis buffer: 6 m*M* Tris-HCl, pH 7.5, 1 *M* NaCl, 100 m*M* EDTA, pH 8.0, 0.5% polyoxyethylene 20 cetylether (Brij 58™), 0.2% deoxycholate, 0.5% *N*-laurylsarcosyl. Store in aliquots at −20 °C.
4. ESP buffer: 30 mL 0.5 *M* EDTA (pH 8.5), 1.5 mL 10% sarkosyl, and 60 mg proteinase K. Store in aliquots at −20 °C.
5. TE buffer: 10 m*M* Tris-HCl, pH 7.5, 1 m*M* EDTA, pH 7.5. Store at room temperature.
6. *SmaI* restriction enzyme and appropriate reaction buffers. Store at −20 °C.
7. 10X TBE buffer, pH 8.3: 0.9 *M* Tris-base, 0.9 *M* boric acid, and 30 m*M* EDTA.
8. Ethidium bromide (10 mg/mL) dissolved in distilled water. Store in a darkened environment.
9. SeaKem HGT agarose (BioWhittaker Molecular Applications, ME, USA).
10. Midrange PFG Marker I.

3. Methods

3.1. Multilocus Sequence Typing

For many species, a crude lysis of bacterial cells is sufficient for the extraction of DNA for MLST.

3.1.1. Crude DNA Extraction for PCR Amplification of MLST Loci

1. Streak the isolate to be studied for single colonies on half of a TSA plate, supplemented with 5% sheep's blood (v/v). Incubate overnight at 37 °C with 5% carbon dioxide.
2. Pipette 50 μL of TE into each well of a PCR plate.
3. Inoculate each well with 1–2 bacterial colonies and seal the plate with adhesive PCR film.
4. Heat each sample to 100 °C for 10 min in a thermal cycler.

3.1.2. PCR Amplification of MLST Housekeeping Loci

1. Amplify internal fragments of the housekeeping genes *aroE, gdh, gki, recP, spi, xpt*, and *ddl* in separate polymerase chain reactions (PCR). The PCR mixture consists of 2 μL of chromosomal DNA, 1 μL of forward and reverse PCR primers (10 p*M*; **Table 11.1**), and 25 μL of PCR Master Mix. Prepare reactions on ice in 48-well PCR plates and seal with adhesive PCR film. Thermally cycle the reactions at 95 °C for 3 min, 35 cycles of 95 °C for 30 sec, 53 °C for 30 sec, and 72 °C for 1 min, followed by a final elongation step of 72 °C for 10 min. The reactions are held in the thermal cycler at 4 °C until their removal.
2. Prepare a 1% (w/v) agarose gel by mixing 1 g of agarose powder in 100 mL of 0.5X TBE and heat until completely dissolved. Allow the gel to cool until safe to touch and add 5 μL of 10 mg/mL ethidium bromide. Pour the gel, which should set in approximately 20 min.
3. Dilute 100 mL of 10X TBE with 1.9 L of distilled water for use as running buffer. Fill the gel tank with running buffer.
4. Once the gel has set, carefully remove the comb(s) and position in the gel tank. Replenish with running buffer, in order to ensure that the gel is submerged.
5. Load 6 μL of each PCR mixture in a well. Include one well of molecular marker for each comb used in the gel. Secure the gel tank cover in place and connect to a power supply, ensuring that the gel runs from the negative to the positive electrode. Run at 150 V for 30 min. Visualize the gel under UV light using a UVIdoc and UVIPhotoMW software (UVItec Ltd., Cambridge, UK), to ascertain whether the reaction was successful and that the amplicon is of the correct size.

3.1.3. Purification of PCR Amplicons

1. Add 60 μL of PEG/NaCl mixture to each sample well and reseal (*see* **Note 2**). Vortex thoroughly and centrifuge at 200 g for 20 sec. Incubate at room temperature for 30 min.

2. Centrifuge at 2,465 g for 30 min at 4 °C in order to pellet any DNA precipitate. Remove the excess PEG/NaCl mixture by removing the adhesive PCR seal and inverting the PCR plate onto tissue and centrifuging at 200 g for 20 sec followed by a separate spin for 1 min on fresh tissue (*see* **Note 3**).
3. Add 150 μL of 70% ethanol (v/v) to each well and reseal with the same adhesive PCR seal. Wash the pellet by centrifuging at 2,465 g for 30 min at 4 °C. Remove excess ethanol by removing the adhesive PCR seal and inverting onto tissue for 30 sec. Place on fresh tissue and centrifuge for 1 min at 200 g.
4. Incubate the open PCR plate at 37 °C for 2 min on a thermal cycler in order to dry the DNA pellet. Pipette 12 μL of sterile distilled water into each well. Reseal the plate with adhesive PCR film and vortex vigorously, before centrifuging for 20 sec at 200 g. Repeat the vortex/centrifuge processes for a total of three times each. Resuspended, purified PCR product may be stored stably at 4 °C for several days, provided that the plate is well sealed. For long-term storage, place the PCR product at −20 °C (*see* **Note 4**).

3.1.4. DNA Sequencing Reaction

It may not be necessary to perform the following steps of the MLST protocol, but they have been included for the sake of completeness. The stages to which the protocol needs to be completed will depend on the requirements of the facility or company employed for DNA sequencing. Some companies only require the unpurified amplicon from the PCR reaction mix, while some companies ask for a purified PCR product.

1. Pipette 2 μL of resuspended, purified PCR amplicon into a new PCR plate. Add 1 μL of either the forward or the reverse (1 p*M*) sequencing primer and 2 μL of BigDye fluorescent terminators. Prepare a second DNA sequencing reaction for each housekeeping gene of each isolate, utilizing the alternative sequencing primer. Seal the PCR plate with adhesive PCR film and centrifuge for 20 sec at 200 g.
2. Thermally cycle the reactions for 24 cycles of 95 °C for 10 sec, 50 °C for 5 sec, and 60 °C for 2 min. The DNA sequencing reactions are held in the thermal cycler at 4 °C until their removal.

3.1.5. Purification of DNA Sequencing Reactions

1. Add 12 μL of sterile distilled water to dilute each sequencing reaction (*see* **Note 5**).
2. Pipette 6 mL of 95% ethanol (v/v) and 240 μL of 3 *M* sodium acetate into a 15 mL falcon tube. Vortex to mix.
3. Add 52 μL of the ethanol/sodium acetate mix to each reaction, reseal, and vortex. Centrifuge for 20 sec at 200 g before incubating the plate for 30 min at 4 °C.
4. Centrifuge the PCR plate(s) for 30 min at 2,465 g in order to precipitate DNA. Remove excess ethanol/sodium acetate mix

by inverting the plate onto tissue. Place on fresh tissue and centrifuge at 200 g for 1 min.

5. Add 150 μL of 70% ethanol (v/v) to each well in order to wash the DNA pellet. Reseal the plate and centrifuge at 2,465 g for 30 min. Remove excess ethanol by inverting the plate onto tissue for 1 min. Transfer to fresh tissue and centrifuge for 1 min at 200 g.
6. Air dry each PCR plate for 15 min at room temperature, before resealing and storing at 4 °C until ready to be run on an automated DNA sequencer.

3.1.6. DNA Sequence Analysis

1. Assemble the forward and reverse trace files for each housekeeping gene into individual contigs using a sequence viewing program such as SeqMan II from the DNAStar package (Lasergene, WI, USA). An example of an allele of each housekeeping gene may be obtained from the MLST site, www.mlst.net, and used as a reference to trim each contig to the appropriate length.
2. Following the generation of sequences, determine the allele numbers by querying the central database at www.mlst.net. Once an allelic number has been obtained for each locus, determine the allelic profile of the isolate and query this against the database at www.mlst.net in order to acquire the ST of the isolate (*see* **Note 6**).
3. Generate a graphical representation of the relatedness of isolates using the program enhanced Based Upon Related Sequence Types (eBURST) *(13)*, available at http://eBURST.mlst.net. For highly recombinogenic species, such as *S. pneumoniae*, the most stringent settings are often used to define the number of loci isolates required to share in order to belong to the same clonal group.

3.2. Pulsed Field Gel Electrophoresis

3.2.1. Preparation of PFGE Plugs

1. Streak half of a TSA plate, supplemented with 5% sheep's blood (v/v), with the isolate to be studied. Incubate overnight at 37 °C with 5% carbon dioxide.
2. Transfer the entire bacterial growth to 1 mL of PET IV buffer in a 1.5 mL microcentrifuge tube, using a sterile swab (*see* **Note 7**).
3. Centrifuge the bacterial suspension at 18,000 g for 3 min. Remove the supernatant by pipetting. Be careful not to disturb the bacterial pellet.
4. Add 100 μL of PET IV buffer to each sample and vortex to resuspend the bacterial pellet.
5. Dissolve 0.4 g of InCert agarose in 50 mL of distilled water by gently mixing and heating for 20 sec intervals until completely dissolved. Allow to cool for 2 min.
6. Add 110 μL of 0.8% (w/v) InCert agarose to each sample and mix by gently pipetting up and down. Pipette 24 μL of

the resulting mixture onto a weighboat and repeat until the entire bacterial agar suspension has been transferred to plugs in the weighboat. Cool at 4 °C for 15 min.

7. Transfer plugs for each isolate to a 1.5 mL microcentrifuge tube.
8. Add 0.5 mL of EC buffer to each tube and incubate at 37 °C for at least 2 h.
9. Remove the EC buffer by pipetting (*see* **Note 8**).
10. Add 0.5 mL of ESP buffer to each sample and incubate overnight at 50 °C.
11. Remove the ESP buffer by pipetting. Add 1 mL of TE buffer and incubate at 37 °C for 30 min to wash the plugs.
12. Remove the TE buffer and replace with a fresh 1 mL of TE buffer. Incubate at 37 °C for 30 min. Repeat the process once more. The plugs may be stored at 4 °C in this final TE buffer wash.

3.2.2. Restriction Enzyme Digest of Agarose Plug

1. Transfer one plug to a new 1.5 mL microcentrifuge tube (*see* **Note 9**).
2. Add 100 μL of 1X reaction buffer, specific to the restriction enzyme to be used, *SmaI*. Incubate at room temperature for 30 min.
3. Remove the 1X reaction buffer by pipette. Add 63 μL of sterile distilled water, 7 μL of 10X reaction buffer, and 1 μL of *SmaI*. A master mix for this step may be prepared during the previous 30 min incubation.
4. Incubate at room temperature for 2 h.
5. Remove the enzyme and reaction buffer mixture by pipette. Add 100 μL of 0.5X TBE buffer and incubate at room temperature for 15 min.

3.2.3. Preparation of the Pulsed Field Gel

1. Prepare 1.3% agarose gel (w/v) by mixing 1.3 g of SeaKem HGT agarose in 100 mL of 0.5X TBE and heating until completely dissolved. Allow the gel to cool until the bottle is safe to touch and add 5 μL of 10 mg/mL ethidium bromide. Pour the agarose into a gel tray with a comb in place. The gel should set in ~30 min. This step may be conducted while the plugs are digesting.
2. Add 2 L of 0.5X TBE to the PFGE chamber. Turn on the pump and cooling module, setting the cooling module to 14 °C. The pump may not immediately function and often requires some attention (*see* **Note 10**).
3. Once the gel has set, carefully remove the comb.
4. Remove the 0.5X TBE buffer from each sample and melt the plugs one at a time by placing in a heat block, set to 90 °C, for 20 sec.

5. Pipette the melted plug into one of the empty wells. This step must be achieved promptly to ensure that the plug does not begin to set again. Two wells of the gel must be reserved for loading with a molecular marker, although this does not require melting. Cut a thin slice of molecular marker and transfer to the appropriate well.
6. Transfer the gel to the centre of the electrophoresis chamber and ensure that the gel is submerged. Replenish the 0.5X TBE buffer if required.
7. Set the required parameters in the control panel of the electrophoresis chamber.
   a. Voltage: 6 V/cm
   b. Included angle: 120°
   c. Initial switch time: 4 sec
   d. Final switch time: 16 sec
   e. Run time: 18 h
8. Visualize the gel under UV light using UVIdoc and UVIPhotoMW software (UVItec Ltd., Cambridge, UK).

3.2.4. Banding Pattern Comparison

The criteria proposed by Tenover et al. *(14)* are those most commonly used in the analysis of PFGE banding patterns. Briefly, these criteria classify banding patterns into four separate categories – indistinguishable, closely related, possibly related, and different – depending on the number of differences between banding patterns. Isolates that have identical banding patterns are categorized as indistinguishable, while isolates that differ by 2–3 bands or 4–6 bands are classified as closely and possibly related, respectively. Isolates differing by seven or more bands are classified as different (*see* **Note 11**).

An example of PFGE analysis of *S. pneumoniae* is shown in **Fig. 11.1**. Using the Tenover criteria, isolates MP1-MEF and MP1-NP are classified as identical, as are isolates MP2-MEF and MP2-NP. Isolate MP3 is classified as possibly related to isolates MP2-MEF and MP2-NP, while MP4 is classified as different from the other five isolates.

During the course of a large epidemiological or clinical study, the banding patterns of a large number of isolates may need to be compared. However, due to the limitations enforced by the number of lanes available in a PFGE gel, it is not always possible to run all samples on the same gel. Thus, several gels may have to be compared. In order to normalize any differences that may be introduced between gels, software such as Bionumerics GelCompare (Applied Maths, Austin, TX) has been developed. Such software utilizes the ladder as a marker, adjusting the banding patterns of each gel to a standard setting so that they might be compared directly.

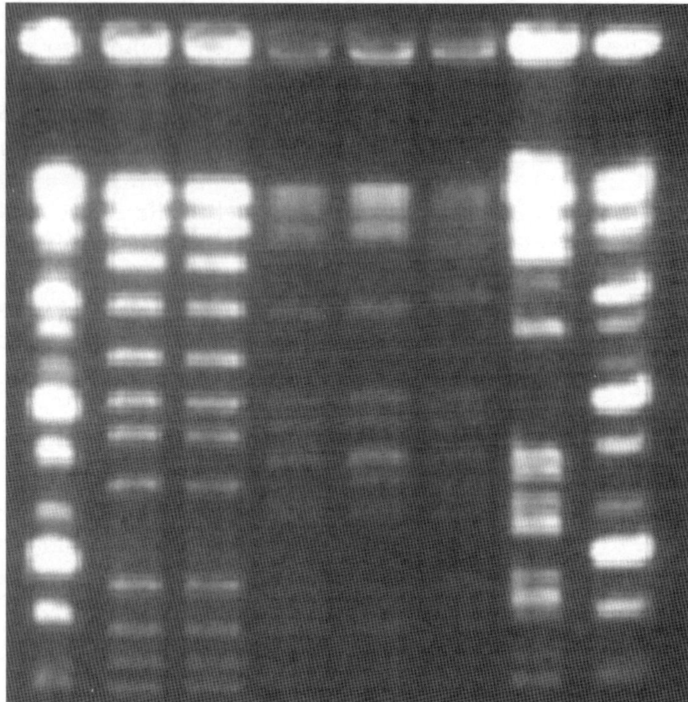

Fig. 11.1. Pulsed field gel electrophoresis image. Lane 1: Ladder; Lane 2: MP1-MEF; Lane 3: MP1-NP; Lane 4: MP2-MEF; Lane 5: MP2-NP; Lane 6: MP3; Lane 7: MP4; Lane 8: Ladder.

4. Notes

1. M13F and M13R tails are added to each of the gene-specific primers to increase the ease of sequencing in a 96-well format. Each of the alleles can be sequenced with the same forward and reverse primer regardless of the housekeeping gene being sequenced, without compromising the accuracy of the sequence data.
2. During the PCR product purification stage and DNA sequencing reaction purification, the same PCR adhesive films may be used after each centrifugation and removal of excess fluid.
3. The sequencing plates and tissue should be wrapped in saran wrap when being centrifuged to remove the excess PEG/NaCl, in order to prevent the PEG/sodium chloride solution from leaking into the centrifuge, as this may impair the functioning of this equipment.
4. While it is possible to purify the PCR products/DNA sequencing reactions of four plates at once, the authors have noted a decrease in quality of the sequences obtained when

more than three plates are processed at once, while the number of repeats required also tends to increase. For PCRs, a plate is considered to be of 48-wells, while for DNA sequencing reactions a plate is considered to be of 96-wells (or the two 48-well plates for forward and reverse reactions of a given PCR plate). The quality of the purified PCR product may be ascertained by running 3 μL on a 1% agarose gel (w/v) at this stage.

5. The addition of sterile distilled water at this point dilutes the excess BigDye fluorescent terminators not utilized during the DNA sequencing reaction, thus reducing the number and size of dye blots obtained when the reaction is run on the automated sequencer.
6. The program Phineus (available at http://www.phineus.org/) has been designed to largely automate this process and has been extensively tested for the MLST schemes of *N. meningitidis* and *S. pneumoniae*, but can also be utilized for the analysis of MLST data of other species.
7. EC and ESP buffers may be removed from the −20 °C freezer at this point, due to the prolonged period of time required for these buffers to thaw.
8. The agarose plugs will be virtually transparent at this stage and during all subsequent washes. Therefore, a 100 μL pipette should be used, to avoid accidentally damaging any of the agarose plugs in the process. Positioning of the pipette at the very bottom of the tube should improve the chance of the pipette tip avoiding an agarose plug.
9. A 10 μL inoculating loop is effective at obtaining an agarose plug from the TE buffer.
10. In order to remove all air bubbles from the system, it is often necessary to reduce the pump speed to nearly zero, before detaching the tubing connecting the pump and chamber from the chamber (ensure that a container is placed in position to collect the TBE buffer that will escape from the chamber at this point). Replenish the TBE buffer level in the pump apparatus by pouring TBE directly into the tubing until nearly full and then reattach to the chamber. Increase the pump speed incrementally. Each step should result in another air bubble being forced into the chamber, and hence being eradicated from the system, until all air bubbles have been removed. The electrophoresis chamber and pump apparatus should be rinsed/flushed regularly with distilled water in order to prevent the build up of salts and excess agarose pieces that may block the pump.
11. The Tenover criteria were established for use with a small number of isolates (≤30) in an outbreak setting. PFGE data should be used in conjunction with epidemiological and/or clinical data to draw appropriate conclusions regarding the

relationship between strains. When typing a much larger number of strains, digital normalization methods should be used.

Acknowledgments

This work was supported by funding awarded to MMP by the National Institute of Allergy and Infectious Diseases (R01 AI068043).

References

1. Maiden, M. C., Bygraves, J. A., Feil, E., Morelli, G., Russell, J. E., Urwin, R., Zhang, Q., Zhou, J., Zurth, K., Caugant, D. A., Feavers, I. M., Achtman, M. and Spratt, B. G. (1998) Multilocus sequence typing: a portable approach to the identification of clones within populations of pathogenic microorganisms. *Proc. Natl. Acad. Sci. U.S.A.* **95**, 3140–3145.
2. Enright, M. and Spratt, B. (1998) A multilocus sequence typing scheme for *Streptococcus pneumoniae*: identification of clones associated with serious invasive disease. *Microbiology* **144**, 3049–3060.
3. Meats, E., Feil, E. J., Stringer, S., Cody, A. J., Goldstein, R., Kroll, J. S., Popovic, T. and Spratt, B. G. (2003) Characterization of encapsulated and noncapsulated *Haemophilus influenzae* and determination of phylogenetic relationships by multilocus sequence typing. *J. Clin. Microbiol.* **41**, 1623–1636.
4. Feil, E. J. and Enright, M. C. (2004) Analyses of clonality and the evolution of bacterial pathogens. *Curr. Opinion Microbiol.* **7**, 308–313.
5. Lee, M., Lee, Y. and Chiou, C. (2006) The suitable restriction enzymes for pulsed-field gel electrophoresis analysis of *Bordetella pertussis. Diagn. Microbiol. Infect. Dis.* **56**, 217–219.
6. Schwartz, D. C. and Cantor, C. R. (1984) Separation of yeast chromosome-sized DNAs by pulsed field gradient gel electrophoresis. *Cell* **37**, 67–75.
7. Birren, B. and Lai, E. (1994) Rapid pulsed field separation of DNA molecules up to 250 kb. *Nucleic Acids Res.* **22**, 5366–5370.
8. Dagan, R., Leibovitz, G., Cheletz, G., Leiberman, A. and Porat, N. (2001) Antibiotic treatment in acute otitis media promotes superinfection with resistant *Streptococcus pneumoniae* carried before initiation of treatment. *J. Infect. Dis.* **183**, 800–806.
9. Yano, H., Suetake, M., Kuga, A., Irinoda, K., Okamoto, R., Kobayashi, T. and Inoue, M. (2000) Pulsed-field gel electrophoresis analysis of nasopharyngeal flora in children attending a day care center. *J. Clin. Microbiol.* **38**, 625–629.
10. Pettigrew, M. M., Foxman, B., Ecevit, Z., Marrs, C. F. and Gilsdorf, J. (2002) Use of pulsed-field gel electrophoresis, enterobacterial repetitive intergenic consensus typing, and automated ribotyping to assess genomic variability among strains of nontypeable *Haemophilus influenzae. J. Clin. Microbiol.* **40**, 660–662.
11. Malchowa, N., Sabat, A., Gniadkowski, M., Krzyszton-Russjan, J., Empel, J., Miedzobrodzki, J., Kosowska-Shick, K., Appelbaum, P. C. and Hryniewicz, W. (2005) Comparison of multiple-locus variable-number tandem repeat analysis with pulsed-field gel electophoresis, *spa* typing, and multilocus sequence typing for clonal characterization of *Staphylococcus aureus* isolates. *J. Clin. Microbiol.* **43**, 3095–3100.
12. Pettigrew, M. M., Fennie, K. P., York, M. P., Daniels, J. and Ghaffar, F. (2006) Variation in the presence of neuraminidase genes among *Streptococcus pneumoniae* isolates with identical sequence types. *Infect. Immun.* **74**, 3360–3365.
13. Feil, E. J., Li, B.C., Aanensen, D. M., Hanage, W. P. and Spratt, B. G. (2004) eBURST: Inferring patterns of evolutionary descent among clusters of related bacterial genotypes from multilocus sequence typing data. *J. Bacteriol.* **186**, 1518–1530.
14. Tenover, F. C., Arbeit, R. D., Goering, R. V., Mickelsen, P. A., Murray, B. E., Persing, D. H. and Swaminathan, B. (1995) Interpreting chromosomal DNA restriction patterns produced by pulsed-field gel electrophoresis: criteria for bacterial strain typing. *J. Clin. Microbiol.* **33**, 2233–2239.

Chapter 12

Fluorescence "In Situ" Hybridization for the Detection of Biofilm in the Middle Ear and Upper Respiratory Tract Mucosa

Laura Nistico, Armin Gieseke, Paul Stoodley, Luanne Hall-Stoodley, Joseph E. Kerschner, and Garth D. Ehrlich

Abstract

Most chronic bacterial infections are associated with biofilm formation wherein the bacteria attach to mucosal surfaces, wound tissue, or medical device surfaces in the human body via the formation of an extracellular matrix. Biofilms assume complex three-dimensional structures dependent on the species, the strain, and the prevailing environmental conditions and are composed of both the bacteria and the extracellular slime-like matrices, which surround the bacteria. Bacteria deep in the biofilm live under anaerobic conditions and must use alternatives to O_2 as a terminal electron acceptor. Thus, the metabolic rates of these deep bacteria are greatly reduced, which renders them extremely resistant to antibiotic treatment, and for reasons not clearly understood, it is often very difficult to culture biofilm bacteria using traditional microbiologic techniques. To directly identify and visualize biofilm bacteria in a species-specific manner, we developed a confocal laser scanning microscopy (CLSM)–based 16S rRNA fluorescence in situ hybridization (FISH) protocol, to find biofilm bacteria in middle ear and upper respiratory tract mucosa, which preserves the three-dimensional structure of the biofilm and avoids the use of traditional culture techniques.

Key words: Middle ear mucosa, upper respiratory tract mucosa, biofilm, FISH, 16S rRNA fluorescent probes, confocal laser scanning microscopy (CLSM).

1. Introduction

Otitis media with effusion (OME) is the chief cause for an office visit to a pediatrician in the United States, and it is an important cause of hearing and speech impairment in children during the early years of life. OME is characterized by recurrent infections,

Bernd Sokolowski (ed.), *Auditory and Vestibular Research: Methods and Protocols, vol. 493*

DOI 10.1007/978-1-59745-523-7_12 Springerprotocols.com

resistance to repeated cycles of antibiotic therapy, and presence of an effusion in the middle ear for more than 3 months. Often the microorganisms causing the infection are unculturable with traditional culture methods. Post et al. *(1)* have shown that PCR can detect the presence of bacterial DNA in culturally sterile middle ear effusions, and Rayner et al. *(2)*, using an RT-PCR-based assay system, detected the presence of bacterial mRNA in 100% of the DNA-positive effusions, establishing that OME results from an active bacterial process often in the absence of culturability.

Recent studies have shown that pediatric OME *(3)* is a biofilm infection, and subsequently biofilms have been shown to be present in numerous upper and lower respiratory infections including adenoiditis, chronic rhinosinusitis (CRS) *(4,5)*, tonsillitis *(6)*, human and experimental cholesteatomas *(7)*, and chronic obstructive pulmonary disease (COPD) *(8)*. Randomized controlled studies have demonstrated that adenoidectomy is effective in the prevention and treatment of OME *(9–11)*, and we found the most common middle ear pathogens also formed biofilms in pediatric adenoidal tissues (manuscript in preparation).

In order to identify bacterial pathogens that are directly associated with host mucosal epithelia in clinical samples, we developed a 16S rRNA fluorescence in situ hybridization (FISH) method to visualize species-specific bacterial biofilms directly in tissues, such as middle ear mucosa (MEM) and upper respiratory tract mucosa, such as adenoids and tonsils. Typically FISH protocols require different fixation methods for Gram-positive and Gram-negative bacteria. However, because data from culture and a priori knowledge of the infectious pathogen are often lacking for clinical specimens and because they may be too small to subdivide, it is not always possible to fix using two different protocols. Therefore, to facilitate detection of both Gram-positive and Gram-negative bacteria simultaneously in clinical tissues, using species-specific 16S rRNA probes, we developed a single fixation process.

FISH is a powerful molecular technique for the identification and three-dimensional characterization of microorganisms in situ, as it provides information about their phylogenetic identity, morphology, and spatial relationship without cultivation. FISH utilizes oligonucleotide probes, designed from 16S rRNA sequences, for the detection of bacterial cells carrying target sequences in their ribosomes. In situ hybridization was first introduced into bacteriology in 1988 by Giovannoni et al. *(12)* using radioactive reporters, which were replaced by fluorescent tags in 1989 by DeLong et al. *(13)*. Over the past two decades, FISH has been applied to the study of bacteria in a variety of complex communities found in natural environments, especially when culturing failed to identify the extent of the microorganisms present, thereby confirming the usefulness of the technique *(14–17)*.

However, FISH does have some problems. These include difficulties in the detection of slow-growing bacteria (with a low number of ribosomes and subsequently low signal fluorescence) and differences in bacterial cell permeability to FISH probes. The latter is particularly problematic since aggressive lysing for a highly impermeable species such as *S. pneumoniae* may lead to complete lysis of more delicate bacteria. Conversely, a less aggressive permeabilization technique may leave some species unstained and, therefore, undetected. To overcome this problem, specimens are often cryosectioned and the sections immobilized on coated slides, so that even if cells are completely lysed, the ribosomes remain relatively localized, providing a strong enough signal for detection. We prefer to view the specimens not in thin section but from an aerial view, which we believe provides a better representation of biofilm distribution and structure. In some cases, biofilms are preserved by embedding in agarose or polyacrylamide gel *(18)*. Also, we suspect that the fixation, permeabilization, and washing steps used in the FISH protocol might cause a detachment of part of the biofilms present on the tissue surface. In fact, in staining fresh tissue with generic fluorescent DNA dyes, we found more bacteria on the epithelial layer than with FISH. However, the latter technique has the advantage of detecting subepithelial and intracellular bacteria often not visible using other methods. In addition to development in fixation, the improvements in brighter fluorochromes, the development of CARD-FISH *(19)*, more effective probe design software, greater availability of commercial probe labeling, and new software for confocal microscopy, all make FISH an excellent technique for the detection and visualization of biofilms in medical specimens.

The 16S rRNA oligonucleotide probes are used to target a range of phylogenetic groups from the species level to the domain. It is common to use a domain probe (eukarya) to identify all bacteria, whereas genus and species probes identify specific bacteria of interest. The specificity of the assay is determined by shifting the probe target from extremely conserved regions (species-specific) to increasingly variable regions in the ribosomal RNA. However, probe accessibility to the target may be limited by the tertiary and quaternary structure of the rRNA, and the distinction of the microorganisms within a species is limited by the high degree of sequence conservation. Schonhuber et al. *(20)* explored the possibility of using tmRNA in in situ hybridization to overcome these limitations.

The fluorescently conjugated 16S rRNA oligonucleotide probes are applied on fixed and permeabilized whole-bacterial cells under stringent hybridization conditions. When these probes specifically hybridize with their complementary target sequences in the ribosomes within the cells, they make the cells fluorescent. If hybridization doesn't occur, or if the degree of hybridization is low, the probes will be removed during washing. Stringency conditions may be adjusted by varying the incubation temperature, the formamide (denaturant) concentration, and the salt content in the hybridization and washing buffers. The present protocol is a modification of the protocol described by Mantz et al. *(21)*, where the stringency conditions are reached by changing the formamide and salt concentrations and leaving constant hybridization and wash temperatures.

FISH combined with confocal laser scanning microscopy (CLSM) allows the visualization of biofilm communities, clearly defining the bacterial species and the three-dimensional structures present in the sample. The great power of the method is the identification and visualization of specific bacteria in the absence of positive bacterial cultures. FISH can also be combined with generic fluorescent stains to assess the percentage and localization of the species in question within the larger bacterial population.

2. Materials

2.1. Sample Fixation

1. 3X phosphate buffered saline (PBS) stock solution: 0.39 m*M* NaCl, 0.70 m*M* NaH_2PO_4, 0.71 m*M* Na_2HPO_4, in 800 mL of Milli-Q (MQ) water, adjust to pH 7.2, using 1 m*M* HCl. Add water to a volume of 1 L and autoclave. Store at room temperature. Prepare 1X PBS from 3X PBS diluting 1:3 with MQ water.
2. Working paraformaldehyde (PFA) solution: Dilute 16% electron microscopy–grade PFA stock solution (Electron Microscopy Science, Hatfield, PA) to 4% with PBS 3X and filter sterilize. Use fresh PFA (stored at 4 °C) for a maximum of 1 week. Concentrated stock solutions of 16% can be kept for long term in the dark at room temperature and diluted with PBS once needed. PFA is a toxic, hazardous material; precautions must be taken in handling it, and it needs to be disposed in accordance with all local, state, and federal regulations (*see* **Note 1**).
3. Working ethanol solutions: 50% and 80% ethanol made from pure ethanol (97%) and 1X PBS. Filter sterilize and store at room temperature.

2.2. Sample Immobilization

1. Gelatin-coated slides: Soak Shandon multi-ppot microscope slides (Thermo Electron Corporation, Waltham, MA) in a solution of 0.075% gelatin (Sigma-Aldrich, St. Louis, MO) and 0.01% $CrK(SO_4)_2$ (Sigma-Aldrich) heated to 70 °C for 1 min. Allow to air dry in a vertical position.

2.3. Additional Permeabilization for Gram-Positive Bacteria

1. 1 *M* Tris-HCl stock solution: Mix 121.1 g of Tris in 800 mL of MQ water and adjust pH to 8.0 with 10 *M* HCl (about 42 mL). Add water to a volume of 1 L, autoclave, and store at room temperature.
2. EDTA stock solution: Mix EDTA to 0.5 *M* in 800 mL of MQ water and adjust pH to 8.0 with 10 *M* NaOH. Add water to a volume of 1 L, autoclave, and store at room temperature.
3. Lysozyme working solution: 0.5 mg/mL of lysozyme (Sigma-Aldrich) in a buffer containing 0.1 *M* Tris-HCl and 0.05 *M* EDTA. Filter sterilize and store in aliquots in freezer at −20 °C.
4. Pure ethanol (97%) (molecular biology grade). Filter sterilize and store at room temperature.
5. Ethanol-PBS solutions: Mix 50% and 80% concentrations in 1X PBS and filter sterilize.
6. Sterile disposable scalpel (Invitrogen, Carlsbad, CA).
7. Petri dishes 60 mm × 15 mm (Thermo Fisher Scientific, Hampton, NH).

2.4. FISH

1. Stock solutions: 5 *M* NaCl and 1 *M* Tris-HCl, pH 8.0. Store at room temperature.
2. Formamide deionized (AMRESCO, Solon, OH). Formamide is a toxic, hazardous material; precautions must be taken in handling, and it needs to be disposed in accordance with all local, state, and federal regulations (*see* **Note 1**).
3. 10% SDS. 100 g SDS (electrophoresis grade), 900 mL of MQ water, heat to 68 °C, adjust pH to 7.2 with 10 *M* HCl, and bring to a volume of 1 L with MQ water. No sterilization. SDS is a toxic, hazardous material; precautions must be taken in handling, and it needs to be disposed in accordance with all local, state, and federal regulations (*see* **Note 1**).
4. In situ hybridization working buffer: Mix 360 μL of 5 *M* NaCl, 40 μL of 1 *M* Tris-HCl, pH 8.0, and *X* μL of formamide deionized (the optimal formamide concentration is determined for each probe; *see* **Section 3.4, step 1**) in a 2 mL snap cap tube. Fill to 2 mL with MQ water. Add 2 μL of 10% SDS (*see* **Note 2**).
5. Washing buffer: Mix 1 mL of 1 *M* Tris-HCl, pH 8.0, *Y* μL of 5 *M* NaCl (optimal concentration depends on the formamide concentration in the hybridization buffer), and *Z* μL of 0.5 *M* EDTA (*see* **Table 12.1**) in a 50 mL conical tube. Fill to 50 mL with MQ water and add 50 μL of 10% SDS (*see* **Note 2**).
6. Microscope slide (Thermo Fisher Scientific).

2.4.1. Fluorescently Conjugated 16S rRNA Probes

1. A short probe consisting of a sequence of ~ 18–24 nucleotides in length that is unique to the target specific group, by showing a perfect base match to the target sequence and more than

Table 12.1
NaCl and EDTA Concentrations in the washing buffer

% form a mide (hybridiz.buffer)	5 M NaCl (=y) μl	0.5 M EDTA (=z) μl
0	9000	–
5	6300	–
10	4500	–
15	3180	–
20	2150	500
25	1490	500
30	1020	500
35	700	500
40	460	500
45	300	500
50	180	500
55	100	500
60	40	500
65	–*	500
70	–	350
75	–	250
80	–	175
85	–	125
90	–	88
95	–	62

*enough NaCl in EDTA

one base mismatch to the same region of the non-target bacteria. Mismatches to non-targeted bacteria should be centralized within the sequence to minimize hybridization to such sites *(22)*. The correct hybridization of two sequences to each other depends on temperature, composition of the hybridization and washing buffers, and the length and G + C content of the oligonucleotide. The G + C content influences the duplex stability and should range between 50 and 60% for probes of 18–24 bases (*see* **Note 3**).

2. Once the probe is designed, several companies are available to synthesize the probe and conjugate the fluorochrome such as 4′, 6-diamidino-2-phenylindole (DAPI),

the sulfoindocarbocyanines (Cy3, Cy5, and Cy7), 6-carboxy fluorescine (FAM), and others to the 5′ end (e.g., Integrated DNA Technologies INC, Coralville, IA).

3. Prepare aliquots of working solution probes at 50 ng/μL and store at −20 °C (*see* **Note 4**).

2.5. Cytology Fluorescent Stains

2.5.1. F-Actin Stain for Visualization of the Cell Cytoskeleton in Tissue

1. 0.1% Triton: Add 100 μL of Triton X100 (Sigma-Aldrich) to 99.9 mL of stock PBS (*see* **Section 2.1, step 1**).
2. Phalloidin (Invitrogen) conjugated with Alexa Fluor fluorescent stain (*see* **Note 5**): Mix 25 μL of the stain in 1 mL of 1X PBS.

2.5.2. DNA Staining for the Visualization of the Cell Nucleus in the Tissue

1. DNA stain such as DAPI (Sigma-Aldrich, St. Louis, MO) at a concentration of 0. 5–1. 0 μg/mL of 1X PBS or 3 μL of Propidium Iodide, Syto 59, or Syto 9 (Invitrogen), in 1 mL of 1XPBS (*see* **Note 5**).

2.6. Microscopy

1. Shandon multi-spot microscope slides (Thermo Electron Corporation, Waltham, MA).
2. Petri dishes 60 mm × 15 mm (Thermo Fisher Scientific).
3. Cover glass, no. $1^1/_2$, 24 mm × 50 mm^2 (Corning Incorporated, Corning, NY).
4. 100% Silicone rubber sealant (*see* **Note 6**).
5. Antibleaching mounting oil (e.g., Citifluor, BacLight mounting oil, Vectashield).
6. Alkaline mounting oil: 50% glycerol in PBS, pH 8.5 (adjust the pH using 1 m*M* NaOH).
7. Immersion oil for fluorescence microscopy Type DF (Cargille Laboratories, Cedar Grove, NJ).
8. CLSM imaging system such as a Leica DM RXE microscope attached to a TCS SP2 AOBS confocal System with high-resolution objectives (Leica Microsystem, Exton, PA).

3. Methods

3.1. Fixation of Various Types of Samples

3.1.1. Fixation of Gram-Negative or Unknown Bacteria in Suspension (Liquid Bacterial Culture for Probe Testing, Ear Effusion, Liquid Samples)

1. Add freshly prepared solution of 4% PFA to liquid sample (2–4% final concentration) with a disposable pipette and incubate for 1–12 h at 4 °C.
2. Spin down biomass, remove supernatant with a disposable pipette (discard in PFA/formamide waste), and resuspend in PBS. Repeat 2X.
3. Spin down biomass, remove supernatant, resuspend in 50% ethanol-PBS, and store at −20 °C until further use.

3.1.2. Fixation of Gram Negative or Unknown Bacteria in Tissue

1. Place the tissue sample in a 2 or 15 mL conical tube. A freshly prepared solution of 4% PFA is added with a disposable pipette to the tissue sample (2–4% final concentration), which is then incubated at 4 °C for 2–12 h, depending on the tissue thickness.
2. Remove the PFA carefully with a disposable pipette and discard in the PFA/formamide waste. Add PBS, using a volume that is 2X the amount of PFA used in **step 1, Section 3.1.2**. Incubate for 10 min at room temperature, and then repeat this step one more time.
3. Remove the PFA, add 50% ethanol-PBS, and store at −20 °C.

3.1.3. Fixation of Gram-Positive Bacteria in Suspension (Liquid Bacterial Culture for Probe Testing, Ear Effusion, Liquid Samples)

1. Spin down biomass in a 2 mL conical tube and remove supernatant.
2. Add 50% ethanol-PBS to the sample in the intended amount (cell concentration can be increased or decreased by adding the appropriate volume. After a few times, you will have a feel for the right amount of 50% ethanol-PBS needed to reach the right cell concentration: After FISH and during microscopy observations, you have to be able to see the morphology well and the brightness of each single cell. Store at −20 °C at least for 24 h or until further use.

3.1.4. Fixation of Gram-Positive Bacteria in Tissue

1. Put the tissue sample in a 2 or 15 mL conical tube. Cover the drained sample completely with 50% ethanol-PBS and store at −20 °C for at least 24 h prior to use.

3.2. Liquid Sample Immobilization for Microscopy

1. The fixed stored liquid sample is added at a 10–20 μL volume per microwell of a gelatin-coated Shandon multi-spot microscope slide.
2. Allow to dry at room temperature in a closed box for a few hours.
3. Wash for a few seconds with MQ water by pipetting the water onto the slide and allowing it to air dry in a closed box.

3.3. Additional Permeabilization of Gram-Positive or Unknown Bacteria (if Needed)

1. Add 20 μL of working lysozyme solution to the immobilized samples on the slide.
2. Place the slide in a 50 mL conical tube containing a wipe in the bottom that is moistened with buffer to prevent the tissue from drying out. Then, incubate at 37 °C for up to 3 h (*see* **Note 7**).
3. Remove the lysozyme solution, rinse 2X with PBS, add 20 μL of 50% ethanol-PBS, and incubate at room temperature for 3 min.
4. Remove ethanol-PBS solution, add 20 μL of 80% ethanol-PBS, and incubate as in **step 3, Section 3.3**, for 3 min.

5. Remove ethanol-PBS solution, add 20 μL of 100% ethanol-PBS, and incubate as before.
6. Remove solution and allow to air dry at room temperature.
7. If tissue contains unknown bacteria, place the fixed and stored tissue in a Petri dish, cut a 0.5–1 mm thick section using a sterile scalpel, and store the residual sample at −20 °C for further analysis. Place the tissue section for use on a slide and treat as in **steps 1–6, Section 3.3**.

3.4. FISH

1. Evaluate the fluorescent probe for its specificity by hybridizing with a pure culture of target organisms and non-target organisms at increasing levels of stringency, changing the formamide concentration from 0 to 50% in steps of 5%, and maintaining the temperature constant *(17)*. The optimal stringency is considered the highest formamide concentration that does not result in loss of fluorescent intensity on the target cells as observed with an epifluorescent microscope (**Fig. 12.1**). At this concentration, hybridization to the non-target bacteria should no longer occur. Probe evaluation includes the following: test probe, domain-level probe (EUB338) *(14)*, nonsense probe (NONEUB338) *(21)*, and a no-probe control to check for sample autofluorescence *(23)*. Once tested on the target bacteria, the probe is ready to be hybridized with the patient sample.
2. Several fluorescent probes can be used simultaneously or sequentially if they are conjugated with different fluorescent dyes and used according to the following procedure (*see* **Note 8**) *(24)*.
   a. Suspend lyophilized probe in 100 μL of MQ water. The probe concentration is often provided by the manufacturer. However, it is useful to check this measurement to assess the quality of the labeling. To determine probe concentration, measure the absorbance of a 1:100 diluted stock solution at 260 nm (1 A_{260} = 20 μg/mL DNA). The Cy3 dye has a maximum absorbance around 550 nm, and the ratio of A_{260}/A_{550} is approximately 1 for a monolabeled 18-mer oligonucleotide.
   b. Use 8 μL of in situ hybridization buffer and 1 μL of fluorescently labeled probe. Two probes can be used simultaneously if they use the same formamide concentration; in this case add 1 μL of each to 8 μL of hybridization buffer. In the case of three probes used at the same time, reduce the total volume of the probes to 2 μL. Probes that need different formamide concentrations have to be used in separate hybridizations: Use the probe with the higher formamide concentration first, followed by the probe with the lower formamide concentration.

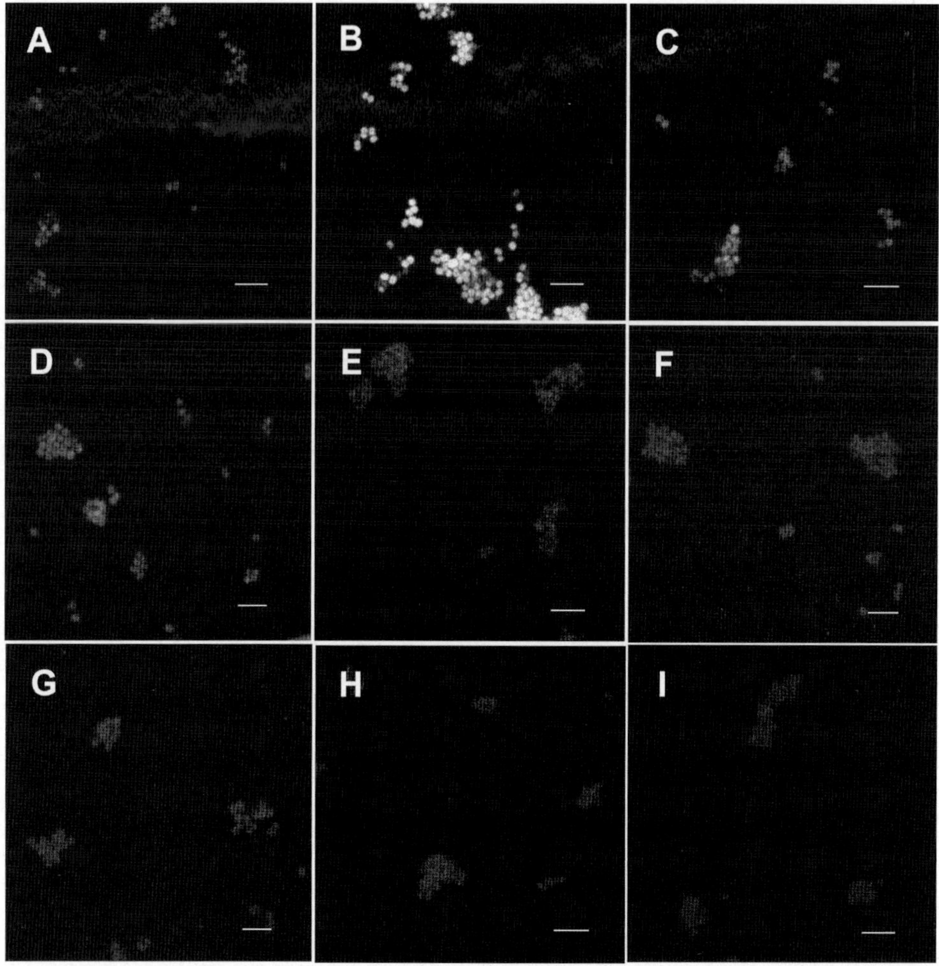

Fig. 12.1. Evaluation of the Mrc88 probe specificity for pure culture of *Moraxella catarrhalis*. FISH was performed at increasing levels of stringency by changing the formamide concentration and maintaining a constant temperature. (**A**) 0%, (**B**) 10%, (**C**)15%, (**D**) 20%, (**E**) 25%, (**F**) 30%, (**G**) 35%, (**H**) 40%, and (**I**) 50%. The optimal stringency is the highest formamide concentration that does not result in a loss of fluorescent intensity, and in this case, it is 10% as shown in **B**. Scale = 5 μm.

c. Place a folded paper towel approximately the size of a slide in a 50 mL tube and pour the remaining hybridization buffer onto the paper towel. This avoids evaporation of buffer in the sample during hybridization. Apply 10 μL of the hybridization buffer–probe mixture onto each well in the slide or onto the MEM. Apply a larger volume of the mixture (same ratio) to cover the sample onto tissue sections. Place the slide with the sample in the tube, screw on cap, place horizontally in the hybridization oven, and hybridize for 90–120 min at 46 °C (*see* **Note 9**).

d. Quickly wash off the hybridization buffer with pre-warmed washing buffer (*see* **Note 10**).

e. Incubate the samples in the remaining pre-warmed washing buffer for 15 min in a water bath at 48 °C (*see* **Note 9**).
f. Rinse the samples with MQ water and allow them to dry at room temperature in the dark.

3.5. Cytology Fluorescent Staining

3.5.1. F-Actin Stain for Visualization of the Cell Cytoskeleton in the Tissue

1. Treat the sample with 0.1% Triton for 3–5 min to increase the permeability of the tissue's cells.
2. Rinse the sample 3X with PBS, then cover it (25 μL for the MEM and 200–250 μL for adenoid and tonsil) with the working solution of Phalloidin/Alexa Fluor (*see* **Note 5**), and stain for 25 min at room temperature in the dark.
3. Rinse the sample 2X with PBS and allow the tissue to stay moist by covering the surface with PBS.

3.5.2. DNA Staining for the Visualization of the Cell Nucleus in the Tissue

1. Cover the sample (25 μL for the MEM and 200–250 μL for adenoid and tonsil) with a working solution of ageneric DNA stain such as DAPI, propidium iodide, or Syto stain for 20 min at room temperature in the dark (*see* **Note 5**).
2. Rinse the sample 2X with PBS and keep the tissue moist by covering the surface with PBS.

3.6. Mounting Specimens for Microscopy

This section provides protocols to examine bacterial cultures, very small ($< 1\,mm^2$) biopsics (in this case MEM) as well as larger (mm^2 to cm^2) tissue biopsies that can have a rough topography varying by as much as 1 cm in thickness.

3.6.1. Mounting Gelatin-Coated Microscopy Slides (*see* Section 3.2)

1. Mount slides in antibleaching oil and cover with a 24 × 50 mm^2 cover-slip. If using fluorescein- or fluorescein-derivate labeled probes, the 1X PBS pH has to be adjusted to 8.5 with 0.1 m*M* NaOH to maximize the brightness of this fluorophore.

3.6.2. Mounting MEM

1. Apply a thin layer of silicone on the edges of a 25 mm^2 glass cover-slip.
2. Place the MEM with two drops of PBS onto the cover-slip and make a chamber by placing another cover-slip over the first.
3. Let the silicone dry for 5–10 min. If using fluorescein- or fluorescein-derivate-labeled probes, the pH of the PBS has to be adjusted to 8.5 with 0.1 m*M* NaOH, to maximize the brightness of this fluorophore.

3.6.3. Mounting Tissue Sections (Adenoid, Tonsils, etc.)

1. Gently absorb the excess liquid from the tissue section by blotting with a paper towel or Kimwipe.
2. Place a smear (~ 0.5 mL) of silicone sealer on the bottom of a 60 mm Petri dish and attach the tissue section by gravity or,

if necessary, by applying some gentle pressure at the edges of the specimen (*see* **Note 6**).

3. Allow to dry for 10 min, and then submerge the sample in PBS. If using fluorescein- or fluorescein-derivate-labeled probes, the PBS has to be adjusted to pH 8.5 with 0.1 mM NaOH, to maximize the intensity of this fluorophore.

3.7. Confocal Laser Scanning Microscopy

3.7.1. Bacterial Culture and MEM Specimens

Place the covered glass chamber with the enclosed MEM on a glass slide. The hybridized samples in the gelatin-coated Shandon multi-spot microscope slide and the covered glass chamber, containing the MEM, can be observed with dry or high-resolution oil immersion objectives.

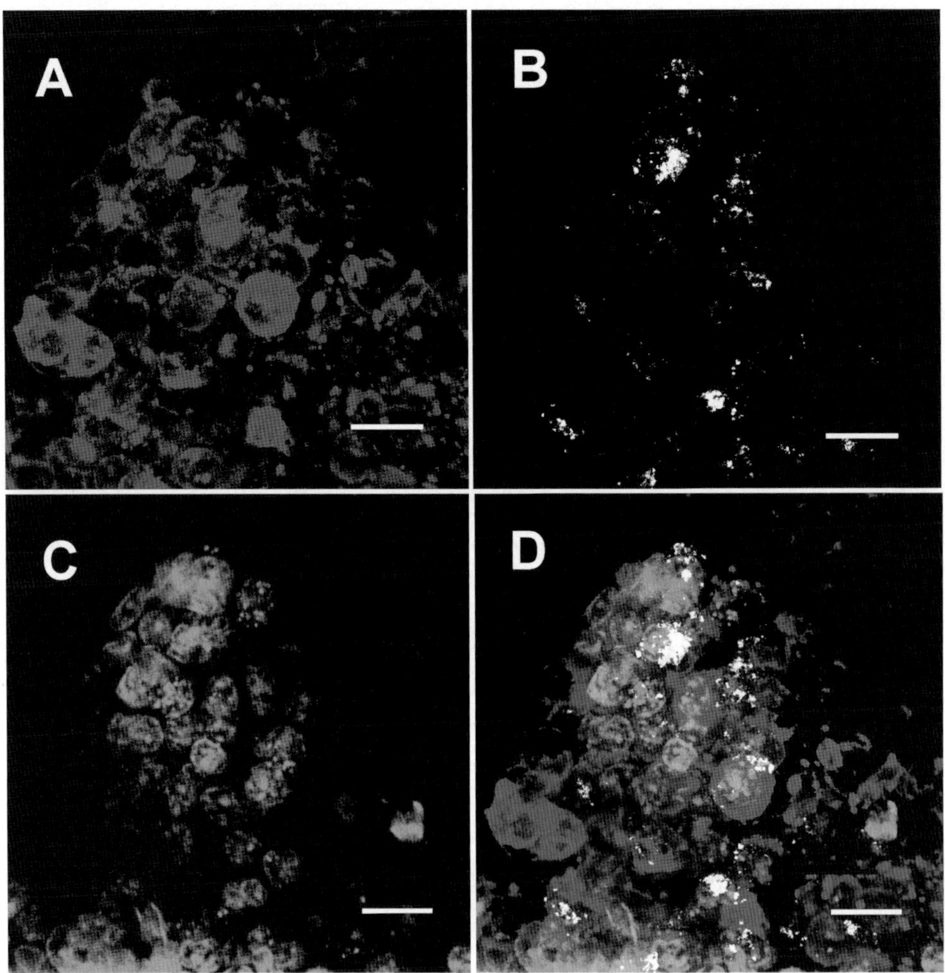

Fig. 12.2. (**A**) Tissue cytoskeleton was revealed by F-actin staining with phalloidin. (**B**) Clusters of bacteria revealed by the eubacterial FISH probe conjugated with Cy3. (**C**) The generic DNA stain Syto 59 stains the nuclei of the adenoid cells and unidentified bacteria. (**D**) Overlay of the three channels showing unidentified extracellular bacteria and intracellular bacterial clusters in the adenoid tissue. *See* color overlay in **Fig. 12.9A**. Scale = 20 μm.

3.7.2. Tissue Mounted in Petri Plate

The tissue attached to the Petri plate and submerged under PBS can be observed with a 63X long working distance, water immersion objective. The advantage of using this objective is that the biofilm can be observed on relatively rough surfaces, as the long working distance minimizes hitting any "high spots" on the specimen with the objective.

Use the appropriate laser line in accordance with the spectral properties of the fluorochrome-conjugated FISH probes used for hybridization. When performing multiple FISH labels simultaneously, it is important to choose fluorophores whose excitation/emission wavelengths have minimal overlap (*see*

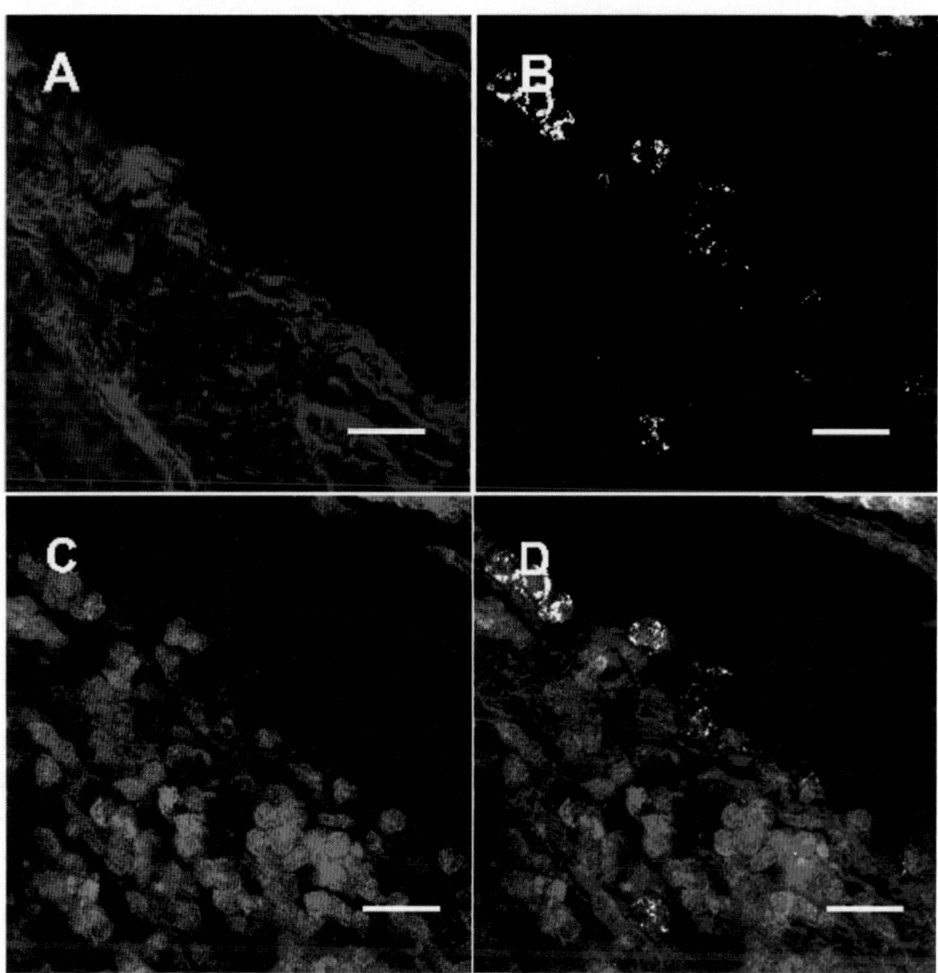

Fig. 12.3. (**A**) Tissue cytoskeleton was revealed by F-actin staining with phalloidin. (**B**) *Staphylococcus* sp. revealed by a staphylococcal FISH probe conjugated with Cy3. (**C**) Tissue nuclei stained with the DNA dye propidium iodide (PI). In the same channel, bacteria stained with FISH eubacterial probe conjugated with Cy5. Usually the fluorescent stain of the tissue nuclei is brighter than the stain of the Cy5 probe, and for this reason it is difficult to see the bacteria when PI is added to visualize the tissue. (**D**) Overlay of the three channels: There is evident presence of *Staphylococcus* sp. associated with the adenoid tissue. *See* color overlay in **Fig. 12.9B**. Scale = 20 μm.

Notes 4, 5). Fluorochromes with different excitation wavelengths can have emission spectra that show a wide overlap, so-called crosstalk, resulting in the recording of several fluorescence signals in a single detection channel, which cannot be separated into individual images. To avoid crosstalk, sequential scanning is recommended. In addition to the detection of fluorescence signals, it is also useful to collect transmitted light, providing the specimen is not opaque, or reflected light, to show the surface of the specimen. For reflected imaging, we use the 488 nm laser line and set a "blue" detector to collect the reflected light. The

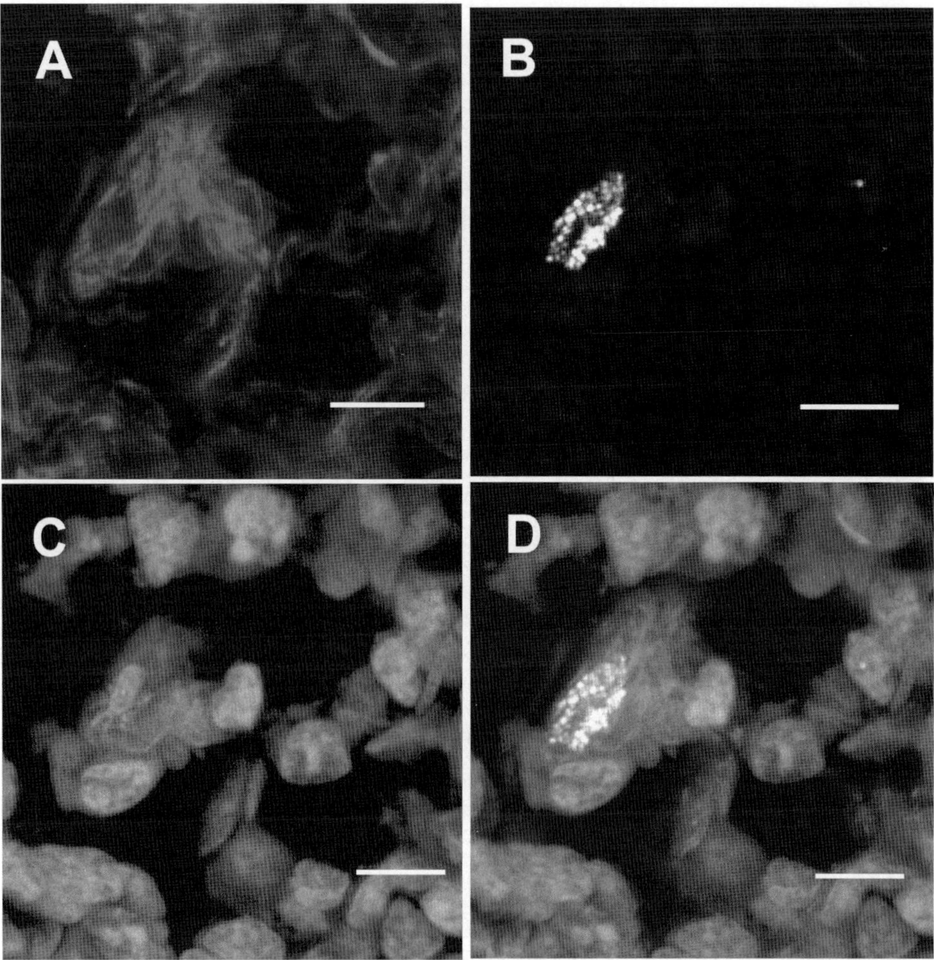

Fig. 12.4. (**A**) Tissue cytoskeleton was revealed by F-actin staining with phalloidin. (**B**) *Streptococcus* sp. revealed by specific FISH probe conjugated with Cy3. **(C)** Tissue nuclei stained with the DNA dye propidium iodide (PI) and all bacteria stained with the generic eubacterial FISH probe conjugated with Cy5 (both PI and Cy5 fluoresce are in the red channel). Usually the fluorescent stain of the tissue nuclei is brighter than the stain of the Cy5 probe, and for this reason it is difficult to see the red bacteria when PI is added to visualize the tissue. (**D**) Overlay of the three channels: There is evident presence of *Streptococcus* sp. associated with the nucleus of an adenoid cell. *See* color overlay in **Fig. 12.9C**. Scale bar = 10 μm.

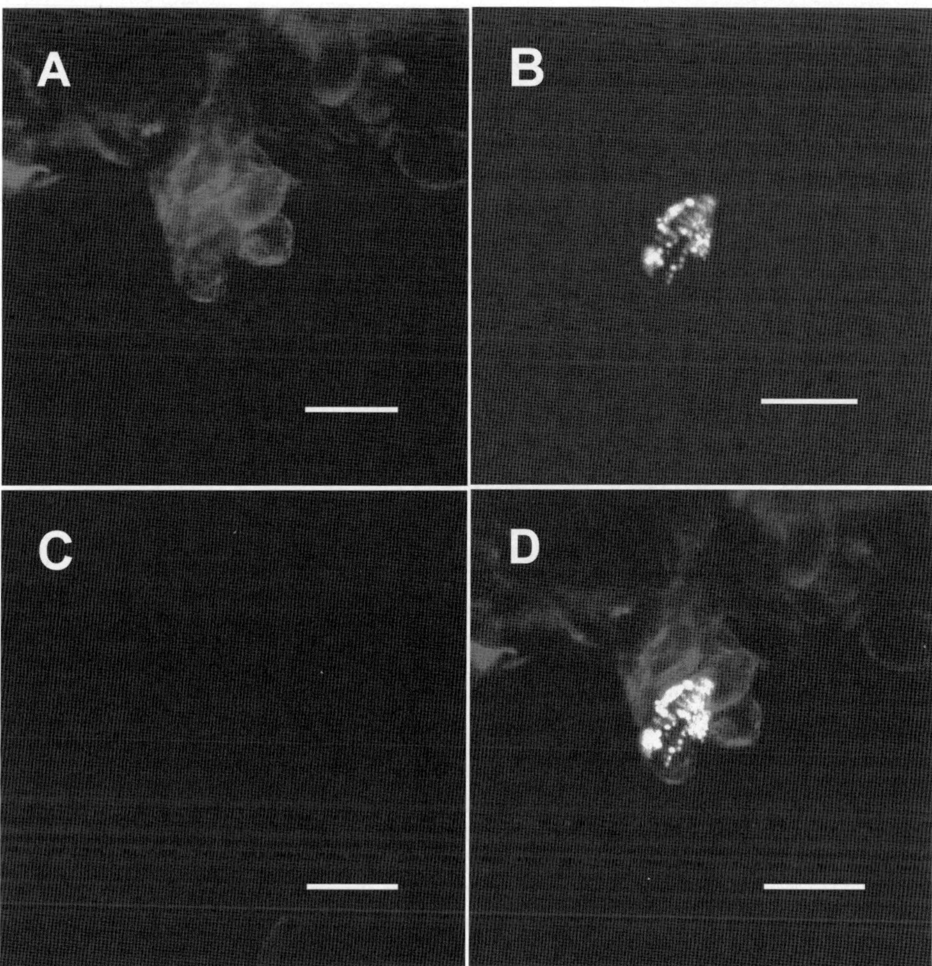

Fig. 12.5. (**A**) Tissue cytoskeleton revealed by F-actin staining with phalloidin. (**B**) Cluster of bacteria revealed by the eubacterial FISH probe conjugated with Cy5. (**C**) No bacteria were revealed using the FISH probe specific for *Streptococcus pneumoniae*. (**D**) Overlay of the three channels showing an intracellular cluster of unidentified bacteria in the adenoid tissue. *See* color overlay in **Fig. 12.9D**. Scale bar = 10 μm.

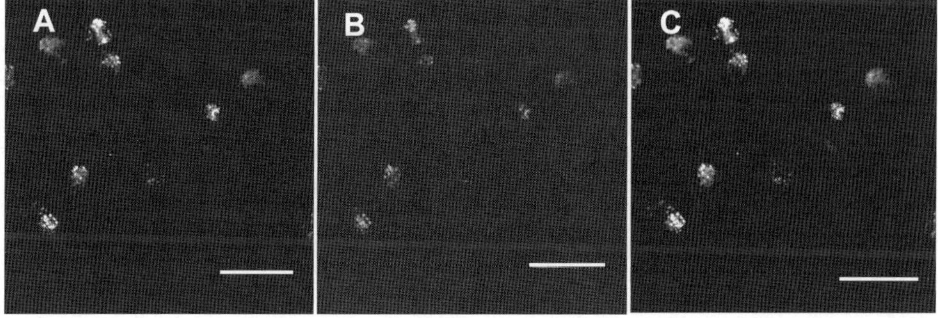

Fig. 12.6. (**A**) Clusters of *Moraxella catarrhalis* revealed by a specific FISH probe conjugated with FAM. (**B**) Eubacterial FISH probes conjugated with Cy5 reveals cluster of bacteria. (**C**) Overlay of the three channels: There is evident presence of bacterial clusters containing *Moraxella catarrhalis* and unidentified bacteria in the adenoid tissue. *See* color overlay in **Fig. 12.9E**. Scale bar = 30 μm.

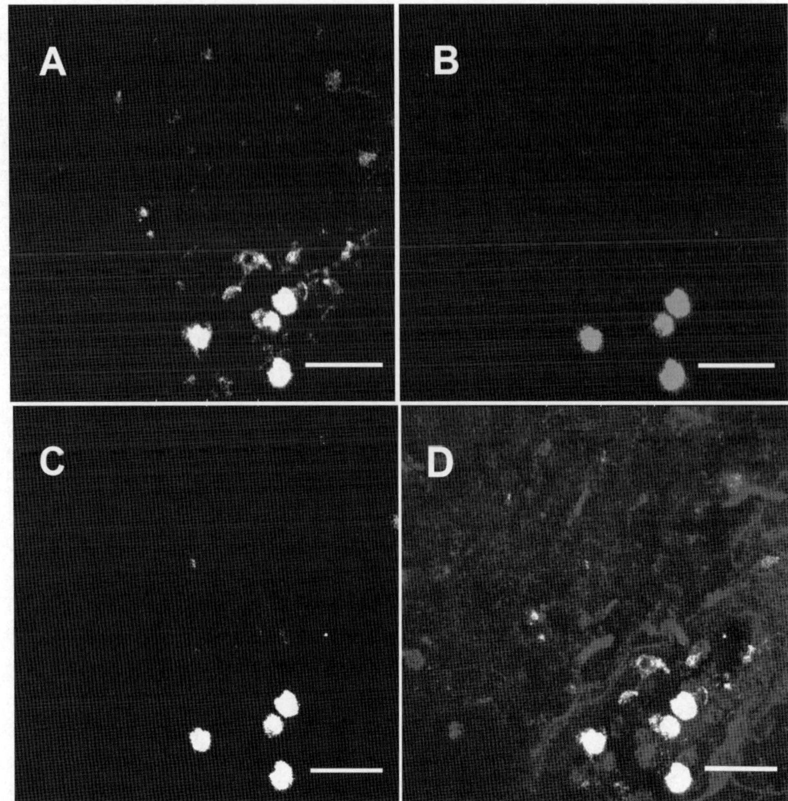

Fig. 12.7. (**A**) Clusters of bacteria revealed by the FISH eubacterial probe conjugated with FAM. (**B**) Clusters of *Staphylococcus* sp. revealed by the specific FISH probe conjugated with Cy3. (**C**) Clusters of *Staphylococcus aureus* revealed by the specific FISH probe conjugated with Cy5. (**D**) Overlay of the three channels: There is evident presence of *S. aureus* clusters revealed by each of the three probes and other bacteria revealed by the eubacterial probe (green channel) in adenoid tissue. Scale bar = 30 μm.

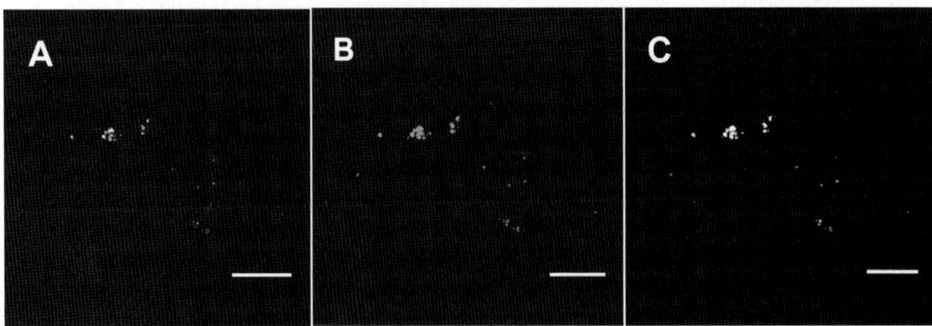

Fig. 12.8. (**A**) *Haemophilus influenzae* revealed in the MEM by the FISH specific probe conjugated with Cy3. (**B**) Bacteria revealed in the MEM by the eubacteria FISH probe conjugated with Cy5. (**C**) Overlay of the two channels showing *H. influenzae* and a few other bacteria in the ear tissue (not visible). *See* color overlay in **Fig. 12.9F**. Scale bar = 15 μm.

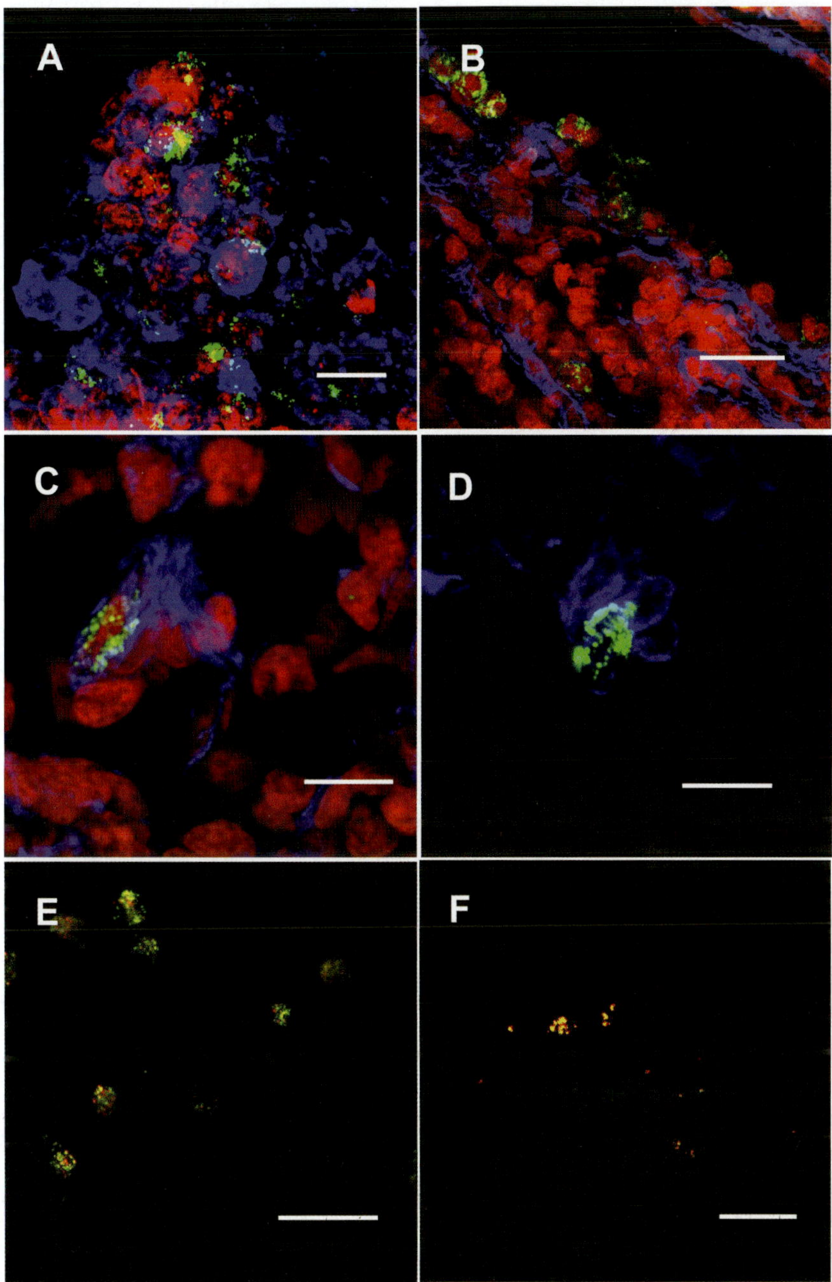

Fig. 12.9. Color overlays from **Figs. 12.2–12.6**, and **12.8**. Tissue cytoskeleton is in *blue*, nuclei of adenoid tissue in *red*, and bacteria in *green-yellow*. (**A**) Unidentified extracellular bacteria and intracellular bacterial clusters in adenoid tissue, revealed by the eubacterial FISH probe. (**B**) Evident presence of *Staphylococcus* sp. associated with adenoid tissue. (**C**) Evident presence of *Sreptococcus* sp. associated with adenoid tissue. (**D**) Intracellular cluster of unidentified bacteria in adenoid tissue revealed by a eubacterial FISH probe. (**E**) Bacterial clusters containing *Moraxella catarrhalis* (*green-yellow*) and unidentified bacteria (*red*) in adenoid tissue. (**F**) *Haemophilus influenzae* (*yellow*) and few other bacteria (*red*) in ear tissue (not visible). Scale bar = 20, 20, 10, 10, 30, and 15 μm, respectively.

Table 12.2
FISH probes, organism specificity, and 16S target sequences

Probe	16S Sequence	Target	Reference
Eub338	GCTGCCTCCCGTAGGAGT	All bacteria	*(15)*
NONEUB338	ACTCCTACGGGAGGCAGC	Nonsense sequence	*(21)*
Hacinf	CCGCACTTTCATCTCCG	*Haemophilus influenzae*	*(33)*
Sta	TCCTCCATATCTCTGCGC	*Staphylococcus* spp.	*(34)*
Str	CACTCTCCCCTTCTGCAC	*Streptococcus* spp.	*(34)*
Spn	GTGATGCAAGTGCACCTT	*Streptococcus pneumoniae*	*(35)*
Mrc88	CCGCCACUAAGUAUCAGA	*Moraxella catarrhalis*	*(3)*
Sau	GAAGCAAGCTTCTCGTCCG	*Staphylococcus aureus*	*(34)*

reflected image is optimized using the Acousto Optical Beam Splitter (AOBS) function, and the best contrast is achieved by using high-gain and high-background detector settings. Examples of CLSM images of adenoid samples and MEM are shown in **Figs. 12.2–12.9**. FISH was performed with several 16S rRNa probes (**Table 12.2**), and other dyes were used in different combinations (**Table 12.3**) to visualize the tissue and localize the bacteria onto and within the sample.

4. Notes

1. It is recommended that chemical goggles, appropriate protective gloves, and masks are worn.
2. It is recommended to add the SDS last in the hybridization and washing buffer to avoid precipitation with the concentrated NaCl and to avoid foam formation.
3. Since the FISH probe design and evaluation process might be fairly time consuming, it is worth searching in the literature for a suitable probe that already exists for your target organism. However, several online molecular sequence databases such as EMBL *(25)* (http://www.ebi.ac.uk/embl/), GenBank *(26)* (http://www.ncbi.nlm.nih.gov/Genbank/), and ribosomal RNA databases such as the Ribosomal Database Project (RDP) *(27, 28)* and the European Ribosomal RNA database *(29)* are available. Moreover, online databases for FISH rRNA probes are available including ProbeBase (http://www.microbial-ecology.net/probebase/) *(30)* and Oligo Retrieval System (ORS, http://soul.mikro.biologie.tu-muenchen.de/ORS/). Probe design can be performed

Table 12.3
Flurorescent probes commonly used showing the maximum excitation wavelength (Ex) the maximum fluorescence emission wavelength (Em), the confocal laser line we use to excite the stain, the nominal color that we commonly assign, and the target and/or conjugate

Fluorescent probe or imaging technique	Ex (nm)	Em (nm)	Laser (nm)	Color	Target and conjugate and for FISH probes the name of the probe, i.e., "Sau"
General Stains					
SYTO9	483	500	488	Green	All bacteria and host nuclei (live bacteria when used in conjunction with propidium iodide)*. In some case extracellular DNA (eDNA) in the EPS can be detected.
Propidium iodide	538	617	488 or 543 or 594	Red	All bacteria and host nuclei (dead bacteria when used in conjunction with propidium iodide)*. In some case eDNA can be detected.
SYTO59	622	644	543	Red	All bacteria and host nuclei. In some case eDNA can be detected.
Transmitted light	NA	NA	Any	Grey	Tissue components and bacteria
Reflected light	NA	NA	488	Blue	Surface of tissue, foreign body surfaces, and bacteria. In some cases cytoskeleton is readily distinguishable.
FISH stains*					
Cy3	550	564	543	Green	Streptococcus genus "Str"

(continued)

Table 12.3 (continued)

Cy3	550	564	543	Green	*S. pneumoniae* "Spn"
Cy3	550	564	543	Green	*H. influenzae* "Haeinf"
Cy3	550	564	543	Green	Stapylococcus genus "Sta"
Cy5	649	666	633	Red	*S. aureus* "Sau"
Cy5	649	666	633	Red	*H. influenzae* "Haeinf"
FAM-5 (pH 9)	492	518	488	Blue or Green	*M. catarrhalis* "Mrc88"
Cy3	550	564	543	Green	Nonsense probe (control) "NonEub"
Cy3, Cy5, FAM-5 (pH 9)					Bacteria universal FISH probe – "Eub388"
Cytology					
Phalloidin-Alexa488	493	519	488	Blue	Cytoskeleton – F-actin

* See Table 12.2 for the probe sequence. We have probes conjugated to different colors, allowing us to select an appropriate fluorescence with respect to other fluorophores being utilized to stain the same specimen.

manually or using computer programs such as Primrose *(31)* or ARB (latin, "arbor" = tree) *(32)*. Primrose (http://www.cardiff.ac.uk/biosi/research/biosoft/Primrose/) is designed to identify 16S rRNA probes using data from RDP, and it runs on the Microsoft Windows 95/NT/2000 operating system. The ARB software package has a PROBE DESIGN tool to search for specific16S rRNA oligonucleotides. The use of this software is limited by the fact that it only runs on certain Unix-based operating systems.

4. It is recommended that small aliquots (50 μL) of probe working solution be prepared. Hybridization signals become dim and the background high when using a repeatedly thawed and frozen probe.
5. Several fluorescent generic DNA dyes can be used to stain the tissue sample. Since the colors that can be observed simultaneously are limited, if the same color is used during FISH and in the tissue staining, it will be difficult to discriminate the bacteria in the tissue. If FISH is performed using different probes conjugated with different colors, it is preferable to observe the sample, annotate what bacteria are present, and then stain the tissue. It is recommended to choose the colors of the generic tissue stains such that it will be possible to see at least the most relevant (specific) probe used during FISH (**Fig. 12.9**).
6. Lubriseal Stopcock grease (Thomas Scientific, Swedesboro, NJ) or a few drops of 5% agarose (Invitrogen) dissolved in water can be used to attach the tissue section to the Petri dish before the observation. It is always important to gently absorb the excess liquid from the tissue section using a paper towel. The tissue can be attached when the agarose is still soft and needs to dry very well before buffer is added for microscope observation with a water immersion objective. The best product to use to obtain a firm attachment will depend on the kind and consistency of the tissue (moistness, texture, etc.).
7. If possible, prior testing should be performed to find the optimal lysozyme incubation time (a shorter time may better preserve the morphology of Gram-negative bacteria that might be present in unknown samples).
8. Several fluorophores can be used during FISH, provided that the microscope is equipped with the appropriate set of excitation and emission filters. Fluorophores that have similar spectral characteristics cannot be used at the same time, but two morphologically distinct organisms can be marked with the same color.
9. The temperature is very critical and must be accurate and steady. It is recommended that the temperature is checked periodically in both the hybridization oven and the water bath with a second mercury thermometer.

10. Transfer the samples rapidly from the hybridization oven to the washing buffer to prevent cooling that can lead to non specific probe binding.

Acknowledgment

We thank J. E. Kerschner, MD, Medical College of Wisconsin, Milwaukee, WI, and J. M. Coticchia, MD, Wayne State University, Detroit, MI, for providing us the MEM and the adenoid tissues.

References

1. Post, J. C., Preston, R. A., Aul, J. J., Larkins-Pettigrew, M., Rydquist-White, J., Anderson, K. W., Wadowsky, R. M., Reagan, D. R., Walker, E. S., Kingsley, L. A., et al. (1995) Molecular analysis of bacterial pathogens in otitis media with effusion. *JAMA* **273**, 1598–1604.
2. Rayner, M. G., Zhang, Y., Gorry, M. C., Chen, Y., Post, J. C., and Ehrlich, G. D. (1998) Evidence of bacterial metabolic activity in culture-negative otitis media with effusion. *JAMA* **79**, 296–299.
3. Hall-Stoodley, L., Hu, F. Z., Gieseke, A., Nistico, L., Nguyen, D., Hayes, J., Greenberg, D. P., Dice, B., Burrows, A., Wackym, P. A., Stoodley, P., Post, J. C., Ehrlich, G. D., and Kerschner, J. E. (2006) Direct detection of bacterial biofilms on the middle-ear mucosa of children with chronic otitis media. *JAMA* **296**, 202–211.
4. Coticchia, J., Zuliani, G., Coleman, C., Carron, M., Gurrola, J., Haupert, M., and Berk, R. (2007). Biofilm surface area in the pediatric nasopharynx: Chronic rhinosinusitis vs obstructive sleep apnea. *Arch. Otolaryngol. Head Neck Surg.* **133(2)**, 110–114.
5. Ramadan, H. H., Sanclement, J. A., and Thomas, J. G. (2005) Chronic rhinosinusitis and biofilms. *Otolaryngol. Head Neck Surg.* **132**, 414–417.
6. Chole, R. A. and Faddis, B. T. (2003) Anatomical evidence of microbial biofilms in tonsillar tissues: a possible mechanism to explain chronicity. *Arch. Otolaryngol. Head Neck Surg.* **129(6)**, 634–636.
7. Chole, R. A. and Faddis, B. T. (2002) Evidence for microbial biofilms in cholesteatomas. *Arch. Otolaryngol. Head Neck Surg.* **128(10)**, 1129–1133.
8. Murphy, T. F. and Kirkham, C. (2002) Biofilm formation by nontypeable Haemophilus influenzae: strain variability,outer membrane antigen expression and role of pili. *BMC Microbiol.* **15(2)**, 7.
9. Wright, E. D., Pearl, A. J., and Manoukian, J. J. (1998) Laterally hypertrophic adenoids as a contributing factor in otitis media. *Int. J. Pediatr. Otorhinolaryngol.* **45(3)**, 207–214.
10. American Academy of Family Physicians; American Academy of Otolaryngology-Head and Neck Surgery; American Academy of Pediatrics Subcommittee on Otitis Media With Effusion. (2004) *Otitis media with effusion. Pediatrics.* **113(5)**, 1412–1429.
11. Tonnaer, E. L., Rijkers, G. T., Meis, J. F., Klaassen, C. H., Bogaert, D., Hermans P. W., and Curfs, J. H. (2005) Genetic relatedness between pneumococcal populations originating from the nasopharynx, adenoid, and tympanic cavity of children with otitis media *J. Clin. Microbiol.* **43(7)**, 3140–3144.
12. Giovannoni, S. J., Turner, S., Olsen, G. J., Barns, S., Lane, D. J., and Pace, N. R. (1998) Evolutionary relationships among cyanobacteria and green chloroplasts. *J. Bacteriol.* **170(8)**, 3584–3592.
13. DeLong, E. F., Wickham, G. S., and Pace, N. R. (1989) Phylogenetic stains: ribosomal RNA-based probes for the identification of single cells. *Science.* **243(4896)**, 1360–1363. Erratum in: (1989). *Science* **245(4924)**, 1312.
14. Amann, R. I., Binder, B. J., Olson, R. J., Chisholm, S. W., Devereux, R., and Stahl, D. A. (1990) Combination of 16S rRNA-targeted oligonucleotide probes with flow cytometry for analyzing mixed microbial populations. *Appl. Environ. Microbiol.* **56(6)**, 1919–1925.

15. Amann, R. I., Krumholz, L., and Stahl, D. A. (1990) Fluorescent-oligonucleotide probing of whole cells for determinative, phylogenetic, and environmental studies in microbiology. *J. Bacteriol.* **172(2)**, 762–770.
16. Amann, R., Ludwig, V., and Schleifer, K. H. (1995) Phylogenetic identification and in situ detection of individual microbial cells without cultivation. *Microbiol. Rev.* **59(1)**, 143–169.
17. Pernthaler, J., Glockner, F. O., Schonhuber, W., and Amann, R. (2001) Fluorescence in situ hybridization with rRNA-targeted oligonucleotide probes, in *Methods in Microbiology* (Paul J. H., ed.), vol. **30**, Marine Microbiology. Academic Press Ltd, London, pp. 207–226.
18. Sekar, R., Pernthaler, A., Pernthaler, J., Warnecke, F., Posch, T., and Amann, R. (2003) An improved protocol for quantification of freshwater Actinobacteria by fluorescence *in situ* hybridization. *Appl. Environ. Microbiol.* **69(5)**, 2928–2935.
19. Pernthaler, A., Pernthaler, J., and Amann, R. (2002) Fluorescence *in situ* hybridization and catalyzed reporter deposition for the identification of marine bacteria. *Appl. Environ. Microbiol.* **68(6)**, 3094–3101.
20. Schonhuber, W., Le Bourhis, G., Tremblay, J., Amann, R., and Kulakauskas, S. (2001) Utilization of tmRNA sequences for bacterial identification. *BMC Microbiol.* **71**, 20.
21. Manz, W., Amann, R., Ludwig, W., Wagner M., and Schleifer K.-H. (1992) Phylogenetic oligodeoxynucleotide probes for the major subclasses of proteobacteria: problems and solutions. *Syst. Appl. Microbiol.* **15**, 593–600.
22. Stahl, D. A. and Amann, R. (1991) Development and application of nucleic acid probes, in *Nucleic Acid Techniques in Bacterial Systematics.* (Stackebrandt, E. and Goodfellow, M., eds.), John Wiley and Sons Ltd., Chichester, UK, pp. 205–248.
23. Hugenholtz, P., Gene, W. T., and Blackall L. (2001) Steroid receptors methods: protocol and assays, in *Methods in Molecular Biology*, vol. **176**, Humana Press Inc., Totowa, NJ, pp. 29–41.
24. Thurnheer, T., Gmür, R., and Guggenheim, B. (2004) Multiplex FISH analysis of a six-species bacterial biofilm. *J. Microbiol. Methods* **56(1)**, 37–47.
25. Kulikova, T., Aldebert, P., Althorpe, N., Baker, W., Bates, K., Browne, P., et al. (2004) The EMBL Nucleotide Sequence Database. *Nucleic Acids Res.* **32**, D27–D30.
26. Benson, A. D., Karsch-Mizrachi, I., Lipman, D. J., Ostell, J., and Wheeler, D. L. (2003) GenBank. *Nucleic Acids Res.* **1**, 23–27.
27. Maidak, B. L., Olsen, G. J., Larsen, N., Overbeek, R., McCaughey, M. J., and Woese, C. R. (1996) The Ribosomal Database Project (RDP). *Nucleic Acids Res.* **24(1)**, 82–85.
28. Maidak, B. L., Cole, J. R., Lilburn, T. G., Parker, C. T., Saxman, P. R., Farris, R. J., Garrity, G. M., Olsen, G. J., Schmidt, T. M., and Tiedje, J. M. (2001) The RDP-II (Ribosomal Database Project). *Nucleic Acids Res.* **29**, 173–174.
29. Wuyts, J., PerrieÁre, G., Van de Peer, Y. (2004) The European ribosomal RNA database. *Nucleic Acids Res.* **32**, D101 D103.
30. Loy, A., Maixner, F., Wagner, M., and Horn, M. (2007) ProbeBase – an online resource for rRNA-targeted oligonucleotide probes: new features 2007. *Nucleic Acids Res.* **35**, D800–D804.
31. Ashelford, K. E., Weightman, A. J., and Fry, J. C. (2002) PRIMROSE: a computer program for generating and estimating the phylogenetic range of 16S rRNA oligonucleotide probes and primers in conjunction with the RDP-II database. *Nucleic Acids Res.* **30(15)**, 3481–3489.
32. Ludwig, W., Strunk, O., Westram, R., Richter, L., Meier, H., Yadhukumar, et al. (2004) ARB: a software environment for sequence data. *Nucleic Acids Res.* **3**, 1363–1371.
33. Hodgart, M., Trebesius, K. Geiger, A. M., Hornef, M., Rosenecker, J., and Heeseman, J. (2000) Specific and rapid detection by fluorescent in situ hybridization of bacteria in clinical samples obtained from cystic fibrosis patients. *J Clin Microbiol.* **38(2)**: 818–25.
34. Trebesius, K., Leitritz, L., Adler, K., Shubert, S., Autenrieth, I. B., and Heeseman, J. (2000) Culture indipendent and rapid identification of bacterial pathogens in necrotising fasciitis and streptococcal toxic shock syndrome by fluorescence in situ hybridization. *Med Microbiol. Immunol.* **188(4)**: 169–75.
35. Kempf, V. A., Trebesius, K., Autenrieth, I. B., (2000) Fluorescent in situ hybridization allows rapid identification of microorganisms in blood cultures. *J. Clin. Microbiol.* **38(2)**: 830–838.

Chapter 13

Positional Cloning of Deafness Genes

Hannie Kremer and Frans P.M. Cremers

Abstract

The identification of the majority of the known causative genes involved in nonsyndromic sensorineural hearing loss (NSHL) started with linkage analysis as part of a positional cloning procedure. The human and mouse genome projects in combination with technical developments on genotyping, transcriptomics, proteomics, and the creation of animal models have greatly enhanced the speed of disease gene identification.

In the present chapter, we first discuss the possibilities for exclusion of known NSHL loci and genes. Subsequently, methods are described to determine the genomic regions that contain the genetic defects. These include linkage analysis with genotyping and statistical evaluation and the determination of copy number variations. In the case of a large genomic region, candidate genes are selected and prioritized using gene expression analysis, protein network data, and phenotypes of animal models. A number of algorithms are described to automate the process of candidate gene selection. The novel high-throughput sequencing techniques might make gene selection and prioritization unnecessary in the near future. Once genetic variants are identified, questions on pathogenicity need to be addressed, which is the topic of the last section of this chapter.

Key words: Positional cloning, linkage analysis, candidate gene selection, mutation analysis, pathogenicity of genetic variants, nonsyndromic hearing loss.

1. Introduction

Nonsyndromic sensorineural hearing loss (NSHL) is extremely genetically heterogeneous. For prelingual NSHL, the predominant patterns of inheritance are autosomal recessive (~75%) and autosomal dominant (~20%). X-linked and mitochondrial inheritance occur in less than 5% of cases *(1)*. Already more than 40 genes are known to be causative for NSHL, but an even larger number remains to be identified (*see Hereditary Hearing Loss Homepage*, by G. van Camp and R. Smith:

Bernd Sokolowski (ed.), *Auditory and Vestibular Research: Methods and Protocols, vol. 493*

DOI 10.1007/978-1-59745-523-7_13 Springerprotocols.com

http://webh01.ua.ac.be/hhh/). The identification of novel deafness genes can be achieved by several strategies, including candidate gene analysis and family studies. In this chapter, we will outline gene identification strategies involving families in which multiple members are affected and families with one or a few patients. In the process of clinical evaluation and sampling of families for genetic analysis, a general physical examination is important to exclude syndromic hearing loss. Also, it is important to evaluate hearing in all members that participate in genetic studies. This evaluation will show the variability of the phenotype in the family and might provide indications for the presence of genocopies or phenocopies that disturb linkage studies. This chapter will address predominantly the analysis of deafness that is inherited in a Mendelian way. Age-related hearing loss is known to be multifactorial, which means that multiple genetic and environmental factors determine the phenotype. Not much is known, as yet, about digenic, polygenic, or complex inheritance in NSHL.

2. Exclusion of Known Loci and Genes

Due to the fact that many genes for NSHL have been identified already, a fast and cost-effective strategy to exclude involvement of one of these known genes is useful before initiation of positional cloning of the causative gene in a family. At this moment, such a strategy is not available for dominantly inherited NSHL and is limited for recessive NSHL. For the latter type of NSHL, ASPER Biotech (http://www.asperbio.com/) offers two assays for the detection of known mutations in a subset of the known deafness genes. In one assay, known mutations are tested in the Usher syndrome genes *CDH23, DFNB3, MYO7A, PCDH15, USH1C, USH1G, USH2A*, and *VLGR1 (2)*. Also, this microarray tests for the presence of mutations in these genes that cause NSHL. The second assay evaluates known mutations in *GJA1, GJB2, GJB6, SLC26A4, SLC26A5*, and the mitochondrial genes for 12S rRNA and tRNA$^{Ser[UCN]}$ *(3)*. This means that only a subset of the known genes can be tested for known mutations. Because of the frequent involvement of the *GJB2* gene in recessive prelingual hearing loss in many populations, this gene should be excluded before further analysis.

2.1. Phenotype-Based Exclusion of Known Deafness Genes in Dominant NSHL

Although the costs for array-based genotyping significantly decreased in the past few years, it is cost-effective to exclude the obvious candidate loci and/or genes before embarking on genome-wide genotyping. This is especially true for dominant NSHL, since in general, the number of family members that have to be tested to obtain statistically significant logarithm of odds (LOD) scores is larger than for recessive NSHL. In addition, the clinical heterogeneity in dominant NSHL allows phenotype-based selection of candidate loci and genes. Here, we will only summarize the most important aspects of phenotypic variability that point toward involvement of a specific locus or gene. A detailed description of phenotype–genotype correlations for dominant NSHL is given by Huygen et al. *(4, 5,* and refs therein*)*. Evaluation of the audiograms of affected family members reveals an audiometric profile, which includes the age of onset, configuration of the audiogram, and presence or absence of progression. Four types of audiogram configurations are discriminated: high-frequency downsloping, flat-to-gently downsloping, mid-frequency or U-shaped, and low frequency. In addition to this audiometric profile, the involvement of the vestibular system can suggest specific DFNA types, such as DFNA9, with an onset after the second decade, as well as DFNA11 and DFNA15. The most robust phenotype–genotype correlations are known for the low-frequency configuration with mutations in the *WFS1* gene (DFNA6/14/38), the mid-frequency configuration with mutations in the ZP domain of *TECTA* (DFNA8/12) or in *COL11A2* (DFNA13), and the high-frequency, downsloping configuration, adult onset, and vestibular impairment with mutations in *COCH* (DFNA9).

2.2. Phenotype-Based Exclusion of Known Deafness Genes in Recessive NSHL

Phenotypic variability of recessive NSHL is limited when compared to that of dominant forms. Most types of recessive NSHL exhibit a severe-to-profound hearing loss for all frequencies with a downsloping configuration. The onset of the hearing loss generally is congenital or in very early childhood. However, a phenotype–genotype correlation enables discrimination of the four autosomal recessive types: DFNB4, DFNB8/10, DFNB13, and DFNB21. *SLC26A4* is the causative gene for DFNB4 and Pendred syndrome *(6, 7)*, and mutations in this gene may account for 5–10% of prelingual hearing loss *(8)*. The differential diagnosis of these disorders can be difficult, especially at a young age. Mutations in *SLC26A4* were associated with phenotypes ranging from mild to profound, either bilateral or unilateral *(9)*. Substantial threshold fluctuations can be observed during progression *(10)*. Inner ear malformations in patients with mutations in the *SLC26A4* gene vary from enlarged vestibular aquaduct to Mondini Dysplasia. For DFNB8/10 (*TMPRSS3*), one Pakistani family was described with an onset between 0 and 12 years of age and progression to profound hearing loss within 4–5 years after onset *(11)*. DFNB13, for which the causative gene has not yet been identified, can also be distinguished based on the progression of the hearing loss from a flat to downsloping configuration, with moderate threshold levels to severe-to-profound hearing loss *(12)*. DFNB21 caused by mutations in *TECTA* is characterized by a flat-to-gently downsloping configuration of the audiogram, with moderate-to-severe prelingual hearing loss *(13, 14)*. Patients with Usher syndrome types I and II present with congenital hearing loss and, therefore, can be among the young children with seemingly NSHL *(15)*. The patients with Usher syndrome type I can, however, be distinguished from those with NSHL, based on delayed motor milestones (ability to walk independently after the age of 18 months) due to vestibular impairment and the severity of the profound hearing loss. The hearing loss in patients with Usher syndrome type II is moderate to severe with a downsloping audiogram.

3. Determination of the Critical Chromosomal Region

By linkage analysis, a genome-wide segregation analysis is performed with polymorphic markers, to determine the chromosomal region that segregates with a trait – NSHL in this case. For statistical significant locus assignment by genome-wide genotyping, a minimum LOD score of 3.3 has to be obtained and a LOD score of 1.86 is suggestive for linkage *(16)*. The LOD score is the $\log_{10}$ of the ratio of the likelihood the disease locus and marker are linked $(0 \leq \theta > 0.5)$ versus the likelihood that they are unlinked. Theta (θ) represents the recombination fraction between two loci – in this case marker and disease. Simulation of linkage analysis allows an estimate of whether the size of the family is sufficient to reach a maximum LOD score of 3.3, and it obtains an indication for linkage at the involved locus. Our rule of thumb is that simulation for a family has to reveal at least a 70% chance to obtain a LOD score > 2 to initiate a genome scan. The regular occurrence of genocopies and/or phenocopies in deafness families can heavily interfere with locus identification in relatively small families. Also, for parametric (assumed pattern of inheritance) simulation and LOD score calculations, age-related penetrance and reduced penetrance have to be taken into account. Therefore, a simulation can best be performed with 95% penetrance for the disease allele and a phenocopy rate of 0.001 (i.e., 0.1% penetrance in individuals that do not carry the disease allele, or one in case of recessive inheritance), to take phenocopies and non-penetrance into account. Liability classes for the disease status can be used to take age-related penetrance into account. When non-penetrance is obvious from the pedigree, the penetrance has to be estimated from that specific pedigree. For single nucleotide polymorphism (SNP)–based genotyping, multipoint LOD scores have to be calculated to identify candidate regions (a LOD score > 1.86 is indicative for linkage). Two-point LOD scores can already reveal candidate regions in variable numbers of tandem repeats (VNTR)–based genome scanning. In two-point LOD scores the information of only one marker is used, whereas in multipoint scores, information is used from multiple markers.

For statistical evaluation of linkage data (LOD score calculation), one could collaborate with researchers in genome centers worldwide that are experienced in linkage analysis. Also, courses are offered by several institutes (e.g., at the Max Delbrűck Centrum für Molekulare Medizin [MDC], Berlin, Germany; Rockefeller University, New York, NY, USA).

Alternatively, commercial genotyping centers can perform both the genotyping and the statistical evaluation. Software for linkage analysis is available as freeware. Programs that are generally used for the statistical evaluation of marker analysis are LINKAGE, Allegro, Genehunter, and SimWalk2. Pedcheck *(17)* and Merlin *(18)* can detect Mendelian and non-Mendelian errors in the genotyping. In case no locus emerges from the analysis, nonparametric evaluation of the data (without pre-assumption of an inheritance pattern) can be performed with, for example, SimWalk2. Information and/or documentation on several programs can be found at http://linkage.rockefeller.edu/soft/and http://watson.hgen.pitt.edu/docs/. Graphical user interfaces for linkage analysis software on Microsoft Windows–based operating systems are easyLINKAGE *(19, 20)* (http://genetik.charite.de/hoffmann/easyLINKAGE/index.html) and ALOHOMORA *(21)*. Also, there are several textbooks on linkage analysis (e.g., *22, 23*).

3.1. Autosomal Dominant Hearing Impairment

3.1.1. Sizeable Families: Linkage Analysis (SNP Array Genotyping)

A genome scan in autosomal dominant NSHL families is generally performed with an array containing about 10,000 SNPs (e.g., Affymetrix 10 K SNP array or Illumina 6 K SNP array). However, VNTR or microsatellite markers are still being used for genome-wide mapping. Applied Biosystems offers mapping sets with an average spacing of 5 or 10 cM, whereas deCODE genetics (http://www.decode.com/genotyping/) offers services for linkage analysis with VNTR markers at 2–10 cM spacing. The informativity of a 10 K SNP array was shown to be higher than that of the ABI mapping set with an average VNTR marker spacing of 10 cM *(24, 25)*. Fewer VNTR markers than SNPs are necessary, due to the greater informativity of VNTR markers. Several genome centers perform array-based or VNTR-based genome-wide scans on a commercial and/or collaborative basis (e.g., deCODE genetics, the Department of Human Genetics, Radboud University Nijmegen Medical Centre, and the GSF National Research Center for Environmental Health).

Generally, one or several candidate regions emerge from a genome scan and confirmation and fine mapping have to be performed. Highly informative VNTR markers (heterozygosity > 0.70), in and flanking the regions can be selected by using the UCSC or ENSEMBL genome browsers and, subsequently, genotyped. Genetic distances between markers based on several genetic maps are available via the NCBI website (Human genome resources; http://www.ncbi.nlm.nih.gov/). In combination with SNPs, the positions of recombinations can be determined that delimit the linkage interval. If necessary, one can select CA repeats or other repeats from the genomic sequence, to determine whether these are polymorphic and informative in the family. Two-point and/or multipoint LOD score calculations, for markers that are used to confirm linkage and for fine mapping, will reveal the maximum LOD score in the interval and thus illustrate the statistical significance of the linkage.

3.1.2. Small Families: Analysis of Copy Number Variations

Small families are not suitable for the identification of novel deafness loci/genes by linkage analysis, due to the lack of statistical power. However, in case of a type of hearing loss that is known to be associated with a limited number of loci, it is possible to determine the involvement of a given locus. VNTR markers are the most likely to be conclusive, because of their higher informativity. It can even be useful to perform linkage analysis in small families, when one or more genes are known to be involved in that specific type of hearing loss, since mutation analysis can lead to false-negative results, due to the presence of mutations outside the protein coding region and exon/intron boundaries. By exclusion of known loci in small families, a patient panel can be built, which is suitable for candidate gene analysis for specific types of hearing loss.

SNP genotyping can be successful in the identification of genetic defects in small families and even isolated cases, when changes in copy number of a specific chromosomal region are causative for the disease. These can be deletions or duplications. The higher the density of genotyped SNPs, the higher the resolution for identifying the copy number defects. Hehir-Kwa and coworkers *(26)* found that with the GeneChip Mapping 250 K Nsp Array (Affymetrix), heterozygous deletions of a region, harboring four or more SNPs, are detected with a power of 95%. Deletions of 50, 100, and 200 kb are thereby detected in 68%, 88%, and 94% of the cases. Duplications are detected with a power of 95%, when seven or more consecutive SNPs are duplicated, and duplicated regions of 100, 200, and 300 kb are detected in 75%, 91%, and 94% of the cases *(26)*, by using the CNAG software package *(27)*.

3.2. Autosomal Recessive Hearing Impairment

3.2.1. Sizeable Families: Linkage Analysis and Homozygosity Mapping

Locus identification in families with autosomal recessive hearing loss can be performed with linkage analysis, and the power of a family can be determined by simulation. However, when there is consanguinity of parents of the affected individuals, locus identification can be performed by homozygosity mapping to find genomic regions that are identical-by-descent (IBD). For this, genome-wide genotyping is only necessary for affected individuals. As was indicated for the identification of loci for autosomal dominant hearing loss, genotyping of both VNTR markers and SNPs (~10 K) is suitable for the identification of the loci in recessive hearing loss. However, for homozygosity mapping with VNTRs, it is important to estimate the expected size of the homozygous region in order to choose a suitable marker density *(28)*. For homozygosity mapping using VNTR markers, candidate regions can be selected by viewing the genotypes for regions for which all affected individuals carry the same allele homozygously. When SNP genotyping is performed, selection of candidate regions can be performed based on the "loss-of-heterozygosity" (LOH) value as calculated by the program CNAG *(27)*. The LOH score is a measure for the likelihood of a stretch of SNPs to be homozygous based on the SNP allele frequencies in the population. An LOH score of ≥ 15 roughly corresponds to a region of ≥ 4 Mb *(29)*. The larger the number of meiosis that separates the consanguineous parents, the smaller the expected LOH value.

Candidate regions determined in a genome-wide scan have to be confirmed and fine mapped as described in **Section 3.1.1**. For families for which homozygosity mapping was performed, the complete family should be included in this phase of the study. A pitfall in the analysis, especially of consanguineous families, is the genetic heterogeneity of recessive hearing loss. Two or three genetic defects in different branches of the family have been reported by several groups *(30, 31)*. Especially, the high carrier rate for mutations in *GJB2*, in populations worldwide, can cause genetic heterogeneity in families, where a genetic defect is segregating in another gene *(30* and Kremer et al., unpublished data).

3.2.2. Small Families: Homozygosity Mapping and Analysis of Copy Number Variation

When there is consanguinity in families with recessive hearing loss, homozygosity mapping can point toward the region with the genetic defect in small families. However, depending on the degree of consanguinity between the parents of patients and the population from which they are derived, the number of candidate regions might still be large *(32)*. Also, since homozygosity for the causative mutation is frequently seen in patients with rare disorders in outbred populations, every genetic type of hearing loss can be regarded as a separate rare disease. Therefore, homozygosity mapping can also be a good strategy to identify the causative gene in small families or isolated cases from this type of population *(33)*. For cases from outbred populations, the homozygous regions are expected to be much smaller than in cases from inbred populations and, therefore, high-density SNP arrays should be used (≥ 250 K). As a rule of thumb, the chromosomal region, containing a homozygous mutation inherited from a common ancestor of the patient's parents, measures 100 cM/number of generations between the common ancestor and the patient *(32)*.

Also, the data from these arrays can be evaluated for copy number variations (CNVs), which can lead to gene identification in cases where one or both mutated alleles of the causative gene involve a change in copy number, as described in **Section 3.1.2**, for autosomal dominant families. The latest generation of arrays (e.g., Affymetrix 5.0 arrays) analyze not only SNPs but also non-polymorphic nucleotides, to cover CNVs in genomic regions that are not well covered by SNPs.

It should be taken into account that isolated cases can also be due to de novo dominant or X-linked mutations.

3.3. X-linked Deafness

X-linked deafness is rare and represents less than 5% of the cases with NSHL. Five loci have been described so far (http://webh01.ua.ac.be/hhh/), one of which, DFN1 (*TIMM8A*), is now known to represent a syndromic form associated with visual disability, dystonia, fractures, and mental deficiency *(34)*. X-linked inheritance can be suspected when only males are affected, which can be the case for DFN3, or when males are more severely affected and/or exhibit an earlier onset of the disease, as described for DFN2, DFN4, and DFN6 *(35–37)*. DFN3 is the most frequent X-linked deafness and is characterized by a temporal bone defect that can be visualized using computerized tomography scanning *(38)*. Deletions of, or mutations in, the intronless gene *POU3F4* are found in ~50% of the DFN3 cases *(39, 40)*. Other DFN3 patients carry variably sized deletions overlapping a 10-kb segment ~900 kb upstream of *POU3F4 (41)*. It is thought that these deletions disrupt an inner ear–specific regulatory element.

3.3.1. Sizeable Families: Linkage Analysis

To determine the locus involved in the disease in a large family (> 5 affected males), SNP arrays (~10 K) and VNTR markers are suitable to be logically confined to markers on the X-chromosome. Genome-wide SNP arrays have the advantage in that disease-associated sizeable CNVs might already be identified. For X-linked inheritance, an LOD score of ≥ 2 is assumed to indicate a statistically significant linkage. Male penetrance is assumed to be 95% for the LOD score calculations, in order to include possible cases of non-penetrance and should, therefore, be adjusted in families where non-penetrance is clearly indicated to be higher. The value for the penetrance in female carriers should be based on the phenotype in the specific pedigree. Since X-linked hearing loss is rare, the frequency of the disease allele should be low, ~0.00001.

3.3.2. Small Families: Analysis of Copy Number Variations

In small families, conclusions will often not be possible with regard to the type of inheritance, since non-penetrance and clinical variation among patients have to be taken into account. Therefore, small families, in which X-linked inheritance cannot be excluded, should be analyzed with all possible forms of inheritance in mind.

3.4. Digenic Inheritance

Digenic inheritance is not frequently indicated by linkage analysis in NSHL. DFNB15 might be inherited digenically, since two chromosomal regions cosegregate with the disease in the family *(42)*. In this study, three NSHL patients inherited the same four marker alleles from the two chromosomal regions, and this combination of alleles was not present in three unaffected sibs. The parents were consanguineous.

4. Identification of Causative Genes in Defined Chromosomal Intervals

4.1. Small Chromosomal Intervals: Sequence Analysis

When a relatively small genomic region (up to 1 Mb) contains an NSHL gene, and there are no known NSHL genes in that chromosomal segment, sequence analysis is the most straightforward technique to find the causative mutation(s). Prioritization of candidate genes (*see* **Section 4.2**), however, should always be considered to save time and resources. The sequence analysis of all genes in a selected chromosomal region might be necessary, if causality of a nucleotide or protein sequence variant is not clear. For example, if protein-truncating mutations or splice site mutations are identified (*see* **Sections 5.1** and **5.2**), the likelihood of pathogenicity is very high. However, if missense variants are found, interpretation is much more difficult and should include an assessment of the degree of evolutionary and chemical conservation of the amino acid residue (*see* **Section 5.3**). Other requirements for pathogenicity are that the variant(s) segregate in the respective families with the disease phenotype, as expected for Mendelian traits, and that the variant is not found in a sizeable cohort of alleles. This means that a polymorphism with a frequency of 0.01 should be absent in 210 alleles of ethnically matched healthy individuals to reach a power of 0.80 *(43)*. However, the absence of a given missense variant in control individuals is no proof for pathogenicity. Vice versa, the presence of a variant in a relatively large percentage (>1%) of control alleles does not rule out its pathologic involvement, as shown for so-called hypomorphic sequence variants in the *ABCA4* and *CFTR* genes *(44, 45)*. These variants do not cause disease in a homozygous situation, but in a compound heterozygous situation together with a severe (in most cases null) mutation; they underlie retinal disease and absence of the *vas deferens* in males, respectively. The identification of a de novo missense mutation can be considered as strong evidence for pathogenicity.

DNA sequence analysis in the last 25 years is based on the chain termination method described by Sanger and coworkers

(46). Basically, truncated complementary DNA sequences are enzymatically synthesized from single-stranded templates using the four bases linked to specific fluorophores *(47)*. The four reactions are mixed and size separated in a denaturing electrophoresis system. The fluorescence signal is recorded as the DNA fragments pass through. The throughput of this technique was significantly increased with the introduction of PCR amplification and capillary electrophoresis. Several electrophoresis systems contain up to 384 parallel capillaries, allowing high-throughput analysis. This method allows the analysis of up to 500–800 bp of sequence and is generally available through in-house facilities in many laboratories. Sequence analysis is also commercially available.

Capillary-based dideoxy sequence analysis has been the "golden standard" for small- and medium-scale sequence analysis. The costs are ~ 0.01 USD/bp, allowing a cost-effective analysis of a selected number of genes. Accurate sequence analysis requires bidirectional analysis, effectively doubling the costs. In a routine DNA diagnostic setting, the number of amplicons analyzed per patient is limited, due to the high costs of materials and personnel. For research purposes, for example, to find a novel disease gene, many more amplicons (sometimes in several patients) can be analyzed, but the material and personnel investments are substantial.

4.2. Large Chromosomal Intervals

When large chromosomal intervals (>1 Mb) must be analyzed for the presence of pathologic mutations, several strategies can help to prioritize the candidate gene analysis.

4.2.1. Candidate Gene Prioritization

4.2.1.1. Expression Characteristics

A specific or abundant presence of an mRNA or protein in inner ear structures renders the corresponding gene a candidate NSHL gene. Cochlear cDNA libraries were previously constructed for guinea pig *(48)*, rat *(49–51)*, mouse *(52, 53)* and, using a subtraction protocol, for mouse *(54)*, chicken *(55, 56)*, gerbil *(57)*, and zebrafish *(58)*. Peters et al. *(59)* constructed MPSS (massively parallel signature sequencing) for different regions of the mouse cochlea and selected genes that are specifically expressed in cochlea or enriched by comparison to the transcriptome of other tissues. Human cochlear cDNA libraries have been constructed without *(60, 61)* and with *(62)* enrichment for inner ear–specific genes. The cDNA fragments generated by Robertson and coworkers can be found at http://www.brighamandwomens.org/bwh_hearing/human-cochlear-ests.aspx.

A cDNA library was constructed using RNA from mouse outer hair cells *(63)*. Also, a mouse inner ear transcriptome *(64)*, a rat vestibular epithelium cDNA library *(65)*, and a zebrafish hair cell transcriptome *(66)* were made. The presence of orthologues of genes from NSHL linkage intervals or IBD regions in these libraries may suggest that the gene is involved in NSHL and has an inner ear–specific function. Reverse transcriptase-polymerase chain reaction analysis, using RNAs from mammalian inner ear and several other tissues, can reveal or confirm the specific or high inner ear expression of a candidate gene. An inner ear protein database was established at the Washington University by I. Thalman and R. Thalman: http://oto.wustl.edu/thc/innerear2d.htm.

Besides inner ear libraries, expression of genes in the inner ear can be evaluated via data available via UniGene (NCBI homepage: http://www.ncbi.nlm.nih.gov), in which transcripts will be included already partially from the above-mentioned cDNA libraries. For mouse genes, data available on the MGI website (http://www.informatics.jax.org) include information on gene expression patterns.

In order to prioritize candidate disease genes, several programs have been developed that use expression characteristics as one of several features. GeneSeeker (http:www.cmbi.ru.nl/geneseeker) *(67, 68)* employs expression and phenotypic data from mouse and man. This program is particularly well suited to predict disease genes for syndromes. eVOC anatomical ontology is used to integrate text mining of biomedical literature and data mining of human gene expression data *(69)*. Disease Gene Prediction (http://cgg.ebi.ac.uk/services/dgpl/) *(70)*, PROSPECTR *(71)* and SUSPECTS *(72)* (http://www.genetics.med.ed.ac.uk/suspects/), and POCUS (http://www.hgu.mrc.ac.uk/Users/Colin.Semple/) *(73)* employ characteristics of known disease genes, such as gene size, evolutionary conservation, as well the candidate genes' expression pattern, phylogenetic extent, and paralogy pattern. The program G2D (genes to diseases) (http://www.ogic.ca/projects/g2d_2/) scores all terms in gene ontology (GO) libraries, according to their relevance to each disease using MEDLINE queries *(74, 75)*. Symptoms are related to GO terms through chemical compounds and G2D performs BLASTX searches *(76)* of the "critical disease region" against all the GO-annotated genes in RefSeq. A comparison of these methods to prioritize type 2 diabetes and obesity candidate genes was performed by Tiffin et al. *(77)*.

It is important to realize that all these methods heavily depend on detailed expression characteristics of genes. Genes expressed in very small inner ear substructures, such as the inner ear hair cells, are underrepresented in inner ear cDNA libraries and may not be scrutinized using the methods listed above. Moreover, all these methods will not reveal genes that encode proteins with very different functional domains that are not yet implicated in NSHL.

4.2.1.2. Animal Model Comparisons

Mutant mouse models that exhibit NSHL, due to inner ear defects, can help identify genes that have a role in the development or function of the inner ear. Many strains of hearing-impaired mice have arisen spontaneously during the last century. In addition, random mutagenesis of mouse genomic DNA by chemicals (e.g., *N*-ethyl-*N*-nitrosourea, ENU), X-ray radiation, or insertion of an extrinsic sequence have been used to create new hearing-impaired mouse strains. Mutations in more than 172 different genes have been reported to be responsible for inner ear defects in mice *(78, 79,* and refs. therein*)*. Many of these are listed in the Jacksons Laboratory's Hereditary Hearing Impairment in Mice database: http://www.jax.org/hmr/map.html and on the site of the Sanger institute: http://www.sanger.ac.uk/PostGenomics/mousemutants/deaf. A systematic search for inner ear defects is carried out by the Mammalian Genetics Unit, Harwell UK (http://www.har.mrc.ac.uk/) and the Institute of Mammalian Genetics (http://www.helmholtz-muenchen.de/en/ieg/group-functional-genetics/enu-screen/mutants-generated/index.html#c5754) *(80, 81)*.

Another excellent model organism for mammalian inner ear defects is the zebrafish (*Danio rerio*). Zebrafish do not possess outer or middle ears but have a typical vertebrate inner ear. They do not contain a specialized hearing organ, but many features are conserved with other vertebrate species. Zebrafish possess four small bones, the Weberian ossicles, linking the swim bladder to the inner ear. These bones enhance hearing and are sensitive to a frequency range of 100–5000 Hz. The inner ear also is used to maintain balance. In addition, zebrafish possess a series of mechanosensory receptors on the surface of the body that are very similar in structure and function to the sensory patches of the inner ear. These are the neuromasts of the lateral line system, arranged in precise lines over the body surface (the so-called lateral lines). Several mutagenesis screens have been carried out *(82)*. ENU mutagenesis screens have generated several hundred lines carrying point mutations *(83, 84)*. Several insertional mutagenesis strategies have been employed in zebrafish, including retrotransposon based and gene trapping (*85* and refs therein).

4.2.1.3. Protein-Protein Interactions

Based on the assumption that common phenotypes are associated with dysfunction in proteins that participate in the same protein complex or pathway, protein sequence and interaction data are employed to predict candidate genes. In the Prioritizer program (http://humgen.med.uu.nl/~lude/prioritizer/) *(86)*, a functional human gene network was developed using known interactions derived from the Biomolecular Interaction Network (BIND; http://www.bind.ca/) *(87)*, the Human Protein Reference Database (HPRD; http://www.hprd.org/) *(88)*, Reactome (http://www.reactome.org/) *(89)*, and the Kyoto Encyclopedia of Genes and Genomes (KEGG; http://www.genome.jp/kegg/) *(90)*. Also, Prioritizer uses gene relationships based on biological processes and molecular function annotations in GO *(91)*, co-expression data from microarray hybridizations from the Stanford Microarray Database *(92)*, the NCBI Gene Expression Omnibus *(93, 94)*, human yeast two-hybrid (YTH) interactions *(95)*, and interactions based on orthologous protein–protein interactions from lower eukaryotes *(96)*. ENDEAVOUR *(97)* uses pathways and protein interactions data combined with text mining of MEDLINE abstracts and LocusLink texts, GO annotations, presence of protein domains, expression data, transcription factor binding sites, cis-regulatory modules, and sequence similarity. In this program the user can select disease genes for training and novel databases can be included. Oti et al. *(98)* achieved a 10-fold enrichment of candidate genes in linkage intervals of unknown disease genes for 10 genetically heterogeneous human diseases employing the HPRD dataset, alongside YTH data from human, fly, worm, and yeast. George et al. *(99)* described a Common Pathway Scanning method similar to the approaches described above. In addition, they use the so-called Common Module Profiling tool based on the hypothesis that disruption of genes of similar function will lead to the same phenotype.

As protein–protein interaction studies in yeast often reveal interactions between (domains of) proteins that do not localize to the same subcellular compartment in the respective organisms, YTH interaction datasets contain many interaction data that do not exist in vivo. Novel protein–protein interaction studies based on in vivo interactions, such as the TAP-tag technology *(100–102)*, likely will yield more accurate protein–protein interaction data that should improve approaches described above.

4.2.2. Medium- and Large-Scale Sequence Analysis

4.2.2.1. Medium-Scale Sequence Analysis: Sequencing by Hybridization

One method to determine up to ~300, 000 nucleotides in a single experiment is based on the hybridization of target (e.g., patient) DNA to an ordered dense array of oligonucleotides (25-mers) *(103)*. Each 25-mer tests the occurrence of the four different nucleotides in its center position. The accuracy of the base calling relies on the discrimination between exact matches and those with single base differences. As both strands are tested, eight different oligonucleotides are used to test the presence of any given single nucleotide variant. The protein-coding exons of up to ~100 genes can be tested in one experiment using this technique. In principle, all 38 known genes for autosomal recessive, autosomal dominant, and X-linked hearing impairment can be analyzed using this technique. Disadvantages of this technology are the relatively high set-up costs, as a minimum of 40 samples need to be analyzed. Estimated costs are USD 2,000–3,000 (~0. 001 USD/bp) per patient. Thus far, sequences are enriched from complex genomes through large-scale parallel PCR amplification reactions, constituting a major bottleneck for high-throughput analysis. Also, repetitive sequences cannot be analyzed, and heterozygous deletions and insertions larger than one nucleotide are not detected. These variants can constitute up to 15% of all variants. Because of the high costs and the time needed for setup, this method is not very useful for the identification of the causative gene from a large linkage interval.

4.2.2.2. Large-Scale Sequence Analysis: Target Enrichment and Sequencing by Synthesis on Arrays

As discussed above, homozygosity mapping and linkage analyses often result in the delineation of novel chromosomal loci for hearing impairment measuring between 1 and 50 Mb. The genes in a 1-Mb region often are amenable to standard capillary sequencing. For larger regions, sequence analysis of all predicted exons, flanking intronic sequences, and promoters is very time consuming and expensive. Recently, several large-scale sequencing technologies have been developed that have increased the throughput another magnitude. Moreover, in comparison to the "sequencing-by-hybridization" technique, these methods also detect small heterozygous deletions and insertions.

Large-scale sequencing techniques can only by applied to sequence analysis of complex genomes, such as the human genome, after the target sequences are selected to fit in one sequencing experiment (see below; maximum of 0.1% of the human genome). Selection of human genomic loci was achieved by microarray hybridization *(104–105)*. First, overlapping 60–90 nt probes, with a positional offset of 20 nt, are linked to Nimblegen microarrays, each containing a maximum of 385,000 unique probes. Seven of these arrays, in principle, capture all predicted exons of the RefSeq *(106)* genes in the human genome, representing 55 Mb (2%) of the euchromatic human genome (2.85 Gb). High molecular weight genomic DNA is fragmented ($\sim$500 bps), ligated to linkers, denatured, and captured with arrayed probes. Selected fragments are recovered by thermal elution and PCR enriched.

In the following two sequencing approaches, single DNA molecules are clonally amplified and used as templates for sequencing-by-synthesis. In the 454 system (http://www.454.com), templates are immobilized and amplified on beads in aqueous-oil emulsions *(107)*. Beads with DNA are purified and placed in picolitre wells for pyrosequencing *(108)*. One deoxynucleoside triphosphate (dNTP) is added per cycle, and the 3′ end of the nascent chain is extended from a primer until a position is reached that requires a different base. A chemiluminescent signal is released in each well and quantified to determine the number of bases incorporated. The process is reiterated with one of four dNTPs typically yielding 100 nt reads after 80 cycles. Up to 40 Mb of raw sequence data are generated in one experiment (i.e., a 10-fold coverage of 4 Mb of sequence). In the Solexa/Illumina system (http://www.solexa.com), single molecules are covalently attached to a planar surface and amplified in situ. Sequencing is carried out by adding a mixture of four fluorescently labeled reversible chain terminators and DNA polymerase. The fluorescent signal is detected for each template, and the fluorophore and reversible block are removed. The terminator–enzyme mix is added for the next cycle, and the process is reiterated. Read lengths are 30–50 bases.

For the application of the large-scale sequencing methods for the identification of novel deafness genes from linkage intervals, an enrichment has to be performed for that specific region, for example for the known and predicted genes in these intervals.

5. Pathogenicity of Sequence Variants

5.1. Protein Truncating Mutations

The two types of mutations that unequivocally result in a premature translation stop are nonsense and frameshift mutations. Nonsense mutations result in the creation of one of three stop-codons (TAA, TAG, TGA), directly at the mutation site. Deletions or duplications often result in a shift of the reading frame, which results in a new, in most cases short, inappropriate coding sequence, downstream of the mutation followed by a stop mutation. The effect of protein truncating mutations in many cases will be complete loss-of-function.

Protein truncating mutations in the majority of cases have an effect on the mRNA. In multi-exon genes, truncating mutations situated in any exon except the last exon and the last 50–55 bp of the penultimate exon in most cases cause nonsense-mediated decay (NMD) of the mRNA *(109* and references therein*)*. NMD probably serves to protect the cells from the putative toxic effect of truncated proteins. As mRNAs are continuously synthesized, NMD will not result in the full absence of mRNA; at least 20% of mRNA will remain. To test whether NMD is acting, somatic cells can be cultured in the presence or absence of agents (e.g., cycloheximide) that inhibit NMD. Subsequently, RNA is isolated and, employing (quantitative) RT-PCR, can be analyzed for the occurrence of NMD. NMD generally does not occur when a mutation resides in a gene in which the protein coding region is not interrupted by introns or when the mutation is located in the last protein coding exon or in the last 50–55 bp of the penultimate protein-coding exon.

In the absence of NMD, the mutant mRNA will encode a truncated protein that in many cases will be misfolded and thereby become unstable. If the truncated protein is stable, critical functional domains at the carboxy-terminus might be lacking, also giving rise to a non-functional or partially functional protein.

5.2. Splice Site Mutations

Efficient splicing requires several recognition sequences within introns: a 5′ splice site, a 3′ splice site, and the branch point sequence (BPS). For the 5′ splice site, the consensus sequence is MAG/GURAGU (M is A or C; R is purine; "/" is the exon–intron boundary; underlined is the canonical, 99% conserved nucleotides). For the 3′ splice site this consensus is Y_nNCAG/R (Y_n is the polypyrimidine stretch; N is any nucleotide; underlined is the canonical, 100% conserved nucleotides), and for the BPS the human consensus is YNCURAY (BP is underlined; Y is pyrimidine). In 0.9% of all splice donor sites in human, the uridine (thymidine in DNA) is replaced by a cytidine nucleotide *(110)*. Any variation in the canonical 5′ and 3′ dinucleotides will have an effect on the proper splicing process. The efficiency of the splice donor site is variably affected if other nucleotides from the consensus sequences are mutated. The −5 through −14 polypyrimidine sequence consists of a preponderance of pyrimidines. The splicing efficiency will be reduced if any of the pyrimidines in a splice acceptor site is mutated to a purine. However, there are also examples of disease-associated mutations in which one pyrimidine is exchanged for another.

Shapiro and Senapathy *(111)* were the first to quantify the efficiency of splice site mutations. Web-based programs can also be used to calculate the effect of any given mutation, for example, NetGene 2 from the Technical University of Denmark (http://www.cbs.dtu.dk/services/NetGene2/), and Splice site prediction from the *Drosophila* Genome Center/Fruitfly (http://www.fruitfly. org/seq_tools/splice.html). In some cases, very subtle splice efficiency changes can have a large effect on the splicing. For example, a mutation in *DFNA5* affects the −6 position (C > G) of the 3′ splice site of intron 7. This mutation was predicted not to affect splicing efficiency but, upon analysis of *DFNA5* transcripts in lymphoblasts, skipping of exon 8 was detected in a significant part of the transcripts *(112)*. The effect of subtle changes in the splice site can depend on the presence or absence of strong cryptic splice sites near the wild-type splice site. In many cases, mutations, other than those affecting the canonical sequences, only have a partial effect on the splicing, leading to mixtures of differentially spliced mRNAs. Apart from these sequences, other intronic nucleotides, upstream of nucleotide −14 of the donor acceptor site and downstream of nucleotide +6 of the splice donor site, play a minor role in the splicing.

The effect of splice site mutations on the splicing process cannot be easily predicted. In many cases the relevant exon will be skipped, but it is also possible that cryptic intronic of exonic splice sites are activated, resulting in the inclusion of intronic sequences or exclusion of exonic sequences, respectively. In the former scenario, the intronic sequence might carry a protein truncating mutation, or, if the inserted sequence does not consist of a multiplex of three, will result in a frameshift and consequent stop mutation further downstream. In the latter scenario, there will be a loss of a discrete segment of the protein if the absent exonic nucleotides measures a multiplex of three; there will be a frameshift and subsequent translational stop if the number of deleted basepairs is not a multiplex of three.

Another group of sequences important for proper splicing are the regulatory elements, that is, splice enhancers and splice silencers. Depending on their location, they are denoted exonic or intronic splice enhancers (ESE/ISE) and exonic or intronic splice silencers (ESS/ISS) *(113)*. Many sequence variants that do not result in an amino acid change, the so called synonymous sequence changes, actually might affect one of these splice regulatory elements. These regulatory elements can be identified using web-based programs (http://rulai.cshl.edu/cgi-bin/tools/ESE3/esefinder.cgi?process=home). In addition, the frequency of these changes in patients versus healthy controls might point to their putative pathologic role.

The ultimate proof that sequence variants affect splicing relies on RT-PCR analysis of the gene of interest, preferably using mRNA from the affected tissue. As a surrogate source of RNA, lymphoblast cells (blood or EBV immortalized cell-lines) or fibroblasts (derived from skin biopsies) can be used. Even genes that are not expressed significantly in these cell types often show illegitimate transcription high enough for the sensitive RT-PCR method. Alternatively, the mutant genomic sequence of interest is cloned into a splicing-test vector, transfected into eukaryotic cells, and tested for its splicing characteristics.

5.3. Missense Mutations

Sequence variants that result in amino acid substitutions (non-synonymous changes) often are the most difficult to interpret. The amino acids are grouped according to charge and hydrophobicity/hydrophilicity (and size) *(114)*. If a sequence variant changes the amino acid to another amino acid from the same "functional" group, this is referred to as a conservative change. In case the mutant amino acid is from another functional group, this is called a non-conservative change, which in most cases has a stronger effect on the structure or function of the protein. The effect of non-synonymous amino acid substitutions can be predicted according to Betts and Russell *(115)*: http://www.russell.embl.de/aas/.

Another criterium that is frequently used is the evolutionary conservation of a given amino acid (comparison between orthologues) or the amino acid homology between similar proteins in the same species (comparison between paralogues). Programs that use the evolutionary conservation for the prediction of the pathogenicity of a specific mutation are SIFT (http://blocks.fhcrc.org/sift/SIFT.html) and PolyPhen (http://coot.embl.de/PolyPhen).

5.4. Copy-Number Variations

With the development of genome-wide SNP array technology, chromosomal deletions and duplications of genomic regions can be detected. Detection sensitivity depends on the number of nucleotides analyzed. These so-called CNVs are found throughout the human genome. CNVs can be benign polymorphic segments, genetic risk factors for multifactorial or polygenic diseases, or pathologic mutations. Data on common benign polymorphisms are increasing *(116)*. If a CNV occurs de novo, this is not proof of pathogenicity, but a strong indication. If an autosomal CNV is identified in a patient, a more detailed analysis of the size of a heterozygous deleted or duplicated segment can be determined using multiplex ligation probe amplification (MLPA) *(117)*. Heterozygous deletions can also be analyzed using genomic Q-PCR *(118–119)*.

5.5. Miscellaneous

5.5.1. Promoter Mutations

Sequence variants in the promoter of a gene can affect transcription, but their interpretation is not straightforward. To test the effect of sequence variants, the transcriptional activity of mutant and wild-type promoters can be analyzed via reporter gene assays. However, promoters vary in size and the functionality of putative promoter elements in reporter assays may not reflect the in vivo situation in specific tissues.

5.5.2. Segregation Analysis

If Mendelian inheritance is expected, sequence variants should segregate accordingly, as expected in families. For autosomal recessive variants, both parents must be heterozygous carriers; affected sibs carry both variants, and healthy sibs do not carry the sequence variant or are heterozygous carriers. For autosomal dominant variants, all affected individuals are expected to carry a heterozygous variant unless penetrance is less that 100%. An exception is phenocopies, that is, individuals apparently afflicted with the disease under study but due to a different cause. Phenocopies are not rare in the field of genetics in view of the large number of non-genetic causes of hearing loss such as infection, exposure to noise, and the high incidence of age-related hearing impairment, which can complicate inherited hearing impairment with a relatively late onset. Also, genocopies can be encountered as already mentioned in **Section 3.2.1**.

5.5.3. Nomenclature

The proper nomenclature of DNA and protein variants can be found at http://www.hgvs.org/mutnomen/ *(120)*. Sequence variants are numbered starting from the A of the ATG start codon and are denoted with the prescript "c."; amino acid variants are denoted with the "p." prescript. Further details can be found at this website.

References

1. Morton, N. E. (1991) Genetic epidemiology of hearing impairment. *Ann. NY Acad. Sci.* **630**, 16–31.
2. Cremers, F. P. M., Kimberling, W. J., Kuelm, M., de Brouwer, A. P., van Wijk, E., te Brinke, H. et al. (2007) Development of a genotyping microarray for Usher syndrome. *J. Med. Genet.* **44**, 153–160.
3. Gardner, P., Oitmaa, E., Messner, A., Hoefsloot, L., Metspalu, A., and Schrijver, I. (2006) Simultaneous multigene mutation detection in patients with sensorineural hearing loss through a novel diagnostic microarray: a new approach for newborn screening follow-up. *Pediatrics* **118**, 985–994.
4. Huygen, P. L. M., Pennings, R. J. E., and Cremers, C. W. R. J. (2003) Characterising and distinguishing progressive phenotypes in nonsyndromic autosomal dominant hearing impairment. *Audiol. Med.* **1**, 37–46.
5. Huygen, P. L. M., Pauw, R. J., and Cremers, C. W. R. J. (2007) Audiometric profiles associated with genetic non-syndromal hearing impairment: a review and phenotype analysis, in *Genes, Hearing and Deafness: From Molecular Biology to Clinical Practice* (Martini, A., Stephens, D., Read, A. P., eds.) Informa Healthcare, New York, NY, pp. 185–204.
6. Everett, L. A., Glaser, B., Beck, J. C., Idol, J. R., Buchs, A., Heyman, M. et al. (1997) Pendred syndrome is caused by mutations in a putative sulphate transporter gene (PDS). *Nat. Genet.* **17**, 411–422.
7. Li, X. C., Everett, L. A., Lalwani, A. K., Desmukh, D., Friedman, T. B., Green, E. D. et al. (1998) A mutation in *PDS* causes nonsyndromic recessive deafness. *Nat. Genet.* **18**, 215–217.
8. Reardon, W., Coffey, R., Phelps, P. D., Luxon, L. M., Stephens, D., Kendall-Taylor, P. et al. (1997) Pendred syndrome-100 years of underascertainment? *Q. J. Med.* **90**, 443–447.
9. Azaiez, H., Yang, T., Prasad, S., Sorensen, J. L., Nishimura, C. J., Kimberling, W. J. et al. (2007) Genotype-phenotype correlations for SLC26A4-related deafness. *Hum. Genet.* **122**, 451–457.
10. Stinckens, C., Huygen, P. L. M., van Camp, G., and Cremers, C. W. R. J. (2002) Pendred syndrome redefined. Report of a new family with fluctuating and progressive hearing loss. *Adv. Otorhinolaryngol.* **61**, 131–141.
11. Scott, H. S., Kudoh, J., Wattenhofer, M., Shibuya, K., Berry, A., Chrast, R. et al. (2001) Insertion of beta-satellite repeats identifies a transmembrane protease causing both congenital and childhood onset autosomal recessive deafness. *Nat. Genet.* **27**, 59–63.
12. Mustapha, M., Chardenoux, S., Nieder, A., Salem, N., Weissenbach, J., El-Zir, E. et al. (1998) A sensorineural progressive autosomal recessive form of isolated deafness, DFNB13, maps to chromosome 7q34-q36. *Eur. J. Hum. Genet.* **6**, 245–250.
13. Mustapha, M., Weil, D., Chardenoux, S., Elias, S., El-Zir, E., Beckmann, J. S. et al. (1999) An α-tectorin gene defect causes a newly identified autosomal recessive form of sensorineural pre-lingual non-syndromic deafness, DFNB21. *Hum. Mol. Genet.* **8**, 409–412.
14. Naz, S., Alasti, F., Mowjoodi, A., Riazuddin, S., Sanati, M. H., Friedman, T. B. et al. (2003) Distinctive audiometric profile associated with DFNB21 alleles of TECTA. *J. Med. Genet.* **40**, 360–363.
15. Smith, R. J. H., Berlin, C. I., Hejtmancik, J. F., Keats, B. J., Kimberling, W. J., Lewis, R. A. et al. (1994) Clinical diagnosis of the Usher syndromes. Usher Syndrome Consortium. *Am. J. Med. Genet.* **50**, 32–38.
16. Lander, E. and Kruglyak, L. (1995) Genetic dissection of complex traits: guidelines for interpreting and reporting linkage results. *Nat. Genet.* **11**, 241–247.
17. O'Connell, J. R. and Weeks, D. E. (1998) PedCheck: A program for identification of genotype incompatibilities in linkage analysis. *Am. J. Hum. Genet.* **63**, 259–266.
18. Abecasis, G. R., Cherny, S. S., Cookson, W. O., and Cardon, L. R. (2002) Merlin–rapid analysis of dense genetic maps using sparse gene flow trees. *Nat. Genet.* **30**, 97–101.

19. Hoffmann, K. and Lindner, T. H. (2005) easyLINKAGE-Plus–automated linkage analyses using large-scale SNP data. *Bioinformatics* **21**, 3565–3567.
20. Lindner, T. H. and Hoffmann, K. (2005) easyLINKAGE: a PERL script for easy and automated two-/multi-point linkage analyses. *Bioinformatics* **21**, 405–407.
21. Ruschendorf, F. and Nurnberg, P. (2005) ALOHOMORA: a tool for linkage analysis using 10 K SNP array data. *Bioinformatics* **21**, 2123–2125.
22. Terwillinger, D. J. and Ott, J. (1994) *Handbook for Human Genetic Linkage*. Johns Hopkins University Press, Baltimore, MD.
23. Nyholt, D. R. (2008) Principles of linkage analysis, in *Statistical Genetics: Gene Mapping Through Linkage and Association* (Neale, B. M., Ferreira, M. A. R., Medland, S. E., and Posthuma, D., ed.) New York: Taylor & Francis Group, pp. 113–134.
24. Schaid, D. J., Guenther, J. C., Christensen, G. B., Hebbring, S., Rosenow, C., Hilker, C. A. et al. (2004) Comparison of microsatellites versus single-nucleotide polymorphisms in a genome linkage screen for prostate cancer-susceptibility Loci. *Am. J. Hum. Genet.* **75**, 948–965.
25. Sellick, G. S., Longman, C., Tolmie, J., Newbury-Ecob, R., Geenhalgh, L., Hughes, S. et al. (2004) Genomewide linkage searches for Mendelian disease loci can be efficiently conducted using high-density SNP genotyping arrays. *Nucleic Acids Res.* **32**, e164.
26. Hehir-Kwa, J. Y., Egmont-Petersen, M., Janssen, I. M., Smeets, D., Geurts van Kessel, A., and Veltman, J. A. (2007) Genome-wide copy number profiling on high-density bacterial artificial chromosomes, single-nucleotide polymorphisms, and oligonucleotide microarrays: a platform comparison based on statistical power analysis. *DNA Res.* **14**, 1–11.
27. Nannya, Y., Sanada, M., Nakazaki, K., Hosoya, N., Wang, L., Hangaishi, A. et al. (2005) A robust algorithm for copy number detection using high-density oligonucleotide single nucleotide polymorphism genotyping arrays. *Cancer Res.* **65**, 6071–6079.
28. Genin, E., Todorov, A. A., and Clerget-Darpoux, F. (1998) Optimization of genome search strategies for homozygosity mapping: influence of marker spacing on power and threshold criteria for identification of candidate regions. *Ann. Hum. Genet.* **62**, 419–429.
29. den Hollander, A. I., Lopez, I., Yzer, S., Zonneveld, M. N., Janssen, I. M., Strom, T. M. et al. (2007) Identification of novel mutations in patients with Leber congenital amaurosis and juvenile RP by genome-wide homozygosity mapping with SNP microarrays. *Invest. Ophthalmol. Vis. Sci.* **48**, 5690–5698.
30. de Brouwer, A. P., Pennings, R. J. E., Roeters, M., Van Hauwe, P., Astuto, L. M., Hoefsloot, L. H. et al. (2003) Mutations in the calcium-binding motifs of CDH23 and the 35delG mutation in GJB2 cause hearing loss in one family. *Hum. Genet.* **112**, 156–163.
31. Lezirovitz, K., Pardono, E., de Mello Auricchio, M. T. B., de Carvalho e Silva, F. L., Lopes, J. J., Abreu-Silva, R. S. et al. (2008) Unexpected genetic heterogeneity in a large consanguineous Brazilian pedigree presenting deafness. *Eur. J. Hum. Genet.* **16**, 89–96.
32. Woods, C. G., Cox, J., Springell, K., Hampshire, D. J., Mohamed, M. D., McKibbin, M. et al. (2006) Quantification of homozygosity in consanguineous individuals with autosomal recessive disease. *Am. J. Hum. Genet.* **78**, 889–896.
33. den Hollander, A. I., Koenekoop, R. K., Mohamed, M. D., Arts, H. H., Boldt, K., Towns, K. V. et al. (2007) Mutations in *LCA5*, encoding the ciliary protein lebercilin, cause Leber congenital amaurosis. *Nat. Genet.* **39**, 889–895.
34. Tranebjaerg, L., Schwartz, C., Eriksen, H., Andreasson, S., Ponjavic, V., Dahl, A. et al. (1995) A new X-linked recessive deafness syndrome with blindness, dystonia, fractures, and mental deficiency is linked to Xq22. *J. Med. Genet.* **32**, 257–263.
35. Tyson, J., Bellman, S., Newton, V., Simpson, P., Malcolm, S., Pembrey, M. E. et al. (1996) Mapping of *DFN2* to Xq22. *Hum. Mol. Genet.* **5**, 2055–2060.
36. Lalwani, A. K., Brister, J. R., Fex, J., Grundfast, K. M., Pikus, A. T., Ploplis, B. et al. (1994) A new nonsyndromic X-linked sensorineural hearing impairment linked to Xp21.2. *Am. J. Hum. Genet.* **55**, 685–694.
37. Del Castillo, I., Villamar, M., Sarduy, M., Romero, L., Heraiz, C., Hernández, F. J. et al. (1996) A novel locus for non-syndromic sensorineural deafness (*DFN6*) maps to chromosome Xp22. *Hum. Mol. Genet.* **5**, 1383–1387.
38. Phelps, P. D., Reardon, W., Pembrey, M., Bellman, S., and Luxon, L. (1991) X-linked deafness, stapes gushers and a distinctive defect of the inner ear. *Neuroradiology* **33**, 326–330.

39. Huber, I., Bitner-Glindzicz, M., de Kok, Y. J. M., van der Maarel, S. M., Ishikawa-Brush, Y., Monaco, A. P. et al. (1994) X-linked mixed deafness (DFN3): cloning and characterization of the critical region allows the identification of novel microdeletions. *Hum. Mol. Genet.* **3**, 1151–1154.
40. de Kok, Y. J. M., van der Maarel, S. M., Bitner-Glindzicz, M., Huber, I., Monaco, A. P., Malcolm, S. et al. (1995) Association between X-linked mixed deafness and mutations in the POU domain gene POU3F4. *Science* **267**, 685–688.
41. de Kok, Y. J. M., Vossenaar, E. R., Cremers, C. W. R. J., Dahl, N., Laporte, J., Hu, L. J. et al. (1996) Identification of a hot spot for microdeletions in patients with X-linked deafness (DFN3) 900 kb proximal to the DFN3 gene *POU3F4*. *Hum. Mol. Genet.* **5**, 1229–1235.
42. Chen, A., Wayne, S., Bell, A., Ramesh, A., Srikumari Srisailapathy, C. R., Scott, D. A. et al. (1997) New gene for autosomal recessive non-syndromic hearing loss maps to either chromosome 3q or 19p. *Am. J. Med. Genet.* **71**, 467–471.
43. Collins, J. S. and Schwartz, C. E. (2002) Detecting polymorphisms and mutations in candidate genes. *Am. J. Hum. Genet.* **71**, 1251–1252.
44. Maugeri, A., van Driel, M. A., van de Pol, T. J. R., Klevering, B. J., van Haren, F. J. J., Tijmes, N. et al. (1999) The 2588G > C mutation in the *ABCR* gene is a mild frequent founder mutation in the western European population and allows the classification of *ABCR* mutations in patients with Stargardt disease. *Am. J. Hum. Genet.* **64**, 1024–1035.
45. Chu, C. S., Trapnell, B. C., Curristin, S., Cutting, G. R., and Crystal, R. G. (1993) Genetic basis of variable exon 9 skipping in cystic fibrosis transmembrane conductance regulator mRNA. *Nat. Genet.* **3**, 151–156.
46. Sanger, F., Nicklen, S., and Coulson, A. R. (1977) DNA sequencing with chain-terminating inhibitors. *Proc. Natl Acad. Sci. U. S. A.* **74**, 5463–5467.
47. Wilson, R. K., Chen, C., Avdalovic, N., Burns, J., and Hood, L. (1990) Development of an automated procedure for fluorescent DNA sequencing. *Genomics* **6**, 626–634.
48. Wilcox, E. and Fex, J. (1992) Construction of a cDNA library from microdissected guinea pig organ of Corti. *Hearing Res.* **62**, 124–126.
49. Ryan, A. F., Batcher, S., Lin, L., Brumm, D., O'Driscoll, K., and Harris, J. P. (1993) Cloning genes from an inner ear cDNA library. *Arch. Otolaryngol. Head Neck Surg.* **119**, 1217–1220.
50. Soto-Prior, A., Lavigne-Rebillard, M., Lenoir, M., Ripoll, C., Rebillard, G., Vago, P. et al. (1997) Identification of preferentially expressed cochlear genes by systematic sequencing of a rat cochlea cDNA library. *Mol. Brain Res.* **47**, 1–10.
51. Harter, C., Ripoll, C., Lenoir, M., Hamel, C. P., and Rebillard, G. (1999) Expression pattern of mammalian cochlea outer hair cell (OHC) mRNA: screening of a rat OHC cDNA library. *DNA Cell Biol.* **18**, 1–10.
52. Beisel, K. W., Shiraki, T., Morris, K. A., Pompeia, C., Kachar, B., Arakawa, T. et al. (2004) Identification of unique transcripts from a mouse full-length, subtracted inner ear cDNA library. *Genomics* **83**, 1012–1023.
53. Ficker, M., Powles, N., Warr, N., Pirvola, U., and Maconochie, M. (2004) Analysis of genes from inner ear developmental-stage cDNA subtraction reveals molecular regionalization of the otic capsule. *Dev. Biol.* **268**, 7–23.
54. Cohen-Salmon, M., Mattei, M. G., and Petit, C. (1999) Mapping of the otogelin gene (OTGN) to mouse chromosome 7 and human chromosome 11p14.3: a candidate for human autosomal recessive nonsyndromic deafness DFNB18 *Mamm. Genome* **10**, 520–522.
55. Killick, R. and Richardson, G. (1997) Isolation of chicken alpha ENaC splice variants from a cochlear cDNA library. *Biochim. Biophys. Acta* **1350**, 33–37.
56. Heller, S., Sheane, C. A., Javed, Z., and Hudspeth, A. J. (1998) Molecular markers for cell types of the inner ear and candidate genes for hearing disorders. *Proc. Natl. Acad. Sci. U.S.A.* **95**, 11400–11405.
57. Zheng, J., Long, K. B., Robison, D. E., He, D. Z., Cheng, J., Dallos, P. et al. (2002) Identification of differentially expressed cDNA clones from gerbil cochlear outer hair cells. *Audiol. Neurootol.* 7, 277–288.
58. Coimbra, R. S., Weil, D., Brottier, P., Blanchard, S., Levi, M., Hardelin, J. P. et al. (2002) A subtracted cDNA library from the zebrafish (*Danio rerio*) embryonic inner ear. *Genome Res.* **12**, 1007–1011.
59. Peters, L. M., Belyantseva, I. A., Lagziel, A., Battey, J. F., Friedman, T. B., and Morell, R. J. (2007) Signatures from tissue-specific MPSS libraries identify transcripts preferentially expressed in the mouse inner ear. *Genomics* **89**, 197–206.

60. Robertson, N. G., Khetarpal, U., Gutiérrez-Espeleta, G. A., Bieber, F. R., and Morton, C. C. (1994) Isolation of novel and known genes from a human fetal cochlear cDNA library using subtractive hybridization and differential screening. *Genomics* **23**, 42–50.
61. Jacob, A. N., Baskaran, N., Kandpal, G., Narayan, D., Bhargava, A. K., and Kandpal, R. P. (1997) Isolation of human ear specific cDNAs and construction of cDNA libraries from surgically removed small amounts of inner ear tissues. *Somat. Cell Mol. Genet.* **23**, 83–95.
62. Luijendijk, M. W. J., van de Pol, T. J. R., van Duijnhoven, G., den Hollander, A. I., ten Caat, J., van Limpt, V. et al. (2003) Cloning, characterization and mRNA expression analysis of novel human fetal cochlea cDNAs. *Genomics* **82**, 480–490.
63. Anderson, C. T. and Zheng, J. (2007) Isolation of outer hair cells from the cochlear sensory epithelium in whole-mount preparation using laser capture microdissection. *J. Neurosci. Methods* **162**, 229–236.
64. Morris, K. A., Snir, E., Pompeia, C., Koroleva, I. V., Kachar, B., Hayashizaki, Y. et al. (2005) Differential expression of genes within the cochlea as defined by a custom mouse inner ear microarray. *J. Assoc. Res. Otolaryngol.* **6**, 75–89.
65. Roche, J. P., Wackym, P. A., Cioffi, J. A., Kwitek, A. E., Erbe, C. B., and Popper, P. (2005) In silico analysis of 2085 clones from a normalized rat vestibular periphery 3′ cDNA library. *Audiol. Neurootol.* **10**, 310–322.
66. McDermott, B. M., Jr., Baucom, J. M., and Hudspeth, A. J. (2007) Analysis and functional evaluation of the hair-cell transcriptome. *Proc. Natl Acad. Sci. U.S.A.* **104**, 11820–11825.
67. van Driel, M. A., Cuelenaere, K., Kemmeren, P. P., Leunissen, J. A., and Brunner, H. G. (2003) A new web-based data mining tool for the identification of candidate genes for human genetic disorders. *Eur. J. Hum. Genet.* **11**, 57–63.
68. van Driel, M. A., Cuelenaere, K., Kemmeren, P. P., Leunissen, J. A., Brunner, H. G., and Vriend, G. (2005) GeneSeeker: extraction and integration of human disease-related information from web-based genetic databases. *Nucleic Acids Res.* **33**, W758–W761.
69. Tiffin, N., Kelso, J. F., Powell, A. R., Pan, H., Bajic, V. B., and Hide, W. A. (2005) Integration of text- and data-mining using ontologies successfully selects disease gene candidates. *Nucleic Acids Res.* **33**, 1544–1552.
70. Lopez-Bigas, N. and Ouzounis, C. A. (2004) Genome-wide identification of genes likely to be involved in human genetic disease. *Nucleic Acids Res.* **32**, 3108–3114.
71. Adie, E. A., Adams, R. R., Evans, K. L., Porteous, D. J., and Pickard, B. S. (2005) Speeding disease gene discovery by sequence based candidate prioritization. *BMC Bioinformatics* **6**, 55.
72. Adie, E. A., Adams, R. R., Evans, K. L., Porteous, D. J., and Pickard, B. S. (2006) SUSPECTS: enabling fast and effective prioritization of positional candidates. *Bioinformatics* **22**, 773–774.
73. Turner, F. S., Clutterbuck, D. R., and Semple, C. A. (2003) POCUS: mining genomic sequence annotation to predict disease genes. *Genome Biol.* **4**, R75.
74. Perez-Iratxeta, C., Bork, P., and Andrade, M. A. (2002) Exploring MEDLINE abstracts with XplorMed. *Drugs Today* **38**, 381–389.
75. Perez-Iratxeta, C., Wjst, M., Bork, P., and Andrade, M. A. (2005) G2D: a tool for mining genes associated with disease. *BMC Genet.* **6**, 45.
76. Altschul, S. F., Madden, T. L., Schaffer, A. A., Zhang, J., Zhang, Z., Miller, W. et al. (1997) Gapped BLAST and PSI-BLAST: a new generation of protein database search programs. *Nucleic Acids Res.* **25**, 3389–3402.
77. Tiffin, N., Adie, E., Turner, F., Brunner, H. G., van Driel, M. A., Oti, M. et al. (2006) Computational disease gene identification: a concert of methods prioritizes type 2 diabetes and obesity candidate genes. *Nucleic Acids Res.* **34**, 3067–3081.
78. Rastan, S., Hough, T., Kierman, A., Hardisty, R., Erven, A., Gray, I. C. et al. (2004) Towards a mutant map of the mouse–new models of neurological, behavioural, deafness, bone, renal and blood disorders. *Genetica* **122**, 47–49.
79. Friedman, L. M., Dror, A. A., and Avraham, K. B. (2007) Mouse models to study inner ear development and hereditary hearing loss. *Int. J. Dev. Biol.* **51**, 609–631.
80. Justice, M. J., Noveroske, J. K., Weber, J. S., Zheng, B., and Bradley, A. (1999) Mouse ENU mutagenesis. *Hum. Mol. Genet.* **8**, 1955–1963.
81. Coghill, E. L., Hugill, A., Parkinson, N., Davison, C., Glenister, P., Clements, S. et al. (2002) A gene-driven approach to the identification of ENU mutants in the mouse. *Nat. Genet.* **30**, 255–256.

82. Patton, E. E. and Zon, L. I. (2001) The art and design of genetic screens: zebrafish. *Nat. Rev. Genet.* **2**, 956–966.
83. Malicki, J., Schier, A. F., Solnica-Krezel, L., Stemple, D. L., Neuhauss, S. C., Stainier, D. Y. et al. (1996) Mutations affecting development of the zebrafish ear. *Development* **123**, 275–283.
84. Whitfield, T. T., Granato, M., van Eeden, F. J., Schach, U., Brand, M., Furutani-Seiki, M. et al. (1996) Mutations affecting development of the zebrafish inner ear and lateral line. *Development* **123**, 241–254.
85. Sivasubbu, S., Balciunas, D., Amsterdam, A., and Ekker, S. C. (2007) Insertional mutagenesis strategies in zebrafish. *Genome Biol.* **8 Suppl. 1**, S9.
86. Franke, L., van Bakel, H., Fokkens, L., de Jong, E. D., Egmont-Petersen, M., and Wijmenga, C. (2006) Reconstruction of a functional human gene network, with an application for prioritizing positional candidate genes. *Am. J. Hum. Genet.* **78**, 1011–1025.
87. Alfarano, C., Andrade, C. E., Anthony, K., Bahroos, N., Bajec, M., Bantoft, K. et al. (2005) The Biomolecular Interaction Network Database and related tools 2005 update. *Nucleic Acids Res.* **33**, D418–D424.
88. Peri, S., Navarro, J. D., Kristiansen, T. Z., Amanchy, R., Surendranath, V., Muthusamy, B. et al. (2004) Human protein reference database as a discovery resource for proteomics. *Nucleic Acids Res.* **32**, D497–D501.
89. Joshi-Tope, G., Gillespie, M., Vastrik, I., D'Eustachio, P., Schmidt, E., de Bono, B. et al. (2005) Reactome: a knowledgebase of biological pathways. *Nucleic Acids Res.* **33**, D428–D432.
90. Kanehisa, M., Goto, S., Kawashima, S., Okuno, Y., and Hattori, M. (2004) The KEGG resource for deciphering the genome. *Nucleic Acids Res.* **32**, D277–D280.
91. Harris, M. A., Clark, J., Ireland, A., Lomax, J., Ashburner, M., Foulger, R. et al. (2004) The Gene Ontology (GO) database and informatics resource. *Nucleic Acids Res.* **32**, D258–D261.
92. Ball, C. A., Awad, I. A., Demeter, J., Gollub, J., Hebert, J. M., Hernandez-Boussard, T. et al. (2005) The Stanford Microarray Database accommodates additional microarray platforms and data formats. *Nucleic Acids Res.* **33**, D580–D582.
93. Barrett, T. and Edgar, R. (2006) Gene expression omnibus: microarray data storage, submission, retrieval, and analysis. *Methods Enzymol.* **411**, 352–369.
94. Barrett, T. and Edgar, R. (2006) Mining microarray data at NCBI's Gene Expression Omnibus (GEO)*. *Methods Mol. Biol.* **338**, 175–190.
95. Stelzl, U., Worm, U., Lalowski, M., Haenig, C., Brembeck, F. H., Goehler, H. et al. (2005) A human protein-protein interaction network: a resource for annotating the proteome. *Cell* **122**, 957–968.
96. Lehner, B. and Fraser, A. G. (2004) Protein domains enriched in mammalian tissue-specific or widely expressed genes. *Trends Genet.* **20**, 468–472.
97. Aerts, S., Lambrechts, D., Maity, S., Van Loo, P., Coessens, B., De, S. F. et al. (2006) Gene prioritization through genomic data fusion. *Nat. Biotechnol.* **24**, 537–544.
98. Oti, M., Snel, B., Huynen, M. A., and Brunner, H. G. (2006) Predicting disease genes using protein-protein interactions. *J. Med. Genet.* **43**, 691–698.
99. George, R. A., Liu, J. Y., Feng, L. L., Bryson-Richardson, R. J., Fatkin, D., and Wouters, M. A. (2006) Analysis of protein sequence and interaction data for candidate disease gene prediction. *Nucleic Acids Res.* **34**, e130.
100. Burckstummer, T., Bennett, K. L., Preradovic, A., Schutze, G., Hantschel, O., Superti-Furga, G. et al. (2006) An efficient tandem affinity purification procedure for interaction proteomics in mammalian cells. *Nat. Methods* **3**, 1013–1019.
101. Tsai, A. and Carstens, R. P. (2006) An optimized protocol for protein purification in cultured mammalian cells using a tandem affinity purification approach. *Nat. Protoc.* **1**, 2820–2827.
102. Gloeckner, C. J., Boldt, K., Schumacher, A., Roepman, R., and Ueffing, M. (2007) A novel tandem affinity purification strategy for the efficient isolation and characterisation of native protein complexes. *Proteomics* **7**, 4228–4234.
103. Chee, M., Yang, R., Hubbell, E., Berno, A., Huang, X. C., Stern, D. et al. (1996) Accessing genetic information with high-density DNA arrays. *Science* **274**, 610–614.
104. Albert, T. J., Molla, M. N., Muzny, D. M., Nazareth, L., Wheeler, D., Song, X. et al. (2007) Direct selection of human genomic loci by microarray hybridization. *Nat. Methods* **4**, 903–905.
105. Hodges, E., Xuan, Z., Balija, V., Kramer, M., Molla, M. N., Smith, S. W. et al. (2007) Genome-wide in situ exon capture for selective resequencing. *Nat. Genet.* **39**, 1522–1527.

106. Pruitt, K. D. and Maglott, D. R. (2001) RefSeq and LocusLink: NCBI gene-centered resources. *Nucleic Acids Res.* **29**, 137–140.
107. Margulies, M., Egholm, M., Altman, W. E., Attiya, S., Bader, J. S., Bemben, L. A. et al. (2005) Genome sequencing in microfabricated high-density picolitre reactors. *Nature* **437**, 376–380.
108. Hardenbol, P., Yu, F., Belmont, J., Mackenzie, J., Bruckner, C., Brundage, T. et al. (2005) Highly multiplexed molecular inversion probe genotyping: over 10,000 targeted SNPs genotyped in a single tube assay. *Genome Res.* **15**, 269–275.
109. Frischmeyer, P. A. and Dietz, H. C. (1999) Nonsense-mediated mRNA decay in health and disease. *Hum. Mol. Genet.* **8**, 1893–1900.
110. Sheth, N., Roca, X., Hastings, M. L., Roeder, T., Krainer, A. R., and Sachidanandam, R. (2006) Comprehensive splice-site analysis using comparative genomics. *Nucleic Acids Res.* **34**, 3955–3967.
111. Shapiro, M. B. and Senapathy, P. (1987) RNA splice junctions of different classes of eukaryotes: sequence statistics and functional implications in gene expression. *Nucleic Acids Res* **15**, 7155–7174.
112. Bischoff, A. M. L. C., Luijendijk, M. W. J., Huygen, P. L. M., van Duijnhoven, G., de Leenheer, E. M. R., Oudesluijs, G. G. et al. (2004) A novel mutation identified in the DFNA5 gene in a Dutch family: a clinical and genetic evaluation. *Audiol. Neurootol.* **9**, 34–46.
113. Amendt, B. A., Si, Z. H., and Stoltzfus, C. M. (1995) Presence of exon splicing silencers within human immunodeficiency virus type 1 tat exon 2 and tat-rev exon 3: evidence for inhibition mediated by cellular factors. *Mol. Cell Biol.* **15**, 4606–4615.
114. Grantham, R. (1974) Amino acid difference formula to help explain protein evolution. *Science* **185**, 862–864.
115. Betts, M. J. and Russel, R. B. (2003) Amino acid properties and consequences of substitutions. In *Bioinformatics for Geneticists.* Barnes, M. R. and Gray, I. C. (ed.) Wiley.
116. Pinto, D., Marshall, C., Feuk, L., and Scherer, S. W. (2007) Copy-number variation in control population cohorts. *Hum. Mol. Genet.* **16 Spec. No. 2**, R168–R173.
117. Sellner, L. N. and Taylor, G. R. (2004) MLPA and MAPH: new techniques for detection of gene deletions. *Hum. Mutat.* **23**, 413–419.
118. Joncourt, F., Neuhaus, B., Jostarndt-Foegen, K., Kleinle, S., Steiner, B., and Gallati, S. (2004) Rapid identification of female carriers of DMD/BMD by quantitative real-time PCR. *Hum. Mutat.* **23**, 385–391.
119. Traverso, M., Malnati, M., Minetti, C., Regis, S., Tedeschi, S., Pedemonte, M. et al. (2006) Multiplex real-time PCR for detection of deletions and duplications in dystrophin gene. *Biochem. Biophys. Res. Commun.* **339**, 145–150.
120. den Dunnen, J. T. and Antonarakis, S. E. (2000) Mutation nomenclature extensions and suggestions to describe complex mutations: A discussion. *Hum. Mutat.* **15**, 7–12.

Part II

Amino Acid Protocols

Chapter 14

Twist-Off Purification of Hair Bundles

Jung-Bum Shin, James Pagana, and Peter G. Gillespie

Abstract

Purification of hair bundles from inner-ear organs allows biochemical analysis of bundle constituents, including proteins and lipids. We describe here the "twist-off" method of bundle isolation, where dissected inner-ear organs are embedded in agarose, then subjected to a mechanical disruption that shears off bundles and leaves them in agarose blocks. With care in the dissection and in clean-up of the isolated bundles, contamination from cell bodies can be kept to a minimum. Isolated bundles can be analyzed by a variety of techniques, including immunocytochemistry, SDS-PAGE, immunoblotting, and mass spectrometry.

Key words: Auditory, vestibular, hair bundles, mechanotransduction, agarose, mass spectrometry.

1. Introduction

1.1. Hair-Bundle Purification Necessity

Hair-cell mechanotransduction, the process underlying hearing and balance, is mediated by hair cells *(1)*. The hair cell's mechanically sensory organelle, the hair bundle (**Fig. 14.1**), responds to deflections by opening cation-selective transduction channels. The apex of a hair cell is dedicated to its bundle, which is made of an array of actin-based stereocilia, accompanied by a single kinocilium in vestibular and developing auditory hair cells. Although there is substantial interest in the identification of proteins involved in mechanotransduction, the small number of hair cells per animal and the scarceness of mechanotransduction components has greatly impeded the discovery of these proteins using traditional biochemical methods.

To enable biochemical characterization, two methods for high-purity isolation of hair bundles were developed, the

Bernd Sokolowski (ed.), *Auditory and Vestibular Research: Methods and Protocols, vol. 493*

DOI 10.1007/978-1-59745-523-7_14 Springerprotocols.com

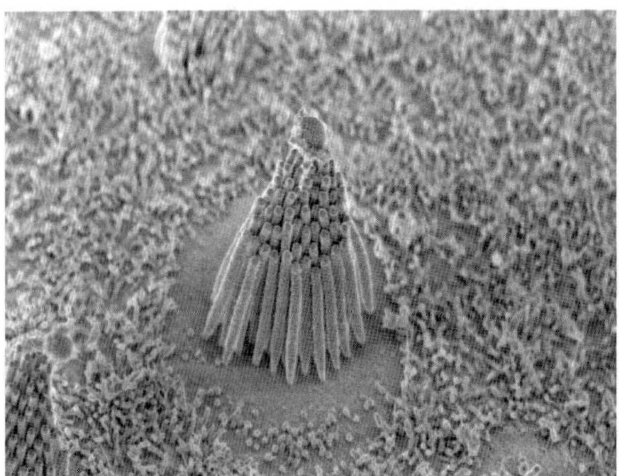

Fig. 14.1. Scanning electron micrograph of a bullfrog sacculus hair bundle. Note the ~60 stereocilia, tapered (and mechanically weak) at their bases. The single kinocilium ends in a bulbous swelling; remnants of the otolithic membrane remain attached here.

bundle-blot *(2)* and the twist-off *(3)*. These protocols require significant effort for dissection and sample processing. For instance, purifying bundles from 100 chicken ears using the twist-off method, yields only ~1.5 μg of total protein and requires about one week of bench work. Although this amount previously was insufficient for extensive biochemical experiments, recent developments in highly sensitive protein-analysis methods, most importantly mass spectrometry-based peptide sequencing *(4)*, now allow characterization of extremely small amounts of protein. The utility of a preparation for purified bundles has, therefore, become immediate. We believe that the twist-off method represents the best method to prepare pure hair bundles, and focus the remainder of the chapter on the detailed description of this method.

1.2. General Principle of the Twist-Off Method

The general principle is as follows. Inner-ear organs are adhered to the bottom of an isolation chamber filled with saline. Molten agarose solution introduced into the chamber displaces the saline then solidifies at 4 °C, capturing the bundles in the hardening gel matrix. A sharp twist or tug on the organs breaks the bundles at their weakest point, the tapered region at the base of the stereocilia. When the organs are lifted away, the bundles remain in the gel matrix and can be excised out. Superfusion of the agarose gel with saline removes soluble contaminants, while careful dissection eliminates cellular debris that otherwise contributes non-bundle proteins to the preparation.

1.3. Purified Hair Bundles from Different Species

The efficiency of the purification varies considerably from organ to organ, with the greatest efficiency occurring for organs with long, thick stereocilia. **Figure 14.2** shows examples of isolated

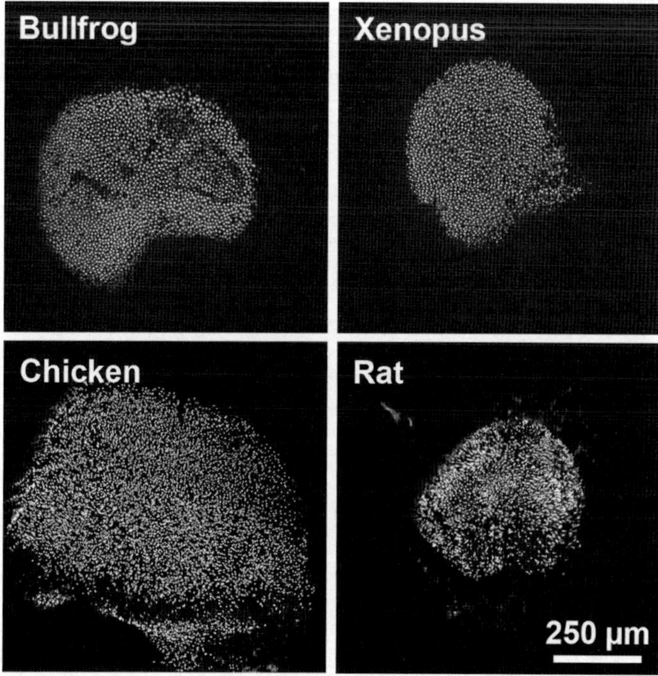

Fig. 14.2. Isolated hair bundles in agarose. Bundles from bullfrog and *Xenopus* saccules, and from chicken and rat utricles were isolated with the twist-off method and stained with fluorescent phalloidin while in agarose. Scale bar applies to all panels.

bundles from bullfrog (*Rana catesbeiana*), African clawed frog (*Xenopus laevis*), chicken (*Gallus domesticus*), and rat (*Rattus rattus*) vestibular organs. Although most bundles appear to be extracted from each organ, counts of isolated bundles reveal that only from the bullfrog or *Xenopus* saccules is it possible to obtain near-100% recovery. Twist-off isolation from chicken and rat utricles produced lower yields, typically between 20–30% and at best 50% (data not shown).

As mentioned above, the hair bundles break at their taper region. Labeling of isolated bundles with or without detergent permeabilization indicates that the bundle membranes have resealed (**Fig. 14.3A–D**), which can be observed directly using transmission electron microscopy imaging of isolated bundles (**Fig. 14.3E–H**) *(3)*. SDS-PAGE analysis of purified bundles (**Fig. 14.4**) demonstrates that bundle samples contain many unique protein bands; when compared to fractions that might contaminate the bundle preparation, bundles clearly contain a distinct set of proteins *(3)*.

1.4. History of Twist-Off Applications in the Literature

The original publication described the principles of bundle isolation from bullfrog saccule, as well as its application for a thorough characterization of hair bundle proteins using biotinylation

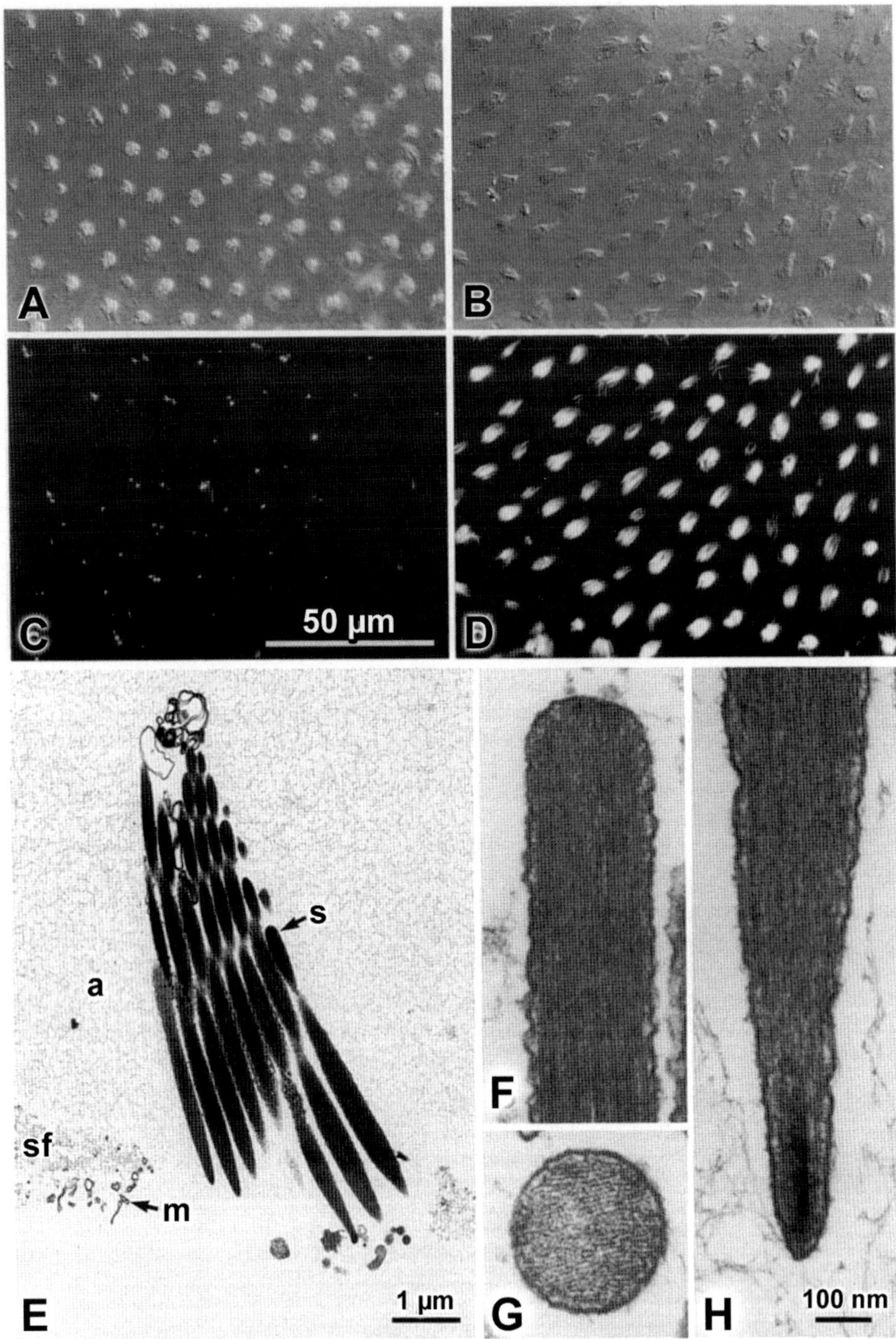

Fig. 14.3. Hair-bundle membranes reseal after isolation. **(A–D)** Light micrographs of bundles in agarose. Bundles were observed with differential interference contrast imaging **(A, B)** or phalloidin staining **(C, D)**; preparations were labeled directly **(A, C)** or following saponin treament **(B, D)**. (**E–H**) Transmission electron microscopy of isolated bundles. Note resealed taper region in (**H**). Labels: a, agarose; s, stereocilia; sf, subotolithic filaments; m, microvillus. All panels reproduced with permission from The Journal of Cell Biology, 1991, **112**, 625–640. Copyright 1991 The Rockefeller University Press.

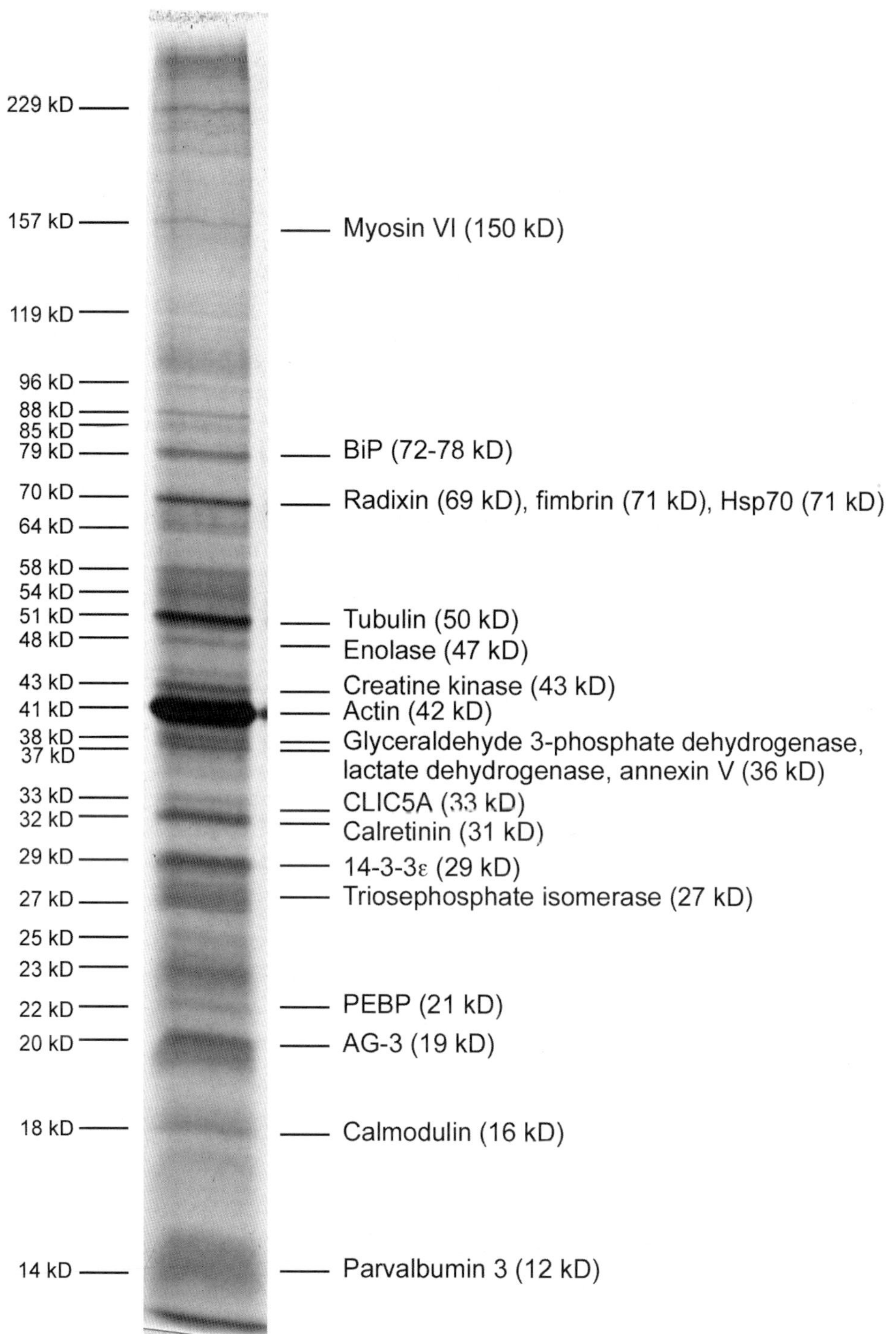

Fig. 14.4. Chicken hair-bundle proteins. Bundle proteins separated by SDS-PAGE and stained with silver. *Left*, apparent molecular masses of major bundle proteins detected in gel. *Right*, expected migration positions of major bundle proteins detected by mass spectrometry.

and SDS-PAGE analysis *(3)*. Although this study detected sub-picogram amounts of bundle proteins, the identity of most protein bands could not be established. Subsequently, the bundle preparation has been exploited for identification and characterization of hair-bundle myosins *(5–9)* and bundle Ca^{2+}-ATPases *(10–12)*.

In the most extensive characterization of bundle proteins so far, we used mass spectrometry (MS) to directly identify proteins in the bundle preparation *(13)*; **Fig. 14.4** shows chicken bundle proteins separated by SDS-PAGE and stained with silver, with major proteins indicated. In several rounds of multidimensional tandem mass spectrometry (MS/MS) experiments, hundreds of putative bundle proteins were identified, among them many known bundle proteins. In addition, several bundle-enriched proteins were newly recognized, including CLIC5 (chloride intracellular "channel" 5) and B-CK (brain-type creatine kinase). As the list of candidate bundle proteins is long, further studies are needed to validate their localization and ultimately to learn their function in the bundle. The study also revealed contaminating proteins, some originating externally (for instance human skin keratin), and others from the cell bodies of hair cells and supporting cells (e.g., mitochondrial, nuclear, and ribosomal proteins). Because the presence of contaminating proteins makes subsequent laborious validation of candidate bundle proteins essential and opens the door for misinterpretation, it is best to minimize contamination in the first place. The use of clean and efficient experimental procedures, as well as selective sampling of bundle preparations, must therefore be emphasized during the entire purification protocol.

2. Materials

2.1. Choice of Organism

2.1.1. Chicken Embryo

Utricles of late-gestation chicken embryos are large (**Fig. 14.2**) and isolated chicken hair bundles are more pure than those from mouse and rat vestibular organs. In addition, the chicken genome is sequenced, enabling accurate MS-based protein identification. A further advantage is that chicken eggs are cheap: fertilized chicken eggs can be purchased from local providers for as little as $2 per dozen, and can be incubated to near hatching in egg incubators (we use the no. 1502 "Sportsman" Egg Incubator; GQF, Savannah, GA). The number and length of hair cells and bundles in the chicken utricle is close to its maximum level at E19-E20, 1–2 days before hatching *(14)*, making that time window the most convenient and efficient for dissection. According to federal policy, chickens are classified as live vertebrate animals only after hatching; chicken embryos are therefore defined as tissue prior to hatching and so IACUC approval for their use is generally not required. For flexible incubation schedules, eggs can be stored at 13–17 °C (we use a small wine cooler) for as long as one week, without noticeable loss of viability of the eggs.

2.1.2. Mouse and Rat

Although hair-bundle purification from vestibular organs of mouse and rat is feasible, the preparations are somewhat less clean than from chicken utricle. Dissections should be performed between P2-P10, when both the utricle and the saccule are accessible (the saccule adheres tightly to the bone in older animals).

2.1.3. Bullfrog and Xenopus Laevis

Our experience is that hair bundle isolation from frog saccules works best in terms of yield and cleanliness. Examination of the remaining epithelium after the twist-off procedure reveals that most bundles of the saccule are successfully extracted, and that contamination from the cell body is virtually nonexistent. Because the *Rana catesbeiana* genome has not been sequenced to date, mass spectrometry analysis is not ideal for identifying bullfrog bundle proteins. However, two *Xenopus* (*laevis* and *tropicalis*) genomes are presently being sequenced. Experimenters should also keep in mind that frogs, especially *Xenopus laevis*, are expensive (~$8/bullfrog, ~$20/*Xenopus*), making large scale experiments difficult if the budget is tight.

2.2. Salines, Buffers, and Other Components

1. Purified water: > 18 MΩ, Milli Q-equivalent.
2. Frog standard saline: 110 m*M* NaCl, 2 m*M* KCl, 2 m*M* $MgCl_2$, 4 m*M* $CaCl_2$, 3 m*M* D-glucose (dextrose), 10 m*M* HEPES, pH 7.4 with NaOH in purified water.
3. Chicken standard saline: 155 m*M* NaCl, 6 m*M* KCl, 2 m*M* $MgCl_2$, 4 m*M* $CaCl_2$, 3 m*M* D-glucose (dextrose), 10 m*M* HEPES, pH 7.4 with NaOH in purified water.
4. Rodent saline: Dulbecco's Modified Eagle Medium, Nutrient Mixture F-12 (Ham's), called DMEM/F12. With L-gultamine, 15 m*M* HEPES; no Phenol Red (cat. no. 11039, Invitrogen- Gibco, Carlsbad, CA).
5. Protease inhibitors: 1:1000 dilutions of 0.2 m*M* phenylmethylsulfonyl fluoride (PMSF; Sigma-Aldrich, St Louis, MO) in 100% ethanol, with sonication; 0.5 mg/mL pepstatin (Sigma-Aldrich) in 100% ethanol; 0.7 mg/mL leupeptin (Sigma-Aldrich) in purified water. PMSF should be made up fresh and used within a day. Pepstatin and leupeptin stocks are kept at −20 °C. Multiple freeze-thaw cycles are fine.
6. O_2, medical grade.
7. Blocking solution: 0.2% saponin, 10 mg/mL BSA (fraction V; Calbiochem, San Diego, CA), 3% normal donkey or goat serum (Jackson ImmunoResearch, West Grove, PA).

2.3. Dissection

1. Stereo dissection microscope (e.g., Stemi SV6, Zeiss, Germany).
2. Surgical scalpel with blades (size 10).
3. Forceps (fine and coarse).
4. Small dissection scissors.
5. Superfine eyelash with handle (Ted Pella, Redding, CA).
6. Dissection dish (35 m*M* plastic Petri dish).
7. Subtilisin, protease type XXIV (Sigma-Aldrich).

2.4. Adhering Organs

1. EASY GRIP Falcon Petri dishes (Becton, Dickinson (BD) and Co., Franklin Lakes, NJ).
2. Glass coverslip (number 1; 22 × 22 mm).
3. Washer ring (nylon ring with outer diameter of 2 cm and inner diameter of 1.3 cm; can be purchased from local hardware store).
4. Nail polish (any brand).
5. Cell-Tak (BD).

2.5. Agarose Solution

1. Low-melting point agarose (Invitrogen).
2. 15 mL and 50 mL conical plastic tubes (Falcon).
3. Pall Acrodisk 32 mm syringe filter, 5 μm filter size (cat. no. 4650, Pall Life Sciences, Ann Arbor, MI).
4. 20 mL plastic syringe.
5. Microwave.
6. Heated water bath (42 °C).
7. Plastic transfer pipette.

2.6. Twist-Off Procedure

1. Microscope stage used to elevate bundle preparations ~7. 5 cm to allow fiber-optic light to be placed underneath **(Fig. 14.5)**. Stage is 30 cm wide, 13 cm deep, and 7.5 cm tall; it is custom-made from aluminum. A 7. 5 × 3. 8 cm opening allows placement of a slide holding a nylon ring.
2. Superfusion system, attached to stage (**Fig. 14.5**). This system consists of a superfusion tube and a suction tube, each with adjustable height, positioned right above the gel surface. Saline is pumped with a peristaltic pump (~1 mL/ min) through ~1. 5 mm inner-diameter Tygon tubing; the tubing is attached to a hollow stainless steel tube (~1. 7 mm outer diameter, ~1. 3 mm inner diameter) that is bent to allow superfusion right on the agarose surface. The tip is flattened to improve flow of saline. Saline, soluble contaminants, and debris are suctioned away via a similar hollow steel tube hooked up to a house vacuum source.
3. Microscope slide (with a washer ring glued on; **Fig. 14.5**).
4. Small surgical scalpel (**Fig. 14.5**).

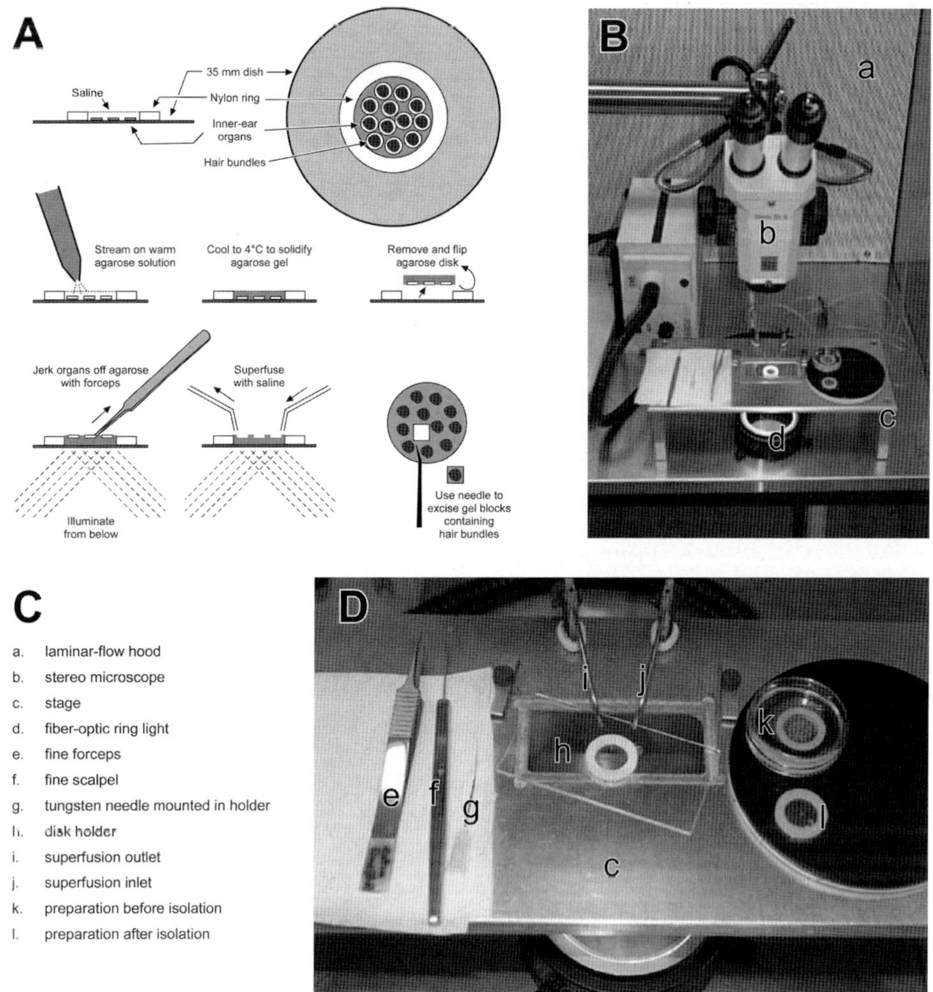

Fig. 14.5. Experimental apparatus. **(A)** Schematic of twist-off isolation. **(B)** Photograph of experimental set-up. **(C)** Key for (**B** and **D**). **(D)** Close-up photograph of tools used for twist-off isolation.

5. Washer ring.
6. Fiber-optic light source (e.g., Schott-Fostec, Elmsford, NY) with ring illuminator.
7. Laminar flow hood (semi-sterile, but anti-dust), with air flow towards the experimenter.
8. Small tungsten needle to pick up agarose blocks (cat. no. 15863, Ernest F. Fullam Inc., Clifton Park, NY); place 30–45° bend in tip and attach to wooden stick or other holder.
9. 1.5 mL microfuge tubes (non-siliconized, e.g., Eppendorf, Eppendorf of North America, Westbury, NY).

3. Method

3.1. Pre-Dissection Preparation and Precautions

1. Set up isolation apparatus in a laminar-flow hood (*see* **Note 1**).
2. Prepare saline (200–300 mL is usually enough), supplemented with protease inhibitor cocktail.
3. Prepare agarose solution (4.5% w/v; *see* **Note 2**) by heating 0.9 g agarose (low-melting point agarose) in 20 mL of appropriate saline for 30 s in the microwave. Allow to cool to ~40 °C in a water bath and filter through a 5 μm syringe filter to remove keratin. Keep the agarose solution at ~40 °C in a water bath.
4. Oxygenate saline for 15 min by bubbling O_2 through 15.2 cm 20 g steel tube; adjust O_2 regulator to produce steady stream of bubbles.
5. Pour 100 mL of saline into a 100-mL Erlenmeyer flask; pump several mL through the superfusion system so that it is ready to go. Place a 0. 45 μm filter in-line to ensure clean superfusion buffer.
6. Prepare the isolation chamber by gluing a nylon washer ring to the bottom of a 35 mm diameter Falcon Petri dish using nail polish as glue.
7. (*Optional*) Prepare Cell-Tak coated isolation chamber (only necessary if organs do not adhere to the untreated Petri dish). Using nail polish, glue a nylon washer ring onto a no. 1 glass coverslip. Spread 10 μL of a 1.8 mg/mL Cell-Tak (BD Biosciences, San Jose, CA) solution over the entire inner area of the washer ring. Some surfaces do not permit spreading of the Cell-Tak; in this case, multiple 0. 5 μL drops of Cell-Tak can be applied to the surface. In any case, the Cell-Tak must dry for ~30 min at room temperature, then rinse once with 100% ethanol, followed by three final rinses in the appropriate saline.

3.2. Dissecting Chicken Embryos

1. Up to 30 utricles can be processed in one isolation procedure.
2. Open E19-20 chicken eggs and sacrifice embryos quickly by decapitation. Cut the head mid-sagitally, remove brain tissue to expose the inner ear, and excise the temporal bone containing entire inner ear.
3. Place the temporal bone containing the inner ear in chicken saline.
4. Chip open the inner ear capsule using coarse forceps and dissect out the utricle (in chicken, the utricle is significantly larger than the saccule) using fine forceps and scissors.
5. Tease the utricular nerve almost completely away from the back of the utricle using forceps (*see* **Note 3**).
6. Remove the otolithic membrane using the eyelash and fine forceps, without protease treatment (*see* **Note 4**).

3.3. Dissecting Bullfrog and Xenopus Laevis Saccules

1. Dissect up to 20 saccules in frog standard saline with protease inhibitors. The saccular nerve should be cut as close as possible to the back side, beveled if possible (*see* **Note 5**).
2. Treat the saccules for 2–3 min in 50 μg/mL subtilisin in frog standard saline, wash three times in standard saline, and carefully remove the otolithic membrane, by lifting the peripheral membrane attachment using an eyelash. Store saccules in frog standard saline at 4 °C with protease inhibitors until further use.

3.4. Dissecting Rat and Mouse Utricles

1. The dissection procedure is very similar to the one described for chicken. Sacrifice rat or mice pups using CO_2 asphyxiation; use decapitation for young animals.
2. Dissect utricles from the exposed ears.
3. Treat utricles for 2–3 min in 50 μg/mL of subtilisin in saline; gently wash with saline three times, then remove the otolithic membrane and the otoconia as in **Section 3.4, step 2** (subtilisin treatment enables more gentle removal).

3.5. Adhering the Organs

The first important step is to adhere the vestibular organs to the bottom of the isolation chamber (*see* **Note 6**).

1. Transfer dissected organs with a small volume of saline into the inside of the washer ring (glued with nail polish to the bottom of a new Petri dish).
2. Using two forceps and under microscopic inspection, stick down the organs (bundles facing up) to the bottom of the dish. It is important not to lesion the central sensory area. Instead, use extra tissue flaps in the periphery as anchor points to adhere the epithelium.
3. If out of unforeseen reasons the organs do not adhere to the bottom of untreated plates, Cell-Tak-coated isolation chambers can be used (*see* **Section 3.1, step 7**).

3.6. Twist-Off

Once the organs have adhered to the dish, they are embedded in agarose and subjected to the "twist-off" procedure.

1. Using a clean plastic transfer pipette (rinsed with filtered water), pipette the agarose solution (4.5%) slowly and carefully onto the organs, displacing saline out of the isolation chamber. Pour off excess agarose; the level of agarose solution should be roughly level with the height of the washer ring.
2. Leave the dish at room temperature for 2 min. This step enables the agarose to engulf the bundles entirely. Do not agitate.
3. Place the dish at 4 °C for 10 min to solidify the agarose. Cover the dish to prevent contamination.
4. Remove excess agarose around the outside of the washer using forceps or a scalpel. Only the agarose disk inside the washer

ring should remain. Using forceps, lift the agarose disk carefully out of the washer ring. The organs will remain embedded in the agarose.

5. Drop the agarose disk upside-down (basal side of organs facing up) into a chamber formed by another nylon washer mounted onto a microscope slide (**Fig. 14.5**). Place the slide onto the dissection stage under the optics of the dissection microscope. Illuminate the agarose disk from below with the fiber optic ring illuminator (*see* **Note** 7).
6. Start the superfusion stream (filtered saline). Superfuse for 10–15 min. During this time, grab the peripheral regions of the individual organs with fine forceps and give them a sharp tug perpendicular to the bundles (more a jerk-off than a twist-off). Discard the organs. Because the agarose disk closely matches the diameter of the new chamber, the disk does not move substantially during the epithelium removal.
7. Cut away any agarose close to bundles that have visible contaminating debris (*see* **Note 8**). Score around and underneath (this can be tricky) the clean part of the bundles using a tungsten needle or fine-tip scalpel, spear the chunk of agarose that contains the bundles with fine forceps or tungsten needle. Transfer to a non-siliconized microfuge tube (pre-weighed to allow measurement of agarose volume).
8. Bundles in agarose can be used immediately or frozen at −80 °C.

3.7. Confirming Purity

Avoiding contamination is the principal challenge in hair-bundle purification (*see* **Note 9**). To assess contamination, perform the following immunohistochemistry control experiment on isolated bundles (**Fig. 14.6**). Note that the goal of this experiment is to give the investigator insight into the correlation between apparent quality of the preparation through the dissection microscope and the actual quality of the bundle preparation. This confirmation step is not carried out for each experiment, but rather is used heuristically to train the investigator in quality control.

1. Carry out bundle isolation to the point where the organs are pulled away from the agarose disk (**Section 3.6, step 6**).
2. Inspect the bundles under the light microscope, choose a few bundle fields that you consider clean and others that you consider contaminated with cell bodies, and cut them out of the gel.
3. Fix the gel blocks containing the two groups of isolated bundles in 3% formaldehyde for 30 min, then wash three times for 5 min.
4. Incubate in blocking solution for 1 h.
5. Incubate at least 2 h with mouse 1:500 dilution of monoclonal α-tubulin antibody (cat. no. T6199, DM1A clone, Sigma-Aldrich) in blocking solution.

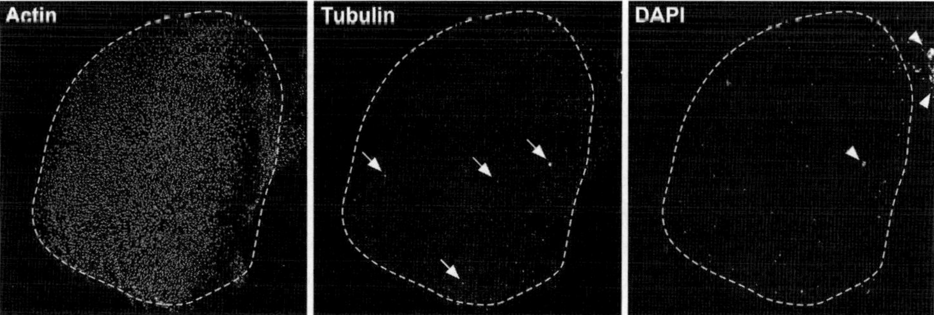

Fig. 14.6. Contamination assessment. Bundles isolated in agarose were labeled with phalloidin (to detect actin), mouse anti-tubulin, and DAPI (to detect nuclei). Although bundle preparations often appear clean when stained with phalloidin, cell-body contamination is better visualized by detecting tubulin (to highlight cell bodies; *arrows*) and nuclei (*arrowheads*). A faint haze in the tubulin image results from kinocilium staining, which is difficult to resolve in the figure. The dashed line indicates the region that would be excised; note that the majority of the contaminating nuclei would not be taken.

6. Wash three times for 10 min in PBS.
7. Incubate for at least 2 h in blocking solution containing 50 μg/mL DAPI (cat. no. D9564, Sigma-Aldrich; from a 5 mg/mL stock in dimethylformamide), FITC-0. 25 μM phalloidin (cat. no. P5282, Sigma-Aldrich) and 5 μg/mL of Cy5 donkey anti-mouse secondary antibody (cat. no. 715-175-150, Jackson ImmunoResearch).
8. Wash 4 times for 10 min in PBS, cut the gel block into a thinner slice with a scalpel (this can be tricky) with the cut running parallel to the length of the bundles, and mount in Vectashield (Vector Laboratories, Burlingame, CA), bundles facing up. It requires some practice to see which side contains the bundles.
9. Cover with a cover glass, and inspect under a fluorescence microscope. A high-quality preparation will have dense coverage of the agarose block with phalloidin signal, representing hair bundles, weak tubulin signal corresponding to kinocilia, and a few bright tubulin spots, and a few nuclei (*see* **Note 10**).

4. Notes

1. Because of the small amount of hair-bundle protein present, contamination by human skin keratin is a particular problem, especially when sensitive detection methods such as mass spectrometry are used for subsequent analysis. Much of the dust in laboratories (and homes) is in fact human skin remnants. If possible, perform the entire procedure under the laminar flow hood. Investigators who shed substantial amounts of dandruff should consider using gloves and hair net.
2. We find that the agarose concentration is critical; a 4.5% gel solution is fluid enough to penetrate between the bundles, but

has sufficient gel strength to apply a strong shearing force to the bundles. Bundle recoveries are much better with this high concentration, as compared to what we have used in the past (3%).

3. A flat utricular back with few nerve remnants helps later when sticking down the utricle, either on clean plastic or a Cell-Tak coated coverslip. Carefully tease away the nerve without damaging the sensory epithelium.
4. Protease treatment disintegrates the chicken otolithic membrane and makes complete removal difficult. In addition, it is crucial that the otolithic membrane is removed gently. Lesioned parts of the sensory epithelium tend to come loose with the bundles, contaminating the preparation.
5. As with the chicken utricle, an important goal of the frog sacculus dissection is to leave as flat a surface as possible. A flat organ sticks much better to the Petri dish. The nerve cannot be teased away as it can in the chicken utricle; with the frog, careful cutting with dissection scissors is necessary.
6. Although the original paper described the use of Cell-Tak as a glue *(3)*, we have been frustrated by poor glue performance of some batches. However, vestibular organs from most organisms (mouse, rat, chicken, and frog) adhere relatively firmly onto uncoated bacteriological Petri dishes (cat. no. 351008 BD Falcon, NJ); attachment requires unused, clean dishes. The entire procedure must be performed in saline, not culture medium, which blocks the sticky coating on these plates.
7. Because stereocilia act as light pipes, oblique illumination (dark-field) allows the isolated hair bundles to be made visible under the dissection scope. We find that one must move the ring light around underneath preparations, perhaps even tilting it slightly, to get optimal contrast for seeing the bundles. Good illumination is very important, as the clean-up step (*see* **Note 8**) is very important for producing clean hair bundles.
8. This step is critical. With proper illumination, both bundles and contaminating cell bodies can be seen through the dissecting microscope. Areas that have few bundles or are rich in contaminating cell bodies should be excised away, leaving only agarose with bundles. In some cases, substantial areas of the agarose have cell-body contamination; these areas should not be harvested. We use a variety of fine tools to carry out this step; typically, one carves the agarose around the regions containing bundles, leaving an agarose pedestal with the bundles. Any part of the agarose that appears to have cellular debris is carved away and discarded.
9. The volume of the cell body is at least twenty times larger than the bundle, and a few contaminating cell bodies can impact the purity of the bundle preparations significantly. It is therefore important, especially when learning the twist-off

method, to train the eye to distinguish between "good" and "dirty" (those with cell body contamination) bundle preparations. The goal is to be able to selectively harvest "good" or pure bundle preparations.

10. Careful examination of different bundle preparations first under the dissecting microscope, then using this labeling procedure, will allow the investigator to appreciate the necessity of cleaning the preparation prior to excision of the bundles from the agarose disk.

Acknowledgments

Supported by NIH grants R21 DC006097, R01 DC002368, and R01 DC007602 (to P.G.G.), and a DRF grant to J.B.S.

References

1. Vollrath, M. A., Kwan, K. Y., and Corey, D. P. (2007) The micromachinery of mechanotransduction in hair cells. *Annu Rev Neurosci* **30**, 339–365.
2. Shepherd, G. M. G., Barres, B. A., and Corey, D. P. (1989) "Bundle-blot" purification and initial protein characterization of hair cell stereocilia. *Proc. Natl. Acad. Sci. U.S.A.* **86**, 4973–4977.
3. Gillespie, P. G. and Hudspeth, A. J. (1991) High-purity isolation of bullfrog hair bundles and subcellular and topological localization of constituent proteins. *J. Cell Biol.* **112**, 625–640.
4. Domon, B. and Aebersold, R. (2006) Mass spectrometry and protein analysis. *Science* **312**, 212–217.
5. Gillespie, P. G., Wagner, M. C., and Hudspeth, A. J. (1993) Identification of a 120 kd hair-bundle myosin located near stereociliary tips. *Neuron* **11**, 581–594.
6. Yamoah, E. N. and Gillespie, P. G. (1996) Phosphate analogs block adaptation in hair cells by inhibiting adaptation-motor force production. *Neuron* **17**, 523–533.
7. Hasson, T., Gillespie, P. G., Garcia, J. A., MacDonald, R. B., Zhao, Y., Yee, A. G., and Corey, D. P. (1997) Unconventional myosins in inner-ear sensory epithelia. *J. Cell Biol.* **137**, 1287–1307.
8. Dumont, R. A., Zhao, Y. D., Holt, J. R., Bahler, M., and Gillespie, P. G. (2002) Myosin-I isozymes in neonatal rodent auditory and vestibular epithelia. *J. Assoc. Res. Otolaryngol.* **3**, 375–389.
9. Holt, J. R., Gillespie, S. K., Provance, D. W., Shah, K., Shokat, K. M., Corey, D. P., Mercer, J. A., and Gillespie, P. G. (2002) A chemical-genetic strategy implicates myosin-1c in adaptation by hair cells. *Cell* **108**, 371–381.
10. Yamoah, E. N., Lumpkin, E. A., Dumont, R. A., Smith, P. J., Hudspeth, A. J., and Gillespie, P. G. (1998) Plasma membrane Ca^{2+}-ATPase extrudes Ca^{2+} from hair cell stereocilia. *J. Neurosci.* **18**, 610–624.
11. Dumont, R. A., Lins, U., Filoteo, A. G., Penniston, J. T., Kachar, B., and Gillespie, P. G. (2001) Plasma membrane Ca^{2+}-ATPase isoform 2a is the PMCA of hair bundles. *J. Neurosci.* **21**, 5066–5078.
12. Hill, J. K., Williams, D. E., LeMasurier, M., Dumont, R. A., Strehler, E. E., and Gillespie, P. G. (2006) Splice-site A choice targets plasma-membrane Ca2+-ATPase isoform 2 to hair bundles. *J. Neurosci.* **26**, 6172–6180.
13. Shin, J. B., Streijger, F., Beynon, A., Peters, T., Gadzala, L., McMillen, D., Bystrom, C., Van der Zee, C. E., Wallimann, T., and Gillespie, P. G. (2007) Hair bundles are specialized for ATP delivery via creatine kinase. *Neuron* **53**, 371–386.
14. Goodyear, R. J., Gates, R., Lukashkin, A. N., and Richardson, G. P. (1999) Hair-cell numbers continue to increase in the utricular macula of the early posthatch chick. *J. Neurocytol.* **28**, 851–861.

Chapter 15

Yeast Two-Hybrid Screening to Test for Protein–Protein Interactions in the Auditory System

Dhasakumar S. Navaratnam

Abstract

We describe a protocol to screen for protein-protein interactions using the Gal-4 based yeast two-hybrid system. In this protocol, we describe serial transformation of bait into an already constructed cDNA library in yeast AH109 cells. We find this method to gives the most number of true interactions. Where a pre-made library in yeast cells is not available, the method outlined can be quickly adapted. AH109 cells can be first transformed with bait containing a vector followed by selection of yeast containing the bait. A second transformation of yeast cells is then accomplished with the cDNA library. The method is quick and can lead to the discovery of significant interactions.

Key words: Protein-protein interactions, Yeast two-hybrid, Gal4 based, cochlea cDNA library, pGBKT7, pGADT7, AH109, serial transformation.

1. Introduction

The yeast two hybrid system was first developed to detect protein-protein interactions between two known proteins *(1)*. There are several different yeast two-hybrid systems available and several key papers and reviews have a detailed accounting *(2–14)*. Yeast two-hybrid systems include the GAL4 based system, the Lex based system and the split ubiquitin based system *(11, 14, 15)*. In this chapter we lay out our laboratory protocol on using the Gal4 yeast two-hybrid system. There are presently many exciting and newer uses of the yeast two-hybrid system including mass screenings to globally define interacting proteins (interactome) *(16–19)*.

A short theoretical description of the GAL4 based system follows, since this method is a recent one. Briefly, the GAL4 based

Bernd Sokolowski (ed.), *Auditory and Vestibular Research: Methods and Protocols, vol. 493*

DOI 10.1007/978-1-59745-523-7_15 Springerprotocols.com

assay described here is based on the fact that eukaryotic transcriptional activators consist of two parts, a DNA binding domain and a transcription activation domain, that need to be in close physical proximity with one another to initiate transcription. The DNA binding domain binds the promoter, whereupon the transcription activation domain "instructs" RNA Polymerase to transcribe a downstream (reporter) gene. The transcripts encoding the protein of interest ("bait") are tagged to the binding domain and a mixture of different unknown proteins are tagged to the activation domain (cDNA). These are then co-transformed into yeast. Transcription ensues when the activation and binding domains are brought into close proximity, as a result of protein-protein interactions between the bait protein and an unknown second protein. By using a selectable marker downstream of the promoter, the yeast two-hybrid screen allows us to choose colonies, where an interaction has occurred between the bait protein and a second unknown protein. These colonies are then picked and grown, and the transcript encoding the protein that interacts with the bait protein is isolated and sequenced.

Described below is our laboratory protocol using the Matchmaker two-hybrid system from Clontech, a commercially available yeast two-hybrid system. This system uses the Gal4 activation and binding domains with several reporter genes (HIS3 and ADE2, α galactosidase and β galactosidase). The HIS3 and ADE2 gene products allow growth in medium lacking histidine and adenine and, also, permit us to screen with different degrees of stringency (i.e., strength of protein-protein interaction). The HIS 3 marker allows for selection with medium stringency while the ADE2 marker allows for selection with high stringency. The galactosidase genes, of which the α form is secreted, allow for easy identification of interacting proteins by blue-white screening. A second feature in using the Matchmaker system is that the inserts and the bait are also epitope tagged, allowing us to confirm protein-protein interaction in-vitro.

In the protocol laid out herein, it is assumed that the user has access to a yeast cDNA library from the tissue of interest in AH109 cells. If not, a commercial library can be purchased and used to serially transform yeast. The yeast are then grown in differentially restrictive conditions (Synthetic dropout media (SD) – His, –Ade, and Synthetic dropout media – His). The plates contain X-alpha-Gal that allows evaluation of galactosidase secretion.

2. Materials

2.1. Hardware

1. 150 mm plates.
2. Benchtop centrifuge and microcentrifuge.

3. 15 and 50 mL conicals.
4. 42 °C and 30 °C water bath.
5. Large platform rocker with speed and temperature controls for growing yeast.
6. Large platform rocker with speed and temperature controls for growing bacteria.
7. 30 °C incubator for yeast plates.
8. 37 °C incubator for bacterial plates.
9. Spectrophotometer.
10. Vortexer.
11. Large platform rocker with speed and temperature controls.

2.2. Solutions

All reagents are prepared in Milli-Q water.

1. Yeast Extract Peptone Dextrose (YPD Medium): Prepare 1 L of medium by mixing 10.0 g of yeast extract (Difco, BD Biosciences, CA), 20.0 gm of Peptone (Difco), 20.0 g of Glucose, in 950 mL of Milli-Q water. Adjust pH to 6.5 and bring volume to 1 L with water. Autoclave at 120 °C for 15 min.
2. Agar Plates: Mix YPD with 10 g of Bacto-agar (Difco), autoclave at 120 °C for 15 min.
3. 2X YPD Medium: Prepare 1 L of medium by mixing 20.0 g of yeast extract (Difco), 40.0 gm of Peptone (Difco), 40.0 g Glucose, and 0.6 g of l-tryptophan in 950 mL of water. Adjust pH to 6.5 with 1 *N* HCl and bring volume to 1 L with water. Autoclave at 120 °C for 15 min.
4. SD (Synthetic Dropout) Medium: Mix 6.7 g of Yeast Nitrogen Base without amino acids, 20.0 g of Glucose, 100 mL of 10X Dropout Supplement in 950 mL of water. Adjust pH to 5.8 with 1 *N* HCl and bring volume to 1 L. Autoclave at 120 °C for 15 min.
5. 10X Complete Dropout Solution. Mix the following components in 1000 mL of Milli-Q water: 200 mg of l-adenine hemisulfate, 200 mg of l-arginine HCl, 200 mg of l-histidine HCl monohydrate, 300 mg of l-isoleucine, 1000 mg of l-leucine, 300 mg of l-lysine HCl, 200 mg of l-methionine, 500 mg of l-phenylalanine, 2000 mg of l-threonine, 200 mg of l-tryptophan, 300 mg of l-tyrosine, 200 mg of l-uracil, 1500 mg of l-valine.
6. 10X Dropout Solution (-Ade/-His/-Leu/-Trp). Mix the following components in 1000 mL of Milli-Q water: 200 mg of l-arginine HCl, 300 mg of l-isoleucine, 300 mg of l-lysine HCl, 200 mg of l-methionine, 500 mg of l-phenylalanine, 2000 mg of l-threonine, 300 mg of l-tyrosine, 200 mg of l-uracil, 1500 mg of l-valine.
7. 10X Dropout Solution (-His/-Leu/-Trp). Mix the following components in 1000 mL of Milli-Q water: 200 mg of l-adenine hemisulfate, 200 mg of l-arginine HCl, 300 mg of l-isoleucine, 300 mg of l-lysine HCl, 200 mg of l-methionine,

500 mg of l-phenylalanine, 2000 mg of l-threonine, 300 mg of l-tyrosine, 200 mg of l-uracil, 1500 mg of l-valine.

8. 10X Dropout Solution (-Leu). Mix the following components in 1000 mL of Milli-Q water: 200 mg of l-adenine hemisulfate, 200 mg of l-arginine HCl, 200 mg of l-histidine HCl monohydrate, 300 mg of l-isoleucine, 300 mg of l-lysine HCl, 200 mg of l-methionine, 500 mg of l-phenylalanine, 2000 mg of l-threonine, 200 mg of l-tryptophan, 300 mg of l-tyrosine, 200 mg of l-uracil, 1500 mg of l-valine.
9. 10X Dropout Solution (-Trp). Mix the following components in 1000 mL of Milli-Q water: 200 mg of l-adenine hemisulfate, 200 mg of l-arginine HCl, 200 mg of l-histidine HCl monohydrate, 300 mg of l-isoleucine, 1000 mg of l-leucine, 300 mg of l-lysine HCl, 200 mg of l-methionine, 500 mg of l-phenylalanine, 2000 mg of l-threonine, 300 mg of l-tyrosine, 200 mg of l-uracil1, 500 mg of l-valine.
10. 50% Polyethylene Glycol (PEG) Stock Solution: Dissolve 50 g of PEG 3350 in 100 mL of water. The solution may need to be heated to 55 °C to dissolve the polyethylene glycol.
11. 10X Li Acetate: Dissolve 6.6 g of LiAc in 85 mL of water and adjust pH to 7.5 with acetic acid. Autoclave at 120 °C for 15 min.
12. 10X TE buffer: Mix 12.1 g of Tris base, 3.72 g of ethylenediaminetetraacetic acid (EDTA, disodium salt) in 950 mL of water and adjust pH to 7.5 with 1 *N* HCl. Autoclave at 120 °C for 15 min.
13. PEG/Li Acetate solution: Mix 8 mL of 50% PEG, 1 mL of 10X Tris-EDTA (TE) and 1 mL of 10X Li Acetate.
14. 3-amino-1,2,4-triazole (3-AT)should be added to SD/-Trp at varying concentrations before plating cells with AH109 cells that are transformed with bait in pGBK T7 plasmid. Double transformants are then grown in SD/-His/-Leu/-Trp medium using a concentration of 3-AT that shows no growth of the bait plasmid transformed AH109 cells.
15. cDNA libraries. Chick cochlea cDNA libraries are available upon request. Please contact Dhasakumar.Navaratnam@Yale.Edu.

3. Methods

There are several different yeast two-hybrid systems available. These include the Gal4-based system, the Lex-based system and the split –ubiquitin system. In this chapter, we describe our protocol for the Gal4 yeast two-hybrid system.

3.1. Two-Hybrid Screening

3.1.1. Preparing the Bait Vector

1. Three methods are available for yeast two-hybrid screens, co-transformation, yeast mating, and serial transformation. Of these, we find serial transformation gives the greatest number of transformants and potential interactions (*see* **Note 1**).
2. Subclone the fragment of cDNA, to be used as bait, into vector pGBK T7. Since the Gal4-based system depends on soluble proteins, it is important that the sequence of interest does not contain potentially membrane spanning regions. Use PCR amplification to generate the cDNA fragments. Design primers complementary to the nucleic acid sequence in the fragment of cDNA with restriction fragments at their 5′ ends. Remember to add an extra 4–5 nucleotides at the 5′ end so that the restriction enzyme can "bite." Also, note that the 5′ end should have the protein of interest in frame with the *myc* epitope.

3.1.2. Testing for Leaky His3 Expression (see **Note 2**)

1. Streak AH109 on YPDA plates and grow until colonies appear (~ 2–4 days).
2. Pick isolated colony and grow in 5 mL of YPD medium overnight at 30 °C.
3. Transfer cells to 50 mL of YPD medium and continue growing at 30 °C.
4. Periodically assess Optical Density (O.D.) at 600 nm to test that the cells are growing.
5. Once O.D. 600 exceeds 0.15–0.3, centrifuge cells at 700 g for 10 min at room temperature.
6. Resuspend cell pellet in 50 mL of 2X YPD.
7. Grow cells at 30 °C for 1–3 h with shaking in a 500 mL flask.
8. Once an O.D. 600 reading of 0.4 is attained, centrifuge cells at 700 g for 10 min at room temperature and discard supernatant.
9. Resuspend cell pellet in 50 mL of water and centrifuge at 700 g for 10 min at room temperature.
10. Resuspend cell pellet in 1.5 mL of 1.1X TE/LiAc solution and centrifuge at 700 g for 10 min at room temperature, and discard supernatant.
11. Resuspend cell pellet in 0.6 mL of 1.1X TE/LiAc solution.
12. In a 1.5 mL microfuge tube, mix 0. 1 μg of bait plasmid in pGBKT7 (>0. 5 μg/μL) and 10 μL of denatured sheared herring testes carrier DNA (10 mg/mL). To denature carrier DNA, add desired amount of DNA from **step 12** to a 0.5 mL PCR tube and set thermal cycler to: 94 °C for 5 min, 4 °C for 5 min, 94 °C for 5 min, and hold at 4 °C.
13. Add 100 μL of AH109 cells from **step 11.** Mix gently by vortexing, then add 0.6 mL of PEG/LiAc solution to 1.5 mL microfuge tube.
14. Mix gently by vortexing, then incubate at 30 °C for 45 min. Mix cells every 15 min.

15. Add 70 μL of DMSO to each 1.5 mL microfuge tube and mix.
16. Place cells in 42 °C water bath for 15 min and mix cells every 5 min.
17. Centrifuge cells at 700 g for 2 min at room temperature.
18. Resuspend in 3 mL of 2X YPD medium and incubate cells at 30 °C for 2 h with shaking.
19. Centrifuge cells at 700 g for 2 min at room temperature.
20. Resuspend each pellet in 0.5 mL of 0.9% NaCl.
21. Spread 50 μL of cells each on 2 SD/-Trp/plates and 2 plates each of SD/-Trp/-His/containing 0, 2.5 m*M*, 5.0 m*M*, 10 m*M*, and 15 m*M* 3-amino-1,2,4-triazole. If there is no growth on SD/-Trp/-His/plates without 3-amino-1,2,4-triazole, the reagent does not needed to be added to SD/-Leu/-Trp/-His in **step 3, Section 3.1.3**. If however, there is leaky His3 expression, add 3-amino-1,2,4-triazole to plates at a concentration where no growth is seen (*see* **Note 2**).

3.1.3. Materials Prepared the Day Before the Y2H Experiment

1. Prepare sufficient subcloned pGBKT7 vector containing cDNA for entire experiment (~ 10 μg). We use Qiaquick purified plasmid with no further purification.
2. Prepare media and plates the day before the planned experiment and set incubator temperature for 30 °C.
3. Prepare the following number of agar plates with specific dropout medium: 200 SD/-Ade/-Leu/-Trp/-His/X-Gal plates, 200 SD/-Leu/-Trp/-His/X-Gal plates, 4 SD/-Leu plates, 4 SD/-Trp plates and 4 SD/-Leu/-Trp/plates. Add 3-amino- 1,2,4-triazole to a final concentration to plates lacking His (-His/) if no growth was seen in previous tests conducted, as described in **Section 3.1.2, step 21**. (*See* **Notes 3 ,4**).

3.1.4. Conducting the Y2H Experiment

The following 10 steps are performed on the morning of the Y2H experiment. We use chemical transformation, because that gives us the most consistent results with good viability (and contrasts favorably with electroporation).

1. Thaw 2 mL of cDNA library (>5×10^7 colony forming units per mL) in AH109 cells at room temperature. The library contains the putative protein-protein partners or "prey."
2. Add entire contents of library into 100 mL of 2X YPD medium.
3. Grow cells at 30 °C for 1–3 h with shaking in a 500 mL flask.
4. Periodically assess O.D. 600 to determine if the cells are growing.
5. Once O.D. 600 exceeds 0.15–0.3, centrifuge cells at 700 g for 10 min at room temperature.

6. Resuspend pellet in 100 mL of 2X YPD and grow cells at 30 °C for 1–3 h with shaking in a 500 mL flask.
7. Once an O.D. 600 reading of 0.4 is attained, centrifuge cells at 700 g for 10 min at room temperature.
8. Resuspend cells in 80 mL of water then centrifuge cells at 700 g for 10 min at room temperature.
9. Resuspend cells in 3 mL of 1.1 x TE/LiAc solution then centrifuge cells at 700 g for 10 min at room temperature.
10. Resuspend pellet in 1.2 mL of 1.1 x TE/LiAc solution. Divide cells into two 0.6 mL aliquots.
11. Prepare two sterile 15 mL conical tubes by cooling and maintaining them on ice, then add in order: 3 μg of bait plasmid in pGBK T7 (>0. 5 μg/μL) and 20 μL of denatured sheared herring testes carrier DNA(10 mg/mL). To denature carrier DNA, add desired amount of DNA to 0.5 mL PCR tube and set thermal cycler to: 94 °C for 5 min, 4 °C for 5 min, 94 °C for 5 min, the hold at 4 °C.
12. Add 600 μL of AH109 cells from **step 10**.
13. Mix gently by vortexing then add 2.5 mL of PEG/LiAc solution to each 15 mL conical tube.
14. Mix gently by vortexing then incubate at 30 °C for 45 min. Mix cells every 15 min.
15. Add 160 μL of DMSO to each 15 mL conical tube and mix.
16. Place cells in 42 °C water bath for 20 min and mix cells every 10 min.
17. Centrifuge cells at 700 g for 10 min at room temperature.
18. Resuspend each cell pellet in 3 mL of 2X YPD medium and incubate cells at 30 °C for 2 h with shaking.
19. Centrifuge cells at 700 g for 10 min at room temperature, then resuspend cells in 30 mL of 0.9% NaCl.
20. Spread 150 μL of cells from conical tube 1 onto each of the 100 150 mm plates containing SD/-Ade/-Leu/-Trp/-His/X-Gal. Spread 150 μl of cells from conical tube 2 onto each of the 100 150 mm plates contining SD/-Leu/-Trp/-His/X-Gal. Spread 150 μl from each tube on two SD/-Leu/-Trp/plate, two SD/-Leu/plate, and two SD/-Trp/plates (*see* **Note 4**).
21. Incubate for 2–6 days at 30 °C or until colonies appear. Colonies should appear by 2 days but no longer than 8 days (*see* **Notes 5–8**).

3.1.5. Picking Colonies and Controlling for Artifacts

1. Count the number of colonies in the SD/-Leu/-Trp/plate, SD/-Leu/plate, and SD/-Trp/plate to determine the number of double transformants (pGADT7 and pGBKT7), and transformants with pGBK T7 respectively. The number of colonies in SD/-Leu/plate will confirm the number of viable cells in the library (titer) (*see* **Note 4**).

2. Pick single colonies from each plate and re-streak on similar selective plates. That is, streak colonies from SD/-Ade/-Leu/-Trp/-His/X-Gal onto SD/-Ade/ -Leu/-Trp/-His/X-Gal plates, and colonies from SD/-Leu/-Trp/-His/X-Gal onto SD/-Leu/-Trp/-His/ X-Gal plates (*see* **Notes 5–8**).
3. Pick individual colonies from each of the new plates and grow in the corresponding restrictive media by inoculating 0.5 mL of medium with one colony each. That is colonies from SD/-Ade/-Leu/-Trp/-His/X-Gal to be grown in SD/-Ade/-Leu/-Trp/-His/medium, and colonies from SD/-Leu/-Trp/-His/X-Gal on SD/-Leu/-Trp/-His/medium.
4. Grow cells at 30 °C for 2 days with shaking at 200 RPM.
5. Centrifuge cells at 700 g for 10 min at room temperature.
6. Resuspend cells in 5 mL of sterile SD/-Trp/medium to isolate only yeast containing the pGADT7-cDNA.
7. Grow cells at 30 °C for 2 days with shaking. Remove 0.85 mL and add 0.15 mL of sterile glycerol. Freeze in nalgene cryovials at −80 °C. These stocks serve as a reservoir in the event of unforeseen errors.
8. Centrifuge remaining cells from **step 7** at 700 g and proceed to isolate DNA by using a yeast mini-prep.

3.2. Yeast DNA Isolation and Transformation of Bacteria

3.2.1. Isolation

We use the Zymoprep I/II™ Yeast Plasmid Minipreparation Kit, Zymoresearch, Orange, CA, to isolate yeast DNA (*see* **Note 9**).

1. Centrifuge 1.5 mL of yeast liquid culture in a 2 mL Eppendorf tube.
2. Add 15 μL of Zymolyase™ for each 1 mL of Solution 1 to make a Solution 1-enzyme mixture.
3. Add 200 μL of Solution 1 to each pellet.
4. Add a further 3 μL of Zymolyase™ to each tube. Resuspend the pellet by mild vortexing and incubate at 37 °C for 60 min.
5. Add 200 μL Solution 2 to each tube. Mix well.
6. Add 400 μL Solution 3 to each tube. Mix well.
7. Centrifuge at 18,000 g for 3 min.
8. Transfer the supernatant to the Zymo-Spin-I column.
9. Centrifuge the Zymo-Spin I column for 30 s at 14,000 g.
10. Discard the flow-through in the collection tube. Make sure the liquid does not touch the bottom part of the column.
11. Add 550 μL of Wash Buffer (ethanol added) onto the column with the collection tube and centrifuge for 1 min at 18000 g.
12. Place column in fresh 1.5 mL microfuge tube and centrifuge at 18,000 g for an additional 1 min.
13. Place column in fresh microtube, add 10 μL of water, and incubate for 5 min at room temperature. Elute plasmid off the column by centrifuging for 1 min at 18,000 g.

3.2.2. Transforming Bacteria with Isolated Yeast Plasmid

The isolated DNA is used to transform bacteria. We find electroporation of the DNA into *E. coli* DH10B (Invitrogen, Carlsbad, CA) gives the best results (*see* **Note 10**). Once *E. coli* are transformed, follow these steps.

1. Spread cells onto ampicillin plates using 5 m*M* glass beads.
2. Pick individual clones using a sterile toothpick to inoculate 3 mL of LB medium, containing 50 μg of Ampicillin/mL. Isolate DNA using a standard mini-prep kit (e.g,. Qiagen plasmid mini-prep kit)
3. Determine the sequence of the inserted cDNA using the T7 primer. We use in-house or commercial sequencing services.

3.3. Confirming Interactions

Interactions demonstrated using the yeast two-hybrid system need to be confirmed by other techniques such as immunoprecipitation (*see* **Note 11**). While an in-depth description of these techniques is beyond the scope of this chapter, we will mention briefly the technique of coimmunoprecipitation. We routinely use *in-vitro* synthesized S35 Methionine-labeled protein, as it is the most useful method. The WGA system from Clontech gives us the most consistent results. We are able to use the sequences in the pGBKT7 and pGADT7 vectors without further subcloning. The vectors can be linearized and used to run off RNA using the T7 promoter. Moreover, both vectors are designed to incorporate a *Myc* and Flag tag at the N-terminus of each inserted cDNA sequence. These tags allow for easy immunoprecipitation and visualization by autoradiography of the corresponding two proteins (confirming interactions).

Also, we follow *in vitro* immunoprecipitation experiments with *in vivo* experiments, using antibodies, where available, to the two interacting proteins. When antibodies to the native proteins are not available, we resort to epitope tagging the entire protein. The proteins are then expressed in HEK cells and separately immunoprecipitated. The converse protein is then detected in the immunoprecipitate by western blotting.

3.4. Where cDNA Libraries in Yeast are Not Available

The method outlined here assumes that you have pGADT7-cDNA libraries in yeast AH109 cells. This method of serial transformation can be used also to test for interactions using commercial non-yeast libraries. In this case, AH109 cells derived from a fresh single colony should be made competent as indicated in **Section 3.1.4, steps 1–10**. The cells should be transformed with bait containing pGBKT7 and grown in SD/-Trp medium. A second round of transformation is then performed using the cDNA library in pGADT7 with the same protocol outlined above (*see* **Section 3.1.4, steps 1–21**). Following the second round of transformation cells are plated on SD/-Ade/-Leu/-Trp/-His/plates and SD/-Leu/-Trp/-His plates to isolate colonies with interacting proteins.

4. Notes

1. We have tried using co-transformation, serial transformation, and yeast mating in trying to detect protein-protein interactions. Of these methods, our data indicate that serial transformations give the greatest number of interacting proteins. The downside to this method is that it is possible that the bait protein could be eliminated by recombination in the event that it is toxic to the cell.
2. At times it is possible that the bait in pGBKT7 can alone activate transcription of the His3 protein. The level of His3 expression is low and can be easily suppressed by the addition of 3-AT. We first test if the bait protein in pGBKT7 activates His3 expression in AH109 cells by transforming these cells with the bait pGBKT7 construct alone. These cells are then plated on SD/-Trp/-His with varying concentrations of 3-AT added to the medium. We then note the concentration of 3-AT at which there is no observed growth. We then add 3-AT to SD/-Trp/-Leu/-His plates used for medium stringency screening of the yeast two-hybrid screen. We do not add 3AT to the high stringency screen (SD/-Trp/-Leu/-His/-Ade).
3. Readers may wonder why we limit ourselves to using relatively stringent conditions for detecting interacting proteins. In our experience, the real difficulty with yeast two-hybrid techniques is not one of sensitivity (number of interactions) but rather one of specificity (number of true interactions). We have, therefore, not had to use low stringency screens. In these screens, yeast, after transformation of bait and prey vectors, are grown in SD/-Leu/-Trp/. After colonies are identified they are re-plated on plates with increasing stringency (SD/-Ade/-His/-Leu/-Trp/and SD/-His/-Leu/-Trp/).
4. Each of the constructs contain different reporter genes, allowing one to select for cells containing that construct. For instance, the pGBKT7 vector contains the tryptophan reporter, allowing cells containing this vector to be grown in medium lacking tryptophan. Likewise, the pGAD T7 vector that contains the cDNA library has the Leucine reporter. Thus, cells containing this plasmid can be grown in medium lacking leucine. Double transformants (that is cells containing both pGBKT7 and pGADT7) can be grown on media lacking both leucine and tryptophan. This must be contrasted with double transformants that show an interaction that will grow in medium lacking leucine, tryptophan, histidine, and adenine. The latter two are reporters that are dependent on an interaction between the GAL4 binding domain and GAL4 activation domain.

5. A feature that must be paid attention to is the shape and the color of yeast in plates incubated in restrictive media. AH109 cells, after 2–4 days of growth, appear as 2–5 mm-sized plaques that resemble a small blob of yoghurt. There are no visible hyphae. Contaminant yeast from ambient air can sometimes grow on these plates but their morphology makes them easy to identify.
6. The times of incubation in restrictive conditions (SD/-His/-Trp/-Leu/; SD/-Ade/-His -Trp/-Leu/) may vary and depends on the strength of the interacting proteins. For instance, weaker interactions may result in growth that is visible after 5–8 days. The criteria that we use is the size of the yeast colony – they must be 2 mm or larger. Often colonies are seen on day 2 that don't grow to be larger than 2 mm.
7. Colonies that express Ade are white or pale pink in color. Colonies that lack Ade turn brown to red in SD/-Ade/-His/-Trp/-Leu/medium.
8. In our screens we use both alpha and beta galactosidase assays only at the point of secondary screening. The reason we do not use this assay in the initial screening is owing to cost. Although we do not use it as a method for detecting interactions, the amounts of alpha and/or beta galactosidase can also be used as a test of the strength of protein-protein interactions. There are several plate and liquid assays (chromogenic and fluorescent) for both these enzymes.
9. We have attempted to isolate yeast DNA using a number of methods with variable success. We have now settled on using the Zymoprep I/II^{TM} Yeast Plasmid Minipreparation Kit, which has given us the most consistent results. The method is similar to the commonly used alkaline lysis method used with bacteria, but has an antecedent enzyme digestion step for breaking down the cell wall.
10. Isolated DNA from yeast mini-preps should be used to transform bacteria. While the commonly used heat shock method can be used, we find the amount of recovered DNA to be variable and at times too low to effectively transform bacteria. We have instead found electroporation (using the manufacturer's instructions) into *E. coli* DH10B (Gibco BRL Life Technologies, Inc.) to give the best results. Spread cells on ampicillin plates.
11. It is now widely recognized in the field that while yeast two-hybrid technology is a sensitive technique, it is susceptible to non-specific artifacts. A number of reasons account for this and are outlined in several of the reviews below. The bait protein too does influence this to some degree. Nevertheless, it is now *de-rigeur* to confirm interactions detected using the yeast two-hybrid interactions by immunoprecipitation. Additionally *in-vivo* and subcellular

interactions can further be determined using fluorescence resonance energy transfer (FRET).

References

1. Brent, R. and Finley, R. L., Jr. (1997) Understanding gene and allele function with two-hybrid methods. *Annu. Rev. Genet.* **31**, 663–704.
2. Bartel, P., Chien, C. T., Sternglanz, R., and Fields, S. (1993) Elimination of false positives that arise in using the two-hybrid system. *Biotechniques* **14(6)**, 920–924.
3. Bartel, P. L., Chien, C.-T., Sternglanz, R., and Fields, S. (1993) Using the two-hybrid system to detect protein-protein interactions, in *Cellular Interactions in Development: A Practical Approach* (Hartley, D. A., ed.) Oxford University Press, Oxford pp. 153–179.
4. Chien, C. T., Bartel, P. L., Sternglanz, R., and Fields, S. (1991) The two-hybrid system: a method to identify and clone genes for proteins that interact with a protein of interest. *Proc. Natl. Acad. Sci. U.S.A.* **88**, 9578–9582.
5. Fields, S. and Sternglanz, R. (1994) The two-hybrid system: an assay for protein-protein interactions. *Trends Genet.* **10**, 286–292.
6. Fritz, C. C. and Green, M.R. (1992) Fishing for partners. *Curr. Biol.* **2**, 403–405.
7. Golemis, E. A., Gyuris, J., and Brent, R. (1994) Interaction trap/two-hybrid systems to identify interacting proteins, in *Current Protocols in Molecular Biology*, John Wiley & Sons, Inc., Hoboken, NJ, Chapters 13 and 14.
8. Golemis, E. A., Gyuris, J., and Brent, R. (1996) Analysis of protein interactions; and Interaction trap/two-hybrid systems to identify interacting proteins, in *Current Protocols in Molecular Biology*, John Wiley & Sons, Inc., Hoboken, NJ, Chapters 20.0 and 20.1.
9. Guarente, L. (1993) Strategies for the identification of interacting proteins. *Proc. Natl. Acad. Sci. U.S.A.* **90(5)**, 1639–1641.
10. Gyuris, J., Golemis, E., Chertkov, H., and Brent, R. (1993) Cdi1, a human G1 and S phase protein phosphatase that associates with Cdk2. *Cell* **75**, 791–803.
11. Johnsson, N. and Varshavsky, A. (1994) Split ubiquitin as a sensor of protein interactions in vivo. *Proc. Natl. Acad. Sci. U.S.A.* **91**, 10340–10344.
12. Luban, J. and Goff, S. P. (1995) *The yeast two-hybrid system for studying protein-protein interactions. Curr. Opin. Biotechnol.* **6(1)**, 59–64.
13. Mendelsohn, A. R. and Brent, R. (1994) Applications of interaction traps/two-hybrid systems to biotechnology research. *Curr. Opin. Biotechnol.* **5(5)**, 482–486.
14. Stagljar I., Korostensky, C., Johnsson, N., te Heesen, S. A. (1998) genetic system based on split-ubiquitin for the analysis of interactions between membrane proteins in vivo. *Proc. Natl. Acad. Sci. U.S.A.* **95(9)**, 5187–5192.
15. Fields, S. and Song, O. A. (1989) novel genetic system to detect protein-protein interactions. *Nature* **340**, 245–246.
16. Fields, S. (2005) High-throughput two-hybrid analysis. The promise and the peril. *FEBS J.* **272**, 5391–5399.
17. Goll, J. and Uetz, P. (2006) The elusive yeast interactome. *Genome Biol.* **7(6)**, 223.
18. Stelzl, U. and Wanker, E. E. (2006) The value of high quality protein-protein interaction networks for systems biology. *Curr. Opin. Chem. Biol.* **10(6)**, 551–558.
19. Suter, B., Auerbach, D., and Stagljar, I. (2006) Yeast-based functional genomics and proteomics technologies: the first 15 years and beyond. *Biotechniques* **40(5)**, 625–644.

Chapter 16

The Use of 2-D Gels to Identify Novel Protein–Protein Interactions in the Cochlea

Thandavarayan Kathiresan, Margaret C. Harvey, and Bernd H. A. Sokolowski

Abstract

Functional proteomics comprises a wide range of technologies for the identification of novel protein-protein interactions and biological markers. Studies of protein-protein interactions have gained from the development of techniques and technologies such as immunoprecipitation, preparative two-dimensional (2-D) gel electrophoresis for peptide mass fingerprinting (PMF), using matrix-assisted laser desorption ionization time-of-flight (MALDI-TOF) mass spectrometry, and liquid chromatography-tandem mass spectrometry (LC-MS/MS). These applications enabled the discovery of putative protein partners without *a priori* knowledge of which one(s) might be relevant. Here, we report the methods by which membrane proteins are isolated from cochlear tissues and prepared for identification by mass spectrometry techniques.

Key words: Cochlea, protein–protein interaction, immunoprecipitation, 2-D gel electrophoresis, isoelectric focusing, in-gel trypsin digestion, peptide extraction, MALDI-TOF-TOF, LC-MS/MS, mass spectrometry.

1. Introduction

Putative interacting protein partners are frequently detected by yeast two-hybrid screenings or *via* the validation of pre-selected candidate proteins. The elucidation of interactomes from large pools of binding proteins, especially those proteins that are weakly or indirectly associated with their targets, are unavoidable limitations of these techniques. More than 100,000 expressed sequence tags (ESTs) from inner ear tissues have been mapped to several vertebrate genomes *(1)*. The combination of

Bernd Sokolowski (ed.), *Auditory and Vestibular Research: Methods and Protocols, vol. 493*

DOI 10.1007/978-1-59745-523-7_16 Springerprotocols.com

immunoprecipitation and proteomics is an effective technique for the identification of interacting partners *(2)*.

One of the most commonly used methods for verification of protein–protein interactions is co-immunoprecipitation (Co-IP) *(3)*. Protein partners are isolated from tissue preparations using an antibody and the immunocomplexes are then precipitated using protein G Sepharose® beads. After a number of washes the target and associated proteins are eluted from the beads. Two-dimensional gel electrophoresis is a robust method to separate complex mixtures of proteins and permits quantitative analysis of protein expression profiles *(4)*. Following separation by isoelectric points (pI), proteins are fractionated by size and stained using a variety of methods that include Coomassie® Brilliant Blue, silver staining, and various fluorescent dyes such as SYPRO® Ruby. Comparison and matching of proteins from different gels is accomplished using software packages. Separations in the first dimension, using immobilized pH gradient strips (IPG), can be performed over a narrow (e.g., 4–5), medium (e.g., 4–7 or 7–10), or wide range (e.g., 3–10) of pH and resolution can be greatly enhanced by permitting migration of proteins over a longer distance (from 7 cm up to 24 cm).

Selections can be made according to the intensity and resolution of protein staining. 2-D gel electrophoresis combined with MALDI TOF-TOF and LC-MS/MS spectrometric peptide identification are powerful tools in the analysis of complex protein mixtures due to their sensitivity and reproducibility *(5)*. This technique has been used widely for determining the identity, quantity and modification states of interacting proteins *(6)*.

The aim of this chapter is to provide insights into protocols for the isolation of cochlear proteins, immunoprecipitation, high quality 2-D gel preparation and analysis, peptide extraction and sample preparation for MALDI-TOF-TOF and LC-MS/MS analysis.

2. Materials

Milli-Q water (minimum of 18 MΩ) is used for the preparation of all solutions. Chemicals are at minimum Analytical Reagent Grade quality, unless otherwise stated.

2.1. Protein Sample Preparation

1. Protease phosphatase inhibitor stock solutions: 50 mg/mL of Pefabloc® SC (4-(2-Aminoethylbenzenesulfonyl fluoride hydrochloride (AEBSF), Fluka/Sigma-Aldrich, St. Louis, MO) in water, stored in 100 μL aliquots; 10 mg/mL of leupeptin (Sigma-Aldrich) in water, stored in 10 μL aliquots; 10 mg/mL of aprotinin (Sigma-Aldrich) in

10 m*M* 4-(2-hydroxyethyl)-1-piperazineethanesulfonic acid (HEPES), pH 8.0, stored in 20 μL aliquots; 100 μg/mL of pepstatin A (Sigma-Aldrich) in 95% ethanol (no need to aliquot), and 500 μ*M* okadaic acid (cat. no. 350-010, Alexis Biochemicals, San Diego, CA) in water, stored in 10 μL aliquots. All inhibitors are stored at −20 °C except for Pepstatin A, which is stable at 4 °C.

2. Tissue Lysis Buffer: 50 m*M* Tris-HCl, pH 8.0, 120 m*M* NaCl, 5 m*M* ethylenediaminetetraacetic acid (EDTA), and 50 m*M* sodium fluoride (NaF). Prepare 100 mL of tissue lysis buffer by dissolving 788 mg of Tris-HCl in 90 mL of water and adjusting to pH 8.0 using 0.1 *N* HCl. Add 701 mg of NaCl, 186 mg of EDTA, and 210 mg of NaF and bring to a volume of 100 mL with water store at 4 °C.
3. Complete Tissue Lysis Buffer: Tissue Lysis Buffer plus 0.5 mg/mL AEBSF, 10 μg/mL leupeptin, 10 μg/mL pepstatin A, 2 μg/mL aprotinin, and 0.5 μ*M* okadaic acid. Prepare 10 mL of complete tissue lysis buffer by adding the following volumes of protease/phosphatase inhibitor stock solutions to 9 mL of Tissue Lysis Buffer: 100 μL of AEBSF, 10 μL of leupeptin, 10 μL of pepstatin A, 20 μL of aprotinin, and 10 μL of okadaic acid. Mix by vortexing briefly and bring up to a volume of 10 mL with Tissue Lysis Buffer store on ice.
4. Solubilization Buffer: Complete Tissue Lysis Buffer containing 0.4% dodecyl β-maltoside detergent (Pierce Biotechnology, Rockford, IL) (*see* **Note 1**). Prepare 1 mL by adding 40 mg dodecyl β-maltoside to 1 mL of Complete Tissue Lysis Buffer and mix.
5. Vortexer, e.g., Eppendorf MixMate equipped with a 1.5–2.0 mL tube holder (cat. no. 022674200, Eppendorf, Westbury, NY).

2.2. Immunoprecipitation

1. rec-Protein G Sepharose® 4B beads (Zymed/Invitrogen, Carlsbad, CA).
2. polyclonal or monoclonal antibody

2.3. Isoelectric Focusing (IEF)

2.3.1. Buffers and Solutions

Prepare 1 mL aliquots and store at −20 °C.

1. Sample Buffer: 7 *M* urea (Bio-Rad, Hercules, CA), 2 *M* thiourea (Sigma-Aldrich), 0.5% Bio-Lyte® pI 3/10 ampholyte (Bio-Rad) (*see* **Note 2**), and 0.4% dodecyl β-maltoside. (*see* **Note 3**). Prepare 50 mL of sample buffer by dissolving 21.02 g of urea and 7.61 g of thiourea in 6 mL of water. Add 250 μL of carrier ampholytes and 200 mg of dodecyl β-maltoside. Mix and bring to a total volume of 50 mL with water.
2. Complete Sample Buffer: Sample Buffer with protease/phosphatase inhibitors (*see* **Note 4**). Prepare a 10 mL volume of complete sample buffer by adding protease/phophatase

inhibitors according to **Section 2.1, step 3**. Mix by vortexing briefly and bring to a final volume of 10 mL with sample buffer.

3. Rehydration buffer: 7 *M* urea, 2 *M* thiourea, 0.2% pI 3/10 ampholyte, 100 m*M* 1,4-dithio-DL-threitol (DTT), and 0.4% dodecyl β-maltoside. Prepare 50 mL of rehydration buffer by dissolving 21.02 g of urea and 7.61 g of thiourea in 6 mL of water. Add 100 μL of carrier ampholyte, 200 mg of dodecyl β-maltoside, and 772 mg of DTT. Bring to a final volume of 50 mL with water.

2.3.2. Supplies

1. The first dimension isoelectric focusing apparatus (IEF) may be obtained from Bio-Rad (e.g., Protean IEF System; Hercules, CA) or Amersham/GE Healthcare (Piscataway, NJ). The second dimension (SDS-PAGE) gel apparatus may be purchased from Bio-Rad (e.g., Mini-PROTEAN® 3) or Amersham. IPG Strips (e.g., cat. no. 163-2000, Bio-Rad) (*see* **Note 5**).
2. Disposable Rehydration/Equlibration Trays (e.g., cat. no. 165-4035, Bio-Rad) (*see* **Note 5**).
3. IEF Focusing Tray with lid, the same size as the IPG strips. This item is supplied with the IEF apparatus.
4. Precut Electrode Wicks (e.g., cat. no. 1654071, Bio-Rad).
5. Mineral oil.
6. Forceps (e.g., cat. no. 1654070, Bio-Rad).
7. Glycerol.
8. Disposable Polyethylene Gloves (e.g., Electron Microscopy Sciences, Ft. Washington, PA).

2.4. SDS-Polyacrylamide Gel Electrophoresis (SDS-PAGE)

1. Tris Base Stock Solution: 0.5 *M* Tris-HCl in water, pH 8.8. Prepare 100 mL by adding 6.06 g of Tris Base (Bio-Rad) to 90 mL of water, mix and adjust pH to 8.8, using 1 *N* HCl. Bring to a final volume of 100 mL with water. Store at room temperature.
2. SDS Stock Solution: 20% SDS in water. Prepare 500 mL by adding 100 g of SDS to ~300 mL of water and mix until completely dissolved. Bring to a final volume of 500 mL with water. Store at room temperature.
3. IPG Strip Equilibration Buffer I: 50 m*M* Tris-base, 6 *M* urea, 30% (v/v) glycerol, 2% SDS, and 1% DTT. Prepare 100 mL by mixing 10 mL of Tris Base stock solution, 36 g of urea, 30 mL of glycerol, 10 mL of SDS stock solution, bring to a volume of 100 mL with water, and then add 1 g of DTT. May be stored at 4 °C for up to 1 week.
4. IPG Strip Equilibration Buffer II: 50 m*M* Tris Base, 6 *M* urea, 30% (v/v) glycerol, 2% SDS, and 4% iodoacetamide. Prepare 100 mL by mixing 10 mL of Tris Base stock solution,

36 g of urea, 30 mL of glycerol, 10 mL of SDS stock solution, bring to a volume of 100 mL with water. Then add 4 g of iodoacetamide and mix. May be stored at 4 °C for up to 1 week.

5. 4X Separating Buffer Stock Solution: 1.5 *M* Tris-HCl, pH 8.8, and 0.4% SDS. Prepare 1 L by dissolving 181.71 g of Tris Base in ~800 mL of water, pH to 8.8, using concentrated HCl, and add 20 mL of SDS Stock Solution. Mix and bring to a total volume of 1 L with water. Store at room temperature.
6. 4X Stacking Buffer Stock Solution: 0.5 *M* Tris-HCl, pH 6.8, and 0.4% SDS. Prepare 1 L by dissolving 60.57 g of Tris Base in ~800 mL of water, pH to 6.8 using concentrated HCl and add 20 mL of SDS stock solution. Mix and bring to a volume of 1 L with water. Store at room temperature.
7. Acrylamide Stock Solution: 30% acrylamide and 0.8% bisacrylamide. Prepare 300 mL by dissolving 87.6 g of acrylamide and 2.4 g of bisacrylamide in ~150 mL of water and mix. Bring to a volume of 300 mL with water. Filter the solution using a 0.2 μM pore size disposable bottle-top filter unit (e.g., Nalgene, Thermo Fisher). Store at 4 °C. This solution is stable for several months. Add N, N, N′, N′-Tetramethylethylenediamine (TEMED) prior to use (*see* **Note 6**).
8. Ammonium Persulfate (APS) Solution: 10% APS in water. Prepare 10 mL by dissolving 1 g of APS in 10 mL of water. Immediately freeze in single use (200 μL) aliquots and store at −20°C.
9. Water-saturated Isobutanol: Prepare 10 mL by combining 5 mL of water with 5 mL of isobutanol in a glass bottle. The separation of a water-saturated isobutanol fraction (layer at the bottom of the bottle) will require approximately 5 min. This can be stored at room temperature for approximately one month.
10. 5X Running Buffer Stock Solution: 625 m*M* Tris base, 960 m*M* glycine, and 2.5% SDS. Prepare 1 L by dissolving 75.7 g of Tris base, 72.06 g of glycine and 125 mL of SDS stock solution in 1 L of water. Do not adjust the pH. Store at room temperature.
11. IPG well comb (e.g., cat. no. 165-3368, Bio-Rad).
12. Prestained molecular weight marker e.g., Precision Plus Protein™ Dual Color Standard (cat. no. 1610374, Bio-Rad) stored in 10 μL aliquots at −20 °C.
13. Agarose Overlay Gel: 1% agarose and 0.03% bromophenol blue. Prepare 100 mL by mixing 1 g agarose in 100 mL of 1× running buffer and add 30 mg of bromophenol blue, molecular biology grade. Store at room temperature.
14. Parafilm® (Pechiney Packaging, Menasha, WI).

15. Coomassie® Brilliant Blue (CBB) G-250 (Bio-Rad) stock solution: 5% CBB in water. Prepare 20 mL by adding 1 g of CBB to 20 mL of water. Mix and maintain in the dark at room temperature.

2.5. Staining of 2-D Gels Using Coomassie® Brilliant Blue

All solutions are stored at room temperature.

1. Gel Fixation solution: 40% methanol and 10% acetic acid. Prepare 100 mL by mixing together 40 mL of methanol and 10 mL of glacial acetic acid in ~40 mL of water. Bring to a volume of 100 mL with water.
2. Gel Staining Stock Solution: 10% ammonium sulfate, 2% phosphoric acid, and 0.1% CBB (*see* **Note 7**). Prepare 500 mL by mixing 10 mL of phosphoric acid in 400 mL of water, add 50 g of ammonium sulfate and 10 mL of CBB stock solution. Bring to a volume of 500 mL with water. This solution should be kept in the dark, such as in a dark brown bottle.
3. Gel Staining Working Solution: 80% Gel Staining Stock Solution and 20% methanol. Prepare 100 mL by adding 20 mL of methanol to 80 mL of Gel Staining Stock Solution and mix. This solution should be prepared on the day of use.
4. Gel Destaining Solution: 10% methanol and 10% acetic acid. Prepare 100 mL by mixing 10 mL of methanol and 10 mL of glacial acetic acid in ~70 mL of water and bring to a volume of 100 mL with water.
5. Glass trays for gel fixation and staining (e.g., Pyrex®, Corning Life Sciences, Lowell, MA).

2.6. Rapid Silver Nitrate Staining

All solutions are stored at room temperature.

1. Gel Fixation Solution: 40% methanol and 35% formaldehyde. Prepare 100 mL by adding 40 mL of methanol and 35 mL of formaldehyde (37% by weight formaldehyde e.g., Thermo Fisher) to 25 mL of water.
2. Stain Enhancer Solution: Sodium thiosulfate (0.025%). Prepare 200 mL by dissolving 50 mg of sodium thiosulfate in 150 mL of water and bring to a volume of 200 mL with water.
3. Gel Staining Solution: Silver nitrate (0.2%). Prepare 100 mL by dissolving 200 mg of silver nitrate in 100 mL with water.
4. Image Developer solution: 3% sodium bicarbonate, 0.025% sodium thiosulfate, 0.004% silver nitrate, and 0.0005% formaldehyde. Prepare 150 mL by dissolving 4.5 g of sodium carbonate in ~100 mL of water, add 3 mL of Stain Enhancer Solution and 75 μL of formaldehyde. Mix and bring to volume of 150 mL with water. (*see* **Note 8**).
5. Stop solution: 380 m*M* citric acid. Prepare 100 mL by dissolving 7.3 g of citric acid in ~70 mL of water. Mix and bring to final volume of 100 mL with water.

2.7. Fluorescent Staining

1. CyDye™ DIGE (Difference Gel Electrophoresis) Fluor dyes, e.g., CY3, CY5, and Sypro™ Ruby (Amersham). Follow manufacturer's instructions.

2.8. Selection and Excision of Spots from 2-D Gels

1. Light box to illuminate film.
2. Pipette tips (1–200 μL): Cut pipet tips to accommodate the individual sizes of spots to be excised.
3. Clean razor blades.
4. Eppendorf Protein LoBind centrifuge tubes (cat. no. 2243108-1).

2.9. Destaining and In-gel Trypsin Digestion

All solutions must be prepared immediately prior to use.

1. Acetonitrile, mass spectroscopy grade.
2. Coomassie® Destaining Solution: 25 m*M* ammonium bicarbonate, pH 8.0, and 50% acetonitrile. Prepare 100 mL by dissolving 198 mg of ammonium bicarbonate in 50 mL of water, adjust pH to 8.0 using 0.1 *N* NaOH. Add 50 mL of acetonitrile and mix.
3. Silver-stain Destaining Solution: 30 m*M* potassium ferricyanide and 100 m*M* sodium thiosulfate. Prepare 50 mL by dissolving 494 mg of potassium ferricyanide and 791 mg of sodium thiosulfate in 50 mL of water.
4. Peptide Digestion Buffer: 25 m*M* ammonium bicarbonate in water. Prepare 50 mL by adding 99 mg of ammonium bicarbonate to ~48 mL of water, adjust pH to 8.0, using 0.1 *N* NaOH. Bring to volume of 50 m with water.
5. Peptide Digestion Solution: 10 μg/mL of Trypsin in Peptide Digestion Buffer. Prepare by reconstituting lyophilized sequencing grade trypsin (e.g., cat. no. 6567, Promega, Madison, WI) in 1 mL of Peptide Digestion Buffer.
6. Vacuum concentrator (e.g., SpeedVac® model SPD121P, Savant/Thermo Fisher).

2.10. Extraction of Peptides from the Gel

All solutions must be prepared immediately prior to use.

1. Peptide Extraction Solution: 50 m*M* ammonium bicarbonate, pH 8.0, 50% acetonitrile, and 0.1% trifluroacetic acid (TFA). Prepare 100 mL by dissolving 395 mg of ammonium bicarbonate in ~48 mL of water, adjust to pH 8.0 with 0.1 *N* NaOH. Add 50 mL of acetonitrile, mix and bring to a volume of 100 mL with water. Add 100 μL of TFA under a fume hood.
2. Peptide Reconstitution Solution: 0.1% TFA in water. Prepare 10 mL by adding 10 μL of TFA to 10 mL of water in a Falcon tube and mix.

2.11. ZipTip® Cleaning of Peptides

1. ZipTip® U-c18 (cat. no. ZTC18 M960, Millipore).
2. ZipTip® Equilibration Solution: 50% acetonitrile in water. Prepare 10 mL by adding equal volumes of acetonitrile and water and mix.
3. Peptide Elution Solution: 0.1% TFA. Prepare 10 mL by adding 10 μL of TFA to 10 mL of water and mix.

2.12. Sample Preparation for MALDI-TOF-TOF and LC-MS/MS

The following materials were supplied by Applied Biosystems, Foster City, CA., except for the LC-MS/MS.

1. Spotter plate.
2. BSA Standard, Spectroscopy Grade.
3. CalMix standard.
4. Hydroxy-3-cyanotranscinnamic acid (HCCA) matrix.
5. 4800 ABI Prism MALDI-TOF-TOF.
6. G4240HPLC chip MS (Agilent Technologies, Santa Clara, CA)

2.13. Bioinformatics Tools for Protein Identification

1. MASCOT Search Engine (Matrix Sciences, London, UK).
2. ProFound (Rockefeller University, New York NY).

3. Methods

Work as quickly as possible during all stages of sample preparation. Keep the tissue to be used for protein extraction on ice during the preparation to minimize proteolytic activity. Protein from 8 cochleae from adult chicks or 15 cochleae from adult mice is sufficient to run 7 cm gels. Use 5 volumes of buffer to one volume of tissue to completely solubilize the proteins. Avoid mouse and chick keratin contamination by rinsing tissues several times following dissection in ice cold saline solution. Avoid human keratin contamination by using a hair net, mask, and by working under a laminar flow hood.

3.1. Protein Sample Preparation

1. Dissect cochleae from 15 day old chicks or adult mice (~30 day-old) by excising tissues under clean conditions and placing them in aliquots of 10 cochleae per cryovial in powdered dry ice. Store at −80 °C or use immediately.
2. Wash tissues the day of the experiment by adding 500 μL of complete lysis buffer to cochlear tissues, vortex briefly, followed by spinning the tissues at 2000 g for 1 min at 4 °C. Discard the buffer and wash tissues in this manner once again.
3. Add 100 μL of Complete Tissue Lysis Buffer to tissues and sonicate three times for 30 s each at a low frequency setting (e.g., between 1–2 using a Sonic Dismembrator Model 100, Thermo Fisher) (*see* **Note 9**), while maintaining the tissues in packed ice. Centrifuge at 750 g for 2 min at 4 °C. Transfer the supernatant to a fresh Eppendorf tube, then add another

100 μL of lysis buffer to the pellet from the first spin and sonicate as before. Combine both crude lysates.

4. Spin the lysate at 100,000 g for 1 h at 4 °C. The supernatant contains the cytoplasmic fraction, while the pellet contains the membrane and cytoskeletal fractions. Transfer the supernatant to a fresh Eppendorf tube and place on ice. Divide the supernatant into two equal volumes in fresh Eppendorf tubes and maintain on ice.
5. Add 130 μL of Solubilization Buffer to the pellet. Vortex at 1500 rpm at 4 °C for 10 min and incubate on a rocking shaker at 4 °C for 1 h. Divide into 2 equal volumes in fresh Eppendorf tubes and maintain on ice. One volume will be used for immunoprecipitations and the other will be used for matrix controls.

3.2. Coimmunoprecipitation

Unless otherwise stated, *all* steps are to be performed either on ice or at 4 °C. (*see* **Note 10**).

1. Washing of Protein G beads: Wash four 25 μL aliquots and two 10 μL aliquots of Protein G beads (packed) by adding 400 μL wash buffer, flicking the tube several times, spinning the beads down at 2000 g for 2 min, and discarding the wash buffer, for a total of three times.
2. Pre-clearing of cytosolic and membrane/cytoskeletal fractions using Protein G beads: Add the cytoplasmic fraction to one of the two 10 μL aliquots of beads and add the membrane/cytoskeletal fraction to the remaining 10 μL aliquot of beads. Incubate with rocking for 15 min. Spin all samples at 2000 g for 2 min and transfer the samples to new tubes, discarding the sedimented beads. Divide the cytosolic fraction equally into 2 tubes and divide the membrane/cytoskeletal fraction equally into 2 tubes.
3. Immunocomplexing of antibody and target protein: Add an appropriate quantity of antibody (this must be determined empirically) to one aliquot of the cytoplasmic fraction and to one aliquot of the membrane/cytoskeletal fraction and incubate with rocking for 1 h. Run control samples alongside.
4. Capture of immunocomplexes: Add the pre-cleared cytoplasmic fraction to each of the two 25 μL aliquots of protein G beads and add the pre-cleared membrane/cytoskeletal fraction to the remaining two 25 μL aliquots of protein G beads. Incubate all samples with rocking for 1 h.
5. Washing of immunocomplexed and non-immunocomplexed (matrix control) beads: Spin each sample at 2000 g for 2 min to collect the beads. Discard the depleted samples by carefully pipeting the fluid overlying the beads. Wash the beads from each sample in 400 μL of wash buffer by flicking the tubes several times, spinning the beads down at 2000 g for 2 min, and discarding the wash buffer. Repeat for a total of 4 times.

6. Elution of immunocomplexed and non-immunocomplexed (matrix control) beads: Immunocomplexes and unspecifically-bound proteins are eluted from the Protein G beads using a combination of equal volumes of IEF sample and IEF rehydration buffers. Add 65 μL of IEF complete sample buffer and 65 μL of IEF Rehydration buffer to each of the four samples of beads. Elute the immunocomplexes/unspecifically-bound proteins from the beads by vortexing at 1500 rpm at room temperature for 10 min (*see* **Note 11**). Sediment the beads in order to recover the immunocomplexes/unspecifically-bound proteins by spinning the samples at 2000 g at 4 °C for 2 min. Carefully remove the fluid above the pelleted beads by pipeting, taking care not to remove any beads, and proceed to IEF. Samples cannot be stored.

3.3. First Dimension: Isoelectric Focusing (IEF)

Use the isoelectric focusing apparatus according to manufacturer's instructions. These instructions apply to the Protean IEF System (Bio-Rad) followed by either Coomassie® or silver staining. Fluorescent detection requires a different set of instructions (*see* **Section 3.5.3**).

1. Remove 4 IPG strips from the −20 °C freezer to thaw and set up one disposable rehydration/equilibration tray of the same size as the IPG strips on a clean, flat surface. Pipet each of the immunoprecipitated samples and the matrix control samples into 4 separate channels by slowly adding the samples along the length of each channel, except for a 1 cm distance at either end. Take care to not introduce any air bubbles. Remove the protective film covering the gel from the back of each IPG strip and place the strip, gel side down, onto the protein solution. At this point, take care to not displace the fluid over the top of the strip and be sure that the full length of the strip is in contact with the sample fluid (*see* **Note 12**).
2. After the strips are completely rehydrated and the samples absorbed into the gels (~1 h), add 1 mL of mineral oil along the surfaces of the strips to prevent dehydration. Maintain the strips in the tray for 12–14 h. (*see* **Note 13**).
3. After rehydration is completed, drain the mineral oil from the strips by gently touching the corner of each strip to a lab wipe (e.g., Kimwipe®, Kimberly-Clark, Roswell, GA).
4. Place one electrode wick on the anode and one wick on the cathode end of each of the 4 channels in an IEF focusing tray in which the IPG strips are to be placed. Wet each wick by adding 8 μL of water by pipeting. Correctly orient each IPG strip to correspond to the anode and cathode ends of the tray. Then place each IPG strip in a separate channel in the tray, add 1 mL of mineral oil and press the strip gently to remove any air bubbles.
5. Place the focusing tray in the IEF cell chamber. Set the Volt-hours (V-h) for 10,000 and set the temperature for 20 °C (this

is the pre-set temperature on the Protean IEF System) (*see* **Note 14**). After completion of the IEF run, remove the strip from the tray and drain the mineral oil. Place each IPG strip in a rehydration tray and continue on to the next portion of the protocol or store at −80 °C.

3.4. Second Dimension: SDS-PAGE

1. Carefully wash glass plates (80 mm × 60 mm) and rinse under running water, followed by two final rinses in water and air-dry. Set up the gel casting apparatus using 1.5 mm spacers. Prepare a mini-SDS gel (7 cm polyacrylamide gel, 12% density).
2. Prepare 10 mL of 12% resolving gel. Mix 2.5 mL of 4X Separating Buffer Stock Solution, 4 mL of Acrylamide stock solution, 3.5 mL of water, 100 μL of APS stock solution, and 20 μL of TEMED. Immediately pour the gel into the glass plates up to a distance of 1.5 cm from the top of the shorter plate and overlay with water-saturated isobutanol. The gel should polymerize in about 30 min. Pour off the isobutanol and rinse the top of the gel twice with water.
3. Prepare 5 mL of 4% stacking gel. Mix 1.25 mL of 4X Stacking Buffer Stock Solution, 680 μL of Acrylamide Stock Solution, 3.1 mL of water, 50 μL of APS stock solution, and 10 μL of TEMED. Use approximately 500 μL of this solution to quickly rinse the top of the resolving gel. Then pour the stacking gel and insert the IPG gel comb. The stacking gel should polymerize within 30 min.
4. Prepare 500 mL of Working Running Buffer Solution. Mix 100 mL of 5X Running Buffer Stock Solution with 400 mL of water in a graduated cylinder. Cover with Parafilm® and invert to mix. Once the stacking gel has set, carefully remove the comb and wash the wells with running buffer.
5. Place the IPG strips in a rehydration tray and add 1 mL of Equilibration Buffer I. Rock gently on a shaker for 20 min. Remove equilibration buffer I, add 1 mL of equilibration buffer II, and rock for 20 min. After incubation remove equilibration buffer II (*see* **Note 15**) and remove remaining equilibration buffer from the strip by gently washing in running buffer.
6. Remove the IPG strip from the tray and gently place it on top of the SDS gel such that the anode end of the IPG strip is oriented toward the left-hand side. The plastic backing of the IPG strip should be placed along the glass plate closest to the operator.
7. Gently press the strip onto the top of the SDS gel with forceps, ensuring that there is full contact between the IPG strip and the entire length of the SDS gel. This will also remove any air bubbles between the strip and the gel.
8. Load 10 μL of the protein standard marker into the small well at the right-hand side of the gel.

9. Melt the agarose overlay gel by microwaving briefly. Allow the agarose gel to cool to approximately 50 °C. Slowly pipet 500 μL of agarose gel along the entire length of the IPG strip, ensuring that no air bubbles are introduced.
10. Transfer the gel cassettes into the running chamber of the gel apparatus. Fill the chamber with running buffer up to the top of the cassettes. Place the lid on the apparatus and connect to the power supply. Set the voltage to 80 V and run the gel for approximately 10 min, then increase the voltage to 100 V. Run the gels until the bromophenol blue tracking dye reaches the bottom of the gels.
11. Gently remove the gels from the glass plates and transfer them into the appropriate fixative corresponding to the staining method of choice.

3.5. Staining of 2-D Gels

Carefully remove each solution by pouring, taking care to not tear the gels.

3.5.1. Coomassie® Staining

1. Place the gels in clean glass trays containing fixative and gently rock for 45 min.
2. Pour off the fixative and replace with Coomassie® Gel Staining working solution for 5–6 h with gentle rocking.
3. Pour off the staining solution and replace with Gel Destaining solution for 1–2 h or until protein spots are clearly resolved (**Fig. 16.1**) (*see* **Note 16**).

3.5.2. Silver Staining

Carefully remove each solution by pouring, taking care to not tear the gels.

1. Place the gel in a clean glass tray containing 100 mL of fixative and gently rock for 10 min.
2. Pour off the fixative and wash the gels with approximately 200 mL of water 2 times for 5 min each.
3. Remove the water and replace with 200 mL of silver stain enhancer solution and gently rock for no longer than 1 min.
4. Wash the gels in 200 mL of water 2 times for 1 min each.
5. Remove the water and replace with 100 mL of silver staining solution and gently rock for 20 min.
6. Wash the gels in 200 mL of water 2 times for 1 min each.
7. Remove the water and replace with 150 mL of developer solution and gently shake the tray by hand until the spots begin to appear. Immediately add 100 mL of stop solution (*see* **Note 17**).

3.5.3. Fluorescent Staining

It is imperative that *all* steps in this protocol be performed in the dark using either a darkroom or covering samples to exclude any light. Perform the same steps for the matrix control samples.

1. After the immunocomplexes have been eluted, combine each control protein sample (the cytoplasmic and the membrane/cytoskeletal fractions) with Cy3® (magenta) and combine

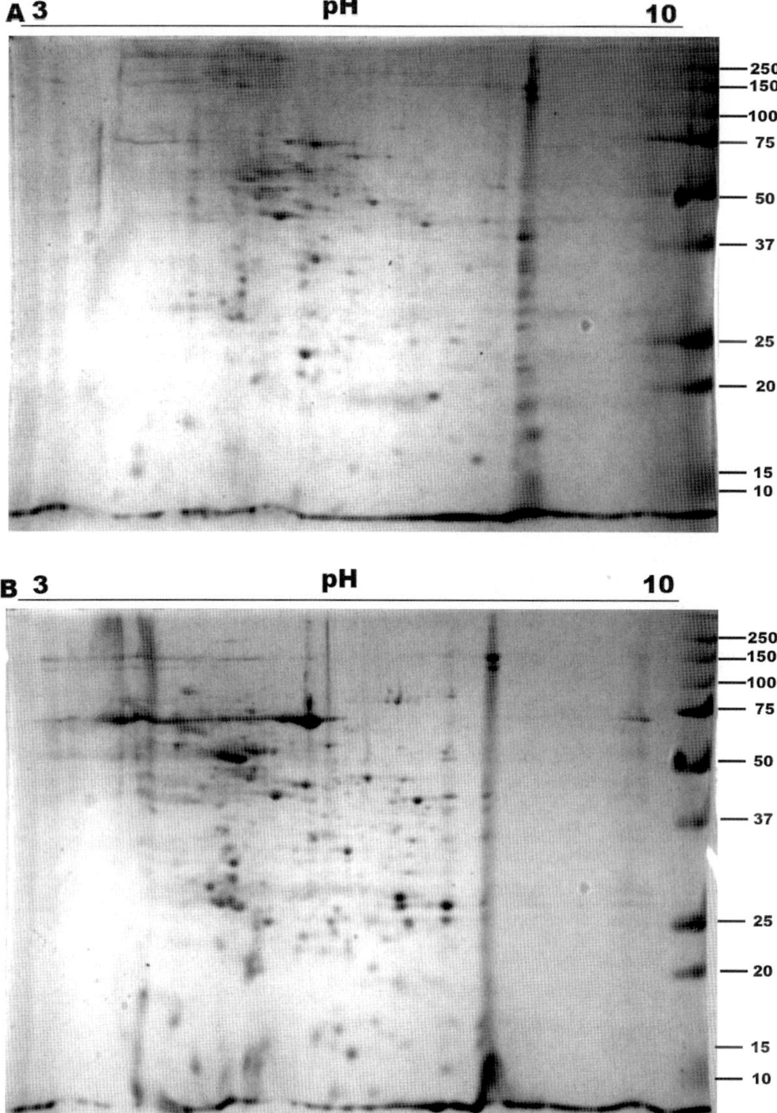

Fig. 16.1. Protein profiles of Coomassie-stained two-dimensional gels. Samples consisted of non-immunoprecipitated mouse cochlear membrane/cytoskeletal proteins (**A**) and cytoplasmic proteins (**B**). Separation in the first dimension was accomplished using IPG strips with a pI range between 3 and 10. The second dimension separated the proteins according to their mobilities under denaturing, reducing conditions using 12% SDS-PAGE.

the immunoprecipitated samples (again, both fractions) with Cy5® (green) and incubate with gentle rocking at room temperature for 30 min.

2. Mix each one of the two control protein sample/Cy3® with one of the two immunoprecipitated sample/Cy5® by briefly vortexing. Continue the 2-D gel protocol through to the finish of the SDS-PAGE portion (*see* **Section 3.3–3.4**) and continue to maintain the gels in the dark in water (*see* **Note 18**).

3.6. Spot Validation and Excision

1. Capture the gel images and print them using a gel documentation system (e.g., The Molecular Imager® VersaDoc™ System, Bio-Rad). The two different fractions can be compared. The spots are identified by pI and molecular weight using PDQUEST software. Spot sets that appear in both the immunoprecipitated samples and the matrix controls must be identified and eliminated from identification. Mark the spots using a numbering system.
2. Wash the gels thoroughly to remove residual acetic acid, methanol, SDS, and buffer salts three times over the course of one day using 500 mL volumes of water. Excise each spot using a fresh pipette tip previously cut to size or an automatic spot cutter (e.g., EXQuest™ Spot Cutter, Bio-Rad).
3. Transfer each gel spot to an individual 1.5 mL Protein Lobind centrifuge tube, label properly and maintain tubes at room temperature.

3.7. Destaining, In-Gel Digestion and Extraction of Peptides

1. To destain a Coomassie®-stained gel spot, add 400 μL of Coomassie® Destaining Solution and vortex at 1500 rpm at room temperature for 4 h. To destain a silver-stained gel spot, add 400 μL of Silver-stain Destaining Solution and proceed in the same manner. Remove the destaining solution, add fresh solution, and repeat these steps two more times.
2. Drain the destaining solution and add 100 μL of acetonitrile and vortex at 1500 rpm at room temperature for 20 min (*see* **Note 19**).
3. Remove the acetonitrile by pipeting and dry the gel spots using a SpeedVac for 15 min at 10 Pa.
4. Add 10 μL of the Peptide Digestion Solution to each tube containing the dried gel spots and incubate in a 37 °C incubator for 14–16 h.
5. The next day, add 100 μL of Peptide Extraction Solution and vortex at 1500 rpm at room temperature for 1 h. Transfer the solution to a fresh Lobind tube. Extract peptides a second time by adding another 100 μL of the Peptide Extraction Solution to the gel spots and vortexing in the same manner.
6. Combine the two extracts and SpeedVac® to completely dry the samples. This will take approximately 1–2 h.
7. Add 10 μL of Peptide Reconstitution Solution and vortex at 1500 rpm at room temperature for 10 min, followed by a brief spin at 2000 g for 1 min. Store samples at −80 °C or proceed for MALDI-TOF/TOF and LC-MS/MS analysis.

3.8. ZipTip® Cleaning of Peptide Samples (*see* Note 20)

Equilibration of ZipTips®: Aspirate and dispense 10 μL volumes of Equilibration Solution through a ZipTip® with a 10 μL pipeter 3–4 times. Repeat with water.

1. Peptide Sample Adsorption: Aspirate and dispense the samples 5–6 times with a ZipTip®. The peptides are now bound to the matrix.
2. Peptide Sample Washing: Aspirate and dispense 10 μL volumes of water 6–7 times with a ZipTip® to de-salt the samples.
3. Peptide Elution: Aspirate 10 μL of Peptide Elution Solution and dispense into a fresh Lobind tube. Aspirate and dispense the sample several times with a ZipTip® to elute all peptides from the ZipTip® matrix. Purified peptides are ready for MALDI-TOF-TOF or LC-MS/MS analysis.

3.9. Sample Preparation for MALDI-TOF-TOF and LC-MS/MS Analysis

1. Place 0.8 μL of each standard (the BSA and the CalMix) on the corresponding labeled sites on the spotter plate. Pipet 0.8 μL of each of the reconstituted peptide samples on the spotter plate. It is imperative that the spots not dry completely.
2. After all the standards and samples are loaded, immediately add 0.8 μL of HCCA matrix to each spot. After approximately 10 min, the spots will be completely dry.
3. Load the spotter plate in the MALDI-TOF-TOF to obtain the spectra for each spot.
4. Inject 2 μL of reconstituted peptide sample into the LC-MS/MS to obtain the spectra for each spot.

3.10. Identification of Proteins by Using Peptide Mass Finger Print (PMF)

1. Proteins are identified by PMF using the NCBI protein database (http://www.ncbi.nlm.nih.gov/). Two search programs are available, namely MASCOT (Matrix Sciences, London, UK) (http://www.matrixscience.com/cgi/search_form.pl?FORMVER=2&SEARCH=PMF) and ProFound (Rockefeller University, http://prowl.rockefeller.edu/prowl-cgi/profound.exe).
2. The following search parameters can be used as follow: bony vertebrate for taxonomy; partial methionine oxidation and carbamidomethylation of cysteine (selections based on the presence of DTT in the IEF Sample Buffer and iodoacetamide in the Equilibration Buffer II, respectively); limit of one missed cleavage (mis-match) and a peptide mass concentration error window of ±50 ppm (both are instrument-specific). Enter the Mr values in the "Query" box and submit the request.
3. The results include a list of significant hits, probability based scores for each hit, and a peptide summary report. Protein identification (ID) is short-listed by score value, the number of peptides that matched, and percentage of sequence coverage (15% or more for MALDI-TOF-TOF and 1% or more for LC-MS/MS). A comparison of the mass values with the PDQuest 2-D gel image analysis is then accomplished and the operator can select a reasonable range (e.g., ±5 kDa).

4. Notes

1. A wide variety of detergents are available to solubilize membrane proteins. The following is a listing of empirically-determined alternatives to dodecyl D-maltoside: decyl D-maltopyranoside, octyl D-glucopyranoside, and the zwitterionic detergent ASB-14 (Amidosulfobetaine-14).
2. The ampholyte pH range directly relates to the pH range of the IPG strip.
3. At the start of a project, when little information is available about the proteome of the tissue to be analyzed, a broad pH range IPG strip (e.g., pH 3.0–10.0) should be selected. This will give an overall impression of the pI range of the proteins present within the sample. In this respect, the use of 7 cm strips permits a rapid assessment of the protein extraction efficiency, as well as information on the pI profiles of the proteins. Then, 18 and 24 cm strips are recommended as these provide maximal resolution. Based on pilot experiments, narrow pH range IPG strips may be chosen in order to improve the resolution in the pH range of interest.
4. Inhibitors might not be necessary due to the strong denaturing activity of urea and thiourea. Some inhibitors, due to their mode of action, can alter the pI of proteins.
5. IPG strips are available in 5 different sizes (7, 11, 17, 18, and 24 cm).
6. TEMED is best stored at room temperature in a desiccator. Purchase small quantities, as TEMED may decline in quality after initial use (gels will take longer to polymerize). This reagent is a neurotoxin and must be handled only under a fume hood.
7. Only after complete solubilization of the ammonium sulfate in the water/acid solution is the CBB stock solution added. This preparation should not be filtered and must be stirred prior to use for even distribution of colloidal particles.
8. Image Developer solution should be freshly prepared to avoid background staining.
9. Sonicate sufficiently well to mechanically disrupt tissues and cell membranes, but not to the point where proteins are denatured and a separation cannot be achieved of membrane proteins from cytoplasmic proteins. After mass spectrometry analysis, one might determine that cytoplasmic proteins appear in the membrane fraction and vice versa. In this case, samples were sonicated too much. If tissues are not sonicated sufficiently, again it is possible that cytoplasmic proteins appear in the membrane fraction.

10. Perform a comparison via 2-D gel analysis of non-immunoprecipitated with immunoprecipitated fractions. This comparison increases the quantity of necessary tissues but might prove useful downstream. Alternatively, comparisons can be made, using only the second dimension (SDS-PAGE), by running a small volume of each fraction on a multi-well gel. Individual bands can be excised and analyzed.
11. Urea is temperature-sensitive, therefore, warming samples above 30 °C must be avoided in order to prevent possible modifications of proteins by carbamylation with subsequent shifts of individual isoelectric focusing points.
12. The volume of protein sample that can be applied to the IPG strips is directly related to the size of the strip. The limits are as follow: 7 cm (125 μL), 11 cm (185 μL), 17 cm (300 μL), 18 cm (325 μL), and 24 cm (450 μL).
13. During the IPG strip rehydration step, the mobility of proteins increases.
14. Protein sample solutions which contain excessive amounts of salt can disrupt the voltage, as indicated by the appearance of a white residue at both ends of the IPG strips. Stop the run and replace the electrode wicks.
15. The equilibration buffer should be added in sufficient volumes such that the IPG strips are floated freely and protein from the IPG strips are sufficiently well equilibrated (*see* **Note 16**). This ensures proper migration of the proteins.
16. Weak staining can be improved by increasing the protein concentration of the samples. Spots that are poorly resolved due to vertical doubling/twinning might be due to improper placement of the IPG strip against the glass plate on the second dimension gel. Vertical streaking of protein on the gel is indicative of insufficient equilibration. Horizontal streaking of the protein on the gel could be an indication that protein samples were not completely solubilized. Consider changing detergents and ensure complete and stable solubilization.
17. It is imperative to remove the stop solution when spots are sufficiently well resolved. Rinse the gels immediately with water to avoid salt deposition as this interferes with downstream analyses.
18. Advantages of 2-D DIGE include significant reduction in gel-to-gel variations and increased sensitivity, e.g., from 0.25–0.95 ηg.
19. Gel spots change in color to a dull white after the addition of acetonitrile. Should this not occur, repeat the destaining step, ensuring that the solution is pH 8.0.
20. Desalting of peptides is imperative in order to avoid interference with mass spectroscopy analyses.

Acknowledgments

This work was supported by NIDCD Grant DC04295 to B.H.A.S.

References

1. Gabashvili, I. S., Sokolowski, B. H., Morton, C. C., Giersch, A. B. (2007) Ion channel gene expression in the inner ear. *J. Assoc. Res. Otolaryngol.* *8*, 305–328.
2. Markham, K., Bai, Y., Schmitt-Ulms, G. (2007) Co-immunoprecipitations revisited: an update on experimental concept and their implementation for sensitive interactome investigations of endogenous proteins. *Anal. Bioanal. Chem.* *389*, 461–473.
3. Duzhyy, D., Harvey, M., Sokolowski, B. (2005) A secretory-type protein, containing a pentraxin domain, interacts with an A-type K+channel. *J. Biol. Chem.* ***280***, 15165–15172.
4. Zheng, Q. Y., Rozanas, C. R., Thalmann, I., Chance, M. R., Alagramam, K. N. (2006) Inner ear Proteomics of mouse models for deafness, a discovery strategy. *Brain Res.* ***1091***, 113–121.
5. Hughes, J. R., Meireles, A. M., Fisher, K. H., Garcia, A., Antrobus, P. R., Wainman, A., et al. (2008) A microtubule interactome: complexes with roles in cell cycle and mitosis. *PLOS Biol.* *6*, 785–795.
6. Cravatt, B. F., Simon, G. M., Yates III, J. R. (2007) The biological impact of mass-spectrometry-based proteomics. *Nature* ***450***, 991–1000.

Chapter 17

Identification of Functionally Important Residues/Domains in Membrane Proteins Using an Evolutionary Approach Coupled with Systematic Mutational Analysis

Lavanya Rajagopalan, Fred A. Pereira, Olivier Lichtarge, and William E. Brownell

Abstract

Structure-function studies of membrane proteins present a unique challenge to researchers due to the numerous technical difficulties associated with their expression, purification and structural characterization. In the absence of structural information, rational identification of putative functionally important residues/regions is difficult. Phylogenetic relationships could provide valuable information about the functional significance of a particular residue or region of a membrane protein. Evolutionary Trace (ET) analysis is a method developed to utilize this phylogenetic information to predict functional sites in proteins. In this method, residues are ranked according to conservation or divergence through evolution, based on the hypothesis that mutations at key positions should coincide with functional evolutionary divergences. This information can be used as the basis for a systematic mutational analysis of identified residues, leading to the identification of functionally important residues and/or domains in membrane proteins, in the absence of structural information apart from the primary amino acid sequence. This approach is potentially useful in the context of the auditory system, as several key processes in audition involve the action of membrane proteins, many of which are novel and not well characterized structurally or functionally to date.

Key words: Phylogenetic, membrane protein, structure-function, site-directed mutagenesis, membrane expression.

1. Introduction

There is usually minimal or no secondary and tertiary structural information for membrane-associated protein families in general, due to the technical difficulties associated with the study of membrane proteins. Attempts to locate specific functional domains

Bernd Sokolowski (ed.), *Auditory and Vestibular Research: Methods and Protocols, vol. 493*

DOI 10.1007/978-1-59745-523-7_17 Springerprotocols.com

and/or residues essential for function in membrane proteins are therefore extremely challenging, and more often than not employ shotgun approaches such as mutating all charged residues, putative phosphorylation or glycosylation sites, consensus/conserved motifs, etc. An alternative approach is to utilize evolutionary relationships in protein families to obtain structural and functional information. The computational method of Evolutionary Trace (ET) ranks individual residues in proteins by correlating amino acid variation patterns with functional divergences in protein families *(1)*. ET analysis thus combines the power of computational analysis with the insights offered by evolutionary and phylogenetic comparison of protein sequences.

The key hypothesis upon which ET analysis rests is that substitutions at residues that are more important to function should necessarily be linked with greater evolutionary divergences than substitutions at residues of lesser importance *(1, 2)*. Therefore, in order to estimate the functional importance of a residue, ET looks for specific patterns of amino acid variations in multiple sequence alignments of a family of ancestrally related proteins (homologs). Specifically, ET looks for conservation *within* subgroup partitions of the entire family, where the partitions are defined by successive divergences at every node of the evolutionary tree. Thus partition 1, or the root partition is the entire family, partition 2 consists of two subgroups, each defined by one of the first two branches, partition 3 consists of three subgroups, each defined by one of the first three branches, and so on until partition n, where each one of the n sequences is in its own singlet subgroup (**Fig. 17.1**). For each residue, ET systematically tests every partition starting from partition 1 to find the first, say partition i, at which a residue's position becomes invariant in everyone of its i subgroups. This partition number, i, is the ET rank of evolutionary importance for that sequence alignment position.

The top-ranked residues (with ranks of 1, 2, 3, ...) are called trace residues and they have important characteristics. First, except for the residues of rank 1 which must be invariant, other residues must vary among some homologs, and the variations may be non-conservative. Second, these variations occur only among *major* evolutionary branches of trace residues, so that top-ranked trace residues are strongly linked with the functional divergences of the corresponding subfamilies. Third, clusters of top-ranked residues are a common feature of protein tertiary structure and they usually indicate functional sites or molecular recognition surfaces in proteins *(3, 4)*. Residues identified by ET analysis can be then subjected to systematic mutational and functional studies, including evaluation of protein stability, trafficking, membrane expression, and appropriate functional assays *(5–8)*. Using these techniques, we have shown that residues in the region of the sulfate anion transporter domain of the outer hair cell membrane

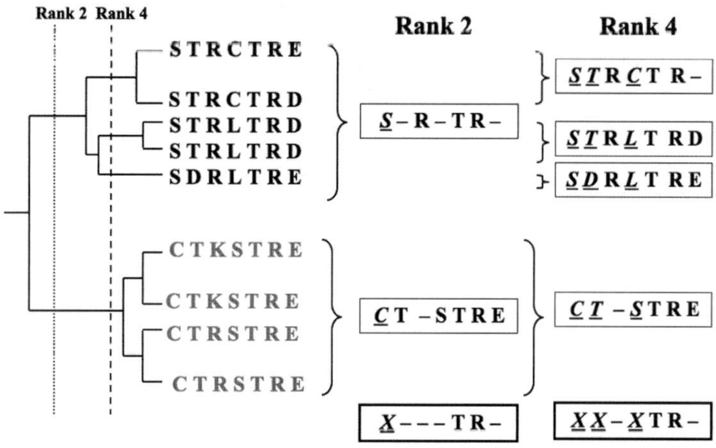

Fig. 17.1. A schematic representation of the residue ranking system used in Evolutionary Trace analysis. Trace residues are invariant within a given branch but may vary between them. Rank 2 residues (position 1, underlined and italicized), defined by the second branchpoint, are invariant within a subgroup (*black*) but vary from the other subgroup (*gray*). Note: the first branchpoint is the root and includes the entire family. Rank 4 residues are invariant within subgroups defined by the fourth branchpoint. Trace residue positions are denoted by Xs, underlined and italicized, at the bottom of the alignment in this schematic.

protein prestin are essential for its function *(9)*. This analysis provided insight into the local structure and packing of the transmembrane helices associated with the sulfate anion transporter domain.

2. Materials

2.1. Evolutionary Trace Analysis

1. Primary amino acid sequence of membrane protein of interest.

2.2. Generation of Mutants

1. Quikchange site-directed mutagenesis kit (Stratagene, La Jolla, CA).
2. Qiagen Spin Miniprep kit (Qiagen, Valencia, CA).
3. Appropriate mutagenesis primers (*see* **Section 3.2 step 2** for primer design). Primers can be obtained from a commercial DNA synthesis service such as Sigma-Genosys (the minimum synthesis scale – 3 OD, and basic purification scale is sufficient. Since primers are supplied as lyophilized powder, dissolve primers in water to a stock concentration of 1 μg/μL, and store at −20 °C until further use.
4. Relevant wild type (*Wt*) gene cloned into a mammalian expression vector.
5. Thin-walled PCR tubes (200 μL capacity).

6. PCR thermocycler.
7. Super Optimal broth – Catabolite repression (SOC) medium (per liter): 20 g Bacto-tryptone, 5 g Bacto-yeast extract, 0.5 g NaCl, pH 7.0. Autoclave and allow to cool. Add glucose to a 20 m*M* final concentration and store at 4 °C.
8. Falcon disposable culture tubes with caps (BD Biosciences, San Jose, CA).
9. Luria Bertani (LB) broth (per liter): 10 g Bacto-tryptone, 5 g Bacto-yeast extract, 10 g NaCl, pH 7.0. Autoclave and store at 4 °C.
10. Plasmid mini-prep kit (Qiagen, Valencia, CA).
11. Appropriate LB agar plates (per liter): 10 g Bacto-tryptone, 5 g Bacto-yeast extract, 10 g NaCl, 1.8 g agar, pH 7.0. Autoclave, cool to about 50 °C, add appropriate selection antibiotic and pour 20–25 ml into each 10 cm plate. Store at 4 °C.
12. XL-1 blue competent *E. coli* cells (Stratagene, San Diego, CA).
13. *Dpn*I restriction enzyme.

2.3. Evaluation of Membrane Expression of Mutants

1. HEK 293 or other appropriate mammalian cell line.
2. 12 mm circular glass cover slips.
3. Dulbecco's Modified Eagle Medium (DMEM; Hyclone, Ogden, UT) supplemented with 10% fetal bovine serum (FBS; Invitrogen, Carlsbad, CA) and 1% penicillin/streptomycin (Invitrogen). Store at 4 °C.
4. Fugene 6 transfection reagent (Roche, Indianapolis, IN).
5. Poly-d-lysine coated 24-well plates (Greiner Bio-One, Monroe, NC).
6. Trypsin (Hyclone).
7. Wild Type (*Wt*) and mutant DNA plasmids.
8. Phosphate-buffered saline (PBS): 137 m*M* NaCl, 10 m*M* Phosphate, 2.7 m*M* KCl, pH 7.4.
9. PCM buffer: Mix PBS with 1 m*M* $CaCl_2$ and 0. 5 m*M* $MgCl_2$ just before use.
10. 4% paraformaldehyde (PFA): Heat 50 mL water to 75 °C using a hotplate or microwave. Add 4 g paraformaldehyde and 10 μl NaOH while stirring. After the PFA dissolves, allow solution to cool to room temperature and filter using Whatman Qualitative Grade 5 filter paper (Maidstone, UK). Add 10 mL PBS and make up to 100 mL using Milli-Q (18 MΩ) H_2O.
11. TritonX-100: 10% stock solution in Milli-Q H_2O.
12. Bovine Serum Albumin (BSA), Immunology grade – IgG free (Jackson ImmunoResearch Inc., West Grove, PA).
13. AlexaFluor350 labeled concanavalin-A (Invitrogen).
14. Appropriate primary and AlexaFluor594-labeled secondary antibodies for protein of interest.

15. Fluoromount™ antifade reagent (Electron Microscopy Sciences, Hatfield, PA); clear nail polish/lacquer.
16. Small forceps; glass microscopy slides (25 × 75 × 1 mm).

3. Methods

3.1. Evolutionary Trace Analysis

1. Identify sequence homologs to the sequence of interest, using a BLAST (Basic Local Alignment Search Tool) search in conjunction with NCBI's (National Center for Biotechnology Information) non-redundant protein sequence database, the blosum62 substitution matrix and default parameters.
2. Retrieve and list the complete sequences of the top 100 homologs (*see* **Note 1**), using the following criteria: e-value >0.05, and covering at least 50% of the sequence of interest. Sequences may be retrieved from NCBI's protein database using the search engine.
3. Generate an initial multiple sequence alignment using the CLUSTALW alignment tool (http://ca.expasy.org/). Use the full alignment option and default parameters (gap open penalty 10, gap extension penalty 0.05).
4. Group the sequences into sub-families based on phylogenetic and evolutionary considerations (i.e., based on the species of the source organism and its position in evolutionary history), using this initial alignment as the starting point (*see* **Note 2**).
5. Refine this initial alignment. Remove evolutionary outliers (protein sequences with remote evolutionary connection to the target sequence; *see* **Note 2**) from the sequence list and delete partial sequence fragments in the multiple alignment.
6. Gaps in the sequence are treated as a 21st amino acid, denoted by an X, in evolutionary trace analysis. This is merely a computational device and not a biophysically relevant assumption. This convention serves to indicate that a deletion or insertion took place that was then conserved in all protein descendants, suggesting some functional importance at the location of those gaps.
7. Assign residues trace ranks based on the minimum number of branches into which the protein family tree must be partitioned for that residue to be invariant within each branch. A residue of rank i is invariant within each of the first i branches of the tree (starting from the root), but variable within one of the first $(i - 1)$ branches. In general, the higher the rank of a trace residue, the more strongly it is likely to be linked to the specific function of the subfamily within which it is conserved. Also, residues can be ranked by measuring the information entropy within each branch and sub-branch as described in Mihalek et al. *(10)*.

8. Identify clusters of highly ranked trace residues in the primary sequence of the protein of interest; such clusters may indicate functionally important regions or domains. These residues may be used for targeting site-specific mutations to identify functional regions, residues or domains in the protein. (*see* **Notes 3–7**).
9. Two ET web servers *(11, 12)* are available at http://mammoth.bcm.tmc.edu/. (*see* **Note 8**)

3.2. Generation of Mutants

1. Design mutations at each trace residue position identified to elicit maximum structural and functional information about the residue at that position. Therefore, create mutants using systematic variations in the size, charge and/or hydrophobicity of each trace residue.
2. Design appropriate mutagenesis primers using the *Wt* nucleotide sequence as a template. Guidelines for primer design are as follows:
   a. Design primers between 25 and 45 bases in length.
   b. Choose a codon for mutation that is approximately in the middle of the sequence, with 10–15 bases on either side; the sequence should incorporate the replacement codon in place of the original *Wt* codon.
   c. Make sure that the annealing temperature (T_m) of the primer is greater than 78 °C.
   d. Design a reverse primer that is complementary to the forward primer.
   e. Design forward and reverse primers with at least 40% GC content and design so that they end in one or more GC bases as a 3′ clamp.
   f. Avoid secondary structure in designed primers, specifically at their 3′ ends.
   g. Analyze primer sequences for secondary structure (hairpins, loops etc), GC content and T_m using free primer analysis software available online at http://www.premierbiosoft.com/netprimer/netprlaunch/netprlaunch. html.
3. Dilute *Wt* plasmid template stock to a working concentration of 10 ng/μL using double-distilled, deionized water. Also, dilute each primer to a working concentration of 125 ng/μl.
4. Assemble a PCR reaction mix in a thin-walled 200 μl microcentrifuge PCR tube. Add each component carefully as individual droplets to the wall of the PCR tube in order to visually verify the addition of each small volume.
   a. 1–3 μL *Wt* DNA template.
   b. 1 μL forward primer and 1 μL reverse primer.
   c. 1 μL dNTPs (supplied with kit).
   d. Add 35 to 37 μL distilled RNAse-free water, followed by 5 μL of the supplied 10X buffer, and 3 μL of the supplied

Quiksolution (Quickchange II kit). Tap or flick the tube gently to facilitate mixing of the components, then place in the heating block of the PCR thermocycler.

5. Program the thermocycler to include the following steps:
   a. Initial 'hot-start' phase (95 °C – 1 min). Add Pfu polymerase at the end of the 'hot-start' phase to ensure complete melting of template DNA and any primer dimers prior to the annealing and extension phases; this improves PCR efficiency
   b. Melting (95 °C – 1 min)
   c. Annealing (60 °C – 1 min)
   d. Extension (68 °C – 1 min/kb of plasmid length).
   e. Repeat steps b through d 17 times, for a total of 18 cycles.
   f. Final extension phase of 68 °C for 7 min.
6. Following PCR amplification, add 1 μL of *Dpn*I restriction enzyme to the PCR tube, and incubate at 37 °C for 60–90 min.
7. Use 1–2 μL of the *Dpn*I-treated amplification product to transform XL-1 blue competent *E. coli* cells (TOP10 cells from Invitrogen may be substituted) by adding the DNA to 50 μL of competent cells in a pre-chilled 15 mL Falcon tube and incubate on ice for 30 min. Heat-shock the cells for 30–35 s at 42 °C followed by incubation on ice for 2 min. Add 250 μL of room temperature SOC medium and plate 50–100 μL onto appropriate antibiotic selection LB agar plates. Incubate overnight at 37 °C.
8. Pick one or two colonies per plate for overnight growth in 2–5 mL of liquid LB medium supplemented with appropriate antibiotics.
9. Perform a plasmid preparation using the Qiagen Spin Miniprep kit (Qiagen) according to the manufacturer's instructions. Verify presence of desired mutations by sequencing.

3.3. Evaluating Membrane Expression of Mutants

3.3.1. Day 1

1. Sterilize glass cover slips under UV light (1 h) and place in wells of a sterile 24-well plate.
2. Add 500 μL of medium (DMEM supplemented with 10% FBS and 1% penicillin/streptomycin) to each well.
3. Add an appropriate amount (12 μL if starting from a confluent 10 cm plate, or 40 μL from a 60 mm plate) of trypsinized HEK 293 cells from a pre-existing cell line into each well for 20% confluence
4. Incubate at 37 °C overnight, to allow the cells to reach 50–70% confluence.

3.3.2. Day 2

1. Mix 225.6 μL of DMEM (without supplements) and 14.4 μL of Fugene 6. Incubate for 5 min at room temperature.
2. Divide the mix into 12 microcentrifuge tubes at a concentration of 20 μL each.
3. Add 400 ng DNA to each tube. Incubate 30 min at room temperature.
4. Add 10 μL into each well. Incubate plate at 37 °C (*see* **Note 8**).

3.3.3. Day 4

1. Remove plate from incubator, remove medium by aspiration.
2. Wash each well 2 times with 500 μL PCM buffer for 5 min at room temperature.
3. Add 300 μL of a 100 μg/mL concentration of Concanavalin-A-Alexa 350 in PCM buffer. Incubate 1 h on ice. (*see* **Note 9**).
4. Wash each well 2 times with 500 μL PCM buffer for 5 min at RT.
5. Fix with 300 μL 4% PFA for 10 min at room temperature.
6. Wash each well 2 times with 500 μL PBS buffer for 5 min at room temperature.
7. Permeabilize with 500 μL PBS containing 0.1% Triton for 3 min at room temperature.
8. Wash 1 time with 500 μL PBS for 5 min at room temperature.
9. Cover with 1% BSA in PBS (500 μL) and incubate for 1 h at room temperature.
10. Incubate cells with 300 μL of the appropriate primary antibody in PBS for 2 h at 37 °C, or overnight at 4 °C. Make sure cells in each well are covered by buffer.
11. Wash each well 2 times with 500 μL PBS buffer for 5 min.
12. Cover cells with appropriate secondary antibody labeled with AlexaFluor594 in PBS for 1 h at room temperature.
13. Wash 3 times with 500 μL PBS.
14. Remove coverslips from wells using forceps, blot excess liquid from the edges onto lint free tissue paper (e.g. Kimwipe, Kimberly-Clark, Dallas, TX).
15. Add a drop of Fluoromount reagent on a clean dry slide. Invert cover slip (cell side down) onto the drop.
16. Seal the edges of the cover slip with clear nail polish. Allow to dry 2–3 h at room temperature before storing slides in the refrigerator at 4 °C.
17. View slides under deconvolution microscope at 63X magnification using appropriate wavelength filters. Analyze images for colocalization of concanavalin-A immunofluorescence with that of AlexaFluor 594, which has been used to label the protein under study.

4. Notes

1. ET analysis can be done for any sample size of sequence homologs. If a structure is available, the statistical significance of structural clustering among top-ranked residues can be computed *(13)* to judge the results reliability. When a structure is not available, a minimum number of 15–20 sequences will typically produce useful evolutionary trace rankings, but this also depends on the evolutionary spread of these sequences.
2. A comprehensive evolutionary and phylogenetic tree encompassing the majority of known organisms can be found at the Tree of Life web project http://www.tolweb.org. This information, combined with the taxonomical information included with the protein sequence in NCBI's database, may be used when ordering the sequences for computational analysis. Evolutionary outliers may also be determined and discarded based on this information.
3. Further information on ET analysis *(1–4)* and its applications to several diverse systems *(5, 6, 14–16)* is readily found in the literature.
4. Other computational methods based on similar evolutionary and phylogenetic considerations have been developed *(17–20)*. Such phylogeny-based computational analyses have been used in at least two studies to predict functional regions in proteins involved in audition *(9, 21)*.
5. Evolutionary Trace residues have been shown to exhibit regular periodicity in transmembrane helices and can, therefore, be used to confirm and/or refine the results of a hydrophobicity-based transmembrane topology prediction *(9)*.
6. We have used Evolutionary Trace analysis, in conjunction with electrophysiological measurements of non-linear capacitance, as a readout of prestin activity, to pinpoint residues in a conserved motif region of the prestin protein that are essential for its function. Further, by analyzing the effects of charge and size substitutions, we have been able to conclude that tight packing of the transmembrane helices in this region is essential for prestin activity. Results of functional assays of individual mutants are shown in **Fig. 17.2** below, and a more detailed account of the study can be found in Rajagopalan *et al.*, 2006 *(9)*.
7. An automated Evolutionary Trace analysis can be now performed at the ET servers at http://mammoth.bcm.tmc.edu/server.html. However, a step-by-step protocol has been provided to facilitate a thorough understanding of the method, and to allow manual refinement of the automated results when necessary.

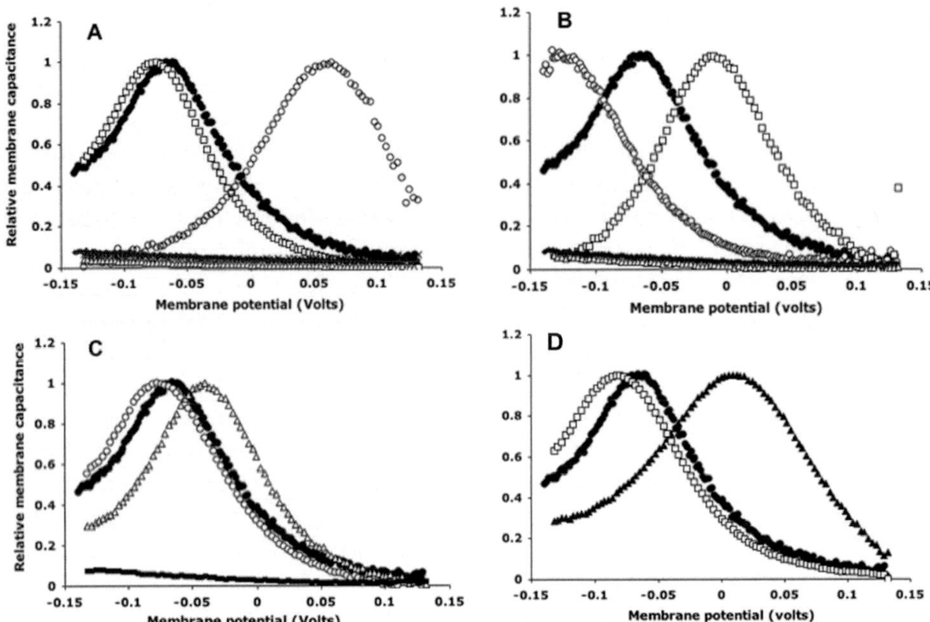

Fig. 17.2. Normalized membrane capacitance of prestin single mutants. (**A**) Five single substitutions at A100: A100S (□), A102G (○), A102V (X), A102L (△) and A102W (□). A100V, L and W (clustered at bottom) have minimal voltage dependence, similar to untransfected or mock-transfected controls. (**B**) Five single substitutions at A102: A102S (□), A102G (○), A102V (X), A102L (△) and A102W (□). A102V, L and W have minimal voltage dependence, similar to untransfected or mock-transfected controls. (**C**) Three single substitutions at L104: L104V (□), L104I (△) and L104W (■). L104W has minimal voltage dependence, similar to untransfected or mock-transfected controls. (**D**) Capacitance plots of L113W (▲) and F117W (□). Representative plots, normalized relative to $C(V_{pkc})$ and C_{lin}, from single cells are shown. In all panels, a representative plot of WT prestin is shown for comparison (•). Differences in magnitude of NLC have been ignored in this representation.

8. These volumes are for an experiment where you use 12 different mutants and all 24 wells, with duplicates of each mutant. Change volumes appropriately depending on your experiment.
9. Concanavalin-A binds to carbohydrates on the cell surface, and is therefore used as a marker for cell membranes. Con-A labeled with AlexaFluor can be used in immunohistochemical analyses to label membranes, which can then be analyzed microscopically for colocalization to various labeled proteins to evaluate membrane expression of these proteins.

Acknowledgments

Preparation of these guidelines was supported in part by fellowship support from the Keck Center for Interdisciplinary Bioscience Training to LR. The authors thank Dr. Ivana Mihalek for helpful comments and suggestions.

References

1. Lichtarge, O., Bourne, H. R., and Cohen, F. E. (1996) An evolutionary trace method defines binding surfaces common to protein families. *J. Mol. Biol.* **257**, 342–358.
2. Lichtarge, O., and Sowa, M. E. (2002) Evolutionary predictions of binding surfaces and interactions. *Curr. Opin. Struct. Biol.* **12**, 21–27.
3. Madabushi, S., Yao, H., Marsh, M., Kristensen, D. M., Philippi, A., Sowa, M. E., and Lichtarge, O. (2002) Structural clusters of evolutionary trace residues are statistically significant and common in proteins. *J. Mol. Biol.* **8**, 139–154.
4. Yao, H., Kristensen, D. M., Mihalek, I., Sowa, M. E., Shaw, C., Kimmel, M., Kavraki, L., and Lichtarge, O. (2003) An accurate, sensitive, and scalable method to identify functional sites in protein structures. *J Mol Biol.* **326**, 255–261.
5. Sowa, M. E., He, W., Slep, K. C., Kercher, M. A., Lichtarge, O., and Wensel, T. G. (2001) Prediction and confirmation of a site critical for effector regulation of RGS domain activity. *Nat. Struct. Biol.* **8**, 234–237.
6. Madabushi, S., Gross, A. K., Philippi, A., Meng, E. C., Wensel, T. G., and Lichtarge, O. (2004) Evolutionary trace of G protein-coupled receptors reveals clusters of residues that determine global and class-specific functions. *J. Biol. Chem.* **279**, 8126–8132.
7. Ribes-Zamora, A., Mihalek, I., Lichtarge, O., and Bertuch, A. A. (2007) Distinct faces of the Ku heterodimer mediate DNA repair and telomeric functions. *Nat. Struct. Mol. Biol.* **14**, 301–307.
8. Shenoy, S. K., Drake, M. T., Nelson, C. D., Houtz, D. A., Xiao, K., Madabushi, S., Reiter, E., Premont, R. T., Lichtarge, O., and Lefkowitz, R. J. (2006) beta-arrestin-dependent, G protein-independent ERK1/2 activation by the beta2 adrenergic receptor. *J. Biol. Chem.* **281**, 1261–1273.
9. Rajagopalan, L., Patel, N., Madabushi, S., Goddard, J. A., Anjan, V., Lin, F., Shope, C., Farrell, B., Lichtarge, O., Davidson, A. L., Brownell, W. E., and Pereira, F. A. (2006) Essential helix interactions in the anion transporter domain of prestin revealed by evolutionary trace analysis. *J. Neurosci.* **26**, 12727–12734.
10. Mihalek, I., Res, I., and Lichtarge, O. (2004) A family of evolution-entropy hybrid methods for ranking protein residues by importance. *J. Mol. Biol.* **336**, 1265–1282.
11. Morgan, D. H., Kristensen, D. M., Mittelman, D., and Lichtarge, O. (2006) ET viewer: an application for predicting and visualizing functional sites in protein structures. *Bioinformatics* **22**, 2049–2050.
12. Mihalek, I., Res, I., and Lichtarge, O. (2006) Evolutionary trace report_maker: a new type of service for comparative analysis of proteins. *Bioinformatics* **22**, 1656–1657.
13. Mihalek, I., Res, I., Yao, H., and Lichtarge, O. (2003) Combining inference from evolution and geometric probability in protein structure evaluation. *J. Mol. Biol.* **331**, 263–279.
14. Geva, A., Lassere, T. B., Lichtarge, O., Pollitt, S. K., and Baranski, T. J. (2000) Genetic Mapping of the Human C5a Receptor. Identification Of Transmembrane Amino Acids Critical For Receptor Function. *J. Biol. Chem.* **275**, 35393–35401.
15. Lichtarge, O., Yamamoto, K. R., and Cohen, F. E. (1997) Identification of functional surfaces of the zinc binding domains of intracellular receptors. *J. Mol. Biol.* **274**, 325–327.
16. Sowa, M. E., He, W., Wensel, T. G., and Lichtarge, O. (2000) A regulator of G protein signaling interaction surface linked to effector specificity. *Proc. Natl. Acad. Sci. U.S.A.* **97**, 1483–1488.
17. Landgraf, R., Xenarios, I., and Eisenberg, D. (2001) Three-dimensional cluster analysis identifies interfaces and functional residue clusters in proteins. *J. Mol. Biol.* **307**, 1487–1502.
18. Armon, A., Graur, D., and Ben-Tal, N. (2001) ConSurf: an algorithmic tool for the identification of functional regions in proteins by surface mapping of phylogenetic information. *J. Mol. Biol.* **307**, 447–463.
19. Hannenhalli, S. S., and Russell, R. B. (2000) Analysis and prediction of functional sub-types from protein sequence alignments. *J. Mol. Biol.* **303**, 61–76.
20. Yang, Z., and Nielsen, R. (2002) Codon-substitution models for detecting molecular adaptation at individual sites along specific lineages. *Mol. Biol. Evol.* **19**, 908–917.
21. Franchini, L. F., and Elgoyhen, A. B. (2006) Adaptive evolution in mammalian proteins involved in cochlear outer hair cell electromotility. *Mol. Phylogenet. Evol.* **41**, 622–635.

Chapter 18

In Vivo Verification of Protein Interactions in the Inner Ear by Coimmunoprecipitation

Margaret C. Harvey and Bernd H. A. Sokolowski

Abstract

Genomics has provided us with vast amounts of data and thus, the challenge to identify and characterize gene products. Proteomics analysis, using methods such as yeast two-hybrid screenings, isoelectric focusing, and mass spectroscopy, generate potentially useful information. To determine functional relationships between and among proteins, however, the initial data for putative protein interactions must first be validated. One technique, which is considered the gold standard, is coimmunoprecipitation.

Key words: Inner ear, ion channels, protein-protein interactions, coimmunoprecipitation.

1. Introduotion

The identification of proteins involved in signal transduction in the inner ear is a daunting task, considering the small amount of tissue with which one is required to work and the low expression levels of many of the most interesting protein species. This difficulty is compounded where the goal is to discover protein-protein interactions. Concentration of the proteins of interest can be achieved by a method of affinity chromatography known as immunoprecipitation *(1, 2)*. Inner ear tissues are isolated quickly, mechanically disrupted by sonication and solubilized under conditions that preserve protein-protein interactions to the greatest extent possible. The whole tissue lysate is used immediately or processed further to isolate a membrane-enriched fraction. Proteins of interest can be separated by employing an antibody directed against one or more epitopes in one protein partner, the caveat being that other closely-associated proteins *in situ* must

Bernd Sokolowski (ed.), *Auditory and Vestibular Research: Methods and Protocols, vol. 493*

DOI 10.1007/978-1-59745-523-7_18 Springerprotocols.com

maintain their association *ex situ*. This "precipitating" antibody is pre-complexed to protein G-coated agarose beads with each molecule of protein G capable of binding two antibody/protein moieties. The immunocomplexes bound to these beads are then eluted in sample buffer under denaturing, reducing conditions and the beads removed by centrifugation. The sample is electrophoretically fractionated by SDS-PAGE and transferred to a membrane. A second antibody targeted against the protein partner is then used for western blot detection of the interaction. The precipitating antibody is used to confirm the presence of the first protein partner in either a tissue lysate or in an immunoprecipitation if the protein exists in low abundance and/or is difficult to capture. Commercially available secondary antibodies that recognize only the heavy chain or the light chain immunoglobulin, be it a monoclonal or polyclonal antibody, are useful in the resolution of protein bands that migrate to the same area as these Ig fragments during electrophoresis.

2. Materials

2.1. Preparation of Stock and Working Solutions for Processing Tissues

All inhibitor stock solutions are stored at −20 °C (*see* **Note 1**).

1. Protease inhibitor stock solutions: These consist of 50 mg/mL of Pefabloc® SC (AEBSF; 4-(2-Aminoethylbenzenesulfonyl fluoride hydrochloride, Fluka, St. Louis, MO) in water stored in 100 μL aliquots, 10 mg/mL of leupeptin (Sigma-Aldrich, St. Louis, MO) in water stored in 10 μL aliquots, 10 mg/mL of aprotinin (Sigma-Aldrich) in 10 m*M* HEPES (4-(2-hydroxyethyl)-1-piperazineethanesulfonic acid), pH 8.0 stored in 20 μL aliquots, 100 μg/mL of pepstatin A (Sigma-Aldrich) in 95% ethanol, and 1 m*M* microcystin-LR (Calbiochem/ EMD Biosciences, San Diego, CA) in DMSO stored in 5 μL aliquots.
2. Lysis buffer stock solutions: 1 *M* Tris-HCl, pH 8.0 (UltraPure™ solution, Invitrogen, Carlsbad, CA) and 0.5 *M* EDTA (ethylenediaminetetraacetic acid, UltraPure™ solution, Invitrogen) pH, 8.0. Prepare 1.5 *M* NaCl in water.
3. Lysis buffer working solution: 50 m*M* Tris-HCl, pH 8.0, 5 m*M* EDTA, pH 8.0 and 120 m*M* NaCl. Prepare 100 mL by mixing 5 mL of 1 *M* Tris-HCl, pH 8.0, 1 mL of 0.5 *M* EDTA, pH 8.0, and 8 mL of 1.5 *M* NaCl. Bring to 100 mL using water and store at 4 °C.
4. Complete lysis buffer working solution: 50 m*M* Tris-HCl, pH 8.0, 5 m*M* EDTA, pH 8.0. 120 m*M* NaCl, 0.5 mg/mL of Pefabloc® SC, 10 μg/mL of leupeptin, 2 μg/mL of aprotinin, 1 μg/mL of pepstatin A and 0. 5 μ*M* microcystin-LR. Immediately prior to use, prepare a 10 mL volume by mixing the following quantities of stock protease/phosphatase inhibitors:

100 μL of Pefablock® SC, 10 μL of leupeptin, 20 μL of aprotinin, 10 μL of pepstatin A, and 5 μL of microcystin-LR. Bring to a final volume of 10 mL with water and maintain in packed ice (*see* **Note 2**).

5. Solubilization buffer(s): Consists of complete lysis buffer (*see* **step 4**) with 1% Triton X-100. Should one of the protein partners reside to a significant extent in lipid rafts, an additional detergent will be required, such as, 1% octyl β-glucoside (Pierce Biotechnology, Rockford, IL) (*see* **Note 3**).

2.2. Preparation of Tissue Samples

1. Cooling agents: Dry ice hammered into powder form, cryovials precooled in crushed ice for each tissue aliquot, and packed crushed ice.
2. Dissection instruments: Due to the density of the temporal bone, several scalpel blades (e.g., No. 11 Feather, Graham-Fields, Atlanta, GA) should be made available. A useful dissection tool is a set of laminectomy forceps (Fine Science Tools, Foster City, CA) and it is imperative that all forceps (e.g., Dumont #5 and Dumont #55, Fine Science Tools) be sharpened and aligned. A small whetstone and emory cloth is sufficient for this purpose
3. Sonicator: (e.g., Sonic Dismembrator Model 100, Thermo Fisher, Waltham, MA).
4. Refrigerated microcentrifuge.
5. Refrigerated shaker equipped to hold microcentrifuge tubes (e.g., Roto Shake Genie, Scientific Industries, Bohemia, NY).
6. Vortexer maintained at 4 °C.
7. 25 gauge 1-1/2″ needle and a 1 mL syringe for trituration.

2.3. Immunoprecipitation

1. rec-Protein G-Sepharose® 4B conjugate beads (Zymed/Invitrogen).
2. Wash buffer: Complete lysis buffer working solution (*see* **Section 2.1, step 4**), containing 0.1% Triton X-100.
3. Elution buffer: 2X Laemmli buffer (Promega, Madison, WI).

2.4. SDS-PAGE and Western Blotting

All reagents should be electrophoresis grade unless otherwise indicated.

1. Polyacrylamide gels: Tris/glycine pre-cast 7.5% (Bio-Rad, Hercules, CA) stored at 4 °C.
2. Markers: The running marker is prestained Precision Plus Protein™ Dual Color (Bio-Rad) and the protein molecular weight marker is peroxidase-labeled Magic Mark™ XP (Invitrogen). Aliquot in volumes of ~10 μL to avoid repeated freeze-thaw cycles and store at −20 °C.
3. 10% SDS (sodium dodecyl sulfide) stock solution: Mix 10 g of SDS (Sigma-Aldrich) in 70 mL of water, then bring to a final volume of 100 mL after SDS has gone into solution (*see* **Note 4**).

4. 10X Electrode Buffer Stock Solution: 2.5 *M* Tris Base and 2 *M* glycine. Add 30.3 g of Tris Base to 700 mL of water and mix until dissolved. Add 150 g of glycine gradually, mix until dissolved and bring the final volume to 1 L with water. Do not pH. Store at 4 °C.
5. Running Electrode Buffer Working Solution: For a working solution add 100 mL of 10X electrode buffer stock solution to 850 mL of water and then add 10 mL of 10% SDS stock solution. Bring to a final volume of 1 L with water, cover with Parafilm®, and mix by inversion several times. Store at 4 °C until use.
6. Working Transfer Electrode Buffer: For a working solution add 100 mL of 10X electrode buffer stock solution to 600 mL of water and then add 200 mL of methanol. Bring to a final volume of 1 L with water, cover with Parafilm® and mix by inversion several times. Store at 4 °C until use.
7. A power supply to run the gel apparatus and transfer apparatus (e.g., Power Pac 3000, Bio-Rad).
8. Blotting membrane: Nitrocellulose (Protran BA85, Schleicher & Schuell/Whatman, Florham Park, NJ), 0.45 μM pore size and filter paper (3 MM chromatography paper, Whatman).
9. Tris-Buffered Saline/Tween 20 (TBST): 50 m*M* Tris-HCl, pH 8.0, 120 m*M* NaCl and 0.05% Tween 20.
10. Blocking buffer and primary and secondary antibody diluents: TBST plus 4% milk buffer (Bio-Rad) (*see* **Note 5**).
11. Polypropylene hybridization bags and apparatus to heat-seal blots during overnight incubation with primary antibody.
12. Horseradish peroxidase-conjugated secondary antibodies: mouse anti-rabbit IgG (mouse monoclonal cat. no. 211-035-109, Jackson ImmunoResearch, Westgrove, PA) at 1:15,000; goat anti-mouse IgG (goat polyclonal cat. no. 12-349, Upstate/Millipore, Danvers, MA) at 1:2000, donkey anti-rabbit IgG (donkey polyclonal cat no. NA934, Amersham/GE Healthcare, Piscataway, NJ) at 1:5000 (*see* **Note 6**).
13. Chemiluminescence detection reagents and supplies: Enhanced Chemiluminescence (ECL)™ kit (Amersham/GE Healthcare), plastic tube for mixing ECL reagents, Saran® wrap (S.C. Johnson & Son, Racine WI), autoradiography cassette with amplifier (e.g., Thermo Fisher), Hyperfilm (Amersham/GE Healthcare), scissors for cutting film™ ECL, a 100–1000 μL pipeter and pipet tips, paper towels, membrane forceps, plastic tape and a timer.
14. Membrane stripping solution: 0.5 *M* acetic acid, pH 2.5 and 1.5 *M* NaCl. Prepare 100 mL by adding 33 mL of 1.5 *M* NaCl to 60 mL of water and add 3 mL of glacial acetic acid.

Mix and pH to 2.5 using glacial acetic acid and bring to 100 mL volume with water.

15. Quantitation: Chemi-imager with densitometry software (e.g., ChemiImager with Spot Denso software, Alpha Innotech, San Leandro, CA).

3. Methods

Inner ear tissues from one animal may be sufficient to obtain a strong signal for each interacting protein (*see* **Note 7**). Design experiments to include a blank bead control or antibody preadsorbed with antigen where available, as a negative control, along with either lysate or immunoprecipitated target proteins as positive controls (*see* **Note 8**). An irrelevant antibody of the same type (isotype-specific for a mAb and source-specific for pAb) and pre-immunized sera, where available, are negative controls as well. Positive controls can consist of using tissue in which reactivity is known and/or using the same primary Abs from other vendors or individuals.

3.1. Preparation of Immunocomplexed Protein G Beads

1. Rinse 50 μL of (packed) protein G beads three times in 500 μL ice-cold wash buffer and centrifuge at 2000 g for 2 min between rinses. Prepare 2 aliquots, one bound to precipitating antibody and another unbound aliquot for control. Uncomplexed washed beads may be used in small quantity (~15 μL) for 30 min should pre-clearing of the samples prove necessary. Empirically determine the amount of antibody required to precipitate the protein of interest. Five μg of either an affinity-purified monoclonal (mAb) or polyclonal (pAb) is a good starting point. Titer mAb tissue culture supernatants and pAb antisera. Allow complexes to form over the course of 1–2 h, minimum, at 4 °C with rocking. Antibody-bound beads can be stored overnight at 4 °C. (*see* **Note 9**).

3.2. Preparation of Tissue Samples

1. Use either freshly excised tissues or an aliquot of tissues from storage at −80 °C. There is no need to thaw frozen tissues. Rinse tissues two times in complete lysis buffer, spinning down briefly each time in a refrigerated microcentrifuge. Add 500 μL of complete lysis buffer to rinsed tissues and sonicate using a low setting (between 1 and 2 on the sonicator described in **Section 2.2, step 2**). Sonicate tissues three times for 30 s, while keeping the tissue cooled in packed crushed ice. Allow one min cool-down period between sonications.
2. Immediately centrifuge crude lysate at 500 g for 10 min at 4 °C to remove debris and nuclei. Use supernatant and discard the pellet.

3. Solubilize the contents of the supernatant over the course of 1 h in solubilization buffer with gentle rocking at 4 °C.

3.2.1. Preparation of Cleared Lysate Sample

1. Centrifuge at 11,000 g for 10 min to clear the solubilized lysate from **Section 3.2, step 3**. If pre-clearing with uncomplexed beads is desired, transfer the supernatant to the tube containing the 15 μL of washed, packed beads from **Section 3.1, step 1** and incubate for 30 min at 4 °C with rocking. Collect the beads and discard, or save and elute bound material to determine any loss of signal. Evenly divide the cleared/pre-cleared solubilized supernatant into the tubes containing the beads (*see* **Section 3.1**), and incubate overnight at 4 °C with rocking.
2. Also, from the solubilized lysate remove and retain ~15 μL in a separate tube, to run on a gel as "lysate," and immediately add an equal volume of 2X Laemmli buffer to this sample, boil for 5 min, and either store on ice or at −20 °C until the gel is run.

3.2.2. Preparation of a Membrane-Enriched Fraction

1. Centrifuge the solubilized lysate from **Section 3.2, step 3** at 20,800 g for 1 h at 4 °C. Meanwhile, prepare a solution of 1% octyl β-glucoside detergent in 750 μL of complete lysis buffer. Emperically determine the amount of protein needed for immunoprecipitation. Test a range in order to conserve tissue. For example, successful precipitations can be performed with ~750 μg of protein. To determine the amount of protein prior to immunoprecipitation, protein assays must be performed using a method compatible with the detergent. For example, the BCA (bicinconchinic acid) kit can be used according to manufacturer's instructions (Pierce Biotechnology).
2. Add the prepared ice-cold detergent solution to the pellet from **step 1** and resuspend the pellet by triturating on ice 20 times, using a syringe, along with intermittent vortexing for a few seconds. Take care to not warm the sample. Finally, solubilize the sample with rocking at 4 °C for 90 min.
3. For pre-clearing with uncomplexed beads, refer to **Section 3.2.1, step 1**.
4. Evenly divide the membrane-enriched sample into tubes containing beads for overnight incubation at 4 °C with rocking (*see* **Note 9**).

3.3. Elution of Immunocomplexes

1. Collect beads by spinning at 2000 g for 5 min at 4 °C. Retain antibody-depleted sample (supernatant overlying the beads) to determine efficiency of precipitation by running a portion out on the gel (*see* **Section 3.2.1, step 2**).

2. Wash beads 5 times in 500 μL of wash buffer, flicking the tube several times during each wash and spinning that sample at 2000 g for 2 min between each wash (*see* **Note 10**).
3. Resuspend beads in 50 μL of 2X Laemmli buffer, vortex briefly, and boil for 5 min.
4. Spin samples at 20,800 g for 5 min at 4 °C in a refrigerated centrifuge. Remove the eluted immunocomplexes overlying the beads, taking care not to draw up depleted beads and immediately maintain sample on ice until ready to fractionate or store at −20 °C.

3.4. SDS-PAGE and Western Blotting

1. These instructions apply to the use of a minigel apparatus (Protean® III) and precast polyacrylamide Tris/glycine gels (both Bio-Rad), but any system will work; follow manufacturer's instructions. Ensure precast gel is not at/near expiration date. After removing comb from gel, rinse wells with freshly prepared electrode buffer and remove the plastic strip at the bottom of the gel. Remove gaskets after each run and rinse thoroughly in water to prevent salt accretion.
2. Place gel in the proper orientation in the assembly stand; use the plastic dam provided by the manufacturer if running only one gel. Boil samples that were stored at −20 °C for 5 min and cool on ice. Vortex and spin down the samples and ladders briefly, and prior to loading, mix each by pipeting up and down several times.
3. Load the following on the gel: lysate, a sample to be used as an IP, and a sample to be used as a co-IP, and ladders. Load according to manufacturer's instructions based on the size of the gel, the maximum volume per well, and the amount of protein required to resolve the species of interest.
4. Run the gel at 60 V until the samples have migrated through the stacking gel. Estimate by monitoring the location of the dye front (the bromophenol blue in the sample buffer) as this migrates a short distance in front of the samples. This takes about 15 min. When the samples have entered the resolving gel, change the voltage to 100 V for a sufficient period of time to resolve the species of interest. Monitor the separation of the bands in the prestained marker.
5. While the gel is running, cut a piece of nitrocellulose membrane that is the size of the running gel. As the run on the gel is near completion, wet the cut piece of membrane by immersing in water for 2 min and then in transfer buffer to equilibrate it until use.
6. When the run is complete, shut off and disconnect the power supply. Remove the gel carefully by first disassembling the cassette. Using a razor blade, remove the wells and the stacking gel, then float the gel off the plate by immersing in a dish of transfer buffer, without methanol, for 15 min with gentle

rocking. Use a sufficient volume of buffer to ensure removal of residual SDS.

7. Electroblot using the tank transfer method. Use a container (e.g., glass casserole dish) to assemble the cassette, while all components are submerged in transfer buffer. Place in the container in the following order, from bottom to top, starting on the cathode portion of the cassette: pre-wet sponges, filter paper, equilibrated gel, equilibrated membrane, filter paper, sponge on the anode side of the cassette, taking care to remove bubbles by gently rolling a small plastic tube across the surfaces. Transfer samples to the membrane at 100 V for 60 min, using either an ice-pack provided by the manufacturer or by packing crushed ice around the outside of the apparatus, or perform the transfer in a chromatography refrigerator (*see* **Note 11**).
8. Rinse membrane in TBS for 15 min with rocking, pour off the TBS and add blocking buffer to block non-specific binding sites on the membrane for 30 min. Emperically determine blocking time and concentration of block by monitoring signal to noise ratio.
9. Pour off the blocking buffer, add the primary antibody diluted in blocking buffer, and incubate overnight with rocking at 4 °C (*see* **Note 12**).
10. Wash blot two times for 10 min each in TBST. Prior to each of the two washes, rinse the blot twice, briefly, in a small volume of TBS without Tween.
11. Incubate in secondary antibody diluted in blocking buffer for 60 min with rocking at room temperature.
12. Wash the blot three times for 10 min each in TBST followed by one wash for 10 min in TBS, with rocking. Again, prior to each wash, briefly rinse the blot twice in a small volume of TBS.
13. While the blot is being washed, bring to room temperature a 1 mL volume of each of the two reagents supplied in the ECL kit. Pour off the final wash and add fresh TBS to the membrane.
14. Assemble the following items together in a bin in preparation for the darkroom: Saran® Wrap, scissors for cutting film, the ECL reagents, a 100–1000 μL pipeter and tips, paper towels, membrane forceps, film, autoradiography cassette, and timer. In the darkroom, mix the two components of the ECL kit in a tube by briefly flicking several times.
15. Remove the blot from the TBS, using membrane forceps, and drain by holding the membrane at an angle on a paper towel, permitting the buffer to be wicked away. Place the membrane on a small sheet of Saran® Wrap and apply the ECL reagents such that the entire blot is covered. Develop for 1 min.

16. Drain the ECL reagents from the membrane by wicking on paper towels, place on a fresh piece of Saran® Wrap and enwrap the membrane. Each of the 4 corners of the wrapped membrane can be maintained in place with small pieces of tape.
17. Place a half-sheet of film over the membrane, close the cassette and expose the film for 1 min. Shorter or longer exposure times might be necessary to resolve the protein bands, progressing to overnight if necessary (*see* **Note 13**).

3.5. Stripping and Reprobing Blots

1. Wash blot two times for 10 min each in TBS and drain.
2. Incubate in stripping solution as prepared in **Section 2.3, item 14** with rocking for 2–5 min and immediately neutralize by adding large volumes of TBS over the course of several buffer changes (*see* **Notes 14,15**).
3. Continue with western blotting as described in **Section 3.4, steps 8–17**.

4. Notes

1. All solutions should be prepared using Milli-Q water (minimum resistance of 18.2 MΩ-cm, unless otherwise stated). All inhibitor stock solutions are stored in single use aliquots with the exception of pepstatin A, which may be stored at 4 °C.
2. The effective range and stability of stock protease/phosphatase inhibitors is noted here. The Pefabloc® SC working solution is effective from 0.1 to 1.0 mg/mL. Leupeptin is effective from 1 to 2 μg/mL, but is stable for only a few hours, therefore a working concentration of 10 μg/mL is used; the stock solution of leupeptin is stable for up to 6 months. Pepstatin A is effective at 1 μ*M* or 1 μg/mL and is stable for about one day; the pepstatin stock solution is stable for several months. Aprotinin is effective from 1 to 2 μg/mL and the stock solution is stable for a couple of months. When purchased from Sigma-Aldrich, aprotinin activity can vary by lot from 3 to 7 TIU (trypsin inhibitor units) per mg. Microcystin-LR is an effective inhibitor of protein phosphatases PP1 and PP2A from low p*M* to n*M* concentrations. However, microcystin is problematic to acquire as it is now on a list of reagents that are being monitored, so a special form must be obtained and submitted prior to receipt.
3. Determine, to the extent possible, in which cell fraction the interacting proteins reside, in order to select the appropriate detergent or set of detergents. For example, if one of the targets is a membrane protein, a question may be, to what extent,

if any, does it reside in lipid rafts? Another question might be, is the target a cytoskeletal protein? For these situations, solubilization with Triton-X is insufficient to liberate membrane proteins or disrupt/solubilize cytoskeletal proteins. Solubilize with a different detergent, that is, one that is nonionic and effective for isolating membrane and cytoskeletal proteins *(3, 4)*.

4. A less pure form of SDS (~95% lauryl sulfate) can contribute to the resolution of variably phosphorylated species. Empirically determine the purity that is optimal.
5. The selection of antibodies is critical. The precipitating antibody must recognize target sequence(s) under native conditions, while the antibody used to probe a blot must bind epitope(s) in proteins that have been denatured. Potential steric hindrance must be avoided by examining conserved domains in both protein partners. Select an antibody against an epitope that does not obviate/preclude binding of antibody to target due to the proximity of the protein-protein binding site(s).
6. All secondary antibodies are HRP-conjugated. Heavy chain (~55 kDa) and light chain (~25 kDa) IgG bands can obscure/prevent the resolution of similar weight species. Presently, 25 kDa-specific secondary antibodies are available for use with polyclonal antibodies, while the 55 kDa-specific species, both for mAb and pAb (Jackson ImmunoResearch) are in development.
7. To prevent degradation, it is imperative to not euthanize more animals than needed to excise the tissues and place them on dry ice in a timely fashion. The dissection of inner ear tissues is time-consuming. Two individuals working in concert at this point is helpful; one euthanizing each animal, bisecting the head and placing it on ice, while the other removes the tissues and parses them into aliquots, each sufficient for conducting an experiment, on dry ice. In this way, freeze-thaw cycles can be avoided. Also, it is imperative to follow your institute's guidelines for the care and use of animals for these procedures.
8. Begin by performing small scale experiments to avoid lag times in processing samples. Lobind protein tubes are useful (e.g., Eppendorf, Westbury, NY). Avoid loss of beads by maintaining an adequate volume of fluid over the beads. Retain all fractions to determine in which fraction the interaction is the most enriched.
9. The gold standard is to use reciprocal coimmunoprecipitations, in which the interaction is tested in both "directions". In order to test this possibility, use the antibody that was used previously to probe the western blot, as the precipitating antibody. Emperically determine if it can bind to its target under native conditions. Using either whole lysates or

membrane-enriched samples, run several lanes out on a gel to test several different concentrations of each antibody and to optimize concentrations. Keep in mind that less sample is required for an immunoprecipitation, since the sample is concentrated for that/those protein(s). A pilot experiment is done at the outset, to determine if the precipitating antibody detects its target protein under native conditions, by including a sample of lysate or immunoprecipitant. The latter is used if the protein is not abundant. The immunoprecipitated sample will demonstrate the efficiency of the precipitating antibody to pull the protein out of the sample. The remaining portion of sample should be run as a coimmunoprecipitation, i.e., by probing the blot using an antibody against the interacting protein. The blot can then be stripped and reprobed, if necessary.

10. The wash stringency after precipitation is critical. Evaluate the loss of signal by considering the potential loss due to the number of washes, the vigor with which they are done, the duration of each, the composition of the wash buffer, and the concentration of detergent. Retain the remainder, that is, the antibody-depleted sample, after removing the immunocomplexed beads, to determine the efficiency. Retain all washes so as to identify potential loss of signal.
11. Samples stored at −20 °C in Laemmli buffer must be boiled for 5 min followed by chilling on ice, vortexing, spinning down, and mixing by pipeting, prior to loading on the gel. Although the samples are denatured and reduced, they must be maintained on ice to avoid degradation, as some enzymes are not rendered inactive at this stage.
12. Prepare working electrode buffers, both running and transfer, from stock solutions immediately prior to use. Wet nitrocellulose membrane in distilled water and equilibrate in transfer buffer for at least 15 min prior to use. Once the gel run begins prepare working transfer buffer and maintain at 4 °C. Adjust to the final volume to be used with methanol immediately prior to use to replace any evaporated methanol.
13. Conserve antibodies by incubating membranes in a pouch fashioned from a hybridization bag (Research Products International, Mt. Prospect, IL) that is heat sealed (e.g., Impulse Heat Sealer, Cleveland Equipment & Machinery, Cordova, TN). Ensure sufficient surface flow by adjusting the rocker and exclude air bubbles prior to sealing. Save the antibody the next day by adding NaN_3 to a concentration of 0.02%, from a 10% stock solution made in water, and clearly mark those aliquots that contain preservative.
14. A lysate signal may take longer to visualize, as it is less concentrated than an immunoprecipitant.
15. Reprobing membrane: ensure blot is stripped by exposing on chemiluminescence film after detection. Expose the film for

the same optimal duration used to resolve the species of interest. If the blot is properly stripped, there will be no signal because the secondary antibody will no longer be present, thereby leaving the detection reagents without an HRP-linked substrate.

Acknowledgments

This work was supported by NIDCD grant DC04295 to B. H. A. S.

References

1. Duzhyy, D., Harvey, M., Sokolowski, B. (2005) A secretory-type protein, containing a pentraxin domain, interacts with an A-type K+channel. *J. Biol. Chem.* **280**, 15165–15172.
2. Markham, K., Bai, Y., Schmitt-Ulms, G. (2007) Co-immunoprecipitations revisited: an update on experimental concept and their implementation for sensitive interactome investigations of endogenous proteins. *Anal. Bioanal. Chem.* **389**, 461–473.
3. Bhairi, S. M. (2001) *A Guide to the Properties and Uses of Detergents in Biology and Biochemistry.* Calbiochem-Novabiochem Corp., La Jolla, CA, 41pp.
4. Carraway, K. L., Carraway, C. A. C. (Eds.) (1999) *Cytoskeleton Signalling and Cell Regulation.* Oxford University Press, New York, NY, 287pp.

Chapter 19

Identification of Transcription Factor–DNA Interactions Using Chromatin Immunoprecipitation Assays

Liping Nie, Ana E. Vázquez, and Ebenezer N. Yamoah

Abstract

Expression of almost every gene is regulated at the transcription level. Therefore, transcriptional factor Transcription factors, consequently, have marked effects on the fate of a cell by establishing the gene expression patterns that determine biological processes. In the auditory and vestibular systems, transcription factors have been found to be responsible for development, cell growth, and apoptosis. It is vital to identify the transcription factor target genes and the mechanisms by which transcription factors control and guide gene expression and regulation pathways. Compared with earlier methods devised to study transcription factor–DNA interactions, the advantage of the chromatin immunoprecipitation (ChIP) assay is that the interaction of a transcription factor with its target genes is captured in the native context of chromatin in living cells. Therefore, ChIP base assays are powerful tools to identify the direct interaction of transcription factors and their target genes *in vivo*. More importantly, ChIP assays have been used in combination with molecular biology techniques, such as PCR and real time PCR, gene cloning, and DNA microarrays, to determine the interaction of transcription factor–DNA from a few potential individual targets to genome-wide surveys.

Key words: Transcription factor, target gene, protein-DNA interaction, inner ear, gene expression, chromatin immunoprecipitation, microarray.

1. Introduction

The ChIP assay is a very powerful technique used to identify regions of a genome associated with specific proteins including but not limited to transcription factors within their native chromatin context *(1)*. In this technique, intact cells are fixed with formaldehyde to cross-link proteins with their associated DNA. Then, the cells are lysed and chromatin is isolated from the nuclei. After fragmentation by sonication, chromatin fragments,

Bernd Sokolowski (ed.), *Auditory and Vestibular Research: Methods and Protocols, vol. 493*

DOI 10.1007/978-1-59745-523-7_19 Springerprotocols.com

containing the protein of interest and their associated DNA, are selectively precipitated using a specific antibody. Eventually, the sequence identities of the precipitated DNA fragments are determined. ChIP has been coupled with many commonly used molecular biology techniques such as PCR and real-time PCR, gene cloning, and DNA microarray to elucidate DNA-protein interactions *(2–4)*. In a standard ChIP procedure, specific DNA sequences are examined by PCR using gene-specific primers. In this case, one must first presume that a specific promoter might be bound by the protein of interest in order to be able to design promoter-specific primers and/or probes. Therfore, this protocol can only be used to identify known target genes for a given protein. Fortunately, techniques have also been developed to identify unknown DNA sequences interacting with known proteins without requiring prior knowledge of such interactions. ChIP cloning is effective for isolating individual target genes, in which the precipitated DNA fragments in a ChIP assay are cloned and subsequently sequenced. In contrast, a combination of ChIP and DNA microarray (ChIP-chip) enables the genome-wide surveys of transcription factor target genes *(5, 6)*.

2. Materials

2.1. Reagents

1. Phosphate Buffered Saline (PBS): 1.06 mM KH_2PO_4, 155.17 mM NaCl, 2.97 mM Na_2HPO_4 in ddH_2O, pH 7.4.
2. 100X Protease inhibitor cocktail (cat no. P8340, Sigma-Aldrich, St. Louis, MO,): 104 mM AEBSF [4-(2-Aminoethyl) benzenesulfonyl fluoride hydrochloride], 0.085 mM aprotinin, 1.53 mM bestatin hydrochloride, 1.40 mM E-64 [N-(trans-Epoxysuccinyl)-L-leucine 4-guanidinobutylamide], 1.90 mM leupeptin hemisulfate salt, 4.22 mM pepstatin in DMSO. Divide into 10 μL aliquots and store at −20 °C.
3. 100X Phenylmethanesulfonyl Fluoride (PMSF) solution: Make a 100 mM solution of PMSF in isopropanol. Divide into 10 μL aliquots and store at −20 °C. Add PMSF immediately before use since PMSF has a short half-life of ~30 min in aqueous solutions.
4. PBS/protease inhibitors (pH 7.4): 1 mM of PMSF and 1X protease inhibitor cocktail in PBS, pH 7.4.
5. 37% Formaldehyde solution (Sigma-Aldrich).
6. 1.25 M Glycine solution in PBS, pH 7.4.
7. Swelling buffer: 5 mM piperazine-N, N′-bis[2-ethanesulfonic acid] (PIPES), pH 8.0, 85 mM KCl, 1% (Octylphenoxy) polyethoxyethanl (IGEPAL® CA-630, Sigma-Aldrich);

before use, add 1 m*M* of PMSF and 1X protease inhibitor cocktail.

8. Nuclei lysis buffer: 50 m*M* Tris-HCl, pH 8.0, 10 m*M* EDTA, 1% SDS; before use, add 1 m*M* of PMSF and 1X protease inhibitor cocktail.
9. 5 *M* NaCl in ddH_2O.
10. 10 mg/mL Proteinase K.
11. 10 mg/mL DNase-free RNase A.
12. Qiaquick PCR Purification kit (cat. no. 28104, Qiagen, Valencia, CA,).
13. Protein A/G Plus Agarose (cat. no. sc-2003, Santa Cruz Biotechnology, Inc., Santa Cruz, CA).
14. IP dilution buffer: 0.01% SDS, 1.1% Triton X-100, 1.2 m*M* EDTA, 16.7 m*M* Tris-HCl, pH 8.1, 167 m*M* NaCl; before use, add 1 m*M* of PMSF and 1X protease inhibitor cocktail.
15. IP washing buffer A: 2 m*M* EDTA, 0.1% SDS, 1% Triton X-100, 20 m*M* Tris-HCl, pH 8.0, 150 m*M* NaCl; before use, add 1 m*M* of PMSF and 1X protease inhibitor cocktail.
16. IP washing buffer B: 2 mM EDTA, 0.1% SDS, 1% Triton X-100, 20 m*M* Tris-HCl, pH 8.0, 500 m*M* NaCl; before use, add 1 m*M* of PMSF and 1X protease inhibitor cocktail.
17. IP washing buffer C: 10 m*M* Tris-HCl, pH 8.0, 1 m*M* EDTA, 1% Igepal®, 1% sodium deoxycholate; before use, add 1 m*M* of PMSF and 1X protease inhibitor cocktail.
18. TE buffer: 10 m*M* Tris-HCl, pH 8.0, 1 m*M* EDTA, pH 8.0.
19. IP elution buffer: 50 m*M* $NaHCO_3$, 1% SDS; before use, add 1 m*M* of PMSF and 1X protease inhibitor cocktail.
20. 10 m*M* Tris-HCl buffer, pH 8.0.
21. 3 *M* sodium acetate, pH 5.2.
22. Phenol, saturated with Tris-HCl, pH 8.0.
23. Chloroform
24. 100% ethanol and 80% ethanol made in ddH_2O.
25. IgG from rabbit serum (cat no. I5006, Sigma-Aldrich).
26. RNA polymerase II 8WG16 monoclonal antibody (cat. no. MMS-126R, Covance, Princeton, NJ).
27. Liquid nitrogen.

2.2. Equipment

1. Medimachine™ (BD Biosciences, San Jose, CA) disaggregation system.
2. 50 μL Medicon (BD Biosciences) disposable polyethylene chambers.
3. Eppendof microcentrifuge 5417R (Eppendorf of North America, Westbury, NY).
4. Eppendof multipurpose centrifuge 5804 R (Eppendorf of North America).
5. Rotating wheel/platform for mixing.
6. Water bath or heat platform.

7. Spectrophotometer.
8. Thermocycler.

2.3. Other Supplies

1. 2 mL Kontes dounce tissue grinder (VWR International, West Chester, PA).
2. 15 mL polystyrene graduated tubes.
3. 18-gauge blunt needle and 1 mL syringe.
4. Cell scraper.

3. Methods

3.1. Preparation of Cross-Linked Cells

3.1.1. Inner Ear Tissue

1. Dissect and collect cochlear tissue (~100 mg) from the mouse inner ear (*see* **Note 1**).
2. Snap freeze the tissue in liquid nitrogen and store at −80 °C for chromatin preparation the next day.
3. Thaw tissue sample on ice.
4. Wash tissue once with 1 mL of ice cold PBS. Centrifuge for 1 min at 500 g. Save tissue pellet and discard supernatant.
5. Cut tissue to ~2 mm small pieces and resuspend in 1 mL of PBS/protease inhibitors (*see* **Note 2**).
6. Cross-link proteins to DNA by adding 27 μL of 37% formaldehyde to the sample and incubate for 15 min at room temperature with shaking (*see* **Note 3**).
7. Stop cross-linking by adding 115 μL of 1.25 *M* glycine solution to the reaction and incubate for 5 min at room temperature with shaking.
8. Centrifuge tissue sample at 500 g for 1 min at 4 °C and discard the supernatant.
9. Wash tissue once with 1 mL of ice cold PBS/protease inhibitors (*see* **Note 4**).
10. Resuspend tissue in 1 mL of PBS/protease inhibitors. Transfer the sample to a Medicone and grind the tissue for 2 min using a Medimachine to disaggregate the tissue. Collect cells from Medicone using an 18 gauge blunt needle and a 1 mL syringe (*see* **Note 5**).
11. Check cell suspension by microscope then centrifuge tube at 500 g for 1 min at 4 °C to collect the cells. The cell pellet can be stored at −80 °C for several months.

3.1.2. Adherent Cells

1. Grow ~3×10^7 cells in four 75 cm^2 culture flasks (*see* **Note 6**).
2. Replace medium with 20 mL of PBS and add 37% formaldehyde directly to flasks to a final concentration of 1%. Incubate cells for 15 min at room temperature with rotation (*see* **Note 3**).

3. Stop cross-linking by adding 1.25 *M* glycine to a final concentration of 0.125 *M* and incubate for 5 min with rotation.
4. Wash cells twice with 20 mL of PBS (*see* **Note 4**).
5. Scrape cells from culture flasks and transfer to a 15 mL conical tube (*see* **Note 7**).
6. Rinse cells once with 5 mL of PBS/protease inhibitors (*see* **Note 2**).
7. Collect cross-linked cells by centrifuging for 5 min at 500 g at 4 °C and discard the supernatant. Cell pellet can be stored at −80 °C for several months.

3.1.3. Suspension Cells

1. Grow ~3×10^7 cells and collect cells by centrifugation for 1 min at 500 g. Resuspend cells in 20 mL of PBS in a 50 mL tube.
2. Add 37% formaldehyde directly to cell suspension to a final concentration of 1%. Incubate cells for 15 min at room temperature with rotation (*see* **Note 3**).
3. Stop cross-linking by adding 1.25 *M* glycine to the reaction to a final concentration of 0.125 *M* and incubate for 5 min at room temperature with rotation.
4. Wash cells twice with 20 mL of PBS (*see* **Note 4**).
5. Resuspend cells in 5 mL of PBS/protease inhibitors and transfer cells to a 15 mL conical tube (*see* **Note 2**).
6. Collect cross-linked cells by centrifuging for 5 min at 500 g at 4 °C and discard supernatant. Cell pellet can be stored at −80 °C for several months.

3.2. Preparation of Sonicated Chromatin

3.2.1. Chromatin Isolation and Sonication

1. Resuspend cell pellet obtained previously from one of the tissue types (*see* **Sections 3.1.1, 3.1.2, or 3.1.3**) in 1 mL of swelling buffer in a 1.5 mL tube. Incubate on ice for 15 min swirling occasionally to resuspend cells.
2. Dounce cells using a 2 mL B dounce to release nuclei and then centrifuge at 2500 g for 5 min at 4 °C to pellet nuclei (*see* **Note 8**).
3. Discard supernatant and resuspend nuclei in 1 mL of nuclei lysis buffer. Incubate on ice for 15 min.
4. Transfer the sample to a 15 mL polystyrene conical tube and sonicate cross-linked chromatin to an average size of 200–1000 bp (*see* **Note 9**).
5. Transfer the sample to a 1.5 mL centrifuge tube and centrifuge at the maximal speed of a microfuge for 15 min at 4 °C.
6. Transfer supernatant to a new 1.5 mL tube and keep it on ice. This supernatant is cross-linked sheared chromatin ready for the ChIP experiment (*see* **Note 10**). However, before performing ChIP, the size and the concentration of sonicated DNA must be checked (*see* **Note 9**). Use 10 μL of this sample to check the quality of the DNA, following the protocol provided in **Section 3.2.2**.

3.2.2. Quality Check of Sonicated DNA

1. Reverse cross-linking: Mix 10 μL of chromatin sample obtained from **step 6 of Section 3.2.1** with 40 μL of IP elution buffer. Add 2 μL of 5 *M* NaCl and boil for 15 min. Cool to room temperature (*see* **Note 11**).
2. Add 1 μL of 10 mg/mL DNase-free RNase A and incubate for 5 min at room temperature.
3. Add 1 μL of 10 mg/mL Proteinase K and incubate at 55 °C for 30 min.
4. Purify sonicated DNA fragments using a Qiaquick PCR Purification kit following the manufacturer's instructions and elute DNA in 50 μL of ddH_2O.
5. Determine concentration of DNA sample using a spectrophotometer.
6. Load ~2–3 μg of purified sheared DNA on a 1% agarose gel. The sonicated DNA fragments should be a smear between ~200–1000 bp.

3.3. Preclear and Immunoprecipitation (IP)

1. Add 20 μL of Protein A/G Plus-agarose beads into each sonicated chromatin sample (~1 mL) and rotate for 2 h at 4 °C.
2. Spin at 3000 g for 5 min at 4 °C and transfer supernatant into a fresh tube. This supernatant is precleared soluble chromatin.
3. Divide the sample into 300 μL aliquots (each aliquot is equivalent to 30 mg of starting tissue and is good for one ChIP experiment) and save 30 μL of chromatin for the preparation of a DNA sample that is 10% of the total input (*see* **Note 12**).
4. Add 10 μg of primary antibody to each 300 μL aliquot and rotate for 2 h in a cold room (*see* **Notes 13** and **14**).
5. Add 40 μL of Protein A/G Plus-agarose beads per IP sample and incubate overnight with constant rotation at 4 °C.
6. Centrifuge at 3000 g for 5 min at 4 °C and discard the supernatant.
7. Wash beads twice with 1 mL of IP washing buffer A (*see* **Note 15**).
8. Wash beads twice with 1 mL of IP washing buffer B.
9. Wash beads twice with 1 mL of IP washing buffer C.
10. Wash beads once with 1 mL of TE buffer.
11. Elute antibody/protein/DNA complexes from protein A/G agarose beads. Add 100 μL of IP elution buffer into each sample, resuspend beads, and incubate for 30 min at room temperature with rotation. Spin down beads and carefully transfer supernatant to fresh tubes. Repeat elution once and combine the two eluates for each sample.
12. Centrifuge samples at the maximal speed of a microfuge for 5 min at 4 °C to remove any remaining beads and transfer supernatant to fresh tubes.

3.4. Cross-Linking Reversal and DNA Purification

1. Add 4 μL of 5 *M* NaCl to the chromatin solution (200 μL) to a final concentration of 0.2 *M*. In parallel, thaw 10% of the total input sample obtained from **step 3 of Section 3. 3** (30 μL) and supplement with 170 μL of IP elution buffer. Add 4 μL of 5 *M* NaCl to the input sample as well. Heat samples 65 °C from 4 h to overnight to reverse the cross-linking (*see* **Note 11**) and then cool to room temperature.
2. Add 1 μL of 10 mg/mL DNase-free RNase A and incubate for 30 min at 37 °C.
3. Add 2 μL of 10 mg/mL Proteinase K and incubate for 2 h at 55 °C and then cool samples to room temperature (*see* **Note 16**).
4. Add 200 μL of phenol to each sample and shake thoroughly by hand. Spin at the maximal speed of a microfuge for 5 min. Transfer the upper aqueous layer to fresh tubes.
5. Add 200 μL of chloroform and shake thoroughly by hand. Spin at the maximal speed of a microfuge for 5 min. Transfer the upper aqueous layer to fresh tubes. Repeat this step once and then transfer the upper aqueous layer to fresh tubes. Combine the two samples.
6. Add 20 μL of 3 *M* sodium acetate (pH 5.2) and 400 μL of absolute ethanol, and mix. Incubate samples for 15 min at −80 °C.
7. Spin down at the maximal speed of a microfuge for 15 min at 4 °C. Discard the supernatant.
8. Wash pellet with 500 μL of 80% ethanol and spin down at the maximal speed of a microfuge for 2 min at 4 °C. Discard the supernatant. Repeat this step once.
9. Vacuum dry the pellet and dissolve each sample in 100 μL of 10 m*M* Tris-Cl buffer (pH 8.0).
10. Determine the concentration of the DNA sample using a spectrophotometer.

3.5. Post-Immunoprecipitation Analysis

Some commonly used molecular biology techniques such as PCR and real-time PCR, gene cloning, and DNA microarrays have been used to analyze the precipitated DNA fragments obtained in the ChIP assay. The detailed protocols for these analyses are out of the realm of this chapter. Here, we will only discuss the general strategies of using these methods in post-ChIP DNA analysis.

3.5.1. PCR Amplification

Real-time PCR is the preferred method for DNA quantification over the Southern slot blot, since ChIP usually yields very small amounts of DNA. The association of a specific DNA sequence with a certain transcription factor is determined by the level of the DNA sequence enrichment in the precipitated DNA compared to the non-precipitated control sample. To get meaningful data, the following control reactions should be included:

1. For PCR effectiveness, use the total input DNA (**step 3 of Section 3. 3)** after the cross-links have been reversed, at various dilutions (1%, 2%, etc.).
2. The mock ChIP reaction, which is performed the same way as the experimental ChIP reaction, except in the absence of primary antibody.
3. Most importantly, an internal control specific to a chromatin region that is not enriched in the ChIP reaction has to be included to verify that the observed enrichment of DNA sequences is genuine.

3.5.2. ChIP Cloning

In this technique, the precipitated DNA fragments in a ChIP assay are first blunted using T4 DNA polymerase and then cloned into a vector. The identities of the cloned DNA fragments are determined by subsequent sequencing and DNA sequence data base search. The main advantage of this technique is that it allows the isolation of individual target genes, regardless of whether the target is known or unknown. However, some drawbacks of this technique should be noted. First, nonspecific sequences have an equal chance of being selected in ChIP cloning. To decrease the amount of nonspecific DNA, the precipitated chromatin sample may be re-immunoprecipitated, which usually results in a lower amount of precipitated DNA. PCR amplification has been used to overcome this problem, but it can cause experimental bias as well. It is recommended to pool DNA samples from multiple immunoprecipitation reactions to achieve a better resolution during the analysis of the protein-DNA interaction in ChIP cloning. Second, the size of DNA fragments is critical for ChIP cloning. To reduce the number of non-specific clones, the chromatin should be sonicated to ~1–2 kb fragments and only clones containing DNA fragments of 500 bp or larger should be analyzed *(7)*. This technique can be used to detect a small number of transcription factor target genes.

3.5.3. ChIP-Chip Assay

In this method, the DNA samples are fluorescent-labeled after ChIP and then analyzed on DNA microarray chips. Two color array platforms are commonly used for this purpose. The enriched DNA is usually labeled with a fluorescent molecule, such as Cy5 or Alexa 647, and the input genomic DNA is labeled with a different fluorescent dye, such as Cy3 or Alexa 555. The two probes are then combined and hybridized to a single DNA microarray. The enrichment levels of DNA fragments can be determined while using the input genomic DNA sample as a reference. Ideally, to achieve a comprehensive and unbiased survey of protein-DNA interactions, the DNA microarrays used in ChIP-chip assays should contain elements that represent the entire genome. This method was used successfully in yeast systems to create high-resolution genome-wide maps of DNA-protein interactions *in vivo (8, 9)*. However, given the high complexity of mammalian genomes, it is technically challenging to design suitable microarrays. Alternatively, CpG island microarrays have been used to investigate genome-wide transcription-DNA *(10, 11)*.

4. Notes

1. We typically collect tissues the day before chromatin purification and store the tissue in a −80 °C freezer after it is snap frozen in liquid nitrogen. We typically prepare chromatin from 100 mg of cochlear tissues (~20 mice) and use the chromatin equivalent of 30 mg starting tissue per antibody in every ChIP. The exact amount of tissue has to be determined empirically, depending upon the abundance of a specific transcription factor in the tissue, the quality of the antibody used, and the cross-linking conditions.
2. PMSF has a short half-life in aqueous solutions. A stock solution of 100 mM in isopropanol should be made and stored at −20 °C and diluted in the buffer immediately before use.
3. In general, cross-linking of DNA and proteins is required for non-histone proteins, such as transcription factors, since they have a weaker association with DNA compared with histones. The aim of cross-linking is to stabilize the association of proteins with DNA at the binding site of chromatin. One percent formaldehyde is commonly used for cross-linking in cells/tissues and 15 min of incubation at room temperature works the best in our experiments. Optimal conditions for formaldehyde cross-linking can be determined using a time-course experiment. Excess cross-linking (>30 min) may lead to difficulties in the chromatin fragmenting to the desired size and to a reduction in antigen availability, thereby producing unsatisfactory results in the ChIP assay *(7)*.
4. Wash tissue or cells thoroughly and try to remove as much solution containing formaldehyde as possible.
5. We use 50 μL Medicones and a Medimachine from Becton Dickinson to obtain a single cell suspension.
6. Use $\sim 1 \times 10^7$ cells per antibody per ChIP. Fewer cells can result in a lower signal-to-noise ratio. Gene expression should be properly induced and cell cultures should be healthy at the time of cross-linking.
7. A brief incubation with trypsin may help in the release of nuclei from cells that do not swell well. It is important to wash trypsinized cells thoroughly with ice cold PBS before continuing with the rest of the protocol.
8. The quality of the nuclei preparation is improved by including an extra step of douncing the cells in swelling solution, which

reduces contamination from cytoplasm. The background in ChIP experiments significantly decreases when using chromatin purified from these nuclei.

9. It is critical to shear chromatin to ~500 bp fragments to pinpoint the DNA binding site of the protein of interest. In general, the larger the fragment size, the more difficult it is to map the exact location of a protein binding site. Optimal conditions for sonication need to be determined empirically to ensure that the majority of chromatin fragments are between ~200–1000 bp. Factors affecting sonication include the volume of the sample, depth of the sonication probe, sonication strength, the duration of sonication, and the condition of cross-linking. One should try to perform sonications under the same conditions to achieve consistent results. We use twelve pulses at 20 s each with a 60 s rest interval at the maximum power of a Bioruptor™ (Wolf Laboratories Limited, Pocklington, York, UK) sonicator. In addition, it is very important to keep samples cold during the entire process.
10. Fresh chromatin usually gives the best ChIP results, even though sonicated chromatin can be stored at −80 °C for a short period of time. In this protocol, sheared chromatin is prepared in nuclear lysis buffer containing 1% SDS, which can be stored at 4 °C for 24–48 h before use without apparent loss of ChIP efficiency.
11. To achieve the best result of reversing cross-linking for ChIP, incubate the sample in 0.2 *M* NaCl from 4 h to overnight at 65 °C. Incomplete reversal of cross-links will result in loss of material. To check the sonication efficiency, however, a short protocol of boiling for 15 min is sufficient.
12. This input sample is equivalent to 10% of the starting genomic DNA for each ChIP assay. It is very useful during post-IP DNA analysis as a control for PCR effectiveness and quantitation of different samples.
13. Antibodies are required in ChIP experiments to capture chromatin regions that contain the protein of interest and their associated DNA. The quality of antibodies is crucial for the recovery of these DNA fragments. The best antibodies are those with high specificity and high affinity, recognizing the epitope after protein-DNA cross-linking. In general, polyclonal antibodies are preferred over monoclonal antibodies to overcome epitope masking. The optimal amount of a specific antibody for the maximum binding of the protein of interest can be determined by performing pilot experiments using increasing amounts of the antibody combined with the cell lysate. However, 10 μg of primary antibody for chromatin obtained from ~30 mg of tissue or ~10^7 cells is a good starting concentration for most antibodies.

14. It is crucial to include proper controls. This protocol is designed to prepare enough sheared chromatin sufficient for three ChIP reactions. Typically, we run an experimental ChIP reaction in parallel with a negative control and a positive control. In the negative control reaction, normal rabbit IgG is used as the primary antibody to eliminate any false positive that may be derived from the system. In the positive control, an RNA polymerase II antibody is used to ensure that each step of the ChIP experiment is working.
15. Serial washings with low-salt and high-salt buffers, lithium chloride buffer, and Tris-EDTA buffer are essential to remove non-specific binding. For each wash, resuspend beads in 200 μL of ice cold buffer by gently tapping the tube and then adding the remaining 800 μL of solution. Rotate for 10 min and collect the beads by centrifuging at 1000 g for 3 min at 4 °C. Orient the cap hinge of tubes on the outside of the rotor to mark the location of the pellet and then pipette out the wash solution carefully.
16. Insufficient proteinase K treatment may cause poor results in the DNA analysis experiments, especially in a ChIP-chip assay.

References

1. Sikder, D. and Kodadek, T. (2005) Genomic studies of transcription factor–DNA interactions. *Curr. Opin. Chem. Biol.* **9**, 38–45.
2. Elnitski, L., Jin, V. X., Farnham, P. J., and Jones S. J. M. (2006) Locating mammalian transcription factor binding sites: A survey of computational and experimental techniques *Genome Res.* **16**, 1455–1464.
3. Weinmann, A. S., Bartley, S. M., Zhang, T., Zhang, M. Q., and Farnham P. J. (2001) Use of chromatin immunoprecipitation to clone novel E2F target promoters. *Mol. Cell Biol.* **21**, 820–6832.
4. Weinmann, A. S. and Farnham, P. J. (2002) Identification of unknown target genes of human transcription factors using chromatin immunoprecipitation. *Methods* **26**, 37–47.
5. Odom, D. T., Zizlsperger, N., Gordon, D. B., Bell, G. W., Rinaldi, N. J., Murray, H. L., Volkert, T. L., Schreiber, J., Rolfe, P. A., Gifford, D. K. (2004) Control of pancreas and liver gene expression by HNF transcription factors. *Science* **303**, 1378–1381.
6. Buck, M. J. and Lieb J. D. (2004) ChIP-chip: considerations for the design, analysis, and application of genome-wide chromatin immunoprecipitation experiments. *Genomics* **83**, 347–348.
7. Das, P. M., Ramachandran, K., van Wert, J., Singal, R. (2004) Chromatin immunopricipitation assay. *Biotechnology* **37**, 961–969.
8. Lee, T. I., Rinaldi, N. J., Robert, F., Odom, D. T., Bar-Joseph, Z., Gerber, G. K., Hannett N. M., Harbison, C. T., Thompson, C. M., and Simon, I. (2002) Transcriptional regulatory networks in *Saccharomyces cerevisiae. Science* **298**, 799–804.
9. Ren, B., Robert, F., Wyrick, J. J., Aparicio, O., Jennings, E. G., Simon, I., Zeitlinger, J., Schreiber, J., Hannett, N., and Kanin, E. (2000) Genome-wide location and function of DNA-binding proteins. *Science* **290**, 2306–2309.
10. Weinmann, A. S., Yan, P. S., Oberley, M. J., Huang, T. H., and Farnham, P. J. (2002) Isolating human transcription factor targets by coupling chromatin. immunoprecipitation and CpG island microarray analysis. *Genes Dev.* **16**, 235–244.
11. Oberley, M. J., Tsao, J., Yau, P., and Farnham, P. J. (2004) High-throughput screening of chromatin immunoprecipitates using CpG-island microarrays. *Meth. Enzymol.* **376**, 315–333.

Chapter 20

Surface Plasmon Resonance (SPR) Analysis of Binding Interactions of Proteins in Inner-Ear Sensory Epithelia

Dennis G. Drescher, Neeliyath A. Ramakrishnan, Marian J. Drescher

Abstract

Surface plasmon resonance is an optical technique utilized for detecting molecular interactions. Binding of a mobile molecule (analyte) to a molecule immobilized on a thin metal film (ligand) changes the refractive index of the film. The angle of extinction of light, reflected after polarized light impinges upon the film, is altered, monitored as a change in detector position for the dip in reflected intensity (the surface plasmon resonance phenomenon). Because the method strictly detects mass, there is no need to label the interacting components, thus eliminating possible changes of their molecular properties. We have utilized surface plasmon resonance to study the interaction of proteins of hair cells.

Key words: Surface plasmon resonance (SPR), hair cell, molecular cloning, protein-protein interaction, fusion polypeptides, interaction kinetics, sensor chip CM5, Biacore 3000.

1. Introduction

Surface plasmon resonance (SPR) binding analysis methodology is used to study molecular interactions *(1, 2)*. SPR is an optical technique for detecting the interaction of two different molecules in which one is mobile and one is fixed on a thin gold film *(1)*. In the work described here, affinity-purified fusion polypeptides are immobilized by an amine-coupling reaction on a sensor chip (Biacore, Piscataway, NJ, USA) inserted into the flow chamber of a Biacore 3000 instrument (Biacore, Uppsala, Sweden). Addition of a second polypeptide, the flow-through *analyte*, to the chamber, results in binding to the immobilized polypeptide *ligand*, producing a small change in refractive index at the gold surface *(3)*, which can be quantified with precision *(4)*. Binding

Bernd Sokolowski (ed.), *Auditory and Vestibular Research: Methods and Protocols, vol. 493*

DOI 10.1007/978-1-59745-523-7_20 Springerprotocols.com

affinities can be obtained from the ratio of rate constants, yielding a straightforward characterization of protein-protein interaction. SPR directly detects mass (concentration) with no need for special radioactive or fluorescent labeling of polypeptides *(5)* before measurement, presenting a great advantage in minimizing time and complexity of the studies.

2. Materials

2.1. Selection of Sensor Chip

1. A CM5 chip, research grade (cat. no. BR-1000-14, Biacore-GE Healthcare, Piscataway, NJ), is commonly used (*see* **Note 1**).

2.2. Production of Ligands and Analytes

1. All reagents are prepared with Milli-Q water. The protein/polypeptide, used as immobilized ligand or mobile analyte, is purified either directly from a tissue source or by recombinant technology with overexpression in bacterial cells (*see* **Section 3.2.4**). Ligand or analyte can be a fusion protein tagged with hexahistidine or glutathione-S-transferase (GST) (as an aid to purification), or can be a synthetic oligopeptide, 20–30 amino acids in length.
2. Mouse brain QUICK-clone cDNA (cat. no. 637301, BD Biosciences-Clontech, Palo Alto, CA, USA) or custom cDNA.
3. Expression vectors pRSET (Invitrogen, Carlsbad, CA, USA) or pGEX (GE Healthcare-Amersham, Piscataway, NJ, USA), for producing fusion proteins/polypeptides by in-frame cloning.
4. Restriction enzymes *BamH1, EcoR1, Hind3, Bgl2, Not1* (Invitrogen).
5. QIAEX-II gel extraction kit (cat. no. 20021, Qiagen, Valencia, CA, USA).
6. *E. coli* Subcloning Efficiency™ DH5α™ cells (cat. no. 18265-017, Invitrogen).
7. One Shot Top 10 cells (cat. no. C4040-03, Invitrogen) for cloning and plasmid isolation.
8. *E. coli* BL 21 Star™ (DE3) cells (cat. no. C6010-03, Invitrogen) for protein expression.
9. LB liquid medium: 1% bacto-tryptone, 0.5% yeast extract, 1% NaCl, adjusted to pH 7.0 with 5 *N* NaOH (cat. no. 244620, BD Biosciences-Difco, San Jose, CA).
10. LB solid medium: 12 g/L agar in LB liquid medium. Autoclave.
11. S.O.C medium: 2% tryptone, 0.5% yeast extract, 10 m*M* NaCl, 2.5 m*M* KCl, 10 m*M* $MgCl_2$, 10 m*M* $MgSO_4$, 20 m*M* glucose (cat. no. 15544-034, Invitrogen).

12. Wizard plus SV Miniprep Kit (cat. no. A1330, Promega, Madison, WI).
13. IPTG (isopropylthio-β-D-galactoside) stock solution: 100 m*M* in water. Sterilize by filtration and store at −20 °C.
14. Ampicillin stock solution: 100 mg/mL in water. Sterilize by filtration and store at −20 °C.
15. Protease Inhibitor Cocktail for use in polyhistidine tagged protein (cat. no. P8849, Sigma-Aldrich, St. Louis, MO).
16. Protease Inhibitor Cocktail for general use (cat. no. P2714, Sigma-Aldrich).
17. Pre-cast 4–12% gradient SDS-PAGE (sodium dodecyl sulfate-polyacrylamide gel electrophoresis) gels (NuPAGE, cat. no. NP0321BOX, Invitrogen).
18. Electrophoresis running buffer (20X NuPAGE SDS Running Buffer, cat. no. NP0001, Invitrogen). Dilute buffer appropriately.
19. Coomassie Blue electrophoresis stain (Simply Blue™ Safe Stain, cat. no. LC6060, Invitrogen).
20. BenchMark™ pre-stained protein ladder (cat. no. 10748-010, Invitrogen).
21. Cell lysis buffer for hexahistidine fusion polypeptide purification: 8 *M* urea in 100 m*M* NaH_2PO_4, 10 m*M* Tris-HCl, pH 8. The wash buffer and elution buffers are the same as the cell lysis buffer, but with the pH adjusted to 6.3 and 4.5, respectively.
22. Nickel-nitrilotriacetic acid (Ni-NTA) spin columns for purification of hexahistidine-tagged proteins (Qiagen).
23. Sonication lysis buffer for protein purification: 1X PBS, 100 m*M* EDTA, 0.5 m*M* DTT, 1X Protease Inhibitor Cocktail for general use, pH 7.4 (Sigma-Aldrich).
24. Glutathione-Sepharose 4B beads for GST fusion tags (cat. no. 17-0756-01, GE-Amersham).
25. L-Glutathione, reduced (cat. no. G4251, Sigma-Aldrich).
26. DispoDialyzer (cat. no. Z368296-10EA, Sigma-Aldrich) or Slide-A-Lyzer Dialysis Cassettes (cat. no. 66107, Pierce, Rockford, IL).
27. Anti-X-Press HRP antibody (cat. no. R911-25, Invitrogen).
28. Anti-GST antibody (cat. no. G7781, Sigma-Aldrich).
29. Coomassie Plus-200 Protein Assay Reagent (cat. no. 23238, Pierce).
30. Quant-iT™ Protein Assay Kit for use with Qubit fluorometer (Invitrogen).
31. Phosphate-buffered saline (PBS): 0.01 *M* PO_4, 0.138 *M* NaCl, 0.0027 *M* KCl, pH 7.4 (cat. no. P3813, Sigma-Aldrich).

2.3. Ligand Immobilization

1. HEPES buffered saline-NaCl (HBS-N buffer): 0.15 *M* NaCl, 0.01 *M* HEPES, pH 7.4 (cat. no. BR-1003-69, Biacore).
2. HBS-P buffer: HBS-N with 0.005% v/v surfactant P20 (cat. no. BR- 1003-68, Biacore).
3. HBS-EP buffer: HBS-N with 3 m*M* EDTA, 0.005% v/v P20 (cat. no. BR-1001-88, Biacore).
4. EDC amine coupling reagent: 0.4 *M* 1-ethyl-3-(3-di-methylamino-propyl) carbodiimide hydrochloride in water (part of cat. no. BR-1000-50, Biacore).
5. Ethanolamine solution: 1.0 *M* ethanolamine-HCl, pH 8.5 (part of cat. no. BR-1000-50, Biacore).
6. NHS amine coupling reagent: 0.1 *M* N-hydroxysuccinimide in water (part of cat. no. BR-1000-50, Biacore).
7. Sodium acetate buffers for immobilization in the acidic range: 10 m*M* sodium acetate, pH 4.0, 4.5, 5.0, 5.5 (cat. no. BR-1003 -49 through -52, Biacore).
8. Sodium tetraborate buffer for immobilization in the basic range: 10 m*M* sodium tetraborate, pH 8.5 (cat. no. BR-1003-53, Biacore).
9. 1.0 *M* NaCl solution.
10. Glycine-HCl solutions: 10 m*M*, pH 1.5, 2.0, 2.5, 3.0. Use for regeneration (cat. no. BR- 1003- 54 through -57, Biacore).
11. 50 m*M* NaOH solution. Use for regeneration (cat. no. BR-1003-58, Biacore).
12. BIAdesorb Solution 1: 0.5% SDS (cat. no. BR-1008-23, Biacore).
13. BIAdesorb Solution 2: 50 m*M* glycine, adjusted to pH 9.5 with 5 *N* NaOH (cat. no. BR-1008-23, Biacore).

2.4. Experimental Binding Measurements

1. HBS-N buffer.
2. HBS-P buffer.
3. HBS-EP buffer.
4. Phosphate -buffered saline with Tween (PBST): 10 m*M* PO_4, pH 7.5, 2.4 m*M* KCl, 138 m*M* NaCl, 0.05% Tween-20.
5. Nonspecific binding reducer (NSB): 10 mg/mL of carboxymethyl dextran sodium salt in 0.15 *M* NaCl, containing 0.02% sodium azide (cat. no. BR- 1006-91, Biacore).
6. Bovine serum albumin (BSA): 2 mg/mL (Pierce).

3. Methods

We perform surface plasmon resonance experiments with a Biacore 3000 instrument. The immobilization involves activation of carboxymethyl groups on a dextran-coated chip by reaction with N-hydroxysuccinimide, followed by covalent bonding of the ligands to the chip surface via amide linkages and blockage of excess activated carboxyls with ethanolamine (6). Reference surfaces are prepared in the same manner, except that all carboxyls are blocked and no ligand is added. During analysis, each cell with an immobilized fusion polypeptide is paired with an adjacent cell on the chip, the latter serving as a reference. The final concentration of bound ligand, expressed in response units (RU), is calculated by subtracting the reference RU from the ligand RU. HBS-N, HBS-P, HBS-EP, or PBST buffer may be used as both running and analyte-binding buffer. Purified fusion polypeptide or protein (analyte), typically at 100 nM, is allowed to flow over the immobilized-ligand surface and the binding response of analyte to ligand is recorded. The chip surface is regenerated by removal of analyte with a regeneration buffer. The maximum RU with each analyte indicates the level of interaction, and reflects comparative binding affinity.

3.1. Selection of Sensor Chip

1. Store the CM5 sensor chip at 4 °C. Bring the chip to room temperature just before the experiment.
2. At the analysis step, dock chip in the instrument. Prime with immobilization running buffer (HBS-N or HBS-EP). The priming process flushes the pumps, the integrated fluidic cartridge, autosampler, and flow cells of the sensor chip with the chosen running buffer. In the *Biacore 3000 Control* program, click the TOOLS tab and then select WORKING TOOLS. Place the running buffer in the buffer shelf, select PRIME, and click START

3.2. Production of Ligands and Analytes

3.2.1. Design of PCR Primers for Production of Insert

1. Design primers so that the target insert will remain in frame when ligated (by adding additional nucleotides to the primer sequence, if necessary).
2. Use a different restriction site for each member of the primer pair to support directional cloning.
3. Amplify desired cDNA message, e.g., from mouse brain cDNA, with the primer pair.
4. Purify the PCR product containing the desired restriction sites by electrophoresis in agarose gels, followed by band excision and sequencing (*see* **Note 2**).

3.2.2. Ligation of Insert into Plasmid Expression Vector

1. Digest the PCR product at 37 °C overnight with appropriate restriction enzymes corresponding to the restriction sites of the PCR product. If the buffer supplied for both restriction enzymes is the same, the two enzymes can be incubated simultaneously with the PCR product. Otherwise, perform the two incubations, as well as **step 2, Section 3.2.2**, for each enzyme consecutively.

2. Extract the digested product (insert) with phenol-chloroform and precipitate it in ethanol (*see* **Note 2**).
3. Digest, extract, and precipitate the expression vector (e.g., pRSET) in a manner similar to that used for the PCR product (*see* **Note 3**).
4. Measure the concentration of the vector and insert by gel analysis using a molecular mass standard of known quantity, by spectrophotometry at 260 nm, or by fluorescence with a Qubit fluorometer (*see* **Note 4**).
5. Perform a standard ligation reaction using T4 DNA ligase (Invitrogen Kit 15224-017). Use a 3:1 insert:vector ratio (*see* **Note 5**).

3.2.3. Transformation of Bacteria by Plasmid Expression Vector

1. Mix 1–2 μL of the vector ligation product with *E. coli* DH5α competent bacterial cells and incubate on ice for 30 min *(2)*.
2. Heat-shock the cell mixture at 42 °C for 45 s.
3. Add 1 mL of S.O.C. medium and mix vigorously in an orbital shaker at 37 °C for 90 min.
4. Spread a 50- to 200-μL aliquot onto an LB agar plate (1.2% agar in LB liquid medium), containing a 75 μg/mL final concentration of ampicillin, and incubate at 37 °C overnight.
5. Select 10 or more colonies from the plate and culture them in 3–10 mL of liquid LB medium containing 75 μg/mL of ampicillin overnight.
6. Pellet the cells by centrifugation at 4,000 g for 5 min.
7. Isolate plasmids with the Wizard plus SV Miniprep Kit.
8. Verify presence of insert by digestion with the two restriction enzymes that were used for cloning (*see* **Note 6**).

3.2.4. Protein Expression from Plasmid Expression Vector

1. Transform *E. coli* BL 21 cells (Invitrogen) with the expression vector construct as in **steps 1–3, Section 3.2.3**.
2. Spread cells onto LB agar plates, containing 75 μg/mL of ampicillin, and allow them to grow overnight.
3. Culture resulting colonies in LB liquid medium, containing 75 μg/mL of ampicillin for 8 h, or until the absorbance at 600 nm reaches an optical density (OD) reading of 0.3.
4. Add IPTG stock solution to yield a final concentration of 1 m*M* in autoclaved LB liquid medium.
5. Culture cells for another 3–4 h in the LB liquid medium.
6. Pellet cells at 4,000 g, add lysis buffer, vortex, and incubate at room temperature for 1 h.
7. Centrifuge at 18,000 g for 25 min and collect the clear supernatant.
8. Electrophoretically analyze 5 μL of the supernatant with a molecular mass standard in a pre-cast SDS-PAGE gel. Stain with Coomassie Blue.
9. Repeat **steps 1–8, Section 3.2.4**, using the expression vector not containing an insert, as a negative control.

10. As a further control, repeat **steps 1–8, Section 3.2.4**, with the omission of the inducer IPTG (*see* **Note 7**).

3.2.5. Purification of Protein (Ligand or Analyte) (see Note 8)

3.2.5.1. Ni-NTA Affinity Purification

1. This purification step is preceded by scaling up the culture (if required), following **steps 1–7, Section 3.2.4**. All procedures are performed at room temperature.
2. Equilibrate a Ni-NTA spin column with 600 μL of cell lysis buffer and centrifuge for 2 min at 700 g at room temperature.
3. Load up to 600 μL of the cleared lysate supernatant from **step 7, Section 3.2.4** onto the Ni-NTA spin column. Centrifuge at 700 g for 2 min and collect the flow-through. Add another 600 μL, if required, and centrifuge again.
4. Wash the column with 600 μL of wash buffer, centrifuge for 2 min at 700 g. Repeat the wash step 3–5 times.
5. Elute the bound fusion polypeptide by adding 200 μL of elution buffer to the spin column and centrifuge at 700 g for 2 min. Repeat the elution two more times.
6. Analyze the protein sample for purity as in **step 8, Section 3.2.4**, on an SDS-PAGE gel, and stain with Coomassie Blue (*see* **Note 9**).

3.2.5.2. Purification by Glutathione-Agarose Beads

1. This purification step is preceded by verification of overexpression and scaling up of the culture (to 50 mL or more), following **steps 1–5, Section 3.2.4** (*see* **Note 10**).
2. Pellet the cells at 4,000 g, wash the pellet by resuspending in sonication lysis buffer, and pellet again. Repeat this wash procedure three times.
3. For subsequent steps, keep the cells on ice.
4. Sonicate cells on ice for 1 min three times, with 30 s pauses between exposures. Add Triton X-100 or Tween-20, to yield a final concentration of 1%, and mix well.
5. Centrifuge at 18,000 g for 25 min at 4 °C and collect the supernatant.
6. Prepare the GST-sepharose 4B beads by washing four times in PBST buffer and preparing a 50% slurry in PBST (*see* also the GE-Amersham manual).
7. Add 100 μL of GST-sepharose slurry for each 1 mL of the supernatant.
8. Incubate at 4 °C overnight with gentle mixing on a rotator. Spin the mix at 600 g, remove the supernatant, and wash the beads 3–5 times with 1X PBST.
9. Elute the bound fusion protein, by mixing the beads with 10 m*M* reduced glutathione, centrifuge at 600 g, and collect the supernatant. Repeat three times and pool the eluted protein.
10. Check the purity of the protein, with SDS-PAGE gels, for the aliquots of lysate, sample flow-through fluid, and buffer washes.

3.2.6. Assay of Purity of Protein Samples (Ligands and Analytes) Used for Binding Studies

1. Dialyze the purified protein extensively (at least 16 h, with 6–7 changes of buffer) at 4 °C, against the running buffer to be used in the SPR binding experiment (*see* **Notes 11 , 12**).
2. To check purity, run protein on SDS-PAGE and stain with Coomassie Blue.
3. Measure spectrophotometric absorbance of the sample at 595 nm (for Coomassie Blue stain) and calculate the concentration from a standard Coomassie Blue absorbance curve constructed from proteins at known concentrations.
4. One may also determine protein concentration by fluorescence methods, such as with the Qubit fluorometer, which is sensitive enough to detect sub-nanogram concentrations of protein (*see* **Note 4**).
5. Store purified protein solutions at −20 °C in separate aliquots to avoid repeated freeze-thaw cycles that may inactivate proteins.
6. Prior to SPR studies (**Section 3.4**), thaw the protein on ice and centrifuge at 4 °C for 30 min at 18,000 g to pellet any precipitated protein.
7. Carefully collect the supernatant.
8. De-gas the sample solution before use (*see* **Note 13**).

3.2.7. Preparation of Cell Lysate Containing Analyte for Binding Studies

1. Pellet bacterial cells that express the analyte protein by centrifuging at 4,000 g.
2. Instead of placing cells in lysis buffer (as in **step 6, Section 3.2.4**), sonicate the cells as described in **step 4, Section 3.2.5.2**. Centrifuge samples at 18,000 g for 25 min at 4 °C.
3. Collect the clear supernatant. As prepared here, non-purified analyte can sometimes be used if the ligand partner is pure, as determined by detection of the ligand as a single band by polyacrylamide gel electrophoresis (*see* **Note 11**).
4. Estimate total protein in the filtered sample by the Commassie Blue spectrophotometric method or the fluorescent method, as described previously.
5. Dilute sample in SPR running buffer (HBS-N or HBS-EP) to the desired concentration, starting with 10 μg/mL of total protein. Typical analyte concentrations are 10–250 μg/mL, depending on the analyte affinity.

3.3. Ligand Immobilization

3.3.1. Preparation for Pre-Concentration: Determining the Ligand Concentration for Optimal Immobilization

1. Utilize 10 m*M* buffer solutions at three to four pH values, e.g., pH 4.0, 5.5 and 6.0, for "pH scouting" (**Fig. 20.1**). Sodium acetate is used at lower pH values, whereas sodium tetraborate is used at pH values greater than 6.0. The pH range is chosen to be below the estimated pI of the fusion protein/polypeptide (*see* **Note 14**).
2. Dilute protein sample in the 10 m*M* buffer solutions prepared at several pH values. The concentration of ligand should be in the range of 2–200 μg/mL (*see* **Note 15**).

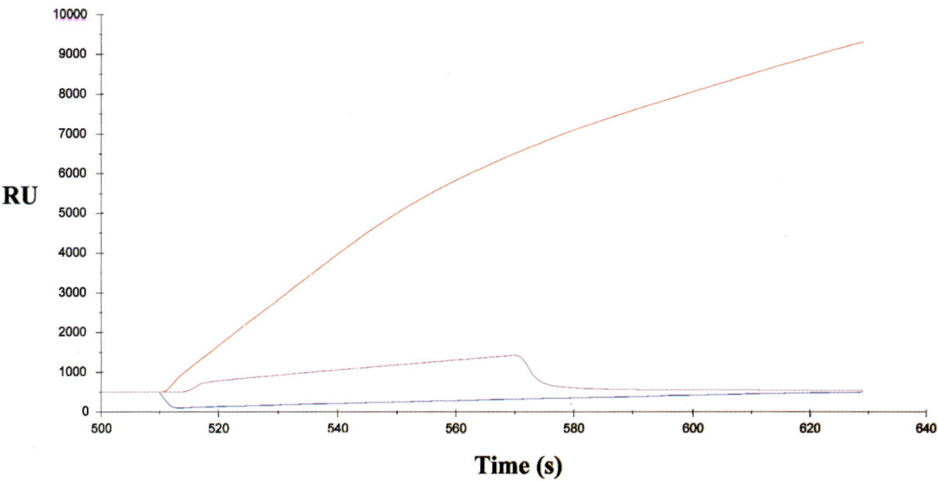

Fig. 20.1. The "pH scouting" of mouse syntaxin 1A hexahistidine fusion polypeptide (model polypeptide of 36 kDa molecular mass and pI 5.67) as ligand. The polypeptide was diluted in 10 m*M* sodium acetate to a final concentration of 50 μg/mL, at three different pH values (4.0, 5.5, and 6.0), prior to the "pre-concentration" testing (*see* **Fig. 20.2**). For pH 4.0 (red curve), the surface concentration of ligand on the chip was much higher than that obtained for pH 5.5 and 6.0 (*pink* and *blue* curves, respectively). Thus, for immobilization of the syntaxin 1A fusion polypeptide via the amine coupling reaction (**Fig. 20.2**), pH 4.0 was chosen (a value less than the pI).

3. Using a sample volume of 100 μL, determine the pH value that gives maximum surface retention for immobilization (**Fig. 20.2**). The Biacore 3000 instrument is automated, so once the program is loaded and samples are placed, the entire reaction proceeds.
4. Open *Biacore 3000 Control* software, click on RUN tab, and then click on APPLICATION WIZARD, then select SURFACE PREPARATION and click on START. In the next panel select pH SCOUTING and click NEXT.
5. In the opening panel, select BUFFERS or add a NEW BUFFER and then click on NEXT, type the LIGAND NAME, specify the INJECTION TIME, FLOW RATE, and FLOW CELL to use.
6. In the next panel, select WASH METHOD, WASH FLOW RATE, SOLUTION TO INJECT (*see* **Note 16**), and INJECTION TIME.
7. In the next panel, assign RACK POSITIONS to samples and solutions, click on NEXT and click START. The result of a pH scouting run is shown in **Fig. 20.1**.

3.3.2. Preparation for Immobilization Via the Amine Coupling Reaction (*see* Note 17)

1. Run a DESORB step with a maintenance chip docked (*see* **Note 18**).
2. Prepare two vials each containing 115 μL of EDC.
3. Prepare two vials each containing 115 μL of NHS.
4. Prepare two empty vials for mixing the reagents in **steps 2 and 3, Section 3.3.2**. (The vials are inserted in the instrument racks for automated mixing of reagents.)

5. Prepare sample (typically 200 μL) by mixing the required amount of ligand with sodium acetate solution at a pH determined by the pH scouting step.
6. Prepare two vials each containing 80 μL of ethanolamine solution, for blocking the unreacted active sites on the chip.
7. Prepare a vial containing 100 μL of 50 m*M* NaOH, for surface regeneration after pre-concentration.
8. Dock the CM5 sensor chip, using the *Biacore 3000 Control* software, and prime the chip with the running buffer (*see* **Note 19**).

3.3.3. Setting Up the Pre-Concentration and Immobilization Steps

1. Open the *Biacore 3000 Control* software (**Fig. 20.2**).
2. Click on the RUN tab, then click on to RUN APPLICATION WIZARD, select SURFACE PREPARATION, and START.
3. When the SURFACE PREPARATION STEPS window opens, select IMMOBILIZATION pH SCOUTING, and then select

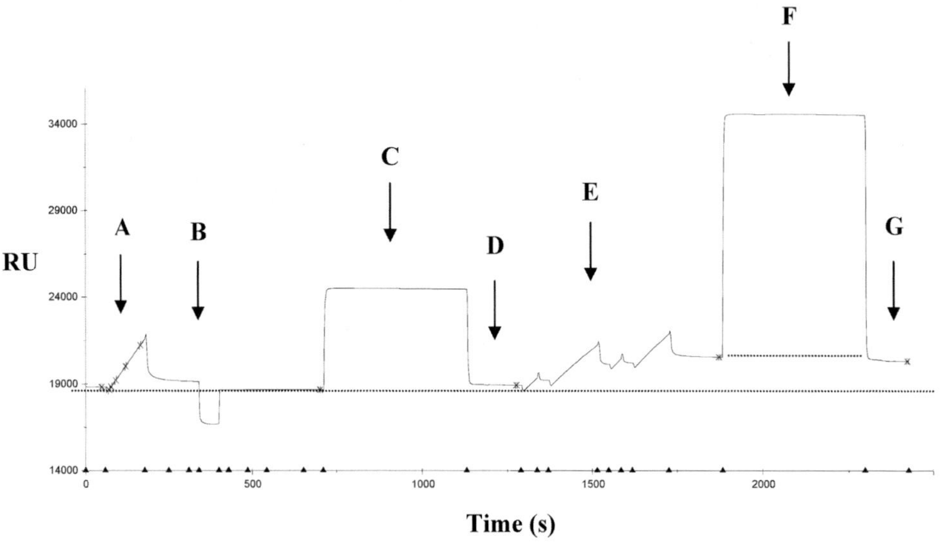

Fig. 20.2. Steps in the immobilization of purified syntaxin 1A fusion polypeptide ligand on a CM5 research-grade sensor chip via the amine coupling reaction. **(A)** "Pre-concentration" test to determine how much of the ligand to inject to reach a targeted (aimed) level of response (RU, response units), here, 1,500 RU. The injection period ends after attainment and overshoot of this criterion value above the baseline of approximately 19,000. The pre-concentration test reflects the proclivity of the ligand for the carboxymethylated dextran surface in its *non-activated* state, maximizing compatibility during the actual covalent binding step, E, and thus providing an estimate of the mass necessary for optimizing the association. **(B)** Washing of the non-covalently bound ligand with 50 m*M* NaOH to remove the ligand completely and obtain baseline. **(C)** Injection of the reaction mix (*see* **Section 3.3.2**) onto the surface, activating the carboxymethyl group by forming a highly-reactive succinimide ester *(7)*, followed by termination of the injection. **(D)** Slight increase in the RU reflecting the activation, compared to the baseline between B and C. **(E)** After surface activation, injection of ligand sample diluted in buffer of the appropriate, pre-determined pH, with continuation of injection resulting in covalent binding of ligand to the reactive surface to yield the targeted RU value of 1,500. This value is the difference between the achieved constant level just before F and the baseline at D. **(F)** Blockage of remaining non-reacted-activated carboxymethyl groups by injection of ethanolamine, followed by cessation of injection. **(G)** Point of attained immobilization of ligand. Small triangles on time line below designate events such as injection, end of injection, and re-filling.

BUFFER (a list of buffers will appear). Click NEXT after selecting the buffer or buffers.

4. Type the NAME of the ligand to be tested, select INJECTION TIME, FLOW RATE, and FLOW CELL (FC; any of the four flow cells, FC 1, 2, 3, or 4 can be used for testing). An injection of 2-min duration, which is the default time, is usually sufficient to test the surface retention of the ligand for the selected pH. The flow rate is set to 5 μL/min (*see* **Note 20**). Click NEXT to continue.
5. For the immobilization reaction, click RUN APPLICATION WIZARD, select IMMOBILIZATION, and click NEXT to select the type of SENSOR CHIP (CM5).
6. Select the IMMOBILIZATION METHOD, then in the next box select AIM FOR IMMOBILIZED LEVEL and click NEXT (*see* **Note 21**).
7. Select the WASH SOLUTION. Use the same wash solution(s) that were used for the pH scouting wash.
8. Select the SAMPLE POSITIONS for the vials to be placed in the instrument rack. Pipette out the required amount of each reagent, including the ligand solution. Manually insert the vials in the rack (*see* **Note 22**).
9. Click NEXT to start the run. At the completion of the run, a table is displayed showing the immobilization achieved (*see* **Note 23**).

3.3.4. Surface Preparation: Regeneration Scouting and Surface Performance Test

1. Open the control software. Perform DESORB, dock the CM5 chip containing the ligand, and prime with running buffer.
2. Click RUN tab, select RUN APPLICATION WIZARD, select SURFACE PREPARATION, and click START when the panel opens.
3. Select REGENERATION SCOUTING and click NEXT. In the panel, enter the ANALYTE NAME, INJECTION TIME, FLOW RATE, and FLOW CELL that contains the ligand. Click NEXT and select the rack positions. Click START and save the file to begin the run.
4. Select SURFACE PERFORMANCE TEST and click NEXT to enter the analyte name. Specify INJECTION TIME, FLOW RATE, AND NUMBER OF CYCLES (2–5 cycles). Select the FLOW CELL (1, 2, 3, 4, 2-1, or 4-3) and click NEXT to set the regeneration conditions.
5. In the panel, select a REGENERATION METHOD (dissociation in buffer, single injection, or two injections), FLOW RATE, SOLUTION/INJECTION (e.g., 10 *mM* glycine HCl, pH 3.0), and INJECTION TIME. Click on the box for STABILIZATION TIME, set the time (2 min or more), and click NEXT.
6. Select the RACK POSITIONS and click START to save and run the experiment (*see* **Note 24**).

3.4. Experimental Binding Measurements

3.4.1. Experimental Binding Run

1. Run DESORB with the maintenance chip docked. Remove the maintenance chip, dock the CM5 ligand chip, and prime.
2. Click RUN, then select APPLICATION WIZARD. Select BINDING ANALYSIS and click START.
3. Choose DIRECT BINDING, click NEXT to select the FLOW CELL, FLOW RATE and the NUMBER OF SAMPLE INJECTIONS. Set the SAMPLE INJECTION TIME and WAIT AFTER INJECTION time.
4. Click NEXT to enter the ANALYTE NAME, NUMBER OF REPLICATIONS, and the ORDER in which the sample is to be analyzed (as entered or random).
5. Select the regeneration (wash) method. Choose SINGLE INJECTION or TWO INJECTIONS.
6. Select FLOW RATE for the regeneration run.
7. Select the TYPE OF SOLUTION (*see* **Note 24**).
8. Click NEXT to reach the next panel to assign the sample positions in the rack. Drag and drop each sample to whichever position is wanted, but make sure the programmed placement corresponds exactly to the subsequent actual placement. This panel also tells one how much is required of each solution.
9. Pipette out the required amount of the solutions and keep in the assigned racks.
10. Press NEXT, and then once again verify if all the samples are in the correct position and in the correct rack.
11. Click START to run the experiment. Begin with a low analyte concentration with a flow rate of 30 μL/min, and choose a reference cell and the ligand cell.

3.4.2. Experimental Kinetic Run (Fig. 20.3)

1. Run DESORB and then dock the CM5 chip with ligand and prime in running buffer.
2. On the *Biacore 3000 Control* Software, click RUN, and select RUN APPLICATION WIZARD.
3. Select KINETIC ANALYSIS and click START.
4. Tick the box for CONCENTRATION SERIES and select as a control experiment MASS TRANSFER.
5. Next, select DIRECT BINDING and click NEXT.
6. In the panel, select the flow cell (2-1, 4-3, 1, 2, 3, or 4), set the FLOW RATE, STABILIZATION TIME, INJECTION TIME and DISSOCIATION TIME. Select the RUN ORDER (random or as entered). Enter ANALYTE NAME, MOLECULAR MASS in Daltons, NUMBER OF REPLICATIONS, and required CONCENTRATIONS. Click NEXT. Select at least five different concentrations and include a buffer blank.
7. In the MASS TRANSFER control panel, select a CONCENTRATION to be analyzed. If using a range of 0–200 nM, use a concentration that is in the range of 100 nM. Click NEXT.

8. Select the REGENERATION METHOD, FLOW RATE, REGENERATION SOLUTION, and the INJECTION TIME. Set the STABILIZATION TIME after regeneration.
9. Select RACK POSITIONS, click NEXT, and click START to begin the run and save the file (*see* **Note 25**).

3.4.3. Data Analysis and Determination of Kinetic Constants

1. Using *BIAevaluation* software (*see* **Note 26**), select the data to be analyzed and create an overlay file by selecting the relevant curves. Align all the curves to the start of the injection. Remove the portions of the plots that lie outside the relevant regions of the association and dissociation curves, since the measurements require only the latter portions.
2. Choose FIT: KINETICS SIMULTANEOUS k_a/k_d to start the kinetic evaluation wizard and perform the curve-fitting process. Set the baseline to zero at a point at the start of the injection in the most stable region. Performing a Y-TRANSFORM command in the wizard panel enables alignment of all the desired curves.
3. Select START TIME and STOP TIME for the injection and select the DATA RANGE, for association and dissociation, to be used in the fitting process. Make necessary adjustments by zooming out and inspecting the selected region visually (*see* **Notes 27, 28**).
4. In the next panel, select the BINDING MODEL. Depending on the binding data, choose the model which best fits the ligand-analyte interaction. (a) 1:1 Langmuir binding, (b) 1:1 binding with shifting baseline, (c) 1:1 binding with mass

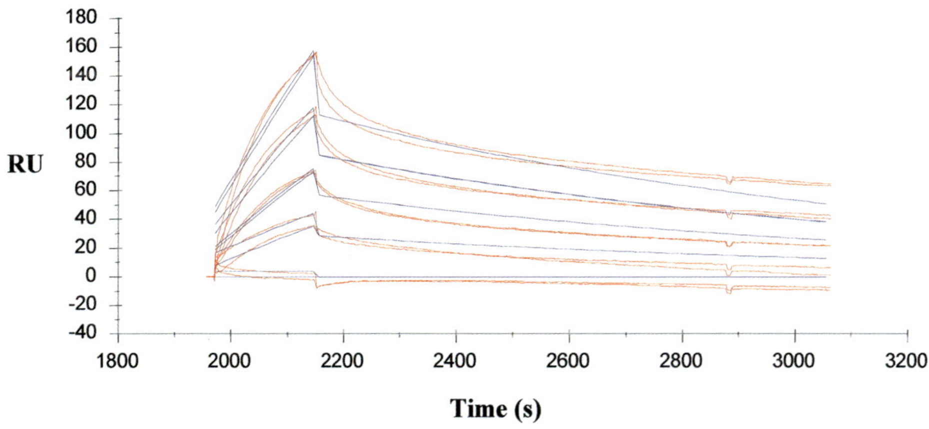

Fig. 20.3. Plots illustrating the experimental curve-fitting methodology for a simple binding model (1:1 Langmuir). Association and dissociation phases can be seen in each plot. Pairs of red traces at each concentration indicate duplicate experimental determinations and blue traces show the corresponding binding model curves. In these studies, a syntaxin 1A fusion polypeptide ligand was immobilized on a CM5 sensor chip by the amine coupling method to a final targeted RU of 707 (immobilization targeting is illustrated in Fig. 20.2). A range of concentrations (0, 10, 20, 40, and 80 nM) of the analyte, a model hexahistidine fusion polypeptide of the calcium channel $Ca_V1.3$ II–III loop with molecular mass 18 kDa, was used for the kinetic evaluation.

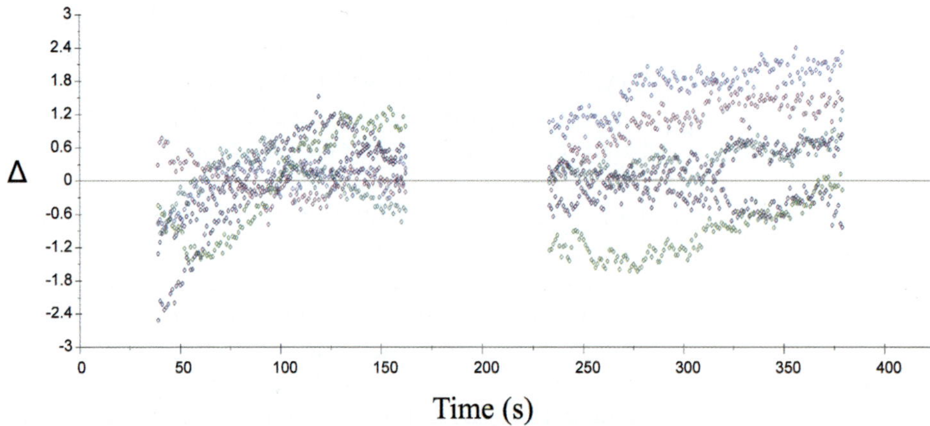

Fig. 20.4. "Residual" plots *(9)* reflecting minimal deviation between experimental and fitted data for a kinetic study (not shown) in which the data match the model. Left and right data sets correspond to chosen association and dissociation portions, respectively, of the kinetic curves. Colors represent different concentrations of analyte (top to bottom, *see* dissociation portion): light blue, 2 nM ; magenta, 0 nM; black, 4 nM; blue-green, 8 nM ; dark blue, 32 nM; green, 16 nM. Note that the scatter in the residual plots is within or around the criterion ±2 RU, indicating that the experimental and model kinetic curves, in this case for a 1:1 Langmuir interaction (*see* **step 4, Section 3.4.3**), show a good fit. Larger deviations from the zero line would indicate that the data should be refitted using another kinetic model.

transfer, (d) bivalent analyte, (e) heterogeneous analyte, (f) heterogeneous ligand, (g) two-state reactions.

5. Fit the data and observe the progress. For simple 1:1 reactions, the computation process is almost instantaneous, but a delay of several seconds to minutes may indicate complex binding or heterogeneity in the data set.
6. After the fit, the curves will appear as an overlay of the experimental and fitted data (**Fig. 20.3**). Close overlap, observed visually, is a good first indication of the validity of the fit.
7. The result is displayed as a report, with all the constants in numerical form. Examine the "residual plot," which is the calculated difference between the experimental and fitted data (**Fig. 20.4**), by clicking on the RESIDUALS tab. If there are systematic deviations between the experimental and fitted curves, the plot will indicate them by displacement from the zero line. Ideally, the noise level in the plot should be on the order of ±2 RU (also *see* **Note 29**).

4. Notes

1. The Biacore CM5 chip is widely employed for SPR interaction studies. Carboxymethylated dextran molecules are attached to a gold-coated surface. The CM5 chip can detect nucleic acids, carbohydrates, and small molecules, in addition to proteins. The chip surface is prepared for binding studies by coupling

ligand molecules to the carboxymethyl group via NH_2, -SH, -CHO, -OH or -COOH linkages. As alternatives to the CM5 chip, CM4, CM3 and C1 chips can be used, which have a lower matrix density of functional groups. Another chip, SA, can be used for immobilization of biotinylated peptides, proteins, nucleic acids, and carbohydrates. HPA and L1 chips are employed for lipid or liposome immobilization. NTA chips are utilized for capturing histidine-tagged molecules *(7)*.

2. If no non-specific PCR bands are present, the product can be purified by phenol-chloroform extraction and alcohol precipitation. Adjust the sample volume to 200 μL by adding deionized water. Add an equal volume of phenol-chloroform (50% phenol saturated with 0.1 *M* Tris-HCl, pH 7.6, 48% chloroform, 2% isoamyl alcohol) to the sample and vortex to mix. Centrifuge at 14,000 g for 5 min. Carefully pipette out the upper phase into a fresh tube and add an equal volume of chloroform, vortex, and centrifuge as before. To the supernatant, add 1/10 volume of 3 *M* sodium acetate, pH 4.5, and 2.5 volumes of ethanol. Incubate at −20 °C for 1 h. Centrifuge at 14,000 g for 10 min at room temperature. Remove the supernatant. Wash pellet in excess 70% ethanol, centrifuge as before, decant the supernatant, and dry pellet in air.
3. The pRSET vector contains sequences for six contiguous histidine residues upstream of the multiple cloning site, and thus the his-tagged fusion protein expressed in bacteria can be purified using a nickel affinity column (Qiagen). Immediately adjacent and upstream to the multiple cloning site is an eight-amino-acid epitope against which antibodies are commercially available (Invitrogen), facilitating antibody-based detection. The pGEX expression vector can also be used (*see* **Note 10**).
4. The Qubit fluorescence spectrometer (Invitrogen) is a convenient instrument for measurement of nucleic acid and protein concentration by fluorescence. In a 0.5 mL tube, pipette 190 μL of the Quant-iT working solution (1/200 dye in buffer; buffer is double-stranded DNA Broad-Range reagent). Add DNA sample (1–10 μL) and bring to a total volume of 200 μL with water. Vortex 2–3 s, incubate for 2 min, and read the fluorescence.
5. Add, to an autoclaved 0.5-mL microcentrifuge tube, 4 μL of 5X ligase reaction buffer, 15–60 fmol vector DNA, 45–180 fmol insert DNA (0. 1–1. 0 μg total DNA), and 1 unit of T4 DNA ligase in 1 μL (Invitrogen). Bring the total volume to 20 μL, mix gently, and centrifuge briefly to collect the contents at the bottom of the tube. Incubate at 4 °C overnight.
6. Further confirmation of the correct insert sequence is accomplished by nucleotide sequencing of the insert before carrying out the protein expression.

7. In verifying the expression of protein, there should be a heavily-stained protein band at the estimated molecular size of the fusion product for the IPTG-induced sample. There may be a low-to-moderately stained band in the non-induced samples. There should be no comparable band in the empty vector sample. Once expression is verified, proceed to the purification step.
8. After establishing protein overexpression for the clonal sequence, the bacterial cells are grown in cultures for scaled-up protein purification. For purification of the fusion protein, generally 50–250 mL of culture is sufficient for small-to-medium-scale protein yield. From a 50 mL culture, one can purify up to 250 μg of protein.
9. For Ni-NTA spin-column (Qiagen) purification, as much as 100 μg of pure protein can be obtained from 1.2 mL of the lysate solution, depending on the level of expression. In some cases, minor bands are visible in the purified sample, and a re-purification step with a fresh spin column is required.
10. For purification using glutathione-agarose beads, PCR primers for domains of interest, containing desired restriction sites, are used in PCR reactions to clone the domains into the pGEX-6P-1 vector. This vector contains a GST coding region followed by a multiple cloning site. The fusion polypeptide product will contain an N-terminal, 26-kDa GST tag, which allows affinity purification. The pGEX-6P-1 vectors are utilized to transform *E. coli* BL21 (DE3) as described in **Section 3.2.3**, and the protein is purified by the glutathione-agarose beads.
11. In general, both ligand and analyte should be as pure as possible, particularly for kinetic analysis of interactions. However, if a well-purified ligand is used, it is possible to determine the concentration of the analyte even if the latter is present in a mixture *(9)*.
12. It is critical that there be no glycerol in the protein samples and buffers.
13. De-gas solutions by fitting a rubber stopper, with inlet tube attached to an aspirator and trap, to the top of a vessel containing the solution and reduce the pressure until air bubbles are expelled.
14. The pH should be adjusted so that the protein has a net positive charge. The pI and molecular mass of the protein can be estimated using the COMPUTE pI/MW tool found at http://au.expasy.org/tools/pi_tool.html.
15. The ligand should be at least 95% pure. Generally, a ligand polypeptide of smaller molecular mass should be paired with a polypeptide/protein of higher molecular mass as analyte, because the SPR response will be higher, due to the more

advantageous (higher) density that can be achieved for the smaller molecular-mass molecule bound to the surface.

16. A single injection of 50 m*M* NaOH at a flow rate of 20 μL/min usually removes all the ligand from the surface (inspect the baseline to see if it returns to the same level as before the run). If the baseline remains elevated, a second wash with the same or another solution is performed to remove the ligand completely.
17. The amine coupling reaction is the reaction generally used for proteins because of readily-available amine groups in proteins. **Fig. 20.2** shows a sensorgram of a typical immobilization experiment using an amine coupling reaction. However, amine coupling is not suitable if the ligand is too acidic (pH<3. 5), if the amine groups are present in the active site, or if there are too many amine groups. In such situations, if thiol groups are present in the ligand, thiol coupling is preferred as an alternative *(7)*. If both are not suitable, capture techniques such as streptavidin-biotin or His-tag nickel affinity methods can be selected. If a well-characterized antibody is available against the ligand, the antibody is immobilized and the ligand captured for binding studies. Non-amine-coupling protocols are not covered here.
18. It is important to perform DESORB before each run in order to clean the micro fluidic tubes, chamber, and injection port. Such recommended instrument maintenance should be carried out on a regular basis. In the *Biacore 3000 Control* program, click the TOOLS tab and then WORKING TOOLS. Place one vial each of BIAdesorb Solution 1 and BIAdesorb Solution 2 in the rack designated by the program. Click DESORB and then START. Use a maintenance chip (chip lacking gold or other surface coatings) during the DESORB; do not use the CM5 chip.
19. As noted in **step 2, Section 3.1**, the sensor chip should be primed before each run. The total duration for the priming is 6.3 min.
20. A slow flow rate (5 μL/min) is best, since the chance for the ligand to adsorb to the surface increases as the flow rate decreases.
21. All the buffers, solutions, and sensor chips are brought to room temperature before the run. De-gas all the solutions. The buffers and regeneration solutions must be filtered through standard 0. 2-μm or 0. 4-μm filters.
22. When analyzing the samples, make sure that there are no air bubbles trapped in the sample solution. Even with prior degassing, bubbles can form during the sample dilution, and sometimes the bubbles may not be clearly visible. Eliminate these bubbles by spinning the tube at top speed (18,000 g) in a benchtop centrifuge for 10–15 s.

23. To minimize the mass transport effect *(8)*, a low level of ligand immobilization is preferred.
24. Regeneration is achieved with the optimal pH and ionic concentration that keep the ligand active. A series of regener-

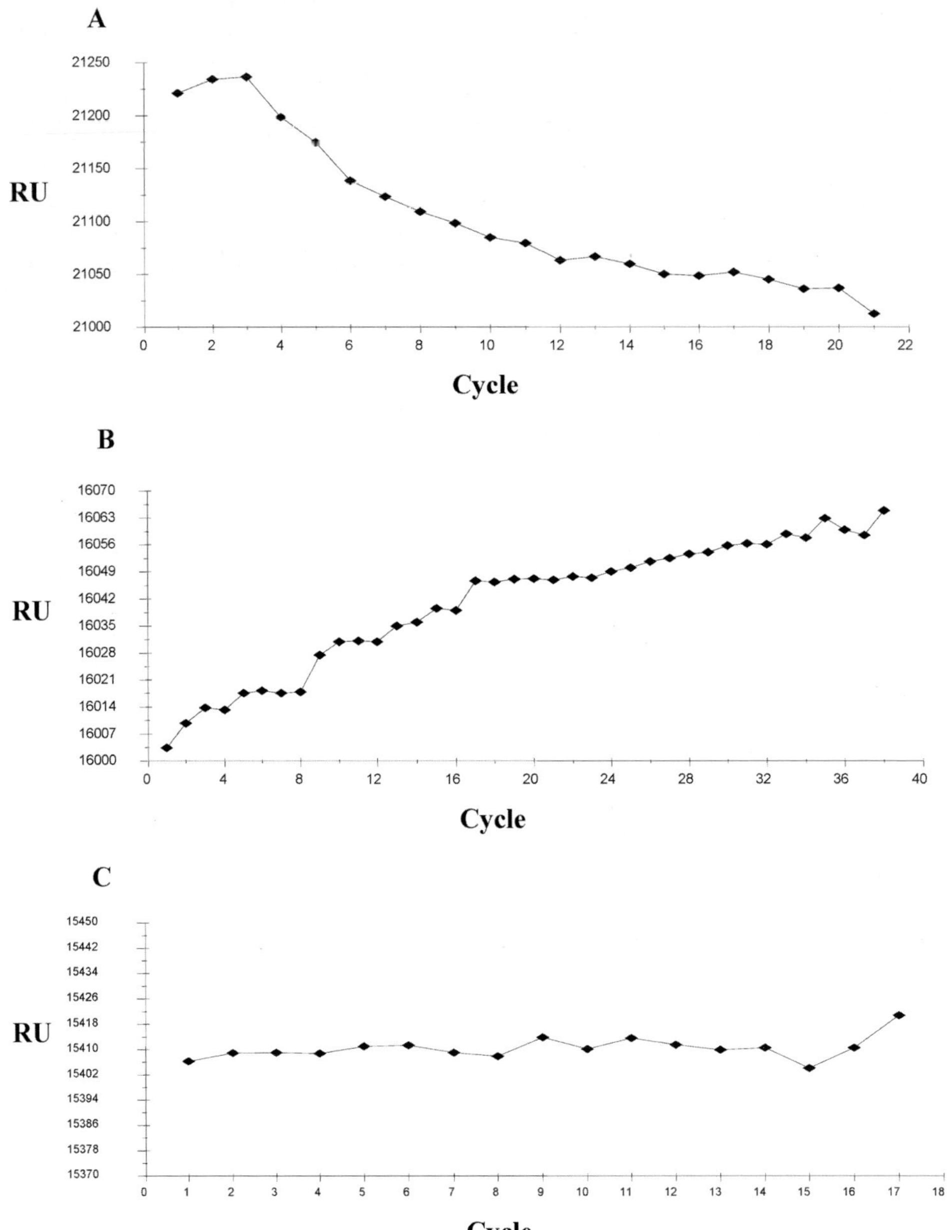

Fig. 20.5. Illustration of unstable vs. stable SPR baselines. Baseline stability is critical to obtaining accurate kinetic data. Plot (**A**) shows a decreasing baseline after each cycle, commonly due to loss of the bound ligand. Plot (**B**) indicates an increase in the baseline after each cycle, reflecting partial dissociation of analyte and poor regeneration. Plot (**C**) shows a stable baseline, indicating minimum loss of bound ligand as well as maximum regeneration. The latter reaction is most likely to yield accurate kinetic data. (Fig. 20.3 illustrates a data set where the baseline was stable and duplicate determinations were similar.)

ation tests should be performed with regeneration buffers, from mild to more stringent conditions of pH and ionic strength. In most cases where the binding affinity is low or moderate, begin with a pH close to neutral (pH 4.5–8.5) and an ionic strength in the range of 0.5–1.0 *M*. 1.0 *M* NaCl should be sufficient to dissociate most of the bound protein (**Figs. 20.5, 20.6**).

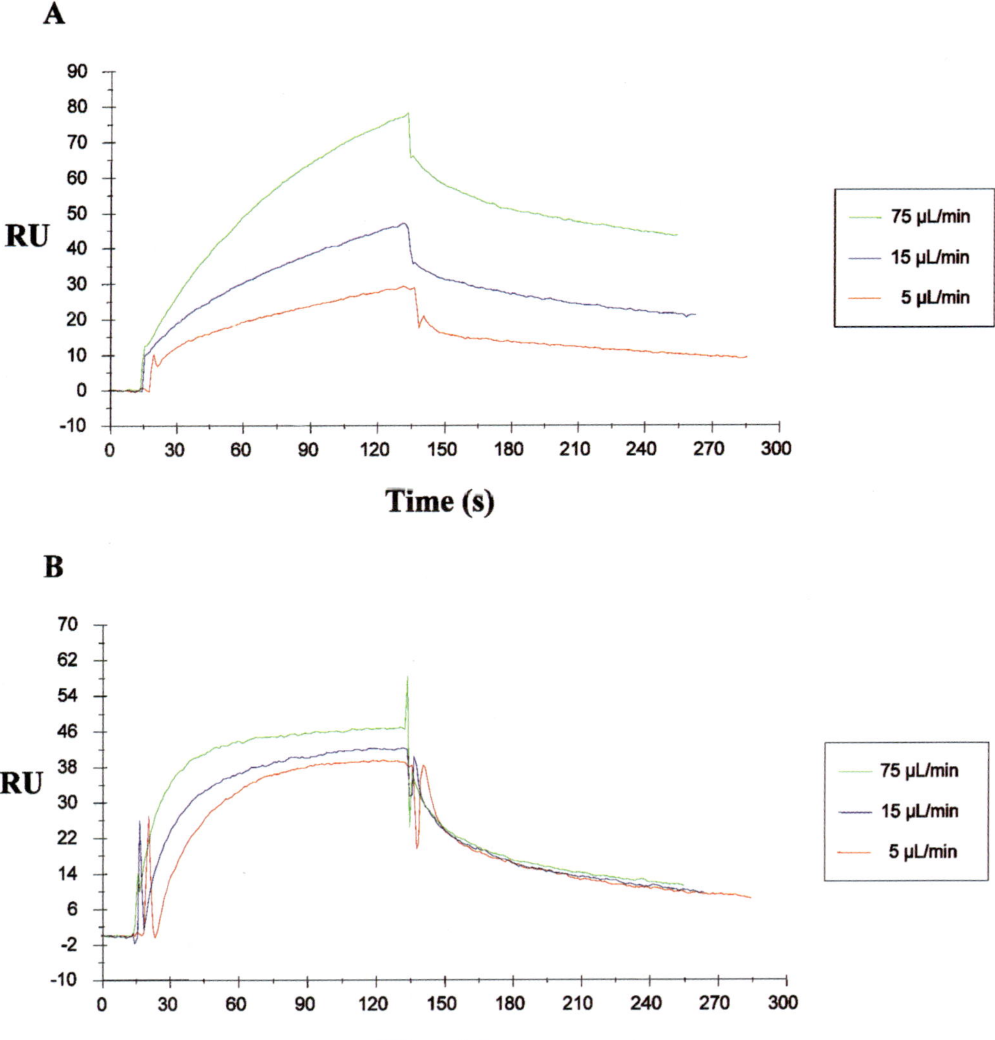

Fig. 20.6. Effect of flow rate on mass transfer. If the association rate constant of an interaction of interest is high, the measured binding rate may reflect the rate of *transfer* of analyte into the matrix (as the limiting rate) rather than the rate of the binding itself *(8)*, and such mass transfer limitations may yield erroneous rate constants. Mass transfer problems, if present, are commonly detected by injecting a certain concentration of analyte sample at different flow rates, as shown. (**A**) The same analyte concentration gives considerably different RUs with different flow rates, indicating the presence of a mass transfer limitation. (**B**) The RUs at different flow rates are more similar, with almost identical dissociation phases, suggesting a minimal mass transfer limitation. If mass transfer limitation is present, as in A, choosing a relatively high flow rate, low ligand density, and experimentally-optimal analyte concentration can considerably reduce the limitation.

25. When undocking the chip from the instrument, it is often advantageous to select *not* to remove the buffer from the flow cells by *un-checking* EMPTY FLOW CELL. The chip can then be wrapped in soft tissue paper or a piece of paper towel and stored in a tight container at 4 °C. The sensor chips with immobilized ligands can be stored for a few days to more than a month.
26. SPR binding measurements (**Fig. 20.7**) are generally performed in duplicate or triplicate. Experimentally-obtained kinetic data can be evaluated with the SPR kinetic evaluation software, *BIAevaluation*. In general, analysis of kinetic data involves the following steps: (a) make overlay plots of several interaction curves, (b) select analysis region, (c) select interaction model and curve fitting, (d) store results of the fitting procedure into an analysis results file.
27. In selecting the data range for the analysis of kinetic plots with *BIAevaluation* software, a short period before the injection start and before the stop (the beginning and end of the association phase of the kinetic plot) should be excluded in the selection to avoid the dispersion effect (the complex response in the refractive index at the beginning and end of injection, not clearly defined mechanistically).
28. Perform separate fits for different regions of the association and dissociation phases to ascertain if the calculated constants are consistent.

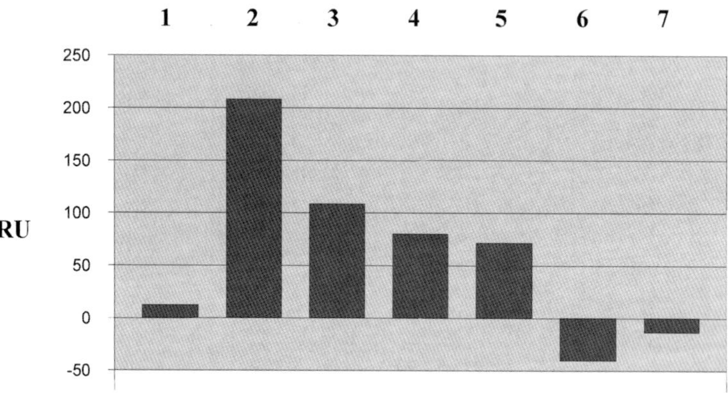

Fig. 20.7. Example SPR data: qualitative analysis of binding between ligand syntaxin 1A fusion polypeptide and otoferlin hexahistidine fusion polypeptide (a model analyte polypeptide of molecular mass 34 kDa), in the presence and absence of calcium. Bar 1 indicates essentially no binding (the slight elevation is a bulk response due to a small increase in the refractive index) when the binding buffer contains 3 mM EGTA. Bar 2 shows a higher response with omission of EGTA and addition of calcium ($100\,\mu M$) to the buffer, indicating a requirement of calcium for the interaction. Bars 3, 4, and 5 show a reduction of the interaction with increasing concentrations of calcium (0.2, 0.5, and 1.0 mM, respectively). Bars 6 and 7 are buffer blanks (negative controls; buffer plus ligand with no analyte). Bar values represent the maximum binding as determined by the RU at the end of the injection of the analyte, relative to baseline just before injection.

29. Although not detailed here, association and dissociation constants can also be obtained directly by determining the concentrations of analyte and ligand under steady-state conditions *(9)*.

Acknowledgments

This work was supported by NIH R01 DC000156, NIH R01 DC004076, and the American Hearing Research Foundation. We thank Dr. Stanley Terlecky, Department of Pharmacology, Wayne State University, for use of the Bia coore 3000 instrument.

References

1. Schuck, P. (1997) Use of surface plasmon resonance to probe the equilibrium and dynamic aspects of interactions between biological macromolecules. *Annu. Rev. Biophys. Struct.* **26**, 541–566.
2. Ramakrishnan, N. A., Drescher, M. J., Sheikhali, S. A., Khan, K. M., Hatfield, J. S., Dickson, M. J., and Drescher, D. G. (2006) Molecular identification of an N-type Ca^{2+} channel in saccular hair cells. *Neuroscience* **139**, 1417–1434.
3. Kretschmann, E. and Raether, H. (1968) Radiative decay of non-radiative surface plasmons excited by light. *Z. Naturforsch. Teil. A.* **23**, 2135–2136.
4. Karlsson, R., Roos, H., Fägerstam, L., and Persson, B. (1994) Kinetic and concentration analysis using BIA technology. *Methods: Companion Methods Enzymol.* **6**, 99–110.
5. Stenberg, E., Persson, B., Roos, H., and Urbaniczky, C. (1991) Quantitative determination of surface concentration of protein with surface plasmon resonance using radiolabeled proteins. *J. Colloid Interface Sci.* **143**, 513–526.
6. Johnsson, B., Lofas, S., and Lindquist, G. (1991) Immobilization of proteins to a carboxymethyldextran-modified gold surface for biospecific interaction analysis in surface plasmon resonance sensors. *Anal. Biochem.* **198**, 268–277.
7. Steffner P. and Markey F. (1997) When the chips are down. *BIA J.* **1**, 11–15.
8. Myszka, D. G., He, X., Dembo, M., Morton, T. A., and Goldstein B. (1998) Extending the range of rate constants available from BIACORE: interpreting mass transport-influenced binding data. *Biophys. J.* **75**, 583–594.
9. BIACORE. (1998) BIAevaluation Version 3 Software Handbook. Chapter 5. Evaluating concentration data. 5.1–5.10.

Chapter 21

Multiplexed Isobaric Tagging Protocols for Quantitative Mass Spectrometry Approaches to Auditory Research

Douglas E. Vetter, Johnvesly Basappa, and Sevin Turcan

Abstract

Modern biologists have at their disposal a large array of techniques used to assess the existence and relative or absolute quantity of any molecule of interest in a sample. However, implementing most of these procedures can be a daunting task for the first time, even in a lab with experienced researchers. Just choosing a protocol to follow can take weeks while all of the nuances are examined and it is determined whether a protocol will (a) give the desired results, (b) result in interpretable and unbiased data, and (c) be amenable to the sample of interest. We detail here a robust procedure for labeling proteins in a complex lysate for the ultimate differential quantification of protein abundance following experimental manipulations. Following a successful outcome of the labeling procedure, the sample is submitted for mass spectrometric analysis, resulting in peptide quantification and protein identification. While we will concentrate on cells in culture, we will point out procedures that can be used for labeling lysates generated from other tissues, along with any minor modifications required for such samples. We will also outline, but not fully document, other strategies used in our lab to label proteins prior to mass spectrometric analysis, and describe under which conditions each procedure may be desirable. What is not covered in this chapter is anything but the most brief introduction to mass spectrometry (instrumentation, theory, etc.), nor do we attempt to cover much in the way of software used for *post hoc* analysis. These two topics are dependant upon one's resources, and where applicable, one's collaborators. We strongly encourage the reader to seek out expert advice on topics not covered here.

Key words: Proteomics, iTRAQ, quantitative mass spectrometry, protein expression.

1. Introduction

A unique watershed moment in biology occurred with the release of the initial drafts of the entire genome sequence derived from the first model organisms. This event led to the completion of whole genomic sequencing of many other organisms, including

Bernd Sokolowski (ed.), *Auditory and Vestibular Research: Methods and Protocols, vol. 493*

DOI 10.1007/978-1-59745-523-7_21 Springerprotocols.com

human. The ability to read genomic sequences and map genes within genomic space is certainly a powerful tool. However, it can also be argued that the vast majority of practicing biologists today actually work on understanding the role of proteins in their chosen sub-discipline, whether to seek an understanding of development, disease states and their progression, or normal cellular processes. Yet the methods used by many biologists to examine the transcriptome only details the state of the transcript and hence the state of gene expression; whereas, it is the protein(s) encoded in each transcript that is more often the actual effector molecule of interest. One assumption tacitly made is that gene expression equals protein expression, but this is clearly not always (some may say rarely!) true *(1, 2)*. Thus, as we push beyond the genomic age, into a post-genomic epoch of biology, one must increasingly come to terms with techniques that are more suited to assess the effectors of biological processes. The transcriptome is a static entity in terms of sequence (save for the uncommon, or experimentally induced, mutation and methylation), whereas proteins have many more degrees of freedom that make their assessment particularly challenging. These degrees can include post-translational modifications, which can be many (phosphorylation, palmitoylation, sumolyation, etc.), the timing of these changes, alternative splicing from the genome, and RNA editing-induced changes to the protein sequence (and therefore function), which is not even revealed in the genomic sequence. Coupling these issues with the observation that many genes can encode more than one protein (and in some cases hundreds of proteins), one begins to get the sense of the enormous task facing the biologist wishing to examine the state of the proteome in their sample of interest. Indeed, while estimates of the number of genes making up the genome generally are settled between 20,000–30,000 genes, estimates of the functional proteome range is in the 100,000's. While biological complexity certainly resides within the genome, a simple comparison of the number and sequence of genes between the worm, *C. elegans*, and humans, illustrates that a major portion of the biological complexity, separating these species, resides in gene function and even more importantly, in the interactions between gene products.

The study of proteins in biological processes crossed a watershed point with the coupling of mass spectrometry (MS) with biological samples, even though techniques used for assessing protein expression and quantification were in use for many years before. The "need" that led to this change in approach is traced to the difficulty of assessing large numbers of proteins simultaneously, as well as assessing proteins for which no discriminatory tag is available (either an antibody, or a fusion protein expressed *in vivo*). The chief advance made possible by MS-based proteomics is the freedom to discover otherwise unanticipated changes in protein expression, post-translational modification, etc. that are involved in particular biological processes. The evaluation of these expression states leads to the establishment of protein interaction networks. Many methods exist for establishing and quantifying interaction networks, however, they all result in testable hypotheses of mechanism(s) that regulate phenotypic changes in an organism. The end result can reveal protein interactions not previously appreciated, as well as altered cellular states observed in disease or experimenter-induced changes in cells/tissues. Thus, an understanding can be attained of how a phenotype occurs when that phenotype is not directly related to an altered gene *(3, 4)*. Inherent in this kind of approach is the movement away from a purely reductionist view of biology, to a more global, interactive view characterized by the systems biology approach.

Many of the most successful proteomics experiments to date (at least related to higher organisms) have used proteomic approaches to study changes in specific structures such as organelles *(5–7)*, synaptic junctional preparations *(8)*, etc. This illustrates a key principle in proteomics – simplifying the sample results in better discrimination of protein changes.

There are a number of issues that must be considered once a decision is made to tackle a proteomics project. A firm commitment must be made to the type of data collection, since in most cases, strategies adopted may be mutually exclusive for other purposes. Thus, if there is an interest in examining phosphorylation states of proteins, a sample preparation/isolation method is used that is significantly different from more routine protein expression studies. Similarly, if interested in glycosylation states of membrane associated proteins, isolation may be significantly different than when assessing the nuclear proteome. Other concerns include whether the sample is pre-fractionated prior to further processing, run on a one- or two- dimensional gel or a liquid chromatography column, tagged for future quantification, or whether the protein remains tagless. One rule of thumb is that each time the sample is manipulated there is a likelihood that protein loss will occur. The threshold at which these loses might alter the outcome of the experiment is difficult or impossible to know *a priori*.

Finally, although space does not allow a full description of mass spectrometry based proteomics here, several review articles can help the uninitiated to navigate the decisions that must be made and to allow a better appreciation of the concerns *(9–13)*. Here, we limit ourselves to detailing the preparation of a protein lysate sample for quantifying expression levels *in vitro* and for giving a quick introduction to the analyses of the resulting data.

1.1. Methods of Labeling for Quantitative Differential Mass Spectrometry

We will outline three of the more common methods of differential labeling in this section and detail one of them further in **Section 3. Figure 21.1** schematically demonstrates the major steps of these methods side by side.

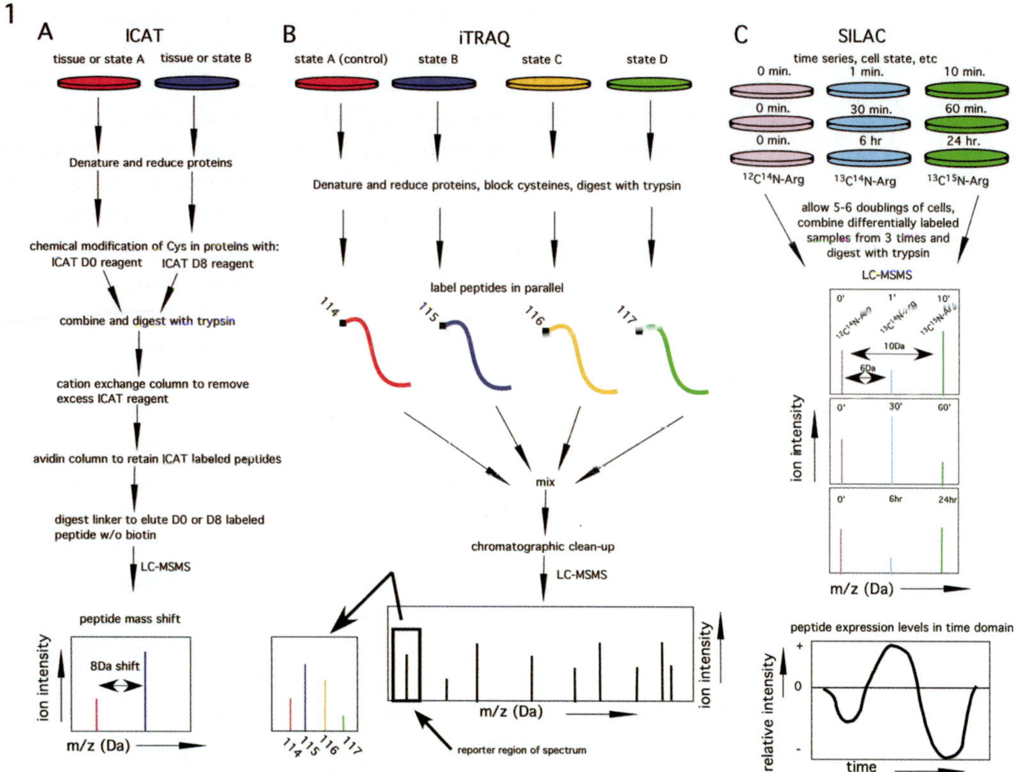

Fig. 21.1. A comparison of the major steps taken to label proteins or peptides for quantitative mass spectrometric analysis. (**A**) In the ICAT procedure, tissues or cells are compared by first denaturing the proteins present in the lysate and then labeling all cysteine residues with one of two labels. The light label contains normal hydrogen at specific sites, while the heavy label contains deuterium substituted in place of the hydrogens. In all, the heavy label is 8 daltons heavier than the light label. The lysates are mixed at this point and digested with trypsin. Following digestion, the resultant labeled peptides are selected for and retained over an affinity column, eluted, and subjected to liquid chromatography and mass spectrometry. Peptides exactly 8Da apart are examined, and when matched by sequence, the ion intensity (or the precursor LC peak area) can be assessed for quantitation purposes. (**B**) In the iTRAQ procedure, cells or tissues of different states are lysed, denatured, and digested. Each pool is labeled separately, then mixed, submitted to LC-MSMS, and assessed for the ion intensity of each label. Labels are freed during peptide bond cleavage and are found in a low mass region of the spectrum, separated by 1Da. (**C**) In SILAC procedures, proteins are metabolically labeled by allowing the cells to incorporate heavy-labeled arginine (depicted here) and/or lysine into the generated proteins. Lysates are generated, mixed, digested, any post-translational modifications desired are selected, and submitted for LC-MSMS. Expression data can be plotted in the time domain to gain an understanding of the temporal nature of modifications. For example, if one is looking at phosphorylation events, data can indicate the time course of phosphorylation and dephosphorylation, as depicted in the bottom panel.

Isotope coded affinity tag (ICAT, **Fig. 21.1A**) was one of the first methods by which peptides could be differentially labeled prior to submission for mass spectrometric analysis *(14)*. Briefly, this technique uses a biotinylated reagent with a specificity toward sulfhydryl groups. The labeling reagent carries either a "heavy" (d_8) or "light" (d_0) tag (**Fig. 21.2A**). This sample is combined with a second, independently labeled sample carrying the other tag, which is then compared to the first. The combined samples are then trypsinized. Labeled peptides are recovered from the mixture via biotin affinity chromatography. Following cleavage of the sample from the biotin moiety and release from the affinity column, the mixture is analyzed via Liquid Chromatography Coupled Mass Spectrometry (LCMS). Analysis consists of examining the ratio of ion intensities of the heavy and light sequence-matched peptides in the MS. The final result yields both sequence information of the peptide following tandem MS procedures and quantification of each peptide. Two potential drawbacks of the ICAT approach are its dependence on the occurrence of cysteines in the proteins of interest (a minor difficulty, as most proteins do contain cysteine residues) and the number of cysteine residues. The latter can be more problematic if the proteins of interest contain few cysteine residues. Unambiguous protein identification from a complex peptide mixture requires good coverage of the parent protein by the recovered peptides. Therefore, if there are few cysteine residues, and thus few labeled peptides, protein identification may be compromised. The advantage of the ICAT system is its relative ease of use, and it's applicability to lysates generated from any source (cell culture, tissues, etc.). A number of protocols have been published concerning ICAT *(15)*.

Isobaric tagging for relative and absolute quantification (iTRAQ, **Fig. 21.1B**) is similar in concept to ICAT, but differs in that amines are modified to carry a label for quantification (**Fig. 21.2B**). With iTRAQ, one also has the ability to multiplex up to four samples for simultaneous differential MS quantification. Additionally, Applied Biosystems is currently (at the time of writing) developing an eight-plex labeling method. Unlike the ICAT system, with iTRAQ, samples are trypsinized first prior to labeling. This approach has the advantage of labeling all peptides, since trypsin cuts at lysine/arginine sequences, and the iTRAQ system labels ε amines of each lysine. In addition, other free amines carry a label as well. The manner by which quantification is accomplished in the MS is similar to the way ICAT- ratios of ion intensities are used to assess changes in peptide expression between samples. In the mass spectra, the location of the quantified peaks is remote from the peptide sequence peaks, due to the cleavage of the label from the peptide during the collision-induced dissociation phase of the tandem MS run. The advantage here is that all label peaks at mass 114, 115, 116, and 117 (labels for masses 113, 118, 119, and 121 are in development) are collected in one place in the spectrum that is relatively "quiet" (**Fig. 21.4**). Thus, suppression by abundant peptide species is generally very low, and the signal more accurately reflects the true expression level. An added advantage to iTRAQ over ICAT is that peptide coverage of protein sequences is more complete due to the greater number of labeled peptides available for analysis. This coverage leads to a better resolution of protein

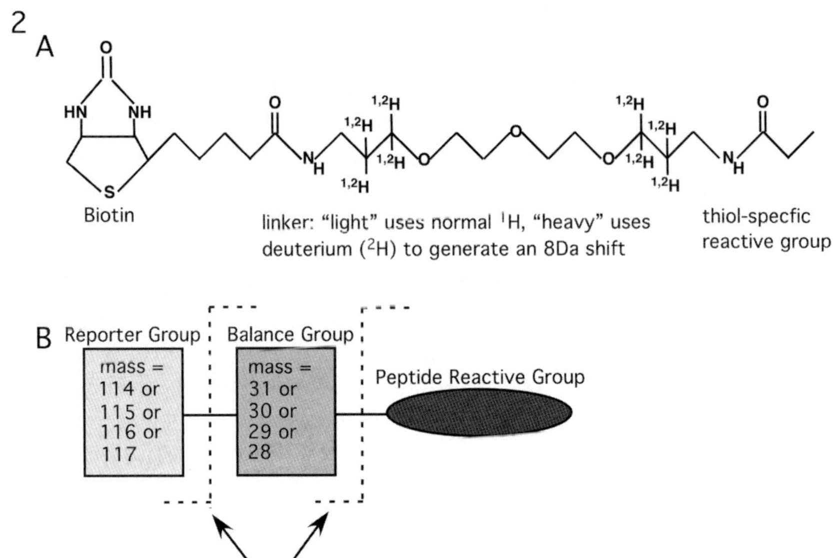

Fig. 21.2. Labels for ICAT and iTRAQ technologies. (**A**) The ICAT label is composed of a biotin moiety used for affinity selection, a linker, which contains either ^{1}H or D (^{2}H) substitutions as indicated, and a thiol reactive group that binds to free cysteines of the protein. The biotin group is cleaved after the affinity column selection step, and the peptide is released. (**B**) The iTRAQ label consists of a reporter group of varying mass and a balance group that is co-varied with the mass of the reporter group such that each labeled peptide carries an isobaric tag (i.e. a tag of the same mass). The label also contains a peptide reactive group that is reactive toward primary amines and ε amines of lysine residues. Upon dissociation in the mass spectrometer, peptide bonds are broken, releasing the peptide, the balance group, and the reporter group. The peptide undergoes further peptide bond breaks to yield the MSMS peptide sequence, while the balance group is lost in the very low mass region of the spectrum. The reporter group is found in a very quiet region of the spectrum and is assessed for total ion count/intensity to yield the quantitative information needed for analysis.

identification, especially with family members that have a highly conserved sequence. While iTRAQ can be used for samples generated from tissue as well as for cell culture, the majority of work thus far is from *in vitro* preparations.

Stable isotope labeling of amino acids in cell culture (SILAC, **Fig. 21.1C**) is significantly different from either ICAT or iTRAQ. With the SILAC approach, all proteins are labeled *metabolically* via incorporation of labeled arginine, lysine, or both *(16, 17)*. Additionally, the label can be light, intermediate, or heavy by using the desired amino acid carrying either ^{13}C, ^{15}N, or double ^{13}C^{15}N substitutions. Because of the number of possible labels (four if one also considers the unlabeled control state), multiplexing is possible with SILAC. The key to successful SILAC labeling is allowing the cells to completely incorporate the labeled amino acids into all the proteins. Therefore, cells are passaged a minimum of five to six times in the presence of the labeled amino acids to accomplish full incorporation *(17)*. SILAC techniques are used to create dynamic temporal maps of protein expression changes in response to various manipulations *(18)*. The strength of the SILAC approach resides in its tagless approach to differential

MS analysis, thus alleviating potential LC problems (co-elution issues, etc.). Additionally, the population of cells analyzed is typically homogeneous, thereby simplifying the proteomic complexity normally inherent in tissue. Thus, signals are cleaner, allowing a deeper probe into the proteome and, thereby allowing access to less abundant and potentially more "biologically significant" proteins. The significant downside to SILAC is its difficulty of use in complex organisms, although reports exist of metabolically labeling *Drosophilia* and *C. elegans* for quantitative proteomic analysis *(19)*. Most recently, the SILAC procedure has been successfully applied to the mouse as well *(35)*. Additionally, a potential arginine to proline interconversion can take place if the arginine is not kept at sufficiently low concentrations in the culture media. This interconversion can lead to artifactual loss of arginine signal. However the most likely problem faced when using SILAC techniques is the issue of cost. Stable isotopes of amino acids can be very expensive and specialized cell culture media is required. Also, serum additives generally cannot be used due to the potential for introducing unlabeled amino acids to the cells. Not all cells grow and thrive in serum-free conditions.

Numerous other methods exist *(20)* for labeling, such as ^{18}O labeling, and peptide acylation, more commonly referred to as a global internal standard technique (GIST). However, ICAT, iTRAQ, and SILAC represent the most commonly used techniques, and it is relatively straightforward to perform the labeling chemistry, as these are available in kit form from vendors.

2. Materials

2.1. Hardware Required

1. Refrigerated bench-top centrifuge capable of 18–20,000 g.
2. SpeedVac® (Thermo Fisher, Inc. Waltham, MA) or similar vacuum concentrator.
3. 60 °C heating block.
4. 37 °C incubator.
5. Vortexer.
6. pH paper.
7. 2.5 ml Hamilton syringe with a blunt 22-gauge needle.
8. Standard lab pipetters.
9. Tissue grinder with ground glass sufaces (Kontes, Vineland, NJ).
10. PepClean C18 columns, (cat. no. 89870, Pierce, Milwaukee, WI,).
11. Probe sonicator.

More sophisticated equipment may be required depending on the level of sample pre-fractionation desired. This can include a system such as:

1. Mini-Rotofor® (BioRad, Hercules, CA).
2. FPLC, or HPLC capable of performing reverse phase chromatography.

These more specialized pieces of equipment will not be covered further, but may be mentioned where appropriate.

2.2. Specialized Reagents/Solutions Required

2.2.1. Lysate Generation

1. Hanks buffered saline solution (HBSS) without calcium chloride, magnesium chloride, or magnesium sulfate (Invitrogen, Carlsbad, CA,).
2. RIPA lysis buffer (or some store bought similar lysis solution such as Pierce's T-Per): 1% NP-40 (or Triton X-100), 1% sodium deoxycholate, 0.1% SDS, 0.15 *M* NaCl, 0.01 *M* sodium phosphate, pH 7.2
3. Protein estimation kit (e.g., cat. no. 23235 Pierce MicroBCA kit,).

2.2.2. iTRAQ Labeling

1. iTRAQ labeling system kit (Applied Biosystems, Foster City, CA).
2. Acetonitrile, (high purity, store at room temperature).
3. 0.5% fresh trifluoroacetic acid in water (preferably made from stock TFA packaged in glass ampoules, store at room temperature).
4. Methanol (high quality, such as HPLC grade, store at room temperature).
5. Absolute ethanol (high quality, such as HPLC grade, store at room temperature).
6. 2% SDS (molecular biology grade) in ddH_2O.
7. 50 m*M* Tris-(2-carboxyethyl) phosphine (TCEP) in ddH_2O.
8. 200 m*M* Methyl methanetiosulfonate (MMTS), or 200 m*M* iodoacetamide, both in isopropanol.
9. 0.5 *M* Triethylammonium bicarbonate, pH 8.5, in ddH_2O.
10. 1 *M* KCl in ddH_2O.
11. Cation exchange loading buffer for chromatographic peptide clean up: 10 m*M* KH_2PO_4 in 25% acetonitrile.
12. Cation exchange chromatography cartridge cleaning solution: 10 m*M* KH_2PO_4, pH 3.0 in 25% acetonitrile/1 *M* KCl.
13. Chromatographic elution buffer: 10 m*M* KH_2PO_4, pH 3.0, (pH using 10 m*M* K_2HPO_4) in 25% acetonitrile/350 m*M* KCl.
14. Trypsin of the highest quality and is treated with L-1-tosylamido-2-phenylethyl chloromethyl ketone (TPCK) (*see* **Section 3.3**). We have successfully used both Applied Biosystems (cat. no. 4352157) and Sigma-Aldrich (cat. no. T6567, St. Louis, MO). Also, *see* **Note 1** on optional use of immobilized trypsin systems. Whatever the source of trypsin, it must be mass spectrometry grade.
15. Acetone of high purity, store at room temp.

3. Methods

Conceptually, the labeling of a sample for iTRAQ analysis is straightforward, and uses well characterized chemistries familiar to most protein chemists. As detailed in the stepwise workflow diagramed in **Fig. 21.3** and explained in detail in **Section 1.1**, the methods describe the following steps:

1. generate protein lysates,
2. reduce disulfide bonds,
3. block the reactive cysteines,
4. digest the proteins to their constituent peptides,
5. differentially label the peptides of each sample with one of the iTRAQ reagents,

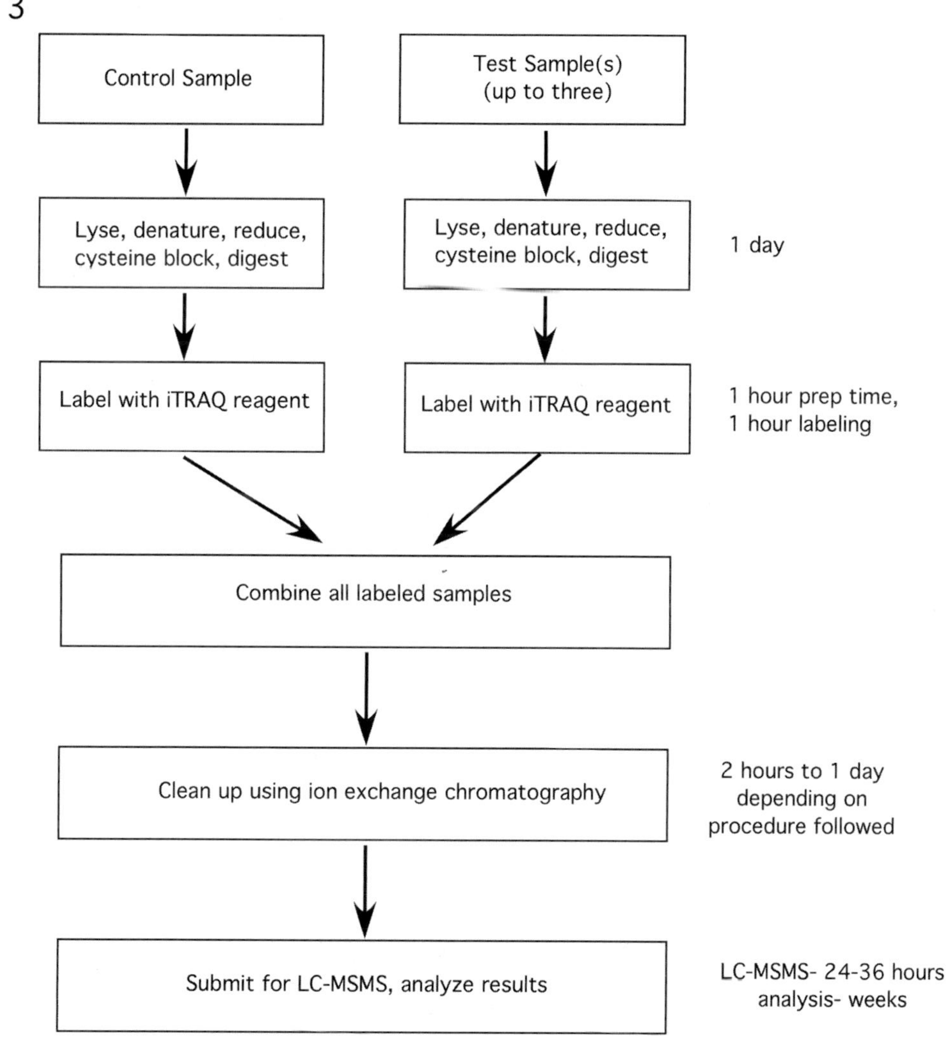

Fig. 21.3. Flow diagram for iTRAQ labeling procedure. See Methods for specific information on each step.

6. combine samples, and
7. submit for mass spectrometer analysis and analyze results.

Sample preparation, processing, and labeling will be covered stepwise. However, only general points of mass spectrometry data analysis will be covered, however, due to the variations and complexities introduced by experiment specific issues. We suggest consultin mass spectrometry core personnel. The goal here is to alert and inform the reader to issues directly under the control of the biologist that can impact downstream analysis so that properly informed decisions can be made where necessary.

3.1. Sample Preparation

3.1.1. Cell Culture

1. Generate samples by manipulating flasks/dishes of cells as required for the experiment (i.e. drug application, activation by bioactive molecules, etc.). Keep samples separate through these steps. We use OC-K3 cells, but the steps detailed here are useful for any cell line or primary cell culture. Numerous cell lines are now available for those interested in cell biological processes of auditory system derived cells *(21, 22)*.
2. Wash cells gently in ice cold Hank's buffered saline solution (HBSS) 2 times.
3. Remove HBSS each time by aspiration, but do not allow the cells to dry. It is vitally important to carry out the lysis steps at 4 °C to inhibit proteolysis.
4. Immediately add the minimal volume of lysis buffer needed to cover the cells, and incubate on ice for approximately 10–20 min. Any formulation of lysis buffer should suffice (RIPA buffer, T-Per or M-Per from Pierce, etc.), and the exact composition of the lysis buffer will depend on the proteins targeted for analysis. For example, isolation and analysis of membrane proteins will demand a different lysis buffer from the standard lysis buffer for cytoplasmic proteins. Usually, the differences between lysis buffers are in detergent and/or osmolarity used. No protease inhibitors should be present, as this will inhibit the tryptic digest in downstream steps.
5. Scrape and transfer cell/lysis buffer solution to an appropriate sized tube. Usually a 1.5–2 ml snap top eppendorf tube will suffice, but this will depend on the size of the dish being lysed.
6. Optional step: Sonicate, on ice, to further disrupt cells, denature genomic DNA, and decrease viscosity of sample.
7. Centrifuge at 18–20,000 g in a refrigerated centrifuge at 4 °C for 20 min to pellet cellular debris
8. Transfer supernatant to fresh tube. Lysate should optimally be used immediately, but can be stored at −80 °C as required.
9. Estimate protein concentration. We have had good success with the Pierce MicroBCA kit (cat. no. 23235). For future use, have on hand approximately 250 μg − 1 mg of protein per sample condition. Adjust the concentration of the sample

to 10 mg/mL. The final protein lysate can be stored at −80 °C for future use.

3.1.2. Tissues

Tissues represent a more difficult sample to deal with, owing primarily to the greater complexity of tissue compared to cells in culture. However, this should not dissuade one from such experiments, but rather alert one to the potential pitfalls associated with working with highly complex material. One major determinant of the success of proteomic analysis of tissue samples is the presence and successful depletion of plasma in the sample. Plasma is in such abundance that it often times masks proteins from detection. Various methods exist for depleting tissue samples of plasma (or at least lessening the contribution of plasma to the protein mixture). For simplicity, we suggest first perfusing ice cold saline transcardially through the animal in an attempt to clear the blood while also slowing proteolysis. Here, the investigator may need to adjust the procedure depending on the requirements for the experiment.

1. Process tissue to remove blood by transcardial perfusion of ice cold saline,
2. Dissect quickly and immerse in lysis buffer at 4 °C in a tissue grinder.
3. Grind tissue to fully lyse the sample. When isolating cochlea, care should be taken to extract any cerebellar tissue from the recess holding the flocculus. Entire cochlear samples can be prepared, or cochlea microdissected without the bone capsule prior to lysis.
4. An optional step utilizing a probe sonicator can be performed. Sonication must be performed at 4 °C. The advantage to sonication is the resultant decrease in viscosity due to the breakdown of the genomic DNA. Experience with the tissue of interest will reveal whether this is a significant problem to be addressed.
5. Follow **steps 6–10 in Section 3.1.1**.

3.2. Sample Processing

Numerous substances in the sample mixture can potentially interfere with the final labeling reaction. Chief among these are amine-containing compounds such as ammonium sulfate, -bicarbonate, -citrate, etc. These will interfere with the labeling process by competing for the iTRAQ label. Other issues to be concerned with are the presence of thiols that are typically introduced into sample mixtures by the addition of DTT or 2-βmercaptoethanol (2βME), which can interfere with the cysteine blocking steps, and high amounts of detergents or denaturants, which will interfere with the tryptic digest step. Should any of these be present in the sample, one must perform a standard acetone precipitation step to isolate the sample from the interfering substance. If DTT or 2βME are present, the acetone precipitation should be performed immediately prior to moving on further in the procedure. If either detergents or primary amines are present, acetone precipitation can be performed either after reducing the sample and blocking the free cysteines (in the case of detergents that might be necessary for maintaining sample solubility), or just prior to tryptic digest (in the case of the presence of primary amines). The best practice is to avoid the need for precipitation at all, as this step represents a major point in the overall procedure where uncontrolled sample loss can occur, thereby introducing bias in the relative amounts of proteins present in each sample. However, precipitation cannot always be avoided. If the trypsin digest will employ an immobilized trypsin (*see* **Note 1**), acetone precipitation will be required, and the sample should be re-suspended only in ddH_2O or a buffer without primary amines. In this case, the acetone precipitation should be performed after denaturing, reducing, and cysteine blocking steps detailed in **Section 3.2.2**. Prior to acetone precipitation, one may also wish to enrich for phosphoproteins (*see* **Note 2 3**).

3.2.1. Acetone Precipitation

Standard acetone precipitation techniques can be used to isolate the proteins of a sample from potentially interfering substances, or to concentrate the sample.

1. Transfer sample to a tube that can hold approximately 10X the sample volume.
2. Chill both the sample and the acetone to 4 °C.
3. Add six volumes of the cold acetone to the sample in the larger tube.
4. Cap the tube, and invert the sample 3–5 times to thoroughly mix the sample and acetone.
5. Incubate the tube at −20 °C for 4 h. A precipitate should become clearly visible.
6. Briefly centrifuge at 4 °C to pellet the precipitate (18,000 g for 2 min).
7. Decant off the acetone from the tube and proceed immediately to the reducing and blocking steps (**Section 3.2.2**). Do not allow the pellet to dry.

3.2.2. Denaturing, Reducing, and Cysteine Blocking the Sample

In order to best label the proteins in the sample, one first must denature the sample to allow equal access to all possible modifiable amino acids carrying primary amines. Because denaturing with chemicals such as SDS will unfold proteins, but not destroy secondary structure induced by the presence of disulfide bonds, the denatured proteins must then be chemically reduced, generally using either DTT or some other reducing agent. The iTRAQ kit uses tris-(2-carboxyethyl)phosphine (TCEP) for this purpose. TCEP is used to denature the sample so that thiols are not introduced to the sample (*see* **Section 3.2**). Finally, the reduced cysteines must be blocked. Typically, this is done with iodoacetamide (at a concentration of 200 m*M* in isopropanol). One may use this, but the iTRAQ kit incudes methylmethane-thiosulfonate (MMTS, 200 m*M* in isopropanol) for cysteine reduction. The advantage of MMTS is that it is a reversible blocker that can be useful should one decide to fractionate the sample by selectively isolating cysteine-containing peptides. We have found the procedure employed by the iTRAQ kit to be easy to follow and to yield excellent labeling.

1. Add 20 μL of dissolution buffer (0.5 *M* triethylammonium bicarbonate, pH 8.5) per tube containing up to 100 μg of protein sample or the acetone precipitated pellet. In order to have ample sample for labeling and processing, one may wish to label 250 μg of protein. Simply scale up the reaction as required. It is critically important, however, to not allow the protein concentration to significantly decrease due to an increase in volume. If the sample is insoluble at a final concentration of 5 mg/mL, *see* **Note 2** for alternatives.
2. Add 1 μL of the denaturant (2% SDS).
3. Check the pH of the solution by spotting some of the sample onto pH paper. The pH should be above 8.0.
4. Vortex to mix well, but avoid formation of bubbles/foam. Spin at low speed if necessary to eliminate foam.
5. Add 2 μL of reducing agent (50 m*M* TCEP) and vortex again to ensure complete mixing.
6. Incubate tubes at 60 °C in a heat block for 1 h
7. Pulse spin briefly to collect sample at the bottom of the tube.
8. Add 1 μL cysteine blocking reagent (200 m*M* MMTS or iodoacetamide).
9. Vortex to mix, and incubate at room temperature for 10 min.

3.3. Trypsin Digest

Trypsin is a pancreatic serine endoprotease derived from pancreatic trypsinogen following a removal of its N-terminal leader sequence. Trypsin is highly selective in its cleavage of peptide bonds, and cleaves only those bonds in which the carboxyl group is contributed by arginine or lysine. When trypsin is treated with L-1-tosylamido-2-phenylethyl chloromethyl ketone (TPCK-treated), any contaminating chymotrypsin activity is irreversibly inhibited. Additionally, when the lysine residues of trypsin are modified by a reductive methylation process, autolytic cleavage of trypsin is also irreversibly inhibited. This results in a loss of trypsin peptide fragments in the mass spectra. Neither treatment alters the trypsin activity toward the proteins present in the lysates. TPCK modification is especially important in iTRAQ labeled mass spectrometry. Contaminating chymotryptic digests result in cleavage of peptides at the carboxyl side of tyrosine, tryptophan, and phenylalanine (i.e. amino acids containing phenyl rings). With longer digests, chymotrypsin will also hydrolyze other amide bonds, particularly those with leucine-donated carboxyls. These digests do not guarantee that a modifiable residue will be present in each resultant peptide. Following tryptic digestion, one can be sure that each peptide generated possesses at least a single modifiable amino acid that can be labeled by the iTRAQ reagents. Thus, TPCK modified trypsin should always be used. Immobilized trypsin (*see* **Note 1**) is also especially useful, since autolysis, and therefore spectral peaks derived from trypsin, is virtually eliminated. However, immobilized trypsin is not provided with the iTRAQ kit. The investigator should determine the need for elimination of trypsin-derived peaks and balance the extra effort/cost with the desired results.

3.3.1. Digesting Samples with Trypsin

1. Dissolve trypsin at 1 μg/μL in ddH_2O or if using trypsin from the iTRAQ kit, use 25 μL Milli-Q standard H_2O per tube. If preparing more than two samples for analysis, prepare 50–60 μg of trypsin, or two tubes from kit.
2. Vortex to mix thoroughly.
3. To each 100 μg of sample in dissolution buffer from **Section 3.2.2, step 10**, add 10 μL of trypsin solution. Keep the final volume below 50 μL, or SpeedVac® as appropriate prior to the addition of trypsin.
4. Vortex to mix thoroughly.
5. Incubate at 37 °C for 16 h (overnight).

3.4. Labeling Samples with iTRAQ Reagents

The heart of the iTRAQ technology lies in the ability of the investigator to efficiently and completely label each peptide in a sample with a traceable, quantifiable tag. Each sample is processed separately to label the peptides.

1. Equilibrate the iTRAQ reagent to room temperature before use.
2. Add 70 μL of absolute ethanol to each iTRAQ label tube that will be used.
3. Vortex approximately 1 min to ensure complete mixing and dissolving.
4. Transfer the entire volume of one iTRAQ label reagent tube to one sample tube such that each sample tube receives one iTRAQ reagent. Usually, controls are labeled with the 114 reagent, but of course this is up to the investigator. It is critical to keep track of which sample received which label, as downstream quantification depends on the proper ratio of control to manipulated state.
5. Vortex to mix, and then pulse spin to collect all solutions to the bottom of the tube.
6. Incubate at room temperature for 1 h.
7. Combine the label reactions into one tube, such that a single tube receives each label singly, without duplication of labels if doing more than four labels (for the 4-plex kit, 8 if using the soon to be released 8-plex kit).

3.4.1. Assessing Completeness of Labeling Reaction

The first time one performs the iTRAQ labeling procedure, it may be worth the added effort to first assess one's success in fully labeling a sample. Without complete labeling, some peptides will not be represented in the reporter region of the spectra (see further discussion below), and therefore the ratio of peptides will be altered due to loss of labeled sample, despite the fact that the peptides are actually present in the sample. A simple test can be performed to assess the completeness with which the sample is labeled with the aid of C18 spin columns. These columns are good general tools for purifying and concentrating peptide samples.

1. Follow **steps 1–6 in Section 3.4**, but only label one digested sample, using the 114 label
2. Dry sample in a SpeedVac® and resuspend in 20 μL of 0.5% trifluoroacetic acid (TFA) in 5% acetonitrile per 30 μg total peptide.
3. Calculate the number of columns required. The number of columns needed depends on the amount of peptide being run. The binding capacity of each column is 30 μg, so calculate the number relative to the binding capacity of the columns and the amount of peptide used.
4. Wet the resin of the spin column with 200 μl of 50% methanol.
5. Spin at 1500 g for one min.
6. Repeat steps 4 and 5, discarding the flow-through each time.
7. Add 200 μL of 0.5% TFA in 5% acetonitrile to each spin column.
8. Spin at 1500 g for one min and discard the flow through.
9. Repeat steps 7 and 8.
10. Apply the labeled sample onto the bed of the spin column.
11. Spin at 1500 g for one min at room temperature.
12. Set flow-through aside in case sample binding needs to be verified.
13. Apply 200 μL of 0.5% TFA in 5% acetonitrile to wash the column and spin at 1500 g for 1 min.
14. Discard flow-through and repeat step 13.
15. Elute the sample from the column by adding 20 μL of 70% acetonitrile to the bed of the column and centrifuging at 1500 g for 1 min.
16. Repeat step 15 with fresh 70% acetonitrile.
17. Dry the sample in a SpeedVac®.
18. Resuspend the sample in 30 μl dissolution buffer.
19. Repeat the labeling reaction as outlined in **Section 3.4, steps 1–6**, using the 117 iTRAQ labeling reagent.

If the initial labeling reaction was not carried to completeness, the sample should contain both the 114 and the 117 labels in the mass spectrum analysis. As long as no 117 label is detected, the initial labeling reaction can be considered complete, and there would be little need to further analyze labeling efficiency in the future unless significant changes to sample preparation are encountered.

3.5. Preparing the Combined Sample for Mass Spectrometry

The entire volume of all samples is combined in a 1:1:1:1 ratio. The most common method of analyzing the iTRAQ sample is via LC-MS/MS. Thus, the sample must be prepared so that it will be efficiently nebulized during the electrospray introduction into the mass spectrometer. At this stage, it is recommended that personnel who run the samples through the mass spectrometer be consulted, because the type of LC one performs will dictate to some degree the final preparation of the sample. A clean up procedure of the sample is included here as a general guide.

Many compounds potentially present in the sample can interfere with the LC-MS/MS analysis. These include, but are not limited to, the dissolution buffer, ethanol, TCEP, any salts, excess iTRAQ labeling reagent, denaturants, and detergents. Thus, a simple method for cleaning the sample needs to be used. Cation exchange chromatography can be used to isolate the sample from interfering compounds and allows one to replace the sample diluent with a more "mass spec friendly" diluent. A variety of cation exchange systems can be purchased. One system that uses spin column and vacuum plate technology for ion exchange chromatography is the VivaPure system (Sartorius Biotech, Inc., Göttingen, Germany). The system appears straightforward to use, but we have had no first-hand experience with this. Included in the Applied Biosystems Methods Development Kit is a system for performing cation exchange chromatography on the sample prior to submission for mass spectrometry. However, this system is useful only for relatively simple samples such as that obtained from phosphopeptide enriched samples, samples that have undergone some other type of affinity enrichment, or samples from cell culture. Complex samples such as those obtained from tissue lysates will require more complicated fractionation, perhaps even over a 2-dimensional LC system, prior to MS analysis. For complex samples, or if more complex fractionation is desired (thus allowing one to uncover signatures of the less abundant proteins in the sample), the appropriate personnel should be consulted.

1. Assemble the cation exchange system as per manufacturer's instructions. Most important is the fitting between the syringe and the needle to port adapter. Be sure that the system is snug to keep solutions from backing up and out of the system.
2. The concentration of undesirable materials such as salts and detergents should be diluted by adding ten volumes (relative to the sample) of 10 m*M* KH_2PO_4 in 25% of acetonitrile (loading buffer).
3. Vortex to mix thoroughly.

4. Ensure the pH of the solution by spotting an aliquot onto pH paper. The pH must be between 2.5 and 3.3. Add more loading buffer as needed to attain the proper pH. The cation exchange column will only work properly when the peptides are carrying a cationic charge. The charge allows them to adhere to the column matrix while inorganics, such as salts, and organics such as acetonitrile, can be flushed away.
5. Prepare the column to accept the sample by injecting 1 mL of 10 m*M* KH_2PO_4 in 25% acetonitrile/1 *M* KCl, pH 3.0 (conditioning buffer). Allow flow through to go to waste. All injections should be made with a slow and steady pace.
6. Inject 2 mL of loading buffer into the cartridge and divert flow-through to waste.
7. Inject diluted sample at a rate of approximately 1 drop/s into the column. Collect the flow-through and save until sure that the peptides were maintained on the column.
8. Inject 1 mL of loading buffer to wash the column. Save in the same tube as used in step 7. This step may be repeated as necessary.
9. Elute retained peptides by slowly (~ 1 drop/s) injecting 500 μL of 10 m*M* KH_2PO_4 in 25% acetonitrile/350 m*M* KCl, pH 3.0 (elution buffer). Collect the eluate into one tube.

3.6. Mass Spectrometric Analysis of iTRAQ Labeled Samples

The iTRAQ system makes use of isobaric tagging reagents (**Fig. 21.2B**) that carry a highly efficient and accurate peptide reactive group, a balance group, and one of a series of reporter groups of differential mass that is capable of maintaining its charge. The tags co-elute during typical reverse phase liquid chromatography performed prior to injection into the mass spectrometer, and are cleavable under MSMS conditions. Upon cleavage in the collision cell of a mass spectrometer, the tag is released from the peptide, thereby giving rise to two important regions of the spectrum: (1) the sequencing portion of the spectrum, where the data for peptide sequence is maintained; and (2) the reporter region of the spectrum, in which the isobaric tags are found. **Figure 21.4** illustrates these regions as viewed in ProteinPilot (*see* **Note 4**). Finally, the reporter region is typically a relatively "quiet" region in the spectrum, and thus one can be confident that the peaks observed are derived from the tag itself. If other peaks are present in the reporter region, this may be an indication of noise resulting from poor sample processing, contaminants, etc.

3.7. What to do with all of Those Data

Both proteomic and genomic analyses are wonderful in that one receives so much data. However, therein also lies the problem (and the danger) – how to verify the results, and what to do with the results once verified. Lists of proteins expressed are useful,

4

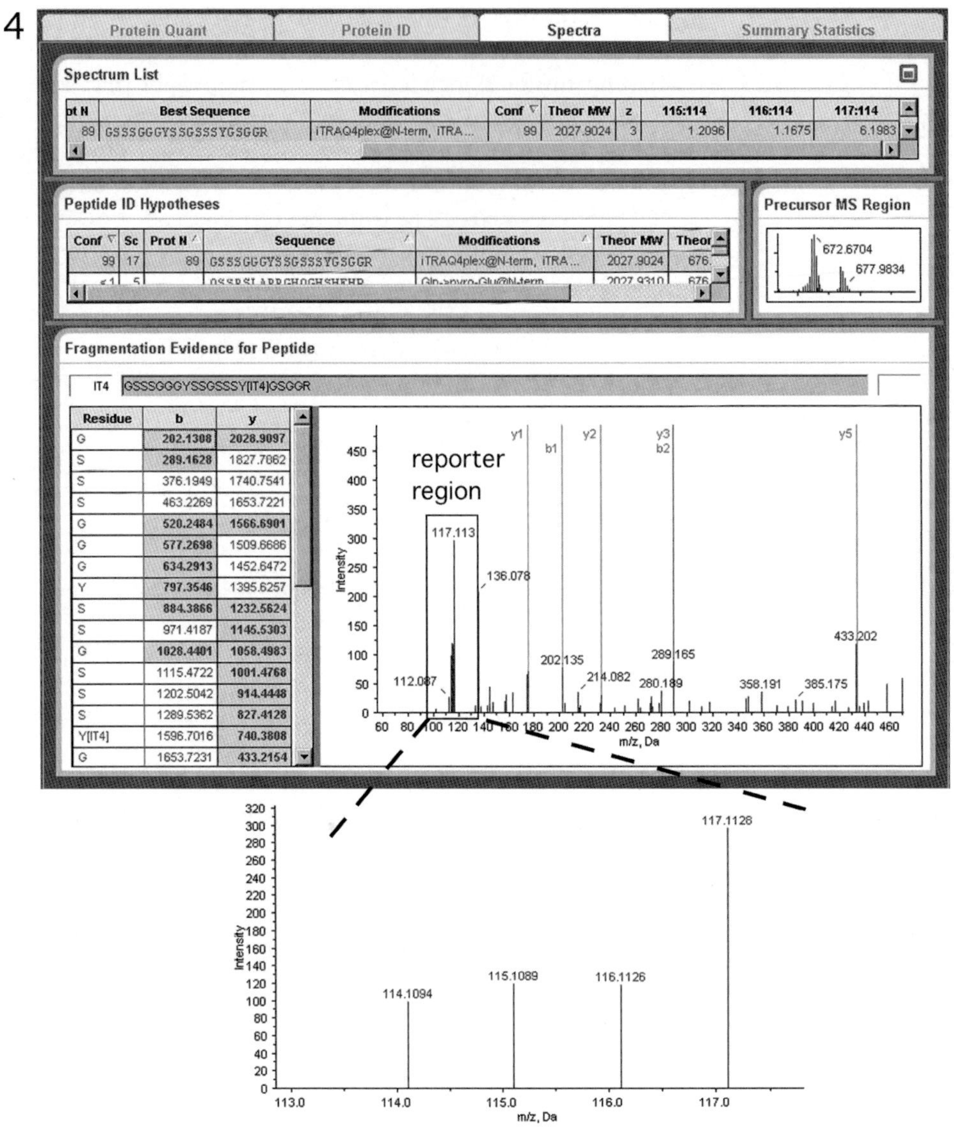

Fig. 21.4. ProteinPilot screen shot of typical results of an iTRAQ experiment. When examining the spectra of a peptide of interest (spectrum list), ratios of labels are indicated. Here, 114 was used as the normal state, and therefore the denominator for the ratio. Additionally, the spectrum can be inspected for manual annotation and analysis if desired in the fragmentation window. Regions can be zoomed in as indicated here to visually inspect the reporter region of the spectrum (*dashed lines*). In this case, this peptide was upregulated only during conditions in state 3 ($\sim$ 6X upregulated), while under conditions 1 and 2 (labeled with 115 and 116 reporters, respectively), no change from baseline (label 114) is detected.

but only at a superficial level. Verification can take the form of western blot analysis, follow-up proteomic analysis, etc. Bioinformatic approaches to proteome analyses are maturing at a steady and ever increasing pace. At this stage, of course, the investigator must decide in which direction to proceed. Listed in **Note 5** are URLs that can be used as a jumping off point to investigate various databases and software. Of particular interest are those that detail methods by which protein:protein interactions are modeled *(23–26)*, and tools with which to generate and investigate interaction pathways. *(27–29)*.

As is clearly evident, protocols that can be advised and published are simply the beginnings of the dive into proteomic space. However, it deserves reiteration here that without proper handling of samples prior to generation of the spectra and quantification of the results, all downstream applications will be compromised. Thus, it is with this in mind that we attempt here to get the investigator off to a good start by informing them of this technology and the major potential pitfalls along the way.

4. Notes

1. Immobilized trypsin can be obtained from Pierce (cat. no. 20230) or Sigma (cat. no. TT0010) in which trypsin is cross-linked to agarose beads and is present as a slurry. To use immobilized trypsin, follow these steps:
   (a) Wash 100–250 μL of the trypsin bead slurry with 500 μl of water or a trypsin digestion buffer without primary amines. Repeat for a total of three times, each time centrifuging the beads down (~ 2000 g) and decanting or pipetting off the supernatant. Be sure to thoroughly re-suspend the beads with each wash and vortex to ensure complete mixing.
   (b) Resuspend the slurry in 200 μL of water or a trypsin digestion buffer without primary amines.
   (c) The sample should be resuspended in water or a trypsin digestion buffer without primary amines at a final concentration of 2 μg/μL.
   (d) Add the sample to the resuspended trypsin slurry.
   (e) Vortex to ensure complete mixing.
   (f) Incubate 16 h (overnight) at 37 °C with end-over-end mixing if possible.
   (g) Isolate the digest by spinning the slurry down and collecting the supernatant.
   (h) SpeedVac® the digest and bring back up in dissolution buffer to 30 μL.
2. On occasion, the sample may be insoluble in the dissolution buffer concentration plus SDS as added. Should this be a problem, the dissolution buffer volume can be increased up to 50 μL without adversely affecting the labeling step. If more is required, one should first SpeedVac® the sample to near

dryness, and add up to 100 μL of dissolution buffer. Should this still not solve a problem of an insoluble sample, two alternatives exist. One may either use a detergent/denaturant or a different buffer. An addition of 1 μL of 2% SDS per 20 μL of the dissolution buffer will keep the concentration of SDS low enough so as to not interfere with the future trypsin digest. Use of denaturants or detergents other than SDS are possible, and included octyl β-D glucopyranoside (OG), NP-40, Tween-20, Triton X-100, and CHAPS (all at less than or equal to a 1% final concentration), or urea at less than 1 *M*. In all cases, the aqueous partition of the sample must not be higher than 40% to avoid problems with final labeling. Any other buffers tried should not carry primary amines, and should buffer at pH 8.0–8.5. These include, but are not limited to, BES, BICINE, CHES, HEPES, MOBS, MOPS, and PIPES. Concentration should be kept at approximately 0.3 *M* such that at the labeling step, the buffer concentration does not fall below 0.06 *M*.

3. Many proteins are regulated by post-translational modifications. One of the most studied modifications is the addition of phosphate groups to distinct residues of proteins. If one is interested in assessing phosphorylation states on proteins in the samples, various enrichment protocols can be followed. The reader is directed to the Pierce Phosphoprotein Enrichment Kit (cat. no. 90003) as a potential starting point, although others exist from different vendors. These columns are simple to use, and yield a significant enrichment of phosphoproteins, but at the expense of losing non-phosphorylated proteins from the sample.
4. A viewer version of Protein Pilot can be obtained as freeware distributed by Applied Biosystems for use with data generated by their QStar and QTrap mass spectrometers. The exact file type the investigator will receive back from the mass spectrometry facility will depend on the type of instrument used, local preferences for software, etc. Here we will only present analysis performed with Protein Pilot software. However, other software will perform in a similar manner. Data represented in ProteinPilot include proteins detected in the sample, the ratio of tags for that protein, individual peptide sequence and tag ratios, and a spectral characterization of tag intensities. From these four simple windows, one can quickly assess relative protein concentrations between the samples analyzed. However, one should also inspect the protein identification window to make investigator-based decisions concerning identifications. In the best cases, one should expect detection of multiple peptides of the protein being identified and these peptides should span numerous regions of the protein. Thus, peptides from N- and C-term regions, as well as internal regions should be

present to allow for the best identification. However, this is not always available, and so further investigation in these cases needs to be done to assess the veracity of the identification in light of incomplete coverage.

5. Useful URLs detailing protein interaction databases, and pathway visualization and structure analysis include:
   (a) Human protein interaction database *(30)*, http://www.hpid.org;
   (b) Human protein reference database, http://www.hprd.org;
   (c) Mammalian protein-protein interaction database (MIPS), htpp:// mips.gsf.de/proj/ppi;
   (d) Molecular interactions database (MINT) *(31)*, http://mint.bio.uniroma2.it/mint;
   (e) Proteomics Identifications database (PRIDE) *(32, 33)*, http://www.ebi.ac.uk/pride;
   (f) Protein-protein interaction network visualization software Cytoscape *(34)* GenePro plug-in for Cytoscape, http://genepro.ccb.sickkids.ca/index.html;
   (g) ProViz http://cbi.labri.fr/eng/proviz.htm

References

1. Anderson, L. and Seilhamer, J. (1997) A comparison of selected mRNA and protein abundances in human liver. *Electrophoresis* **18**, 533–537.
2. Gygi, S., Rochon, Y., Franza, B., and Aebersold, R. (1999) Correlation between protein and mRNA abundance in yeast. *Mol. Cell Biol.* **19**, 1720–1730.
3. Grant, S. and Blackstock, W. (2001) Proteomics in neuroscience: From protein to network. *J. Neurosci.* **21**, 8315–8318.
4. Gavin, A., Aloy, P., Grandi, P., Krause, R., Boesche, M., Marzioch, M., et al. (2006) Proteome survey reveals modularity of the yeast cell machinery. *Nature* **440**, 631–636.
5. Coughenour, H., Spaulding, R., and Thompson, C. (2004) The synaptic vesicle proteome: A comparative study in membrane protein identification. *Proteomics* **4**, 3141–3155.
6. Andersen, J., Lam, Y., Leung, A., Ong, S., Lyon, C., Lamond, A., and Mann, M. (2005) Nucleolar proteome dynamics. *Nature* **433**, 77–83.
7. Yates, J., Gilchrist, A., Howell, K., and Bergeron, J. (2005) Proteomics of organelles and large cellular structures. *Nat. Rev. Mol. Cell Biol.* **6**, 702–714.
8. Langnaese K., Seidenbecher C., Wex H., Seidel B., Hartung K., Appeltauer U., Garner A., Voss B., Mueller B., Garner C. C., and Gundelfinger E. D. (1996) Protein components of a rat brain synaptic junctional protein preparation. *Brain Res. Mol. Brain Res.* **42**, 118–122.
9. Yates, J. (1998) Mass spectrometry and the age of the proteome. *J. Mass Spectrometry* **33**, 1–19.
10. Yates, J. (2000) Mass spectrometry. From genomics to proteomics. *Trends Genet.* **16**, 5–8.
11. Fountoulakis, M. (2004) Application of proteomics technologies in the investigation of the brain. *Mass Spectrometry Rev* **23**, 231–258.
12. Ferguson, P. and Smith, R. (2003) Proteome analysis by mass spectrometry. *Ann. Rev. Biophys. Biomol. Struct.* **32**, 399–424.
13. Mann, M., Hendrickson, R., and Pandey, A. (2001) Analysis of proteins and proteomes by mass spectrometry. *Ann. Rev. Biochem.* **70**, 437–473.
14. Gygi, S., Rist, B., Gerber, S., Turecek, F., Gelb, M., and Aebersold, R. (1999) Quantitative analysis of complex protein mixtures using isotope-coded affinity tags. *Nature Biotechnol.* **17**, 994–999.
15. Shiio, Y. and Aebersold, R. (2006) Quantitative proteome analysis using isotope-coded

affinity tags and mass spectrometry. *Nature Prot.* **1**, 139–145.

16. Ong, S. and Mann, M. (2006) A practical recipe for stable isotope labeling by amino acids in cell culture (SILAC). *Nature Prot.* **1**, 2650–2660.
17. Amanchy, R., Kalume, D., and Pandey, A. (2005) Stable isotope labeling with amino acids in cell culture (SILAC) for studying dynamics of protein abundance and posttranslational modifications. *Sci. STKE* **267**, pl2.
18. Blagoev, B., Ong, S., Kratchmarova, I., and Mann, M. (2004) Temporal analysis of phosphotyrosine-dependent signaling networks by quantitative proteomics. *Nature Biotechnol.* **22**, 1139–1145.
19. Krijgsveld, J., Ketting, R. F., Mahmoudi, T., Johansen, J., Artal-Sanz, M., Verrijzer, C. P., Plasterk, R. H., and Heck, A. J. (2003) Metabolic labeling of *C. elegans* and *D. melanogaster* for quantitative proteomics. *Nat. Biotechnol.* **21**, 927–931.
20. Julka, S. and Regnier, F. (2004) Quantification in Proteomics through Stable Isotope Coding: A Review. *J. Proteome Res.* **3**, 350–363.
21. Rivolta, M., Grix, N., Lawlor, P., Ashmore, J., Jagger, D., and Holley, M. (1998) Auditory hair cell precursors immortalized from the mammalian inner ear. *Proc. Biol. Sci.* **265**, 1595–1603.
22. Rivolta, M. and Holley, M. (2002) Cell lines in inner ear research. *J. Neurobiol.* **53**, 306–318.
23. Bork, P., Jensen, L. J., von Mering, C., Ramani, A. K., Lee, I., and Marcotte, E. M. (2004) Protein interaction networks from yeast to human. *Curr. Opin. Struct. Biol.* **14**, 292–299.
24. Sharan, R., Suthram, S., Kelley, R. M., Kuhn, T., McCuine, S., Uetz, P., Sittler, T., Karp, R. M., and Ideker, T. (2005) Conserved Patterns of Protein Interaction in Multiple Species. *Proc. Natl. Acad. Sci. U.S.A.* **102**, 1974–1979.
25. Sharan, R., Ulitsky, I., and Shamir, R. (2007) Network-Based Prediction of Protein Function. *Mol. Systems Biol.* **3**, 88.
26. Fernandez-Ballester, G. and Serrano, L. (2006) Prediction of Protein-Protein Interaction Based on Structure. *Meth. Mol. Biol.* **340**, 207–234.
27. Iragne, F., Nikolski, M., Mathieu, B., Auber, D., and Sherman, D. (2005) Proviz: Protein interaction visualization and exploration. *Bioinformatics* **21**, 272–274.
28. Yan, W., Lee, H., Yi, E. C., Reiss, D., Shannon, P., Kwieciszewski, B. K., Coito, C., Li, X. J., Keller, A., Eng, J., Galitski, T., Goodlett, D. R., Aebersold, R., and Katze, M. G. (2004) System-based proteomic analysis of the interferon response in human liver cells. *Genome Biol.* **5**, R54.
29. Suderman, M. and Hallett, M. (2007) Tools for Visually Exploring Biological Networks. *Bioinformatics* **23**, 2651–2659.
30. Han, K., Park, B., Kim, H., Hong, J., and Park, J. (2004) Hpid: The human protein interaction database. *Bioinformatics* **20**, 2466–2470.
31. Chatr-aryamontri, A., Ceol, A., Palazzi, L. M., Nardelli, G., Schneider, M. V., Castagnoli, L., and Cesareni, G. (2007) Mint: The molecular interaction database. *Nucleic Acids Res.* **35**, D572–574.
32. Jones, P., Cote, R. G., Cho, S. Y., Klie, S., Martens, L., Quinn, A. F., Thorneycroft, D., and Hermjakob, H. (2008) Pride: New developments and new datasets. *Nucleic Acids Res.* 36, D878–D883.
33. Jones, P., Cote, R. G., Martens, L., Quinn, A. F., Taylor, C. F., Derache, W., Hermjakob, H., and Apweiler, R. (2006) Pride: A public repository of protein and peptide identifications for the proteomics community. *Nucleic Acids Res.* **34**, D659–D663.
34. Shannon, P., Markiel, A., Ozier, O., Baliga, N. S., Wang, J. T., Ramage, D., Amin, N., Schwikowski, B., and Ideker, T. (2003) Cytoscape: A software environment for integrated models of biomolecular interaction networks. *Genome Res.* **13**, 2498–2504.
35. Krüger, M., Moser, M., Ussar, S., Thievessen, I., Luber, C.A., Fomer, F., Schmidt, S., Zanivan, S., Fässler, R., Mann, M. (2008) SILAC mouse for quantitative proteomics uncovers Kindlin-3 as an essential factor for red blood cell function. Cell 134, 353–364.

Part III

Imaging Protocols

Chapter 22

Fluorescence Microscopy Methods in the Study of Protein Structure and Function

Heather Jensen-Smith, Benjamin Currall, Danielle Rossino, LeAnn Tiede, Michael Nichols, and Richard Hallworth

Abstract

As more and more proteins specific to hair cells are discovered, it becomes imperative to understand their structure and how that contributes to their function. The fluorescence microscopic methods described here can be employed to provide information on protein-protein interactions, whether homomeric or heteromeric, and on protein conformation. Here, we describe two fluorescence microscopic methodologies applied to the outer hair cell-specific membrane protein prestin: the intensity and fluorescence lifetime (FLIM) variants of FRET (Fluorescence Resonance Energy Transfer), used in the study of protein-protein interactions, and the Scanning Cysteine Accessibility Method (SCAM), used for the determination of protein conformation. The methods are readily adaptable to other proteins.

Key words: Prestin, FRET, FLIM, SCAM, confocal microscopy, cyan fluorescent protein, yellow fluorescent protein.

1. Introduction

An abundance of new proteins has been revealed in hair cells by biochemical and genetic strategies *(1–3)*. Frequently, the mechanisms and functions of these new proteins are as yet unclear. Here, we describe fluorescence microscopic methods applied to the study of a novel protein specific to the outer hair cell, prestin *(4)*. This membrane protein is believed to be the agent of outer hair cell motility, which in turn mediates the outer hair cell contribution to cochlear amplification *(5)*. In the first method, hypotheses concerning putative multimerization by prestin are tested using Fluorescence (sometimes, Förster) Resonance Energy Transfer

Bernd Sokolowski (ed.), *Auditory and Vestibular Research: Methods and Protocols, vol. 493*

DOI 10.1007/978-1-59745-523-7_22 Springerprotocols.com

(FRET) *(6)*. FRET is the non-radiative exchange of energy between fluorophores and occurs only when two fluorophores with overlapping emission and excitation spectra, referred to as the donor and the acceptor, are sufficiently close (in practice, about 5 nm). If FRET is occurring, excitation of the donor leads not only to donor photon emission but also to photon emission by the acceptor in it's wavelength range. The two modified fluorescent proteins (FPs), cyan fluorescent protein (CFP) and yellow fluorescent protein (YFP), make an excellent FRET pair and can be introduced into a cell-based expression system as FP constructs of the proteins of interest *(7)*. In our experiments, FRET between prestin-CFP and prestin-YFP constructs in co-transfected HEK 293 cells is used to demonstrate multimeric self association of prestin *(8, 9)*. Two variants of FRET are used, an intensity-based procedure that separates CFP and YFP emissions by linear un-mixing *(10)*, and a Fluorescence Lifetime IMaging (FLIM) method *(11–13)* that examines the lifetimes of excited CFP fluorophores. For both procedures, the acceptor photobleach method is used, which reduces concerns about cross-talk between donor and acceptor channels *(9, 14)*. In the second method, hypotheses concerning the structure of prestin in the plasma membrane (PM) are examined. Competing and mutually incompatible 10 and 12 transmembrane (TM) models for prestin have been advanced *(15, 16)*. We test these models using the Scanning Cysteine Accessibility Mutagenesis method (SCAM) *(17)*. Prestin, coupled C-terminal to enhanced GFP (eGFP), is expressed in HEK 293 cells. Cysteines, whether naturally occurring or introduced by mutagenesis, are alkylated by a membrane-impermeable thiol-reactive reagent (maleimide) coupled to biotin in living cells. Binding to the biotin by streptavidin coupled to a fluorophore (in this case, Alexa Fluor 568, chosen because its emission band is easily separable from that of eGFP) is used to determine whether a given amino acid position is present on the extracellular surface of prestin.

2. Materials

2.1. Cell Culture and Transient Transfection (FRET/FLIM, SCAM)

1. Dulbecco's Modified Eagle's Medium (DMEM) supplemented with 10% fetal bovine serum (FBS) (Biomeda, Burlingame, CA).
2. 0.25% trypsin, 1X.
3. Lipofectamine 2000™ reagent (Invitrogen, Carlsbad, CA).
4. 25 cm^2 cell culture flask for cell culture maintenance.
5. 35 mm glass-bottomed culture dishes (MatTek, Ashland, MA) for experiments.

6. Human embryonic kidney (HEK) 293 cell line (ATCC, Manassas, VA).
7. eGFP-N2 plasmid (Clontech).
8. Cerulean-CFP (cCFP) plasmid, obtained by material transfer agreement from the D. Piston laboratory, Vanderbilt University *(18)*.
9. Venus-YFP (vYFP) plasmid, obtained by material transfer agreement from the A. Miyawaki laboratory, Brain Science Institute, RIKEN, Saitama, Japan *(19)*.

2.2. Mutagenesis (SCAM)

1. QuikChange II XL Site Directed Mutagenesis Kit (Stratagene, La Jolla, CA).
2. Template DNA - cDNA (gerbil prestin), obtained as a gift from Jian Zuo, cloned into plasmid vector EGFP-N2 (Clontech, Mountain View, CA) to generate a gerbil prestin-GFP fusion protein (*see* **Note 1**).
3. Mutagenic primers are designed according to Stratagene primer guidelines. Primers should be 25–45 base pairs in length, not directly overlapping. The mutated base pairs should also be located in the middle of the primers. Primers should be phosphorylated at the 5' end.
4. MyCycler Personal Thermal Cycler (Bio-Rad, Hercules, CA).

2.3. Maleimide Labeling (SCAM)

1. Dulbecco's Phosphate Buffered Saline (DPBS) (Invitrogen).
2. N^{α}-(3-maleimidylpropionyl) biocytin (biotin maleimide) (Molecular Probes, Eugene, OR). Prepare a 1–10 m*M* stock solution in dimethyl sulfoxide (DMSO).
3. 2-mercaptoethanol.
4. Phosphate Buffered Saline (PBS): 22 m*M* $NaH_2PO_4.H_20$, 78 mM anhydrous Na_2HPO_4, 1 L distilled water, 0.1 g thimerosal (as an antimicrobial).
5. Paraformaldehyde (PFA): Prepare as a fresh 4% (w/v) solution in PBS.
6. Wash buffer: 0.1% bovine serum albumin (BSA) in PBS.
7. Streptavidin-Alexa Fluor 568 conjugate (Molecular Probes): Prepare a 1:200 working solution in wash buffer from a 1:1 stock in distilled water.

2.4. Confocal Microscopy (FRET/FLIM, SCAM)

1. A Zeiss LSM 510 META NLO scanning confocal microscope (Zeiss MicroImaging, Thornwood, NY) with a 63X 1.4 numerical aperture oil objective, modified by addition of a R3809U photomultiplier detector (Hamamatsu, Bridgewater, NJ) coupled to a single photon counter board (SPC, Becker & Hickl, Berlin, Germany).
2. A META detector (available as a component of the confocal microscope), consisting of a band of 32 narrow band (10.2 nm) detectors covering the range 360–700 nm.

3. An argon laser with lines at 488 nm for excitation of GFP, 458 nm for excitation of cCFP, and 514 nm for bleaching of vYFP.
4. A Helium-Neon laser at 543 nm for excitation of Alexa Fluor 568 (standard equipment on many confocal microscopes).
5. A Chameleon ULTRA titanium:sapphire laser (Coherent, Santa Clara, CA) for two-photon excitation of cCFP (800 nm).
6. An SPC-40 computer board and SPCImage software for acquisition and analysis of photon lifetimes (Becker & Hickl, Berlin, Germany).
7. LSM Image Browser software for off-line image analysis (Zeiss MicroImaging).
8. ImageJ software for image analysis (open source, obtained from http://rsb.info.nih.gov/ij).
9. Mounting Medium A (MMA): solution of equal parts PBS and glycerol, 1% (w/v) n-propylgallate, and 1.5 μg/mL 4', 6-diamidino-2-phenylindole, dilactate (DAPI) (Molecular Probes), used in SCAM studies.
10. Mounting Medium B (MMB): solution of equal parts PBS and glycerol, used in FRET/FLIM studies. N-propylgallate is omitted because it may interfere with FRET, and DAPI is omitted because its emission band overlaps that of cCFP.
11. Round glass cover slips (30 × 0.15 mm).
12. Rubber cement.

3. Methods

3.1. Construct Generation for FRET and FLIM

1. Generate constructs containing protein coding sequences for the specific protein constructs of interest. Also, construct or obtain from another investigator a plasmid vector expressing a construct consisting of cCFP and vYFP linked by a short linker sequence (e.g., GGGPGGG) containing a central proline. This construct will bring cCFP and vYFP into proximity and will thereby serve as a positive control for FRET. As a negative control, consider a fusion construct of vYFP and a related protein not found in the same tissue.
2. Human embryonic kidney cells (HEKs) are plated onto 35 mm glass-bottomed culture dishes. Once cells reach 80–90% confluency (2–3 days) they are transfected with various cCFP and/or vYFP coupled constructs using Lipofectamine™ 2000 reagent. Our optimum transfection efficiency is achieved with a DNA: Lipofectamine™ 2000 ratio of 4 μg (total):10 μl respectively (*see* **Note 2**). After an 18 h incubation period, the medium is replaced with DMEM

culture medium (*see* **Section 2.1, step 1**). Transfected cells are ready for fixation 24–48 h post addition of the transfection reagents.

3. HEK cells are fixed with 4% PFA in PBS for 30 min, mounted using MMB, and covered with glass coverslips. To preserve fluorescence, dishes are sealed with rubber cement (*see* **Note 3**) and stored at 4 °C.

3.2. Acceptor Photobleaching Intensity Fluorescence Resonance Energy Transfer (apFRET)

1. Linear unmixing is used to separate spectra of overlapping fluorophores when both are excited by the same excitation wavelengths. Both cCFP and vYFP are excited by 458 nm. Spectra are acquired at each pixel of an image using the META detector. The spectra at each pixel is considered to consist of the weighted sum of the spectra of the two fluorophores present, i.e.,

$$\text{for wavelengths } \lambda = \lambda_1, \lambda_2 \ldots\ldots\ldots\ldots \lambda_n,$$

$$S(\lambda) = A \times Donor(\lambda) + B \times Acceptor(\lambda),$$

where $S(\lambda)$ is the spectrum observed at the pixel, $Donor(\lambda)$ and $Acceptor(\lambda)$ are the spectra for the donor and acceptor over the same wavelength range, and A and B are weighting constants proportional to the quantity of each fluorophore in each pixel. The goal is to determine A and B from the set of linear equations above. This can be readily achieved by least squares fitting in Excel or in software supplied by Zeiss (*see* **Note 4**).
2. Donor and acceptor spectra are obtained from cells transfected with the cCFP construct only or the vYFP construct only, using the 458 nm laser line and the META detector set to collect 472–643.2 nm fluorescence (*see* **Note 5**).
3. A pre-acceptor photobleaching cCFP/vYFP image from a cell with both cCFP and vYFP labeled protein(s) is acquired under the above acquisition parameters.
4. Linear un-mixing. The cCFP and vYFP specific portions of this image are assigned to separate profiles using Zeiss LSM software or Excel to unmix the overlapping spectra using the spectra acquired in **Section 3.2, step 2** (**Fig. 22.1A–C**).
5. Acceptor (vYFP) photobleaching using a 514 nm laser line and band pass filter of 535–590 nm is performed in one or more regions of interest (ROIs) containing cCFP/vYFP fluorescence. Consecutive scans are used to reach at least an 80% decrease in vYFP intensity.
6. A post-acceptor photobleaching cCFP/vYFP image is acquired and unmixed as described above (*see* **Section 3.2, step 3**) (**Fig. 22.1D–F**).
7. FRET is detected as an increase in cCFP (donor) fluorescence intensity after photobleaching of vYFP (acceptor). FRET

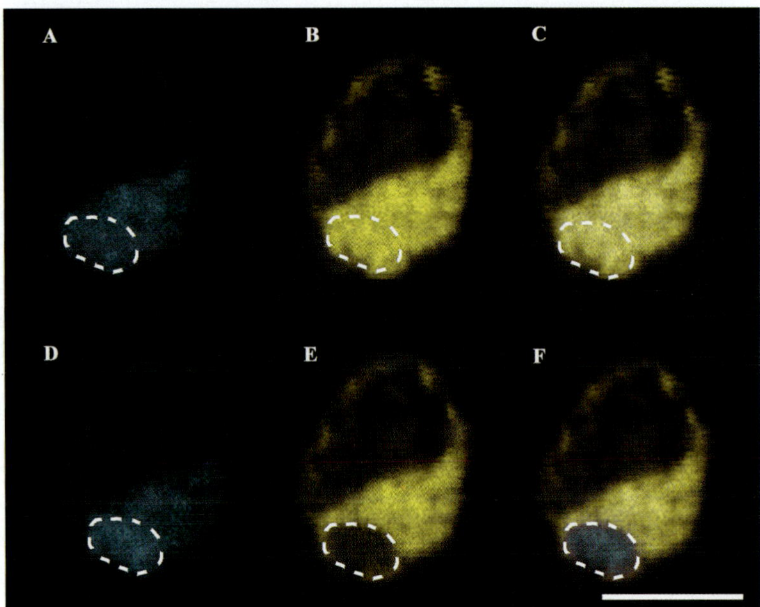

Fig. 22.1. Intensity-based measurements of apFRET in a specific ROI (*dashed lines*) of cells transfected with with plasmids expressing prestin-FP C-terminal constructs. Prestin-cCFP (donor) and prestin-vYFP (acceptor) fluorescence intensity before (**A, B**) and after (**D, E**) prestin-vYFP photobleach (C, F merged images). The observed increase in prestin-cCFP fluorescence after acceptor photobleach indicates intermolecular FRET between prestin molecules and in turn indicates prestin multimerization. Scale bar = 20 μm.

efficiency E is calculated as:

$$E = \left[\frac{(A_1 - A_0)}{A_0}\right] \times 100\%,$$

where A_1, A_0 are respectively the cCFP fluorescence intensities after and before vYFP photobleaching in the ROI. cCFP fluorescence intensities of non-photobleached ROIs in the same scan window are used to calculate potential baseline shifts in cCFP intensity occurring during the imaging process. These values can be used to normalize the data set.

3.3. Fluorescence Lifetime Imaging (FLIM) apFRET

1. A pre-acceptor photobleaching cCFP fluorescence lifetime image is acquired with 800 nm fs-duration pulses from a tunable Chameleon ULTRA Titanium:Sapphire laser with a pulse repetition rate of 80 MHz using the Zeiss confocal microscope (**Fig. 22.2A**). Emitted photons are filtered by a 500 nm short-pass filter and detected with a R3809U photomultiplier detector. A model SPC-40 board, synchronized with the laser pulses and the confocal scan signal, is used to measure photon emission latency. Photons are accrued for 60–120 s (*see*

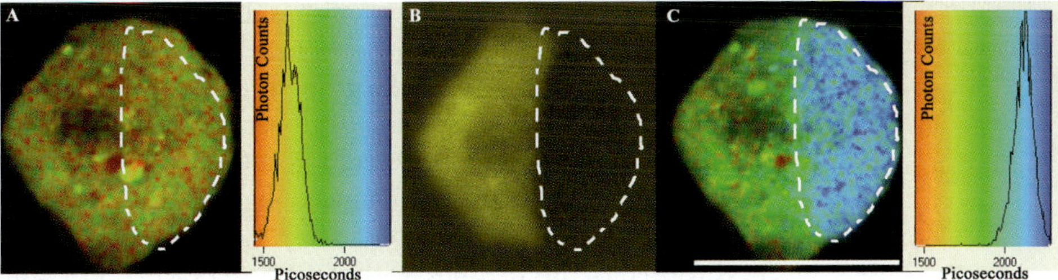

Fig. 22.2. cCFP lifetime measurements in an apFRET experiment in a specific region of interest (*dashed lines*). cCFP fluorophore lifetimes from an HEK cell transfected with a plasmid expressing a cCFP-vYFP linked construct, before (**A**) and after (**C**) vYFP (acceptor) photobleaching (**B**, shown as fluorescence intensity). Scale bar = 20 μm. Lifetime color pseudoscale to the right of each lifetime figure.

Note 6). Photon count rates are kept well below 1 photon per pulse to avoid double counting artifacts.

2. Acceptor (vYFP) photobleaching is performed as previously described in **Section 3.2, step 3** (**Fig. 22.2B**).
3. A post-acceptor photobleaching lifetime image is acquired as previously described in **Section 3.3, step 1** (**Fig. 22.2C**).
4. Photon latency histograms, displaying accumulated pixel information from specific ROIs before and after acceptor photobleaching, are generated for each pixel in the ROI with SPCImage. Mean lifetime distributions (in picoseconds) weighted by intensity for each ROI are fit with two time constants with a χ^2 value between 1.0 and 2.0. FRET efficiency E is calculated as:

$$E = \left[1 - \frac{\tau_{DA}}{\tau_D}\right]\%,$$

where τ_D and τ_{DA} are the lifetimes of the donor alone (post-vYFP photobleaching) and the donor in the presence of the acceptor (pre-vYFP photobleaching), respectively.

3.4. Mutagenesis

1. A predicted extracellular loop residue in the protein of interest should be mutated to a cysteine residue with the QuikChange II XL Site Directed Mutagenesis kit. A mutation to a cysteine residue in a predicted intracellular loop can act as a negative control (*see* **Note 7**).

3.5. Preparation of Samples for SCAM

1. HEK 293 cells are maintained in DMEM supplemented with 10% FBS on 25 cm^2 tissue culture flasks. After cells reach 80% confluency, they are removed from the cell culture flask with trypsin and split 1/10 on 35 mm culture dishes for the experimental procedure. A 1/10 split will provide experimental dishes that are ready for transfection in 48 h. One experimental dish is sufficient for each wild type and

mutant construct. Make two dishes for negative controls (*see* **Note 8**).

2. Transfect each dish with 4 μg of either Wt or mutated DNA using 10 μl Lipofectamine 2000™ according to Invitrogen's recommendations. Check cells for GFP synthesis in 48 h. Mock transfect (Lipofectamine only, no DNA present) the negative control dishes.
3. Rinse cells three times with DPBS. The medium must be completely aspirated.
4. Incubate the cells with biotin maleimide (obtained from the 1–10 m*M* stock solution in DMSO) dissolved in PBS to a final concentration of 50 μ M for 10 min at room temperature. The concentration of DMSO in PBS must not exceed 1% (v/v).
5. Treat cells for 1 min with 2-mercaptoethanol to a final concentration of 14 m*M* to remove excess maleimide.
6. Aspirate the solution from the cells. Rinse cells thoroughly with PBS three times for 5 min each.
7. Fix cells in fresh 4% PFA for 30 min at room temperature.
8. Remove the PFA and rinse cells in wash buffer three times for 5 min each.
9. Under light restricted conditions at room temperature, incubate cells in streptavidin-Alexa Fluor 568 (1:200) for 1 h.
10. Discard the streptavidin-Alexa Fluor 568 and rinse the cells with wash buffer three times for 5 min each.
11. Mount the cells with MMA and cover with a round glass coverslip. Remove excess mounting media. Seal the cover slip with rubber cement. Store dishes at 4°C in the dark.

3.6. Confocal Microscopy for SCAM

1. The samples are viewed on a Zeiss LSM 510 META NLO scanning confocal microscope equipped with a 63X oil objective. GFP is excited using a 488 nm laser line and collected with a band pass filter of 500–530 nm. Streptavidin-Alexa Flour 568 is excited using a 543 nm laser line and collected with a band pass filter of 565–615 nm.
2. For integral membrane proteins such as prestin, the PMs of 10–12 cells are imaged in each experimental group to determine the optimal imaging parameters. These parameters are averaged within and between each experimental group, and are used to acquire all of the subsequent images. For each experimental group, images of 20 transfected cells are acquired for data analysis.

3.7. Analysis of Fluorescence for SCAM

1. Raw pixel intensity data exported from GFP (green channel) and biotin maleimide-streptavidin-Alexa Flour 568 (red channel) are used to compare fluorescence intensities within and between experimental groups.

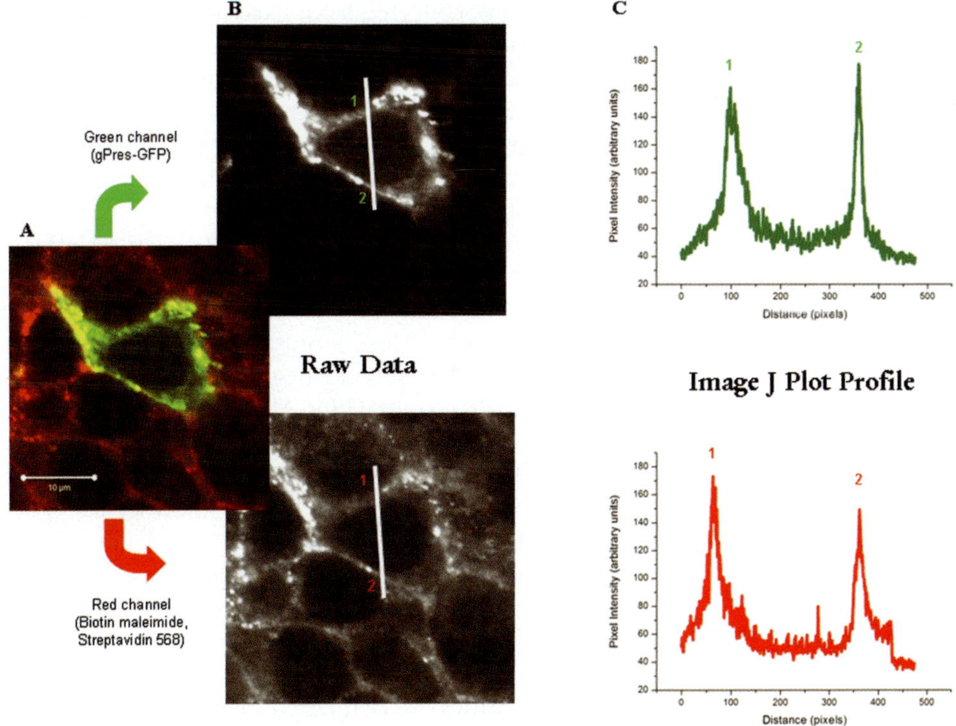

Fig. 22.3. Collected images and analysis of peak fluorescence intensity amplitudes. (**A**) Confocal image of combined GFP (*green* channel) and biotin maleimide-strepavidin-Alexa Flour 568 (*red* channel) which is exported into (**B**) raw pixel intensity data. (**C**) The peak fluorescence intensity amplitudes in the plot profile are obtained from the white line as in (**B**).

2. Fluorescence intensities are compared by averaging peak fluorescence amplitudes from two or more representative PM regions using ImageJ software (**Fig. 22.3**) (*see* **Note 9**).
3. The ratio of Alexa Flour 568 intensity to GFP intensity (*see* **Note 10**) is compared between experimental groups. Addition of a cysteine to the extracellular loop should increase the ratio of Alexa Flour 568 fluorescence intensity to GFP fluorescence intensity, while additions to the intracellular loops or transmembrane regions will not.

4. Notes

1. Gerbil (or rat) prestin are the preferred prestin models. Mouse prestin, for unknown reasons, does not reliably incorporate into the membrane of HEK 293 cells.
2. Optimum transfection efficiency may require alteration of the DNA (μg): Lipofectamine™ 2000 ratio.
3. Use of acetone-based sealants will reduce or eliminate fluorescence from fluorescent proteins.

4. A third spectrum, collected from fixed untransfected cells, may be used for background subtraction by unmixing.
5. Detector settings and fluorophore absorptions will vary with the FRET pair.
6. Although increasing photon collection can increase the data available for calculating lifetimes, prolonged imaging can also lead to fluorophore bleaching.
7. Mutated proteins must be functional (structural effects are minimal).
8. The first negative control is incubated with biotin maleimide and then streptavidin-Alexa Flour 568, in the absence of plasmid transfection. The second negative control is transfected but incubated with streptavidin-Alexa Flour 568 alone.
9. In the first negative control, in which biotin maleimide and streptavidin-Alexa Flour 568 is added and there is no plasmid transfection, there should be minimal fluorescence. The first negative control consists of the background fluorescence of native HEK 293 integral membrane proteins with cysteines in their extracellular loops. In the second negative control (Streptavidin-Alexa Fluor 568 alone), there should be no fluorescence in the red channel.
10. The GFP intensities are compared between groups to ensure there are no significant differences in protein insertion in the PM between experimental groups.

Acknowledgments

Supported by DC 02053 (to RH), RR17417-01 (to Creighton University), and NSF-EPSCoR EPS-0346476 (to RH).

References

1. Steel, K. P. and Kros, C. J. (2001) A genetic approach to understanding auditory function. *Nat. Genet.* **27**, 143–9.
2. Frolenkov, G. I., Belyantseva, I. A., Friedman, T. B., and Griffith, A. J. (2004) Genetic insights into the morphogenesis of inner ear hair cells. *Nat. Rev. Genet.* **5**, 489–98.
3. Bork, J. M., Peters, L. M., Riazuddin, S., Bernstein, S. L., Ahmed, Z. M., et al. Usher syndrome 1D and nonsyndromic autosomal recessive deafness DFNB12 are caused by allelic mutations of the novel cadherin-like gene CDH23. (2001) *Am. J. Hum. Genet.* **68**, 26–37.
4. Zheng, J., Shen, W., He, D. Z., Long, K. B., Madison, L. D., and Dallos, P. (2000) Prestin is the motor protein of cochlear outer hair cells. *Nature* **405**, 149–55.
5. Zheng, J., Madison, L. D., Oliver, D., Fakler, B., and Dallos, P. (2002) Prestin, the motor protein of outer hair cells. *Audiol. Neurootol.* 7, 9–12.
6. Gordon, G. W., Berry, G., Liang, X. H., Levine, B., and Herman, B. (1998) Quantitative fluorescence resonance energy transfer measurements using fluorescence microscopy. *Biophys. J.* **74**, 2702–13.
7. Pollok, B. A. and Heim, R. (1999) Using GFP in FRET-based applications. *Trends Cell Biol.* **9**, 57–60.
8. Wu, X., Currall, B., Yamashita, T., Parker, L. L., Hallworth, R., and Zuo, J. Prestin-prestin and prestin-GLUT5 interactions in HEK293T cells. (2007) *Dev. Neurobiol.* **67**, 483–97.
9. Hallworth, R., Currall, B., Nichols, M. G., Wu, X., and Zuo, J. (2006) Studying inner ear

protein-protein interactions using FRET and FLIM. *Brain Res.* **1091**, 122–31.

10. Dickinson, M. E., Bearman, G., Tille, S., Lansford, R., and Fraser, S. E. (2001) Multispectral imaging and linear unmixing add a whole new dimension to laser scanning fluorescence microscopy. *Biotechniques* **31**, 1272, 74–6, 78.
11. Chen, Y. and Periasamy, A. (2004) Characterization of two-photon excitation fluorescence lifetime imaging microscopy for protein localization. *Microsc. Res. Tech.* **63**, 72–80.
12. Elangovan, M., Wallrabe, H., Chen, Y., Day, R. N., Barroso, M., and Periasamy, A. (2003) Characterization of one- and two-photon excitation fluorescence resonance energy transfer microscopy. *Methods* **29**, 58–73.
13. Murata, S., Herman, P., and Lakowicz, J. R. (2001) Texture analysis of fluorescence lifetime images of AT- and GC-rich regions in nuclei. *J. Histochem. Cytochem.* **49**, 1443–51.
14. Karpova, T. S., Baumann, C. T., He, L., Wu, X., Grammer, A., Lipsky, P., et al. (2003) Fluorescence resonance energy transfer from cyan to yellow fluorescent protein detected by acceptor photobleaching using confocal microscopy and a single laser. *J. Microsc.* **209**, 56–70.
15. Navaratnam, D., Bai, J. P., Samaranayake, H., and Santos-Sacchi, J. (2005) N-terminal-mediated homomultimerization of prestin, the outer hair cell motor protein. *Biophys. J.* **89**, 3345–52.
16. Deak, L., Zheng, J., Orem, A., Du, G. G., Aguinaga, S., Matsuda, K., and Dallos, P. (2005) Effects of cyclic nucleotides on the function of prestin. *J. Physiol.* **563**, 483–96.
17. Zhu, Q. and Casey, J. R. (2007) Topology of transmembrane proteins by scanning cysteine accessibility mutagenesis methodology. *Methods* **41**, 439–50.
18. Rizzo, M. A., Springer, G. H., Granada, B., and Piston, D. W. (2004) An improved cyan fluorescent protein variant useful for FRET. *Nat. Biotechnol.* **22**, 445–9.
19. Nagai, T., Ibata, K., Park, E. S., Kubota, M., Mikoshiba, K., and Miyawaki, A. (2002) A variant of yellow fluorescent protein with fast and efficient maturation for cell-biological applications. *Nat. Biotechnol.* **20**, 87–90.

Chapter 23

Ion Imaging in the Cochlear Hair Cells

Gregory I. Frolenkov

Abstract

Regulation of important cellular functions via signaling pathways is a fundamental property of the cell. Intracellular Ca^{2+} is probably a best known second messenger in cell biology. In mechanosensory cells of the inner ear, the hair cells, intracellular Ca^{2+} participates in a variety of functions including mechano-electrical transduction, synaptic transmission, and efferent regulation of the outer hair cells, one of two types of hair cells in the mammalian cochlea. The outer hair cells are responsible for the amplification of sound-induced vibrations within the cochlea, which determines the sensitivity of mammalian hearing. Besides Ca^{2+}, another intracellular ion, Cl^- may have very specific function in the same outer hair cells. Intracellular Cl^- is required for the motor function of prestin, a unique plasma membrane molecular motor of these cells. The goal of this article is to review practical aspects of the techniques suitable for imaging of Ca^{2+} and Cl^- in live mammalian cochlear hair cells.

Key words: Calcium, chloride, fluorescence, organ of Corti, hair cells.

1. Introduction

There are approximately 15,000–30,000 sensory hair cells in the mammalian inner ear. Mammalian hair cells exit the cell cycle in embryonic development and never proliferate after differentiation. Biochemical characterization of the hair cell proteins is usually a daunting task due to very limited amounts of material that could be obtained from one ear. Therefore, most of the proteins essential for the sensory function of hair cells have been and most likely will be identified through the alternative approaches, such as positional cloning of the genes responsible for different forms of hereditary deafness in humans or the analysis of mouse models of hereditary hearing loss.

Bernd Sokolowski (ed.), *Auditory and Vestibular Research: Methods and Protocols, vol. 493*

DOI 10.1007/978-1-59745-523-7_23 Springerprotocols.com

The next steps after identification of a new protein essential for hearing is usually to determine where it is expressed, to generate knockout or transgenic mouse models, and to determine the function of the protein. In many cases, the last step represents a "bottleneck", because the complexity of cell physiology does not allow using any stereotyped experimental technique for determining the function. However, some techniques are used more widely than other ones. An example of a widely used technique is ion imaging in living cells.

Regulation of important cellular functions via signaling pathways is a fundamental property of the cell. Intracellular Ca^{2+} is one of the most widely used second messengers in cell physiology. In mammalian hair cells, intracellular Ca^{2+} participates in a variety of cellular functions including mechano-electrical transduction *(1)*, synaptic transmission *(2)*, and efferent regulation of outer hair cell electromotility *(3)*. Besides Ca^{2+}, another intracellular ion, Cl^{-} may have very specific functions in outer hair cells. Intracellular Cl^{-} is required for the motor function of prestin, a unique plasma membrane molecular motor that is critical for the amplification of sound-induced vibrations within the cochlea *(4, 5)*. The goal of this article is to review the practical aspects of techniques suitable for imaging Ca^{2+} and Cl^{-} in live mammalian cochlear hair cells. The article describes critical considerations in the hardware design, protocols for preparing the specimens, rationale for choosing specific fluorescent probes, and step-by-step instructions for loading the specimens with fluorescent indicators as well as performing both ratiometric and non-ratiometric ion imaging. Finally, I describe the artifacts that are most common in the studies of the cochlear hair cells.

2. Materials

2.1. Imaging Setup

The success of ion imaging depends almost entirely on the available hardware. A system customized for a particular application usually provides better or at least the same performance as the general-purpose expensive equipment that one can find in core facilities. The most critical pieces of the hardware are listed below. The considerations that need to be kept in mind, while building a new or customizing the existing Ca^{2+}-imaging setup, are discussed in **Notes 1–5**. Additional information on live cell microscopy can be found at http://micro.magnet.fsu.edu/.

1. Mechanically stable inverted advanced research microscope (e.g. Axiovert 200, Zeiss, Germany; Eclipse TE2000, Nikon, Japan; or IX71, Olympus, Japan). Upright microscopes (e.g. Axioskop 2 FS, Zeiss; Eclipse FN1, Nikon; or BX51WI, Olympus) may be used for special applications such as imaging of

hair cell stereocilia. However, the best spatial and temporal resolutions can be achieved only with oil-immersion objectives that are typically used on an inverted microscope (*see* **Note 1**). The microscope should be equipped with a basic stage for mounting Petri dishes. In these stages, the specimens are usually mounted on a removable round insert that can be easily rotated. More expensive universal microscope stages that are designed for rectangular slides are not very convenient because they do not allow rotating the sample.

2. Anti-vibration table (e.g. 63–500, Technical Manufacturing Corporation, Peabody, MA; or 9100, Kinetic Systems, Boston, MA). Damping of mechanical vibrations is more critical in live cell imaging as compared to other fluorescent imaging applications (*see* **Note 2**).
3. Confocal attachment (e.g. CARV II, BD Biosciences, San Jose, CA), if needed. Please, note that most ion imaging applications do not require confocal mode (*see* **Note 3**).
4. Microscope objective lens is the most critical element of any ion imaging setup. This lens assembly focuses the excitation light beam of a particular wavelength to the specimen and simultaneously collects the light of a longer wavelength that is emitted by fluorescent ion-sensitive probe (**Fig. 23.1**). The objective must have the highest practically possible numerical aperture (*see* **Note 1**).
5. A digital camera with the quantum efficiency of no less than 50% and active cooling down to at least −20 °C (*see* **Note 4**).
6. A light source that is capable of producing substantial UV illumination and reasonably fast switching of excitation wavelengths (*see* **Note 5**).
7. An image acquisition and processing software (e.g. MetaMorph or MetaFluor, Molecular Devices, Downingtown, PA; or Simple PCI, Hamamatsu Corp., Sewickley, PA).

2.2. Hair Cell Isolation

1. Leibovitz L-15 cell culture medium without phenol red (Invitrogen, Carlsbad, CA). L-15 is the best known bath solution to prolong viability of isolated hair cells.
2. Collagenase type IV (Invitrogen, Carlsbad, CA) dissolved in Ca^{2+} and Mg^{2+} containing Hanks' Balanced Salt Solution (HBSS, cat. no. 14025, Invitrogen, Carlsbad, CA) at 10 mg/mL. This stock solution is stored for not more than 10 days at 4 °C.
3. Hamilton syringe (blunt end, 50 μL, cat. no. 80565, Hamilton Co., Reno, NV), a very helpful tool for cell isolation.
4. Microsurgical rongeur (e.g., cat. no. 14292, World Precision Instruments, Sarasota, FL).
5. Gold Seal® thin, no. 0, rectangular (24 × 60 mm) cover slip glass (cat. no. 63751, Electron Microscopy Sciences, Hatfield, PA) are used as a bottom for the imaging chamber. Similarly

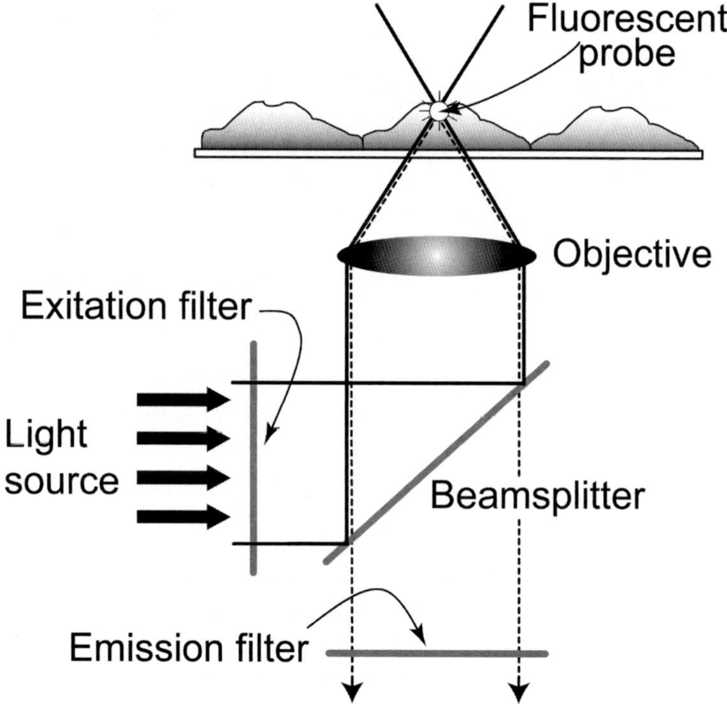

Fig. 23.1. Epifluorescent imaging of ion-sensitive indicators. Broad-spectrum illumination passes a narrow-band excitation filter that is transparent only to the wavelengths that correspond to the absorption spectrum of the fluorescent indicator. These wavelengths are not transparent to the beamsplitter, which reflects the illumination beam to the objective lens. The lens focuses the beam, thereby exciting fluorescence. The emitted fluorescent light has a different spectrum with the maximum in the longer wavelength region. These longer wavelengths are transparent to the beamslitter, thus, passing the emitted light to the eye of the observer or to a digital camera. The narrow-band emission filter blocks any wavelengths that are outside the spectral region of the indicator's maximal fluorescence. A combination of an excitation filter, a beamsplitter, and an emission filter is known as a "filter set", which is generally specific for each ion-sensitive indicator (**Table 23.1**).

thin cover slips may also work, but their transparency to UV illumination must be determined.

2.3. Culturing Organ of Corti Explants

1. Dulbecco' Modified Eagle's Medium (DMEM) with high glucose content (4.5 g/L) and 25 m*M* of HEPES (Invitrogen, Carlsbad, CA). Supplement DMEM with 7% fetal bovine serum (FBS, cat. no. 16000, Invitrogen, Carlsbad, CA).
2. Ready to use stock solution of collagen type I in 0. 02 *N* acetic acid, pH 3.67 (cat. no. 08–115, Upstate, Lake Placid, NY).
3. Ammonium hydroxide (Sigma-Aldrich, St. Louis, MO).
4. Ampicillin stock solution: Rehydrate ampicillin salt (Invitrogen, Carlsbad, CA) in 20 mL of sterile distilled water producing a stock solution of 10 mg/mL that is stored at 4 °C.

5. Glass-bottom Petri dishes (Electron Microscopy Sciences, Hatfield, PA). We found this low-profile model (7 mm height, 50 mm diameter, glass opening diameter of 30 mm) to be the most convenient due to very good access for any perfusion or recording pipettes.
6. Hand-held manual pipette pump (cat. no. 378980000, Bel-Art, Pequannock, NJ). A handy tool for transferring pieces of the organ of Corti.

2.4. Fluorescent Probes

1. Select an appropriate fluorescent probe for the particular application (*see* **Note 6**). Consult **Table 23.1** for available probes, their excitation/emission wavelengths, and recommended filter sets for each fluorescent probe.
2. Stock solutions: Membrane impermeable dyes are generally dissolved in distilled water filtered through $0.22\,\mu m$ syringe filter. Cl^- indicators are typically dissolved in methanol. Cell-permeable Ca^{2+} indicators should be reconstituted using high quality, anhydrous dimethylsulfoxide (DMSO). Once prepared, stock solutions of fluorescent indicators should be stored at $< -20\,°C$, well sealed, frozen, and desiccated. Consult **Table 23.1** for concentrations of stock solutions and solvents.
3. Intrapipette solution for loading a cell via patch pipette (mM): 12.5 KCl, 131 KGlu (potassium gluconate), 2 $MgCl_2$, 8 K_2HPO_4, 2 KH_2PO_4, 2 Mg_2ATP, 0.5 EGTA (ethylene glycol-bis(2-aminoethylether)-N, N, N', N'-tetraacetic acid), 0.2 Na_4GTP. Adjust to pH 7.4 with KOH and osmolarity to 325 mM/kg with D-glucose. Filter the solution with $0.22\,\mu m$ syringe filter twice and make 1 mL aliquots. Store aliquots at $-20\,°C$. To make a working solution, an aliquot of the stock solution of the dye is diluted with the intrapipette solution to a final concentration (*see* **Table 23.1**).
4. Cell-permeable Ca^{2+} indicators for loading the cells: Pre-mix an aliquot of dye stock solution with an equal amount of dispersing agent Pluronic F127 (P-3000MP, Invitrogen, Carlsbad, CA) in order to facilitate cell loading. Then dilute this mixture with L-15 to a final concentration that is dependent on the dye in use (*see* **Table 23.1**).
5. Prepare all working solutions of the fluorescent probes just prior to the experiment and keep on ice. Cover the tubes with aluminum foil to prevent photodecomposition.

2.5. Loading the Cell Via a Patch Pipette

1. Patch clamp amplifier (e.g., Axopatch 200B, Molecular Devices, Sunnyvale, CA; or EPC 10, HEKA Elektronik, Germany). Less expensive amplifiers (e.g., PC-501A, Warner Instruments, Hamden, CT) can be used if the experiments

Table 23.1
Fluorescent probes for imaging Ca^{2+} and Cl^- in cochlear hair cells

	Catalog No[1]	Solvent	M.W.[2]	Qty (mg)	Dissolve (μL)	Stock (mM)	Aliquot volume (μL)	Add F-127 (μL)	Dilute (mL)	Final conc. (μM)	Final solvent (%)	Incubate	Ex[3]	Em[3]	Filter set[4]	Comment
Ca^{2+} indicators (cell-impermeable)				*Stock solution*					*Working solution*							
Fura-2 (K^+ salt)	F-1200	H_2O	832	1	**240**	5	10	N/A	1	**50**	1	N/A	340 380	510	XF04-2	A practical ratiometric probe for Ca^{2+}
Indo-1 (K^+ salt)	I-1202	H_2O	840	1	**238**	5	10	N/A	1	**50**	1	N/A	355	405 495	XF07	An alternative ratiometric probe for low-resolution Ca^{2+} imaging
Fluo-3 (K^+ salt)	F-3715	H_2O	960	1	**208**	5	10	N/A	1	**50**	1	N/A	500	545	XF105-2	Low background fluorescence, very large responses
Oregon Green, BAPTA-1, (K+ salt)	O-6806	H_2O	1114	0.5	**90**	5	10	N/A	1	**50**	1	N/A	475	535	XF115-2	Prominent background fluorescence, good responses
Oregon Green, BAPTA-5N (K^+ salt)	O-6812	H_2O	1159	0.5	**86**	5	10	N/A	1	**50**	1	N/A	475	535	XF115-2	Low affinity indicator for Ca^{2+} levels above 1 μM.

(continued)

Table 23.1 (continued)

Ca^{2+} indicators (cell-permeable)																
Fura-2 (AM)	F-1201	DMSO	1002	1	**100**	**10**	5	5	10	**5**	0.1	20–30 min				
Indo-1 (AM)	I-1203	DMSO	1010	1	**99**	**10**	2	2	4	**5**	0.1	30–40 min				
Fluo-3 (AM)	F-1241	DMSO	1130	1	**89**	**10**	5	5	10	**5**	0.1	20–30 min	These indicators have the same spectral			
Oregon Green, BAPTA-1 (AM)	O-6807	DMSO	1258	0.05	**20**	**2**	2.5	2.5	2	**2.5**	0.25	20–30 min	properties as the cell-impermeable forms			
Oregon Green, BAPTA-5N (AM)	O-6813	DMSO	1303	0.05	**20**	**2**	2.5	2.5	2	**2.5**	0.25	20–30 min	above.			
Chloride indicators																
MQAE	E-3101	Methanol	326	100	**307**	**1000**	10	N/A	2	**5000**	0.5	5–6 h	350	460	XF02-2	A practical Cl^- indicator for hair cell studies
SPQ	M-440	Methanol	281	100	**355**	**1000**	10	N/A	2	**5000**	0.5	5–6 h	350	460	XF02-2	Less sensitive indicator for very large changes of Cl^-

[1]Catalog numbers for Invitrogen products are provided. Similar products may be obtained from other sources (e.g. TEF Labs, Austin, TX).
[2]M.W. – Molecular weight.
[3]Ex – the middle of transparency window of the excitation filter (in nm). Em – the middle of transparency window of the emission filter (in nm).
[4]Catalog numbers for Omega Optical (Brattleboro, VT) products are provided. Equivalent filter sets may be obtained from other sources (e.g. Chroma Technology, Rockingham, VT).

involve ion imaging only and extensive recordings of the ion channel activity are not planned.

2. Micropipette puller (e.g., P-97, Sutter Instruments, Novato, CA).
3. Micromanipulator (e.g., MHW-3, Narishige, Japan; or PCS-5000, Burleigh, Canada).
4. Soda glass or borosilicate capillaries with an embedded filament (e.g. cat. no. 1B100F-4, World Precision Instruments, Sarasota, FL).
5. Non-metallic needle for filling capillaries (cat. no. MF28G-5, World Precision Instruments, Sarasota, FL).

3. Methods

3.1. Isolation of Guinea Pig Cochlear Hair Cells

1. The maximal yield of viable isolated outer and inner hair cells is obtained from guinea pig (*see* criteria of viability below, **Section 3.1, step 5**), when compared to other rodents. Therefore, this protocol assumes the isolation of guinea pig hair cells. However, it can be used also for isolating hair cells from rats or mice but the number of viable cells will be substantially reduced.
2. Adult guinea pigs (200–400 g) are euthanized with CO_2 and decapitated using a guillotine for large rodents (World Precision Instruments). The temporal bones are removed from the skull with a rongeur and placed in L-15. The bulla is opened to expose the cochlea, and the otic capsula is chipped away with a surgical blade (no. 11) starting from the base.
3. Dissect strips of the organ of Corti from the modiolus with a fine needle (gauge no. 26 or 27). Using a Pasteur pipette connected to a hand-held pipette pump, transfer the strips to a 100 μL drop of L-15, containing a membrane-permeable form of an ion-sensitive indicator (e.g. Fura-2/AM). Incubate for a specified time (*see* **Table 23.1**) at room temperature. Please, keep in mind that Table 23.1 provides only approximate incubation times. The optimal incubation times have to be determined experimentally to provide a maximal fluorescent signal without significant compartmentalization of the dye into intracellular organelles (*see* **Note 7**). Approximately 15–20 min before the end of the incubation period, add 1 mg/mL of collagenase to the same drop. If membrane-impermeable forms of a fluorescent indicator will be used, pre-incubate strips of the organ of Corti only with collagenase for 15–20 min.
4. While cells are incubating, affix the cover glass to the inset of the microscope stage (*see* **Section 2.1, step 1**) and mount the inset to the stage of an inverted microscope. After incubation, gently collect all strips of the organ of Corti into a Hamilton

syringe and transfer them to a 100 μL drop of L-15, which is placed on the top of the coverslip lying above the microscope objective. Then gently dissociate the cells by reflux through the needle of a Hamilton syringe and allow cells to settle. If clean cover glass is used, the cells will adhere to the glass and will be ready for imaging after 5–10 min. To maintain the osmolarity during a long period of observation, a laminar flow of L-15 (about 5 mL/h) through a drop of L-15 medium can be arranged using a micro-perfusion system (e.g. BPS-4, ALA Scientific Instruments, Westbury, NY).

5. The viability of isolated hair cells is typically determined by high resolution differential interference contrast (DIC) imaging. The signs of cell damage are: apparent changes in the shape of the cell, extensive Brownian motion of intracellular organelles, damaged stereocilia, and mislocalization of the nucleus.

3.2. Cultured Organ of Corti Preparation

1. Cultured organs of Corti from early postnatal (postnatal days 0 through 5, P0-P5) rats or mice are probably the most physiologically viable *in vitro* preparations for hair cell studies. In this preparation, hair cells exhibit apparently normal morphology and mechanosensitivity for several days, whereas acutely isolated cells typically die within several hours. The protocol below is based on the original technique of organotypic culturing *(6, 7)* that was adapted to commercially available low-profile (7 mm height, 50 mm diameter) glass-bottom Petri dishes.
2. To coat glass-bottom Petri dishes with collagen, dilute the stock solution of collagen in sterile distilled water (1:1) and cover a central portion of the dish with ~ 15 μL of diluted solution. Then, incubate the dishes in ammonium hydroxide vapor for 20 min at room temperature. This step forms the collagen gel at the bottom of the dish and should be performed in a chemical hood. After incubation, wash dishes 2–3 times in distilled water and fill with the cell culture medium (HEPES-containing DMEM). Prepare the dishes at least 6 h before dissecting the organ of Corti and keep them at 37 °C in 5% CO_2 to equilibrate the ionic environment within the collagen gel.
3. Euthanize the animal, according to Protocols established by the Institutional Animal Care and Use Committee, and dissect strips of the organ of Corti. Aspirate the cell culture medium from the Petri dish and place dissected organs of Corti on the collagen layer with the hair cell bundles facing up. Add 200–250 μL of fresh medium and spread the medium in such a way that the surface tension will hold the organ of Corti explant in the middle of the dish. Seal the Petri dishes with Parafilm® and keep them at 37 °C for 6–12 h. Then, remove Parafilm and add additional fresh medium. Keep the cultures at

37 °C in 5% CO_2 thereafter. Experiments are performed within 1–5 days after culturing.

4. Usually, there is no need for an antibiotic to keep organ of Corti cultures sterile. However, 10 mg/L of ampicillin can be added to the cell culture medium, if necessary. There is no observable effect of this antibiotic on Ca^{2+}-signals, mechanotransduction currents or electromotility.
5. Prior to the experiment, replace the cell culture medium with L-15. If the experiment requires loading with cell-permeable indicator, add this indicator to the L-15 medium and incubate for a specified time (*see* **Table 23.1**). Please, keep in mind that **Table 23.1** provides only approximate incubation times. The optimal incubation times have to be determined experimentally to provide a maximal fluorescent signal without significant compartmentalization of the dye into intracellular organelles (*see* **Note 7**). After incubation, wash the dish 2–3 times with fresh L-15 medium and mount the dish to the stage of a microscope. If the experiment requires loading the cell with cell-impermeable indicator, the surface of a hair cell needs to be cleaned by removing the neighboring cells with a suction pipette.

3.3. Loading the Cell Via a Patch Pipette

1. There are two common ways of loading the cell with a membrane-impermeable dye: microinjection and diffusion from a patch pipette. During microinjection, an injection pipette makes a transient hole in the plasma membrane that needs to be resealed after pipette withdrawal. Resealing the membrane is a relatively rare event with outer hair cells older than P5-P6, perhaps due to massive incorporation of prestin molecules into the plasma membrane *(8–10)*. Therefore, loading of a cell-impermeable fluorescent indicator through a patch pipette is strongly preferred, at least for outer hair cells.
2. Thaw an aliquot of intrapipette solution and add cell-impermeable indicator to produce a final working concentration (see **Section 2.4, step 3** and **Table 23.1**). Unless seal formation is really problematic, do not filter this solution because filters may absorb the dye. Fill a small tuberculin syringe with dye-containing intrapipette solution, cover the syringe with aluminum foil, and keep on ice throughout the experiment.
3. Produce a patch pipette with a tip diameter of $\sim 1\,\mu m$. Fill the pipette from the back using a non-metallic syringe needle. Mount the pipette to the holder of the patch clamp amplifier and make a seal on the basolateral surface of the hair cell. This step is identical to sealing in the conventional patch clamp technique.
4. Break the plasma membrane using an electric pulse ("zap" feature of the amplifier) under sustained but very gentle negative pressure in the pipette ($< 0.5\,kPa$). Immediately

after breaking the membrane, release the pressure or even apply a gentle positive pressure. The dye will rapidly diffuse into the cell without apparent changes of the cell shape. A manual microinjector (e.g. IM-5A, Narishige, Japan) and a pressure monitor (e.g. PM-01, World Precision Instruments) significantly facilitate all the above manipulations. A commonly used rupture of the membrane patch with a negative pressure usually flushes the very tip of the pipette with cytoplasm, resulting in less reproducible loading time. In addition, it is harder to control the cylindrical shape of an outer hair cell after negative pressure. Rupture by "zap" may not work if the seal resistance is less than $\sim$ 1 GΩ. If the seal is poor, it is better to do another patch with a fresh pipette and a new cell rather than continue the experiment with a bad seal.

3.4. Non-Ratiometric Ion Imaging

1. First, using *minimal intensity* of fluorescent illumination, adjust the camera exposure time to obtain non-saturated clear images of the hair cell. Check the absolute values of the resting fluorescence of the cell. Keep in mind that for an N times increase in fluorescence following stimulation, the resting signal should be about 1/Nth of the total dynamic range of the recording system (i.e. for a camera with a 12 bit digitizer, the maximum intensity would be $2^{12} = 4096$ and the optimal resting fluorescence would be 4096/N). If the expected fluorescence increase is unknown and/or the optimal level of the signal requires unrealistically long exposures, adjust the illumination to obtain an $\sim$ 10% signal. Record a test series of images for the same time as planned for an actual experiment.
2. Select the region of interest. Plot the changes of the average intensity of fluorescence in this region over time using image processing software. This plot will provide an estimate of photobleaching (light-induced decomposition of the fluorescent molecules) and the stability of the light source. A gradual decrease in intensity represents photobleaching, while spontaneous sparks may represent instability of the light source. The exposure time and the intensity of illumination can now be adjusted to produce optimal recording conditions with an overall signal decrease due to photobleaching within $<$ 10% of the resting fluorescence (*see* **Note 8**). If instability of the light source is suspected, plot the changes of intensity over time in the region without any cells. If similar sparks but with smaller amplitude are still discernible in the cell-free image, change the light source.
3. Using optimal illumination and exposure time, record the actual experimental series of image frames. The first frames of the series ($\sim$ 20% of the total number of frames) should be taken before the start of any stimulation to ensure the stability of the baseline.

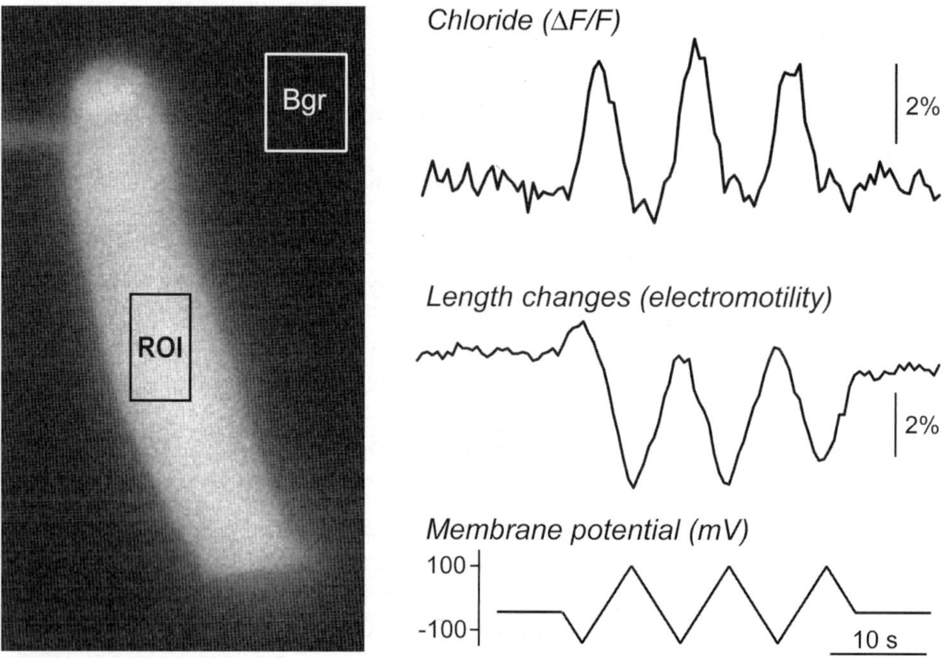

Fig. 23.2. Non-ratiometric imaging of Cl^- in an isolated outer hair cell. The cell was loaded with 5 mM of MQAE via a patch pipette. A region of interest (ROI) covers the portion of the cell that was attached to the glass. A region of background fluorescence (Bgr) was chosen at the right top corner of the image. Records show (from top to bottom): voltage-driven changes of intracellular Cl^-, changes of the length of the cell, and command potential. Length changes were measured as previously described *(17)*. Intrapipette solution contained no Cl^-.

4. Select a region of interest covering the whole body of the cell or a selected portion of the cell (**Fig. 23.2**). This region should not include any moving portion of the cell (*see* **Note 9**). Construct the "background" region (**Fig. 23.2**) around an area without any cells (make sure that there are no cells in the out-of-focus planes, too). Subtract the average background fluorescence from the signal in the region of interest for each frame. Please note that an increase in Ca^{2+} causes an increase in fluorescence, while an increase in Cl^- produces a *decrease* in fluorescence. This difference is because Cl^- ions are detected via diffusion-limited collisional quenching, while Ca^{2+} ions bind indicators directly. For Ca^{2+} indicators, compute the ratio $\Delta F/F = [F(t) - F(0)]/F(0) \times 100\%$. For Cl^- indicators, compute another ratio $\Delta F/F = [F(0) - F(t)]/F(0) \times 100\%$. In these equations, t is time, $F(t)$ is fluorescence at each consecutive frame, and $F(0)$ is pre-stimulus (baseline) fluorescence. In theory, these values should correctly monitor the changes of intracellular Ca^{2+} and Cl^- concentrations *(11–13)*, but only if the cell does not change shape. In experiments with cochlear hair cells, the shape of the cell may change substantially, especially in outer hair cells.

Therefore, we usually refrain from drawing quantitative conclusions from non-ratiometric ion measurements.

3.5. Ratiometric Ca^{2+} Imaging with Fura-2

Presently, we are not aware of any practical method for ratiometric imaging of Cl^- in hair cells. Two most common ratiometric probes for Ca^{2+} imaging are Fura-2 and Indo-1. Fura-2 exhibits changes in the absorption wavelength on Ca^{2+} binding, while Indo-1 exhibits a shift of the emission wavelength. At first glance, Indo-1 is more attractive than Fura-2, because the practical excitation wavelength of Indo-1 (355 nm) is somewhat longer than the shortest excitation wavelength of Fura-2 (340 nm). Therefore, imaging with Indo-1 may not require special UV-transparent optics and may be potentially performed even in confocal systems. However, the emitted fluorescence of Indo-1 should be detected at the wavelengths of 405 and 495 nm. At these wavelengths, the objective lenses are usually not well corrected, which results in the shift of the focal plane between these wavelengths as much as about a micrometer. Such a shift might be tolerable for low magnification imaging of brain slices but is certainly intolerable for high resolution imaging of cochlear hair cells. Thus, this section describes only the ratiometric imaging of Ca^{2+} with Fura-2, as the most practical choice.

1. Using the protocol described in **Section 3.4, steps 1** and **2**, determine the optimal illumination intensity and exposure time at both wavelengths of Fura-2 excitation (340 and 380 nm). Although ratiometric techniques can correctly monitor Ca^{2+} even in presence of substantial photobleaching, it is still desirable to control photobleaching in order to reduce potential phototoxicity (*see* **Note 8**).
2. Similar to **step 3 in Section 3.4**, record the actual experimental series of image pairs (340/380 nm). Correct each image individually for background fluorescence, which is determined as the average intensity of fluorescence in a manually selected reference region, common to the whole sequence and devoid of any obvious cellular structures. Then, construct a ratio (F_{340}/F_{380}) of each consecutive pair of 340 nm/380 nm images using image processing software. The intensity of each pixel in this image will reflect Ca^{2+} concentrations at the corresponding point of the cell (**Fig. 23.3**). However, if the cell is moving, substantial artifacts may appear at the edge of the cell (*see* **Note 9**).
3. The F_{340}/F_{380} values averaged over a region of interest or measured at each point of the image can be converted into the concentration of free Ca^{2+} $[Ca^{2+}]_i$. Although the most accurate calibration should be performed in a live cell *(11)*, such calibration requires determination of the background fluorescence at zero $[Ca^{2+}]_i$, which is a complicated task for cochlear hair cells. Fortunately, $[Ca^{2+}]_i$ in these cells is typically higher

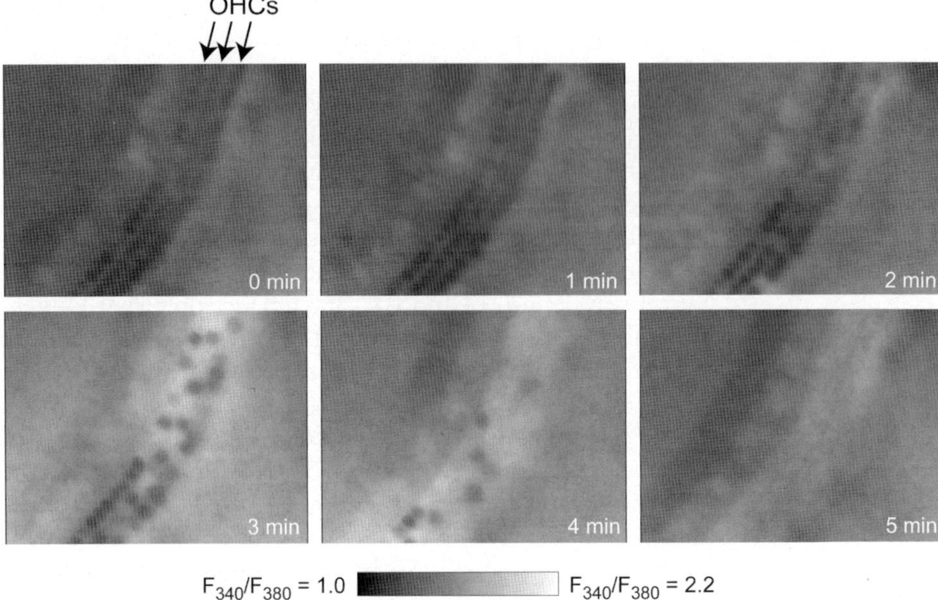

Fig. 23.3. Continuous ratiometric Ca^{2+} imaging in a cultured organ of Corti preparation. The sample was loaded with a cell-permeable form of Fura-2. In the beginning of the recording, three rows of outer hair cells (OHCs) exhibited distinct low levels of Ca^{2+}. Starting from approximately the second minute of continuous UV illumination, the outer hair cells became loaded with Ca^{2+} one by one and exploded. By the end of the fifth minute of observation, all outer hair cells died. The age of the mouse organ of Corti specimen was postnatal day 3 plus 4 days *in vitro*.

than 100 n*M* and a simplified *in vitro* procedure for Fura-2 calibration, using a commercially available Ca^{2+} imaging calibration kit (F-6774, Invitrogen), gives reasonable estimations of $[Ca^{2+}]_i$. For calibration, use exactly the same imaging parameters (illumination intensity and exposure times) as were used to capture the experimental series.

4. Notes

1. The ability of the objective lens to collect as much light as possible is critically important for fluorescent imaging. This property is characterized by a number, numerical aperture (N.A.) that is marked on any objective barrel right after magnification (e.g. "100X/1.40" means that this objective has a magnification of 100X and a N.A. of 1.40). The intensity of a fluorescent signal is proportional to the forth power of the numerical aperture and inversely proportional to the second power of magnification. In theory, the numerical aperture (but not magnification) determines the maximum resolution that could be achieved with a given objective. In practice, objectives with high magnification are still needed to obtain the best

resolution, albeit the effect of numerical aperture on the resolution is the most dramatic one. Therefore, high numerical aperture (N.A. = 1.30–1.45) oil immersion objectives with a magnification of 40 to 100X and a short working distance (0.17–0.2 mm) are commonly used on an inverted microscope, when the specimen is observed through a thin cover glass. On an upright microscope, water immersion objectives (40X–100X) with smaller numerical apertures (N.A. = 0.95–1.00) and longer working distances (~ 1–3 mm) are typically used. In studies of cochlear hair cells, the upright microscope is needed for observations of hair cell stereocilia. Air objectives (N. A. < 0. 9) are never used for ion imaging. The most expensive objectives have multiple optical corrections that are usually achieved by adding additional lens elements, which inevitably results in the loss of fluorescent signal. Thus, the simplest design of the objective is usually preferable for ion imaging. The objective also does not need any phase rings (marked as "Ph1", "Ph2" or "Ph3" on the objective barrel). These rings also absorb the light. Finally, ratiometric imaging of Ca^{2+} with Fura-2 indicator and Cl^- imaging with MQAE (N-(ethoxycarbonylmethyl)-6-methoxyquinolinium) both require excitation with ultra-violet (UV) illumination. Therefore, the objective lens must be able to pass illumination with wavelengths down to 340 nm. Such objectives are usually marked as "UV", "Fluor", "Ultrafluor", "Fl", "Fluar", "Neofluar", or "Fluotar".

2. In contrast to imaging of fixed cells, when the cells are immobilized in a medium with high viscosity and covered with a cover glass, live cells are usually situated in a small imaging chamber or a Petri dish that contains substantial amounts of liquid. Mechanical vibrations of the microscope stage and/or submerged micropipettes produce bulk oscillations of the bath medium. These oscillations entrain the cells and debris. Therefore damping of mechanical vibrations is an important consideration for any live-cell imaging setup. The most common source of vibrations is the vibration of the building. Anti-vibration tables usually damp these sorts of vibrations very effectively. The second common source of vibrations is the vibrations that are generated by the cooling system of the imaging camera, by the cooling system of the light source, by the moving parts of an excitation filter wheel (if available), and by the moving parts of the motorized microscope. Some of this equipment can be mechanically decoupled from the table (light source and filter wheel), while the other equipment cannot (camera and motorized parts of a microscope). It is advisable to avoid using, in live-cell imaging, any motorized microscopes and digital cameras with substantial levels of vibrations.

3. Conventional or confocal microscopy? The answer depends on the spatial size of the investigated events. A major advantage of confocal microscopy is the removal of "fluorescent haze", the out-of-focus signals from the optical planes below and above the focal plane. When the changes of Ca^{2+} or Cl^{-} throughout the cytosol of the hair cell are investigated, there is no obvious need for confocal microscopy even in the organ of Corti preparation. However, when the sub-micrometer-size microdomains of Ca^{2+} are studied either in the stereocilia *(14, 15)* or in the cell body *(16)*, the use of confocal microscopy is essential. The "payoff" for the high spatial resolution of confocal microscopy is a significant loss of emitted light, which limits the investigation only to strong signaling events. Additionally, currently available confocal systems do not allow UV illumination with wavelengths down to 340 nm, which makes the use of more successful ratiometric fluorescent probes for Ca^{2+} problematic (e.g., Fura-2).
4. A light detector (digital camera) is the next most critical element of the imaging setup after the objective lens. Due to the very fast advancement of technology, it is hard to provide any advice regarding particular manufacturers and/or models. However, there is "a rule of thumb" that can be used in evaluating the camera. The camera has to be able to visualize a cell that is barely visible to the human eye. A test sample can be prepared by loading commonly used HEK293 cells with a limited amount of non-ratiometric Ca^{2+} indicator, such as Fluo-3, which has minimal fluorescence at resting conditions. When these cells are grown to monolayer, they exhibit background spontaneous Ca^{2+} waves and sparks that should be easily detected by the camera. The image of all cells (including inactive cells) has to be clear, free of noise, and obtained in a reasonably fast exposure time (on the order of tens of milliseconds). As of today, only a few cameras with a price tag of $\sim \$40,000$ will meet these criteria. These cameras utilize "on chip" multiplication of the photon-generated charges and deep thermoelectric cooling of the imaging sensor (down to $\sim -70\,^{\circ}C$). The multiplication boosts the sensitivity of the camera while deep cooling is essential for the best signal-to-noise ratio. A variety of commercially available cameras without multiplication, but with cooling to at least $-20\,^{\circ}C$, represents a less expensive but slower alternative. These cameras produce an image of similar quality with a longer exposure time ($\sim$ 1 s). When imaging speed is the priority, an affordable solution may be provided by somewhat older "intensified CCD" cameras that also multiply photoelectrons. The cameras without multiplication and without thermoelectric cooling that may be offered for routine fluorescent microscopy are unlikely to be useful in ion imaging of

live cells due to their low sensitivity and excessive noise at low signal intensities.

5. The light source has to emit a substantial amount of UV light at wavelengths of 340–350 nm. This requires a light source with a special quartz condenser that is optional for most research microscopes. The xenon light source is preferred over the mercury source due to a more favorable ratio of 340/380 nm lines in the spectrum of the emitted light. For ratiometric Ca^{2+} imaging with Fura-2, the wavelengths of excitation should be switched between successive imaging frames. In the imaging setups with the slow light detectors (*see* **Note 4**), switching of the wavelength can be accomplished by a relatively simple filter wheel (e.g., Lambda-10, Sutter Instruments, Novato, CA). For faster wavelength switching, a special light source is needed (e.g., Polychrome V, TILL Photonics, Munich, Germany; or Lambda DG-4, Sutter Instruments, Novato, CA).
6. A variety of fluorescent probes for Ca^{2+} and Cl^- is available. **Table 23.1** lists probes that have been successfully used in our studies of cochlear hair cell physiology. For Ca^{2+} probes, both cell permeable and cell-impermeable forms were tested. Among Cl^- probes, only the probes that slightly permeate the membranes were tested in cochlear hair cells. These Cl^- indicators, as well as membrane-impermeable Ca^{2+} indicators were loaded into the cell via a patch pipette, which provides the best control over the intracellular concentration of the dye, but also results in washing out important intracellular messengers. It is still to be determined whether slightly permeable Cl^- indicators could be loaded into the hair cells through the plasma membrane by prolonging (several hours) the incubation of the cultured organ of Corti preparations. Membrane-permeable indicators preserve the cell signaling pathways better, but their concentration inside the cell is usually unknown. Cell-permeable forms of Ca^{2+} indicators are the acetoxymethyl (AM) ester derivatives. Once inside the cell, the lipophilic AM ester groups are cleaved by nonspecific esterases and the florescent probe is released into the cytosol.
7. Ion-sensitive indicators are retained in the cytoplasm for a limited time only. With time, or if the sample was overincubated with a membrane-permeant indicator, the dye will compartmentalize into intracellular organells, producing a distinct pattern of non-uniform fluorescence within the cell body. Compartmentalization renders some fraction of the indicator insensitive to changes of the ion level in the cytoplasm, which represents a problem for both ratiometric and non-ratiometric imaging. A sample with significant compartmentalization is not useful for any further studies of cytosolic Ca^{2+} or Cl^-.

8. Photo-induced decomposition of a fluorescent indicator (photobleaching) generates free radicals inside the cell, whose accumulation is toxic to the cell. In the organ of Corti, the outer hair cells are the most susceptible to phototoxicity (**Fig. 23.3**). Unfortunately, accumulation of free radicals inside an outer hair cell develops silently until $[Ca^{2+}]_i$ suddenly increases, the cell becomes swollen and eventually breaks. Switching off the illuminator after noticing an $[Ca^{2+}]_i$ increase is usually too late because the cell will die anyway. Therefore, care should be taken to avoid photobleaching as much as possible.
9. Ratiometric Ca^{2+} imaging depends on the ratio of the two consecutively captured images. If the cell is moving, the edge of the cell on the subsequent image does not overlap with the same edge on the previous image. A strong artifact of movement occurs at the edge after calculating the ratio. Movement artifact is a problem in studies of outer hair cell electromotility. This problem can be resolved by using shorter exposure times and/or slower electric stimulation.

Acknowledgments

I thank Dr. Guy P. Richardson for teaching me his technique of culturing organ of Corti explants. The author is supported by the Deafness Research Foundation, the National Organization for Hearing Research Foundation, and the University of Kentucky.

References

1. Fettiplace, R. and Hackney, C. M. (2006) The sensory and motor roles of auditory hair cells. *Nat Rev. Neurosci.* 7, 19–29.
2. Moser, T., Neef, A., and Khimich, D. (2006) Mechanisms underlying the temporal precision of sound coding at the inner hair cell ribbon synapse. *J. Physiol.* **576**, 55–62.
3. Frolenkov, G. I. (2006) Regulation of electromotility in the cochlear outer hair cell. *J. Physiol.* **576**, 43–48.
4. Dallos, P. and Fakler, B. (2002) Prestin, a new type of motor protein. *Nat. Rev. Mol. Cell Biol.* **3**, 104–111.
5. Dallos, P., Zheng, J., and Cheatham, M. A. (2006) Prestin and the cochlear amplifier. *J. Physiol.* **576**, 37–42.
6. Russell, I. J. and Richardson, G. P. (1987) The morphology and physiology of hair cells in organotypic cultures of the mouse cochlea. *Hear. Res.* **31**, 9–24.
7. Russell, I. J., Richardson, G. P., and Cody, A. R. (1986) Mechanosensitivity of mammalian auditory hair cells in vitro. *Nature* **321**, 517–519.
8. Kalinec, F., Holley, M. C., Iwasa, K. H., Lim, D. J., and Kachar, B. (1992) A membrane-based force generation mechanism in auditory sensory cells. *Proc. Natl. Acad. Sci. USA* **89**, 8671–8675.
9. Belyantseva, I. A., Adler, H. J., Curi, R., Frolenkov, G. I., and Kachar, B. (2000) Expression and localization of prestin and the sugar transporter GLUT-5 during development of electromotility in cochlear outer hair cells. *J. Neurosci.* **20**, RC116.
10. Zheng, J., Shen, W., He, D. Z., Long, K. B., Madison, L. D., and Dallos, P. (2000) Prestin is the motor protein of cochlear outer hair cells. *Nature* **405**, 149–155.
11. Kao, J. P. (1994) Practical aspects of measuring [Ca2+] with fluorescent indicators. *Methods Cell Biol.* **40**, 155–181.
12. Rudolf, R., Mongillo, M., Rizzuto, R., and Pozzan, T. (2003) Looking forward to

seeing calcium. *Nat. Rev. Mol. Cell Biol.* **4**, 579–586.

13. Inglefield, J. R. and Schwartz-Bloom, R. D. (1999) Fluorescence imaging of changes in intracellular chloride in living brain slices. *Methods* **18**, 197–203.
14. Denk, W., Holt, J. R., Shepherd, G. M., and Corey, D. P. (1995) Calcium imaging of single stereocilia in hair cells: localization of transduction channels at both ends of tip links. *Neuron* **15**, 1311–1321.
15. Lumpkin, E. A. and Hudspeth, A. J. (1995) Detection of Ca2+ entry through mechanosensitive channels localizes the site of mechanoelectrical transduction in hair cells. *Proc. Natl. Acad. Sci. USA* **92**, 10297–10301.
16. Tucker, T. and Fettiplace, R. (1995) Confocal imaging of calcium microdomains and calcium extrusion in turtle hair cells. *Neuron* **15**, 1323–1335.
17. Frolenkov, G. I., Mammano, F., Belyantseva, I. A., Coling, D., and Kachar, B. (2000) Two distinct Ca(2 +)-dependent signaling pathways regulate the motor output of cochlear outer hair cells. *J. Neurosci.* **20**, 5940–5948.

Chapter 24

Atomic Force Microscopy in Studies of the Cochlea

Michio Murakoshi and Hiroshi Wada

Abstract

The high sensitivity of mammalian hearing is achieved by amplification of the motion of the cochlear partition. The origin of this cochlear amplification is the elongation and contraction of outer hair cells (OHCs) in response to acoustical stimulation. This motility is made possible by a membrane protein embedded in the lateral membrane of OHCs. The gene of this protein has been identified and termed prestin. We, herein, present a method for observation by atomic force microscopy (AFM) of prestin expressed in the Chinese hamster ovary (CHO) cell plasma membrane. To obtain a stable sample for AFM imaging in liquid, we used as an example in the protocol provide here, CHO cells transfected with prestin or FLAG-tagged prestin, and untransfected CHO cells *(1)*. The cells attached to a substrate were subjected to ultrasonic waves generated from a sonicator probe so that the inside-out plasma membranes remained on the substrate. Prestin was immunostained with mouse anti-FLAG primary antibody and FITC-conjugated goat anti-mouse IgG secondary antibody. The lipid of the plasma membrane was labeled with fluorescence probes. The cytoplasmic faces of the cells were then observed in liquid by the tapping mode of AFM at low and high magnifications. More particle-like structures 8–12 nm in diameter were observed in the plasma membranes of the prestin-transfected CHO cells than in those of the untransfected CHO cells. Since the difference between these two types of cells is due to the existence of prestin, such particle-like structures in the prestin-transfected CHO cells are possibly constituted by prestin.

Key words: Prestin, transfection, Chinese hamster ovary cell, plasma membrane isolation, sonication, atomic force microscopy, fluorescence microscopy.

1. Introduction

Sound is transformed into mechanical vibration at the tympanic membrane and then transmitted into the cochlea via the ossicles. In the cochlea, the transmitted sound is converted

Bernd Sokolowski (ed.), *Auditory and Vestibular Research: Methods and Protocols, vol. 493*

DOI 10.1007/978-1-59745-523-7_24 Springerprotocols.com

into mechanical vibration of the basilar membrane, and this mechanical signal is finally translated into electrical signals by inner hair cells (IHCs). During this translation process, outer hair cells (OHCs) elongate and contract in sync with the mechanical signals. This reciprocal somatic motion of the OHCs amplifies the basilar membrane vibration and thereby the input signal to the IHCs is amplified, thus, the term "cochlear amplifier." The high sensitivity of mammalian hearing is achieved due to this amplification. OHC somatic motility, which was thought to be associated with a membrane protein in the lateral wall of OHCs *(2)*, was later found to be a motor protein *(3–8)*. The gene coding this protein was identified in the gerbil cochlea and termed prestin *(9)*. Since its identification, prestin has been intensively studied and found to be a direct voltage-to-force converter that uses cytoplasmic anions as extrinsic voltage sensors that can operate at microsecond rates *(10)*.

In early studies, using an electron microscope, the lateral membrane of the OHCs was found to be densely covered with particles about 10 nm in diameter *(11–14)*. More recently, the microstructure of the lateral wall of the OHC was examined by atomic force microscopy (AFM; *15–17*) and the plasma membrane was confirmed to contain many particles with a diameter of about 10 nm. These particles are believed to be the motor protein. However, since there are many different kinds of intrinsic membrane proteins other than prestin in the plasma membranes, it has been impossible to clarify which structures observed in such membranes are prestin.

Recently, prestin was reported in insect cells and was purified from these cells. Using the purified prestin, its 3-D structure was clarified as a bullet-shaped particle by transmission electron microscopy (TEM; *18*). However, the structure of purified prestin is not necessarily identical to that of prestin expressed in the plasma membrane of the native cell. In addition, TEM-based analysis of the protein structure requires time-consuming and cumbersome techniques to prepare a sample that contains sufficient amounts of purified protein for experimentation.

Here, we report a simple method for observation by AFM of membrane proteins expressed in the native plasma membrane under quasi-physiological conditions. We describe how to prepare CHO cells that are already prestin-transfected or untransfected, (do not express GLUT-5 *(19)*), and their plasma membranes for observation with AFM *(1)*. Descriptions include how CHO cells on a substrate are sheared open by exposure to ultrasonic waves so that only the flat inside-out plasma membranes remain on the substrate and how the cytoplasmic faces of the isolated membranes are observed by AFM in liquid.

2. Materials

2.1. Cell Culture

1. Prestin-transfected CHO cells, FLAG-tagged prestin-transfected CHO cells and untransfected CHO cells *(20, 21)*.
2. RPMI-1640 medium (Sigma-Aldrich, St. Louis, MO) supplemented with 10% fetal bovine serum, 100 U penicillin/mL and 100 μg streptomycin/mL.
3. Phosphate-buffered saline (PBS): 137 m*M* NaCl, 8.1 m*M* Na_2HPO_4, 2.68 m*M* KCl, 1.47 m*M* KH_2PO_4, pH 7.4.
4. 100 μ*M* ethylenediamine tetraacetic acid (EDTA) in PBS (EDTA-PBS).
5. 25 cm^2 culture flasks (Nunc, Roskilde, Denmark).

2.2. Isolation of Inside-Out Plasma Membranes from CHO Cells

1. External solution: 145 m*M* NaCl, 5.8 m*M* KCl, 1.3 m*M* $CaCl_2$, 0.9 m*M* $MgCl_2$, 10 m*M* HEPES, 0.7 m*M* Na_2HPO_4 and 5.6 m*M* glucose, pH 7.3. Stored in single use aliquots at 37 °C (*see* **Note 1**).
2. Hypotonic buffer: 10 m*M* piperazine-N, N′-bis[2-ethanesulfonic acid] (PIPES), 10 m*M* $MgCl_2$, 0.5 m*M* ethyleneglycotetraacetic acid (EGTA), pH 7.2. Stored in single use aliquots at 4 °C (*see* **Note 2**).
3. High-salt buffer: 2 *M* NaCl, 2.7 m*M* KCl, 1.5 m*M* KH_2PO_4, 1 m*M* Na_2HPO_4, pH 7.2. Stored in single use aliquots at room temperature.
4. 0.05% (w/v) trypsin in PBS (*see* **Note 3**).
5. PBS as in **Section 2.1 step 3**.
6. Sonicator (e.g., XL-2000, Misonix, Farmingdale, NY).

2.3. Hybrid Imaging by AFM and Fluorescence Microscopy

1. Imaging buffer: 28 mL of 0.2 *M* NaH_2PO_4 is combined with 72 mL of 0.2 *M* Na_2HPO_4 buffer and the mixture is then diluted with 200 mL of Milli-Q water, resulting in 300 mL of 0.1 *M* phosphate buffer with a pH of 7.2 (*see* **Note 4**).
2. PBS as in **Section 2.1 step 3**.
3. 0.22 μm pore-size syringe filters (Millipore, Billerica, MA).
4. 35 mm glass-bottomed dishes (Asahi Techno Glass, Chiba, Japan).
5. 35 mm plastic dishes (Nunc).
6. AFM system (NVB100, Olympus, Tokyo, Japan), in which the AFM scanning unit is mounted on an inverted fluorescence microscope (IX70, Olympus, Tokyo, Japan).
7. V-shaped silicon nitride cantilevers (OMCL-TR400PSA-2, Olympus; *see* **Note 5**).

2.3.1. Prestin Labeling for Isolated Plasma Membranes

1. 4% paraformaldehyde phosphate buffer solution (Wako Pure Chemical Industries, Osaka, Japan).

2. Blocking reagent for immunostaining: Block Ace (Dainippon Pharmaceutical, Osaka, Japan) or equivalent.
3. Anti-FLAG antibody (Sigma-Aldrich, St. Louis, MO).
4. FITC-conjugated goat anti-mouse IgG antibody (Sigma-Aldrich).

2.3.2. Lipid Labeling for Isolated Plasma Membranes

1. Chloromethyl Moiety- Diiodide (CM-DiI™; Invitrogen, California).

2.4. High-Magnification Imaging by AFM

1. 1% (v/v) glutaraldehyde: Prepared from 25% glutaraldehyde solution (Kanto Chemical, Tokyo, Japan) in PBS for each experiment.
2. CM-DiI (Invitrogen).
3. PBS as in **Section 2.1 step 3**.
4. Imaging buffer: 0.1 *M* phosphate buffer (*see* **Note 4**).
5. 35 mm plastic dishes (Nunc).
6. AFM system (NVB100, Olympus), in which the AFM scanning unit is mounted on an inverted fluorescence microscope (IX70, Olympus).
7. V-shaped silicon nitride cantilevers (OMCL-TR400PSA-2, Olympus; *see* **Note 5**).

3. Methods

3.1. Isolation of Inside-Out Plasma Membranes from CHO Cells

1. Culture prestin-transfected, FLAG-tagged prestin-transfected, and untransfected CHO cells in 25 cm^2 flasks at 37 °C with 5% CO_2. When the cells become confluent, remove the culture medium from the flask and detach the cells by incubating with 5 mL of 100 μM EDTA-PBS solution (*see* **Note 6**).
2. Put the EDTA-PBS solution containing the cells into a 15-mL tube and centrifuge at 250 g for 5 min.
3. Remove the supernatant and add 5 mL of the external solution to the tube.
4. Agitate the external solution containing the cells by pipetting in the tube using an automatic pipettor equipped with a 5 mL pipette.
5. Put 1 mL of the external solution into glass-bottomed dishes or plastic dishes in accordance with the intended purpose (*see* **Note 7**).
6. Deposit 1 mL of the cell-containing external solution on each of the dishes containing the external solution (*see* **Note 8**)
7. After ~ 10 min, most of the cells reattach to the dish substrate (*see* **Note 9**).

8. Discard the external solution and wash the cells twice with external solution warmed to 37 °C (*see* **Note 10**).
9. After removing the external solution, immerse the dish in 60 mL of ice-cold hypotonic buffer in a 100 mL glass beaker using forceps. Shear the cells open by exposure to ultrasonic waves for ~ 100 ms in the buffer, i.e., sonication (*see* **Note 11**).
10. After sonication, take the dish out of the beaker and remove the remaining hypotonic buffer.
11. Wash the dish three times with PBS (*see* **Note 12**).
12. Incubate the dish, on which the isolated plasma membranes are attached, in 1 mL of high-salt buffer for 30 min at room temperature to remove the cytoskeletal materials and the peripheral proteins *(22)*.
13. Remove the high-salt buffer from the dish using a pipette and wash the dish three times with PBS. Incubate the isolated membranes in 1 mL of 0.05% trypsin for 5 min at room temperature to remove the remaining materials (*see* **Note 13**).
14. Gently remove the trypsin solution and wash the dish three times with PBS.
15. Choose isolated plasma membranes whose locations on the substrate remain unchanged after sonication as samples for viewing, to ensure that the cytoplasmic face is up (*see* **Note 14**).

3.2. Hybrid Imaging by AFM and Fluorescence Microscopy

1. In this experiment, use the isolated membranes of the FLAG-tagged prestin-transfected CHO cells and those of the untransfected CHO cells.
2. Employ glass-bottomed dishes to prevent autofluorescence of a substrate (*see* **Note 7**).
3. Fix the isolated membranes in 1 mL of 4% paraformaldehyde phosphate buffer solution for 30 min at room temperature (*see* **Note 15**).
4. After fixation, remove the fixative and rinse the plasma membranes three times with PBS and incubate in 1 mL of Block Ace for 30 min at 37 °C (*see* **Note 16**).
5. Discard the blocking solution and wash the membranes three times with PBS. Incubate the sample in 300 μL of mouse anti-FLAG primary antibody in PBS (1:240) for 1 h at 37 °C (*see* **Note 16**).
6. Remove the primary antibody and wash the plasma membranes three times with PBS. Prepare the FITC-conjugated goat anti-mouse IgG secondary antibody at a dilution of 1:150 in PBS. Add CM-DiI to the antibody solution to a final concentration of 2 m*M* and incubate the sample

with 300 μL of this solution for 30 min at 37 °C (*see* **Note 16**).

7. Set up an AFM system, while incubating the sample in the secondary antibody. Attach a cantilever to a cantilever holder using forceps and attach the holder to the AFM scanning unit.
8. Mount the plastic dish, containing 7 mL of imaging buffer filtered previously with a 0. 22-μm pore-size syringe filter, on the stage of the AFM (*see* **Note 4**).
9. Place the cantilever into the buffer and align so that the laser irradiates the back of the cantilever and is reflected into the center of the laser detector.
10. Leave the cantilever aligned in the buffer for ~ 30–60 min at room temperature (*see* **Note 17**). The AFM system is then ready for imaging.
11. Remove the secondary antibody and wash the plasma membranes three times with PBS. Immerse the sample in 7 mL of imaging buffer filtered with a 0. 22-μm pore-size syringe filter. The sample is ready to be placed on the AFM stage.
12. Remove the plastic dish containing imaging buffer that was previously placed on the AFM stage and replace with the dish containing the sample.
13. Position the cantilever above the isolated plasma membranes to be imaged.
14. Fluorescence images are obtained by an inverted fluorescence microscope equipped with a Plan Apo 100X (numerical aperture = 1.40) oil-immersion objective and a cooled CCD camera (DP70, Olympus; *see* **Notes 18** and **19**).
15. Oscillate the cantilever near its resonance frequency in liquid (3.6–4.7 kHz) so that AFM images are obtained using the oscillation imaging mode (Tapping mode™, Digital Instruments, Santa Barbara, CA), which minimizes sample damage during scanning (*see* **Note 20**).
16. Move the cantilever in the vicinity of the sample either manually or automatically, depending on the AFM system employed. This approach is termed engagement (*see* **Note 21**). The sample is now ready to be scanned by the AFM.
17. Obtain low-magnification AFM images from the isolated membranes at a scan rate of 0.1 Hz, i.e., the scan speed is 1. 0 μm/s when the scan size is 5. 0 μm. Each scan line has 512 points of data and an image consists 512 scan lines (*see* **Note 22**).
18. An example of fluorescence images and a low-magnification AFM image of the isolated plasma membrane of the FLAG-tagged prestin-transfected CHO cell is shown in **Fig. 24.1**.

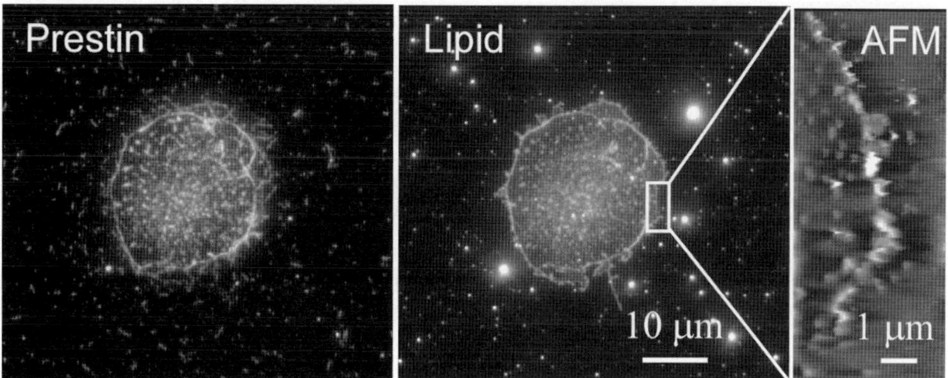

Fig. 24.1. Fluorescence images of prestin (*left*) and lipid-labeled (*center*) isolated membranes and a corresponding AFM image (*right*) of the isolated plasma membrane of a FLAG-tagged prestin-transfected CHO cell. A low-magnification AFM image was obtained from the boxed area shown in the center panel.

3.3. High-Magnification AFM Imaging

1. In this experiment, the isolated membranes of the prestin-transfected CHO cells and those of the untransfected CHO cells are used (*see* **Note 23**).
2. Use plastic dishes as the substrate (*see* **Note 7**).
3. Set up the AFM system first and follow the above-mentioned procedure (*see* **steps 7–10 in Section 3.2.**).
4. Fix the isolated membranes in 1 mL of 1% glutaraldehyde in PBS for 20 min at room temperature (*see* **Note 15**).
5. Remove the fixative and wash the sample three times with 1 mL of PBS.
6. Incubate the membranes with 300 μL of 2 m*M* CM-DiI in PBS for 5 min at 37 °C and subsequently for 15 min at 4 °C (*see* **Note 16**).
7. Remove the CM-DiI solution and wash the sample three times with 1 mL of PBS.
8. Immerse the isolated membranes in 7 mL of imaging buffer previously filtered with a 0. 22-μm pore-size syringe filter (*see* **Note 4**).
9. Remove the plastic dish containing the imaging buffer from the AFM stage and replace with the dish containing the sample.
10. Set up the cantilever and the inverted fluorescence microscope as described in **steps 13–16, Section 3.2.**).
11. Obtain high-magnification images (scan size of 1.0 × 0. 5 μm) from the isolated membranes at a scan rate of 0.3–0.4 Hz (scan speed is 0. 6–0. 8 μm/s). Each scan line contains 512 points of data, and an image consists of 256 scan lines.
12. Examples of the high-magnification AFM images of the prestin-transfected and untransfected CHO cells are shown in **Fig. 24.2**.

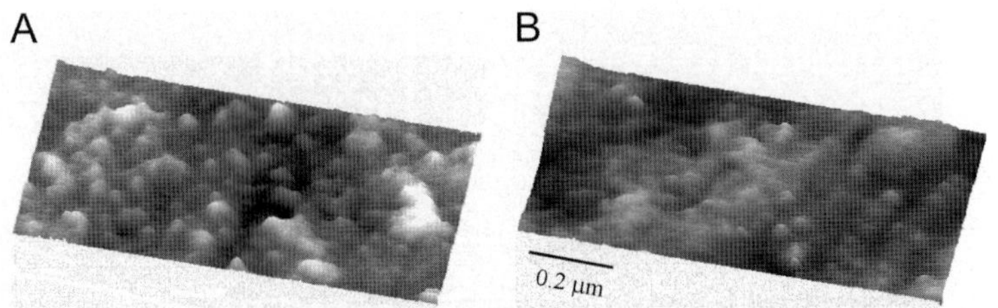

Fig. 24.2. Three-dimensional AFM images of the isolated plasma membranes of CHO cells. (**A**) Prestin-transfected CHO cell and (**B**) Untransfected CHO cell. More particle-like structures 8–12 nm in diameter exist in the plasma membrane of the prestin-transfected CHO cell than in that of the untransfected CHO cell.

4. Notes

1. Unless stated otherwise, all solutions should be prepared using Milli-Q water.
2. Hypotonic buffer should be cooled to 4 °C. Otherwise, the extraction rate of the inside-out plasma membrane on the substrate of the dish decreases.
3. To prevent autolysis, store 0.05% trypsin in PBS at −20 °C and do not leave at room temperature. May only be stored at −20 °C for up to one week.
4. Alternatively, Hanks' Balanced Salt Solution (HBSS: 5.33 mM KCl, 0.44 mM KH_2PO_4, 137.93 mM NaCl, 0.34 mM Na_2HPO_4, 0.56 mM glucose) can also be used.
5. The cantilever has a pyramidal tip. The typical radius of its curvature is 16 ± 1 nm (personal communication with Olympus).

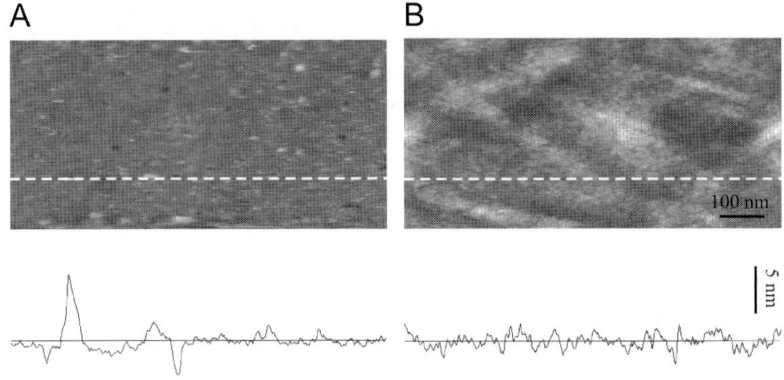

Fig. 24.3. The surface topography of different dish substrates. (**A**) Glass-bottomed dish. (**B**) Plastic dish. The section along the dotted line is presented at the bottom of each panel. Mean surface roughness of the glass-bottomed dish substrate and the plastic dish substrate are 0.59 nm and 0.48 nm over the scan length of 1.0 μm, respectively.

6. This protocol can be adapted for many other experimental applications, such as immunofluorescence experiments, electron microscopic experiments, etc.
7. A glass-bottomed dish is used for hybrid imaging by AFM and fluorescence microscopy to prevent autofluorescence. A plastic dish is used for high-magnification AFM imaging because the surface of a plastic dish substrate is much flatter than that of a glass-bottomed dish (*see* **Fig. 24.3**).
8. The cells should be sparsely scattered on the dish substrate to minimize the unwanted contamination of the substrate when the cells, containing many proteins, are sheared open by sonication. Therefore, when the number of the cells in the external solution is large, a lesser amount of the cell-containing external solution should be added to the dish, and vice versa. An example of a differential interference contrast (DIC) image of CHO cells sparsely spattered on a dish is shown in **Fig. 24.4**.
9. A 10 min incubation time is sufficient for cell reattachment. Further incubation does not result in better adhesion of cells; instead, it causes the cells to deteriorate. Hold the dish to the light and observe it from the bottom. If the cells are successfully attached to the substrate, the bottom of the dish should be translucent.

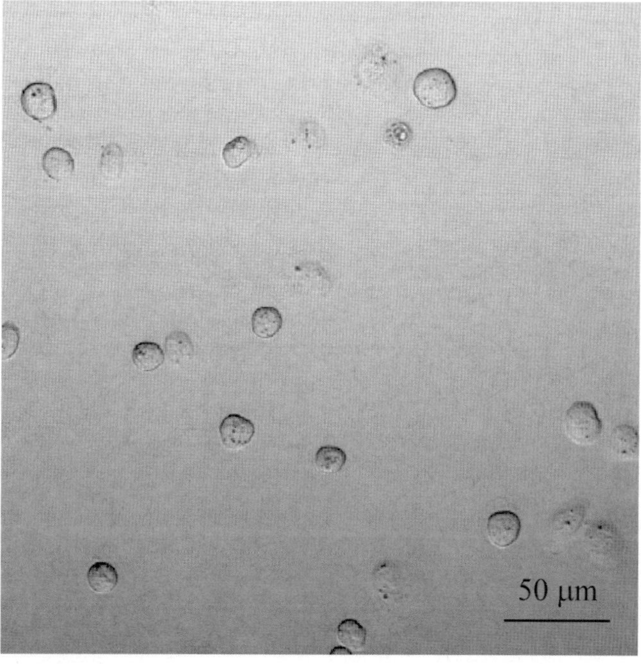

Fig. 24.4. Differential interference contrast (DIC) image of CHO cells. The cells should be sparsely scattered on the dish to prevent contaminating the dish substrate with debris from sonicating the cells.

10. An external solution warmed to 37 °C is more effective than one kept at room temperature for the removal of unwanted materials attached to the surface of the isolated plasma membranes, such as cell fragments, proteins contained in the culture medium, etc. Do not remove the solution in the dish completely when changing solution, as the cells deform due to the effect of surface tension when they come in contact with air. Unless stated otherwise, such careful attention should be maintained throughout the experiment.
11. If you use an XL-2000 sonicator, set the power dial to 3 (max. is 20, corresponding to 100 W) or less. Set the tip of the probe about 10 mm above the cells. Rhythmically push the thumb pulsing switch once for a few moments, generating a pulse of ultrasonic waves from the probe. This pulse is likely to produce a few 100 ms ultrasonic waves.
12. After sonication and PBS washing, hold the dish to the light and view it from the bottom. The substrate that was beneath the sonication probe should be transparent as a result of the isolated plasma membrane. On the other hand, the substrate that was far from the sonication probe is often not transparent; rather, it remains translucent because the cells are not exposed to the ultrasonic waves and, therefore, the plasma membranes are not isolated. Thus, if the isolation procedure of the plasma membrane is successfully done, the bottom of the dish is transparent at its center and translucent at its periphery. Empirically, however, if the diameter of the transparent part of the dish is much bigger than 10 mm, the cells located in that part have been removed completely. This result indicates that the plasma membrane isolation was unsuccessful and you must go back to **step 1, Section 3.1**. If there is no transparency whatsoever, most of the cells are not sheared open. In this case, bring the tip of the sonication probe close to the substrate and repeat the procedure. Do not adjust the output power of the sonicator.
13. Trypsin-induced proteolysis progresses in a time-dependent manner. Do not immerse the sample in this solution for more than 5 min.
14. To check the position of the cells before and after sonication, arbitrarily draw a grid consisting of sub-millimeter squares on the bottom of the glass-bottomed dish with a fine-tipped pen before sonication. When using a plastic dish, the bottom of the dish is scratched with a fine needle (e.g., 26-gauge), instead of a fine-tipped pen. Although several kinds of dishes with gridded substrates are commercially available, the above-mentioned methods are easy to use.
15. Glutaraldehyde fixation gives an AFM image of better quality than paraformaldehyde fixation. On the other hand, glutaraldehyde fixation sometimes reduces the antigenicity of the target protein. Thus, it is not applicable for AFM imaging

when using immunofluorescence. In this case, paraformaldehyde fixation is a simple alternative. Experimenters have to find the most appropriate fixation method for each experimental design.

16. While incubating the sample in buffer at 37 °C, the closed dish is completely covered with Kimwipes® soaked in distilled water to prevent the sample from drying out, especially when using a small amount of buffer (*e.g.*, 300 μL).
17. The silicon nitride cantilever used in this study is coated with thin gold/chromium film on its back side to reflect the laser beam effectively. Due to this composition, the cantilever is bent when its back is irradiated by the laser, producing the so-called bimetal effect. This effect causes thermal drift, i.e., misalignment of the laser. Although thermal drift is an inevitable phenomenon, incubation of the cantilever with the imaging buffer for ~ 30–60 min prior to imaging minimizes this drift.
18. Differential interference contrast images of isolated plasma membranes cannot be obtained because they are transparent. Thus, the membranes are stained with a lipophilic dye in order to see if the cells are sheared and the plasma membranes are isolated on the substrate.
19. This system enables us to record fluorescence and AFM images of the same sample simultaneously. A mercury lamp is used as a light source. FITC fluorescence is detected with a narrow-band filter cube (U-MNIBA, Olympus, Tokyo, Japan), which consists of a 470–490-nm excitation filter, a 505-nm dichroic mirror, and a 515–550-nm emission filter. CM-DiI fluorescence is detected with a wide-band filter cube (U-MWIG, Olympus, Tokyo, Japan) composed of a 520–550-nm excitation filter, a 565-nm dichroic mirror, and a 580-nm emission filter.
20. High-quality images tend to be obtained when the drive frequency is higher than 4 kHz in this AFM system. Therefore, find a resonance frequency higher than 4 kHz.
21. Use of the automatic stop function on an AFM system may be problematic, i.e., the cantilever may hit the sample, causing deformation and/or contamination of the tip, resulting in loss of image quality. Thus, we recommend using the manual stop function if available. When using the manual stop function, the distance between the cantilever tip and the sample surface can be estimated by the following procedure.
    (a) Oscillate the cantilever at its resonance.
    (b) Engage the cantilever manually.
    (c) When the oscillating cantilever comes to within about 10 μm of the sample surface, a fluid damping effect occurs (*i.e.*, the film damping phenomenon). This effect dampens the oscillating motion of the cantilever, resulting in a

reduction in its amplitude. This amplitude is usually displayed by the AFM indicator and/or the AFM monitor. Therefore, when you notice an amplitude reduction, stop the engagement, because this reduction indicates that the cantilever is ~ 10 μm above the sample. Engage the cantilever again at the minimal increments allowed by your AFM system. Monitor the force curve and find the point where the cantilever starts to deflect. This point is the contact point between the cantilever tip and the sample surface.

22. Although adjustments of AFM scan parameters are quite important to obtain high-quality AFM images, the parameters substantially depend on the AFM system you use. Therefore, please refer to the users' manual of your AFM system for details of parameter settings.
23. The FLAG-tagged prestin-transfected CHO cells are not used to obtain high-magnification AFM images because three FLAG peptides (~ 3 kDa) are used to tag the C terminus of prestin. This tag size affects the image size of prestin when done at high magnification.

Acknowledgments

The authors are grateful to Professor M. Sokabe (Nagoya University, Japan), Dr. D. Furness (Keele University, U.K.), Mr. A. Yagi (Olympus, Japan), and Professor D. He (Creighton University, U.S.A.) for their helpful suggestions and comments. This work was supported by a Grant-in-Aid for Scientific Research on Priority Areas 15086202 from the Ministry of Education, Culture, Sports, Science and Technology of Japan, by a Health and Labour Science Research Grant from the Ministry of Health, Labour and Welfare of Japan, and by a grant from the Human Frontier Science Program to H.W., and by a Grant-in-Aid for JSPS Fellows 19002194 from the Japan Society for the Promotion of Science and by Special Research Grants 11170012 and 11180001 from the Tohoku University 21st Century COE Program of the "Future Medical Engineering Based on Bio-nanotechnology" to M.M.

References

1. Murakoshi, M., Gomi, T., Iida, K., Kumano, S., Tsumoto, K., Kumagai, I., Ikeda, K., Kobayashi, T., and Wada, H. (2006) Imaging by atomic force microscopy of the plasma membrane of prestin-transfected Chinese hamster ovary cells. *J. Assoc. Res. Otolaryngol.* 7, 267–278,
2. Dallos, P., Evans, B. N., and Hallworth, R. (1991) Nature of the motor element in electrokinetic shape changes of cochlear outer hair cells. *Nature* **350**, 155–157.
3. Brownell, W. E., Bader, C. R., Bertrand, D., and de Ribaupierre, Y. (1985) Evoked mechanical responses of isolated

cochlear outer hair cells. *Science* **227**, 194–196.

4. Kachar, B., Brownell, W. E., Altschuler, R., and Fex, J. (1986) Electrokinetic shape changes of cochlear outer hair cells. *Nature* **322**, 365–368.
5. Ashmore, J. F. (1987) A fast motile response in guinea-pig outer hair cells: the cellular basis of the cochlear amplifier. *J. Physiol.* **388**, 323–347.
6. Santos-Sacchi, J., and Dilger, J. P. (1988) Whole cell currents and mechanical responses of isolated outer hair cells. *Hear. Res.* **35**, 143–150.
7. Liberman, M. C., Gao, J., He, D. Z., Wu, X., Jia, S., and Zuo, J. (2002) Prestin is required for electromotility of the outer hair cell and for the cochlear amplifier. *Nature* **419**, 300–304.
8. Dallos, P. and Fakler, B. (2002) Prestin, a new type of motor protein. *Nat. Rev. Mol. Cell Biol.* **3**, 104–111.
9. Zheng, J., Shen, W., He, D. Z., Long, K. B., Madison, L. D., and Dallos, P. (2000) Prestin is the motor protein of cochlear outer hair cells. *Nature* **405**, 149–155.
10. Cheatham, M. A., Huynh, K. H., Gao, J., Zuo, J., and Dallos, P. (2004) Cochlear function in Prestin knockout mice. *J. Physiol.* **560**, 821–830.
11. Arima, T., Kuraoka, A., Toriya, R., Shibata, Y., and Uemura, T. (1991) Quick freeze, deep etch visualization of the 'cytoskeletal spring' of cochlear outer hair cells. *Cell Tissue Res.* **263**, 91–97.
12. Forge, A. (1991) Structural features of the lateral walls in mammalian cochlear outer hair cells. *Cell Tissue Res.* **265**, 473–483.
13. Kalinec, F., Holley, M. C., Iwasa, K. H., Lim, D. J., and Kachar, B. (1992) A membrane-based force generation mechanism in auditory sensory cells. *Proc. Natl. Acad. Sci. U.S.A.* **89**, 8671–8675.
14. Souter, M., Nevill, G., and Forge, A. (1995) Postnatal development of membrane specialisations of gerbil outer hair cells. *Hear. Res.* **91**, 43–62.
15. Le Grimellec, C., Giocondi, M. C., Lenoir, M., Vater, M., Sposito, G., and Pujol, R. (2002) High-resolution three-dimensional imaging of the lateral plasma membrane of cochlear outer hair cells by atomic force microscopy. *J. Comp. Neurol.* **451**, 62–69.
16. Wada, H., Usukura, H., Sugawara, M., Katori, Y., Kakehata, S., Ikeda, K., and Kobayashi, T. (2003) Relationship between the local stiffness of the outer hair cell along the cell axis and its ultrastructure observed by atomic force microscopy. *Hear. Res.* **177**, 61–70.
17. Wada, H., Kimura, K., Gomi, T., Sugawara, M., Katori, Y., Kakehata, S., Ikeda, K., and Kobayashi, T. (2004) Imaging of the cortical cytoskeleton of guinea pig outer hair cells using atomic force microscopy. *Hear. Res.* **187**, 51–62.
18. Mio, K., Kubo, Y., Ogura, T., Yamamoto, T., Arisaka, F., and Sato, C. (2008) The motor protein prestin is a bullet-shaped molecule with inner cavities. *J. Biol. Chem.* **283**, 1137–1145.
19. Inukai, K., Katagiri, H., Takata, K., Asano, T., Anai, M., Ishihara, H., Nakazaki, M., Kikuchi, M., Yazaki, Y., and Oka, Y. (1995) Characterization of rat GLUT5 and functional analysis of chimeric proteins of GLUT1 glucose transporter and GLUT5 fructose transporter. *Endocrinology* **136**, 4850–4857.
20. Iida, K., Konno, K., Oshima, T., Tsumoto, K., Ikeda, K., Kumagai, I., Kobayashi, T., and Wada, H. (2003) Stable expression of the motor protein prestin in Chinese hamster ovary cells. *JSME Int. J.* **46C**, 1266–1274.
21. Iida, K., Tsumoto, K., Ikeda, K., Kumagai, I., Kobayashi, T., and Wada, H. (2005) Construction of an expression system for the motor protein prestin in Chinese hamster ovary cells. *Hear. Res.* **205**, 262–270.
22. Ziegler, U., Vinckier, A., Kernen, P., Zeisel, D., Biber, J., Semenza, G., Murer, H., and Groscurth, P. (1998) Preparation of basal cell membranes for scanning probe microscopy. *FEBS Lett.* **436**, 179–184.

INDEX

E

F

G

R

S

T